T0399230

Shield Tunnel Engineering

Shield Tunnel Engineering

From Theory to Practice

SHUYING WANG
School of Civil Engineering, Central South University,
Changsha, P.R. China

JINYANG FU
School of Civil Engineering, Central South University,
Changsha, P.R. China

CONG ZHANG
School of Civil Engineering, Central South University of
Forestry and Technology, Changsha, P.R. China

JUNSHENG YANG
School of Civil Engineering, Central South University,
Changsha, P.R. China

ELSEVIER

Elsevier
Radarweg 29, PO Box 211, 1000 AE Amsterdam, Netherlands
The Boulevard, Langford Lane, Kidlington, Oxford OX5 1GB, United Kingdom
50 Hampshire Street, 5th Floor, Cambridge, MA 02139, United States

British Library Cataloguing-in-Publication Data
A catalogue record for this book is available from the British Library

Library of Congress Cataloging-in-Publication Data
A catalog record for this book is available from the Library of Congress

ISBN: 978-0-12-823992-6

For Information on all Elsevier publications
visit our website at https://www.elsevier.com/books-and-journals

Publisher: Glyn Jones
Acquisitions Editor: Glyn Jones
Editorial Project Manager: Naomi Robertson
Production Project Manager: Anitha Sivaraj
Cover Designer: Mark Rogers

Typeset by MPS Limited, Chennai, India

Contents

Foreword

The shield tunnelling method is a fully mechanized tunnel construction method that is notable for its safety, speed, relatively low cost, and minimal environmental impact. With the result of nearly 200 years of development, shield tunnelling has been widely used for constructing tunnels for urban rail transit, underwater transportation, highways, railways, municipal pipe systems, hydraulic engineering, gas transmission, and other fields. It has become one of the most important construction methods for tunnels and underground spaces. This method provides strong contributions to the construction and technical development of modern tunnel and underground engineering.

In recent years, with the rapid development of shield tunnel engineering, the length, diameter, and buried depth of shield tunnels have been increasing, and their geological conditions, structural forms, and surrounding environments have become more and more complex. This situation not only poses new requirements and challenges, but also provides a platform and opportunities for the growth of technology. Moreover, the demand for engineers and technicians who are proficient in the knowledge of shield tunnelling is increasing with the continuous growth of the scale and quantity of tunnel and underground space construction. Therefore a clear, comprehensive, and illustrative textbook or reference book for shield tunnel engineering is particularly important for the education and training of professional technicians.

This book has been edited by a team of university faculty members who are fully experienced in teaching and research. The editors systematically reviewed the theoretical and technical knowledge points in shield tunnel engineering and integrated a large number of practical engineering cases and scientific research results. The aim of the book is not only to enable beginners to comprehensively learn about design, construction, and maintenance of shield tunnels, but also to help readers to deeply understand the related problems as illustrated by actual cases. This book provides a comprehensive discussion of shield tunnel engineering and is worthy of careful reading by university students and researchers in the relevant professional fields. The book also can be used as a textbook for teachers and students majoring in civil engineering, water conservancy, urban underground space, and rail transit and as training material and a technical reference for design, construction, management, and scientific research personnel engaged in shield tunnelling and underground engineering, shield machine manufacturing, and other fields.

We hope that the publication of this book will contribute to the education of technical personnel and the progress of shield tunnelling technology.

Prof. Xiangsheng Chen
Member of the Chinese Academy of Engineering, Beijing, P.R. China

Preface

Since the beginning of the 21st century, there have been great opportunities and challenges in tunnel engineering construction with the rapid development of high-speed railways, subways, and other types of rail transportation. As one of the main construction methods of tunnel engineering in weak ground, shield tunnelling is an important area of study. Many new theoretical advances and technological developments have been seen in their design, construction, and operation management. In particular, shield tunnel engineering has developed quickly in China, with its diverse geological conditions, and many practical experiences have been accumulating in the development process.

This book systematically lays out the basic professional knowledge of shield tunnel engineering and incorporates advanced scientific research results such as soil conditioning technology, slurry treatment technology, backfill grouting technology, and shield tunnel environmental impact and control. The editorial team has many years of working experience to obtain a deep understanding of the shield tunnelling method. Their knowledge on shield tunnel engineering is introduced in depth in this book through reviews of the current states of technology development and many typical engineering cases. The main content of this book includes the development of shield tunnel engineering in the world, geological investigation for shield tunnelling, shield machine configurations and working principles, segmental lining structure design, shield machine selection, shield tunnelling and segment assembling, backfill grouting, muck conditioning and recycling, slurry treatment and waste slurry recycling, ground settlement and control, defects of shield tunnel lining and their treatments.

Participants in the writing of this book include Shuying Wang, Jinyang Fu, and Junsheng Yang from Central South University, and Cong Zhang from Central South University of Forestry and Technology. The specific division of the organization and writing work for this book is as follows: Shuying Wang is responsible for the writing of Chapters 1, 3, 4, 6, 7, and 9, Jinyang Fu is responsible for the writing of Chapters 2, 5, 11, and 12, Junsheng Yang is responsible for the overall planning of the book and compilation of part of Chapter 10, and Cong Zhang is responsible for the writing of Chapter 8, and a part of Chapter 10.

Professor Wei Wang from Central South University and Professor Fei Ye from Chang'an University acted as the chief reviewers of this book. They reviewed the writing outline and all the manuscripts in detail, and they put forward many valuable opinions. Professor Xuemin Zhang and Professor Mingfeng Lei from Central South University also provided review opinions. Many thanks to them.

In the process of writing, many graduate students from the Department of Tunnel Engineering, School of Civil Engineering, Central South University participated in part of the material collection, illustration drawing, and text sorting. Thanks equally to all.

This book was financially supported by the National Natural Science Foundation of China (No. 51778637, 52022112). Additionally, during the preparation of this book, the editors received favors from China Railway Construction Heavy Industry Corporation Limited, China Railway Engineering Equipment Group CO., LTD, China Railway No.5 Engineering Group CO., LTD, China Railway Tunnel Group CO., LTD, and other companies. They provided valuable materials including some figures and data. Thanks a lot for their help.

This book can be used as a textbook for postgraduates and undergraduates in civil engineering, urban underground space engineering, and other majors and fields in general colleges and universities as well as for professional and technical personnel engaged in the design, construction, and scientific research of shield tunnel engineering. It can also be used as a reference book for teachers, students, and short-term trainees in vocational and specialized higher education institutions.

Instinctively, the editors seek to describe the specialized knowledge exactly but the complexity of the shield tunnel engineering confounds them. There may exist errors and deficiencies in the book, so readers are urged to read the book critically. However, it is hoped that the book contains sufficient information to guide readers to an appropriate answer or solution.

Shuying Wang
Jinyang Fu
Junsheng Yang
Cong Zhang

CHAPTER 1

Introduction

1.1 Concepts of shield tunnel engineering

Shield tunnelling method is an underground excavation method in which a shield machine with a metal shell is used to excavate ground and install lining supports under the protection of the metal shell. The shield machine completes the operations, including ground excavation, muck removal, segment assembly, machine advancement, and others (Fig. 1.1). The shield tunnelling method is generally suitable for tunnel construction in weak formations. The shell acts as a temporary support for the excavated cavern, bears the pressure of the surrounding ground, and keeps groundwater out of the shield machine.

Internationally, the shield tunnelling method is generally put in the category of the construction methods using tunnel boring machines (TBMs). In contrast, engineers in China assign the category of TBM specifically to rock tunnelling machines. Unlike shield tunnelling, the TBM construction method is suitable for hard rock stratum tunnels, mainly for construction of mountain tunnels. In some Chinese cities with hard rock,

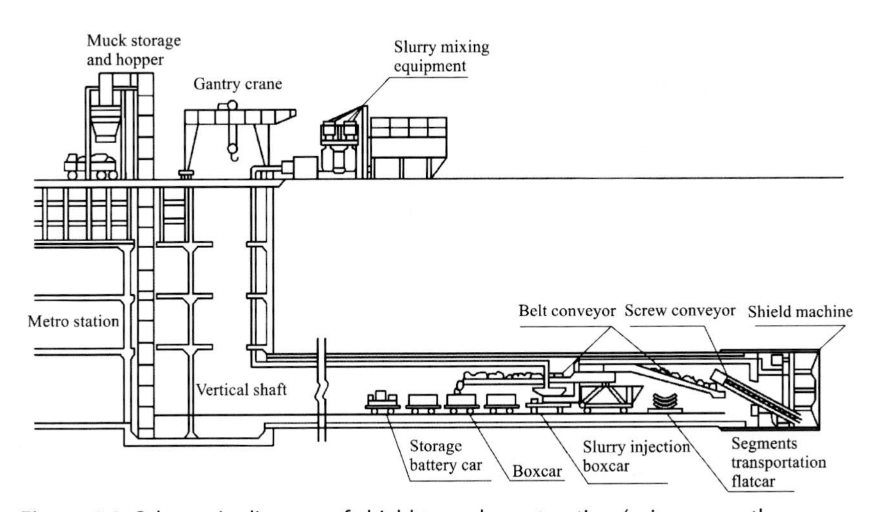

Figure 1.1 Schematic diagram of shield tunnel construction (using an earth pressure balance shield as an example).

Shield Tunnel Engineering
DOI: https://doi.org/10.1016/B978-0-12-823992-6.00001-1

such as Qingdao, Shenzhen, and Chongqing, the TBM method has also been used to construct municipal tunnels such as metro tunnels. The TBM tunnelling method is similar to the shield tunnelling method in that the cutters equipped in the shield cutterhead excavates the stratum. The difference is that most TBMs are open in the front and there is no need to have a closed chamber to provide pressure to support ground. In addition, there are also big differences between a shield machine and a TBM in the type of protective shell, thrust system, and lining type.

The basic requirements of the shield tunnelling method include the ability to perform full-face excavation, to ensure the safety of the workers in the shield, to avoid the ground large deformation and instability, to install lining in the tail of the shield, and to support the shield machine at the end of the segment through a force transmission system such as jacks [1]. Additionally, the shield method includes a shield tail sealing system to prevent water leakage from the shield tail and grouting systems (including synchronous grouting and secondary grouting) to fill the gap between the shield and the surrounding ground for reducing ground settlement. The shield tunnelling method is highly integrated with advanced disciplines such as computer technology, automation, informatization, system science, and management science. It adopts advanced technology such as electronics, informatics, telemetry, and remote control to guide and monitor all operations to ensure that the shield machine is always in the excellent working conditions.

With the development of urban tunnels, the shield tunnelling method has been widely used to construct urban underground projects such as metro intervals, municipal highways, water culverts, and underground pipe corridors. The shield tunnelling method has become one of the leading construction methods for tunnel construction. Compared with traditional tunnel construction methods, the shield tunnelling method has the following advantages:

1. The construction personnel and equipments work under the protection of the shield shell, and the shell can smoothly pass through a weak stratum.
2. The excavation, segment assembly, and other operations of the shield machine can be automated, and remotely controlled and informationized. The speed of tunnelling is high, the construction labors required as few, and the economic and social benefits are greatly improved.
3. There are few field operations, excellent concealment, low environmental impact from noise and vibration, and the effect of construction on the surrounding buildings and underground pipelines can be controlled well.

Inevitably, the shield tunnelling method also has certain disadvantages:

1. Shield machines are expensive. Once the shield machine enters the stratum, it must complete the excavation. It cannot be reconstructed extensively in the cave, so the shield machine and its parts must be selected carefully.
2. The construction technologies, such as equipment manufacturing, pneumatic equipment supply, segment assembly, and lining structure waterproofing, must be closely coordinated with the requirements of civil engineering, geological engineering, and mechanical engineering.
3. The earth pressure balance and slurry shield machines are commonly used for tunnel construction, and the stratum conditions at the excavation face cannot be seen, requiring the shield machine to have wide adaptability.
4. For tunnels with small-radius curves and large slopes, it is difficult to control shield attitude during their constructions.
5. The engineering benefit for building short tunnels is poor with the shield tunnelling method.

As a branch of tunnel and underground engineering, shield tunnel engineering has developed into an applied science and engineering technology that are used in the planning, survey, design, construction, and maintenance of shield tunnels. The issues involved include the shield tunnel engineering survey, shield machine configuration and working principles, shield machine selection, shield lining structure type and design, shield launching and receiving, shield advancing and segment assembly, backfill grouting, muck conditioning, slurry treatment, environmental impact and its control during shield tunnel construction, structural damage and remediation of the shield tunnel, and others.

1.2 Types of shield machines

Shield machines can be classified according to section shape, tunnelling principle, shield diameter, and partition structure between the excavation face and the operating room. In terms of section shape, in addition to the commonly used circular shield machines, there are horseshoe-shaped, rectangular, and multicircular shaped shields. Compared with circular shield machines, tunnel construction using noncircular shield machines are technically challenging, and some problems such as disadvantages in structural strength and positioning control must be solved [2]. According to the classification of tunnelling principles, shield machines are generally divided into open types and closed-chamber types, as shown in Fig. 1.2.

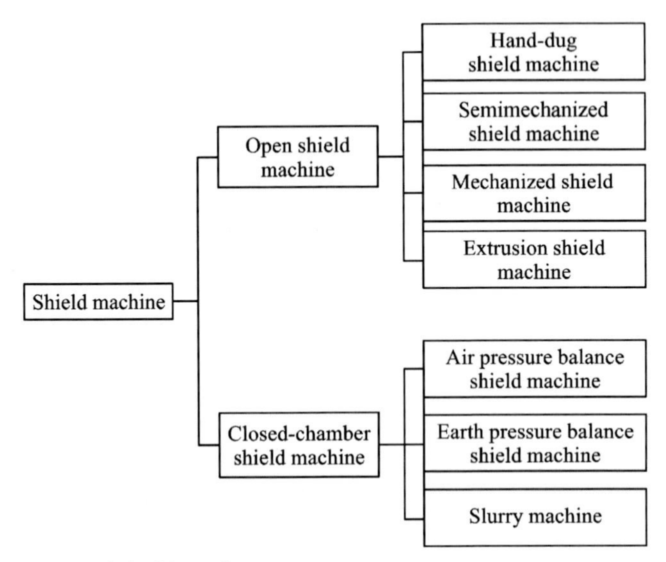

Figure 1.2 Types of shield machines.

From the perspective of excavation methods, the open shield machines can be classified as hand-dug shield machines, semimechanized shield machines, or (full) mechanized shield machines. As shown in Fig. 1.3, the hand-dug shield machine requires use of traditional tools such as sharp picks, cross picks, or shovels for manual excavation and uses movable front eaves, baffles, and other facilities to retain soil. In many cases it is necessary to prereinforce the ground to stabilize the excavation face before smooth tunnelling. As shown in Fig. 1.4, the semimechanized shield machine uses single-bucket excavators, backhoes, and outrigger rock drills for partial excavation. Sometimes tools such as shovels are needed to excavate the hard-to-reach parts of the excavator, and the soil-retaining method is similar to that used for the hand-dug shield machine. However, to provide enough working space for single-bucket excavators and other facilities, the excavation face is seperated into parts as large as possible. In many cases, it is also necessary to prereinforce ground to stabilize the excavation face.

As is shown in Fig. 1.5, the mechanized shield machine uses a rotatable cutterhead mounted in the front of the shield for full-face excavation, which is usually done in strata with strong self-stability. For soft strata, it is also necessary to use prereinforcement technology to stabilize the excavation face.

In terms of the supporting medium in the closed chamber, the closed-chamber shield machines are divided into air pressure balance shield machine, earth pressure balance (EPB) shield machine, and slurry shield machine.

Figure 1.3 Hand-dug shield machine.

Figure 1.4 Semimechanized shield machine.

The EPB shield machine is used for a full-section excavation of the strata with a cutterhead, as shown in Fig. 1.6. It is provided with a muck storage room between the bulkhead and the excavation face, which is called the excavation chamber (or soil chamber, muck chamber), and additives are added to the muck in the chamber to improve its workability. Then a jack cylinder system is used to push the muck through the bulkhead, and a screw conveyor is used to discharge it to generate a

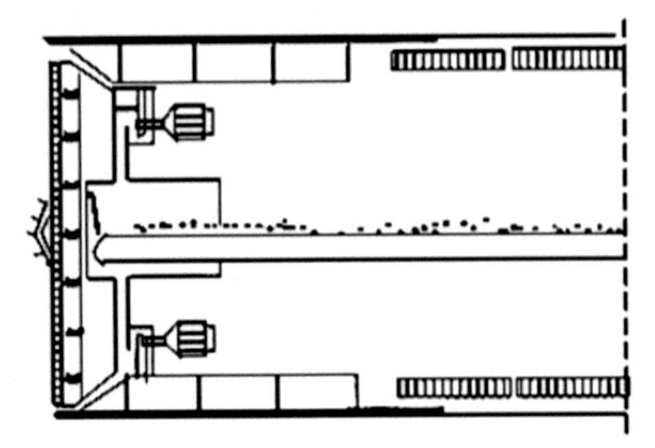

Figure 1.5 Mechanized shield machine.

Figure 1.6 EPB shield machine.

pressure in the muck, balancing the soil-water pressure and thereby retaining the soil at the excavation face.

As shown in Fig. 1.7, the slurry shield machine also uses a cutterhead to excavate the ground. It sets up a slurry chamber between the bulkhead and the excavation face (the German slurry shield also contains an air chamber, and the slurry pressure is adjusted by air pressure) and stabilizes the ground by applying pressure to the slurry in the chamber.

In general, the EPB and slurry shield machines can use the closed chambers to build pressure to stabilize the excavation face. By contrast, the stability of the excavation face of the tunnel constructed by hand-dug, semimechanized, and mechanized open-shield machines mainly depends

Figure 1.7 Slurry shield machine.

on the strength of the ground itself, or prereinforced ground is required to achieve the stability of the excavation face. However, for all closed shield machines, the ground conditions at the excavation face cannot be visually observed as they can when open shield machines are used, and a complete muck conditioning or slurry treatment system is required.

1.3 Development history of shield machine and tunnelling method

As a highly mechanized method of tunnelling, the shield tunnelling method depends largely on the development of the shield machine. It is particularly necessary to understand the development history of the shield machine.

1.3.1 Overview of development in countries other than China

1.3.1.1 The birth of shield machine

In 1806 French-born engineer Marc Isambard Brunel observed that a shipworm could drill holes in a wooden board [3–6]. The shipworm is a bivalve mollusk with a shell on its anterior end. When drilling through wood, it secretes liquid and coats the wall of the hole to form a tough protective shell, which resists the expansion of the wood when it is wet and prevents the shipworm from being flattened. Inspired by the drilling of shipworms and their use of secretions as protective shells, Brunel proposed the shield tunnelling method in 1818 and registered a patent in the United

Kingdom. In 1825 Brunel and his son, Isambard Kingdom Brunel, used the shield tunnelling method for the first time in an underwater tunnel under the Thames River in England. The shield machine had a rectangular cross-section with a height of 6.78 m and a width of 11.43 m, and it used brick segments as a support structure. Because the front of the shield was open and thus was not effective in blocking water, the project was shut down several times by water inrush accidents, which caused heavy casualties. With summing up the failure lessons, the Brunels built an improved shield machine with a square cast-iron frame and started construction again in 1834. After unremitting efforts, the underwater tunnel was completed in 1841 and put into use in 1843. The project took about 20 years in total. Although the construction lasted a long time and caused big losses due to the water inrush accidents, it provided rich experiences that could be used to improve the shield machine and shield tunnelling method. The shield was initially called a cell or cylinder. Moulton used the term "shield" for the first time in a patent application in 1866.

1.3.1.2 The evolution of excavation-face support method

In 1830 Lord Cochrane invented a pneumatic shield machine for crossing saturated water-bearing formations and obtained a patent. In 1869 Peter Barlow and James Henry Greathead constructed the second tunnel beneath the Thames River, using the round shield and fan-shaped cast-iron segments. Compared with the previous project using shield tunnelling method, this second project progressed relatively smoothly. Later, in 1887, Greathead used the compressed-air method to stabilize the excavation face for the first time in the South London Railway Tunnel, which laid the foundation for the modern shield tunnelling method. After a series of improvements, the shield tunnelling method was widely adopted in France, Germany, the United States, Russia and the Soviet Union, and other European and American countries in the late 19th to mid-20th centuries. The method was adopted to build tunnels for different functions in Baltimore, Paris, Berlin, Moscow, and other cities.

Previously, manual excavation was used for shield tunnelling method. In 1876 British inventors John Dickinson Brunton and George Brunton applied for the first patent for a mechanized shield machine. As shown in Fig. 1.8, the hemispherical rotating cutterhead is used to excavate the ground, and the soil slag falls on the hopper of the cutterhead [7]. The dirt is transported to the belt conveyor. In 1896 the British inventor J. Price developed a mechanized excavating shield machine with a spoke cutterhead, as shown in

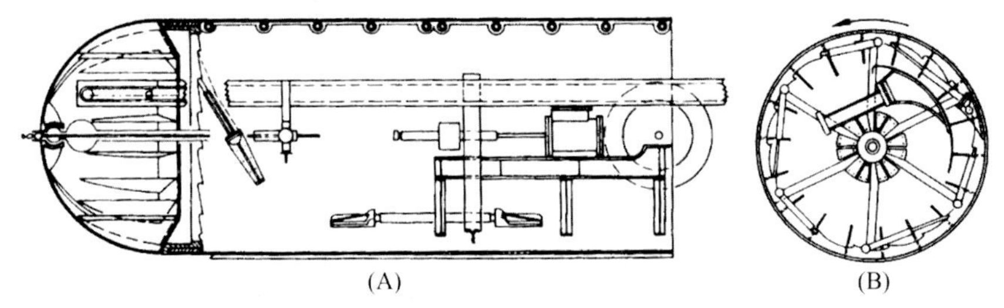

Figure 1.8 The first mechanized excavation shield machine, as a UK patent proposed by J. D. and G. Brunton in 1876. (A) Side view. (B) Front view. *Stack, B. Handbook of mining and tunnelling machinery. Chichester: Wiley; 1982.*

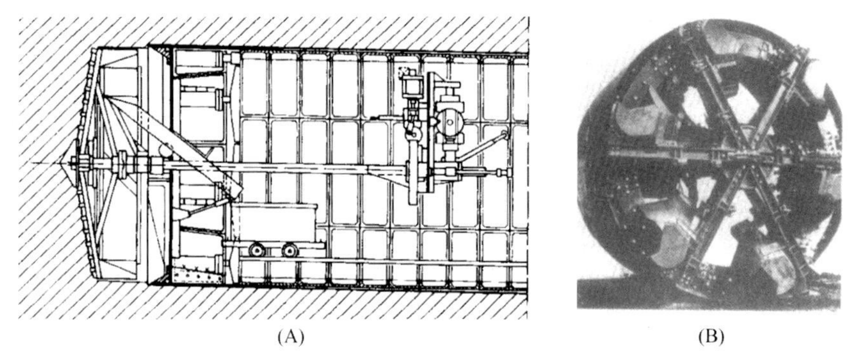

Figure 1.9 The first spoke-type cutterhead mechanized shield machine, as a UK patent proposed by J. Price in 1896. (A) Design drawing. (B) Shield machine with a spoke cutterhead. *Stack, B. Handbook of mining and tunnelling machinery. Chichester: Wiley; 1982.*

Fig. 1.9, and a circular shield was combined with the rotating cutterhead [7]. The four spokes of the cutterhead are equipped with cutters, and the cutterhead is driven by a long-axis motor. The shield machine was successfully used in the construction of a clay tunnel in London in 1897.

In 1896 the German inventor Haag proposed a shield machine that used liquid to support the tunnel excavation face and sealed the excavation chamber as a pressure chamber. He applied for the first German patent for the shield machine, as shown in Fig. 1.10.

More than 100 years after Brunel proposed the patent of the shield tunnelling method, Japan, the first country in Asia to adopt the shield tunnelling method. It was used in the construction of the National Railway Uetsu Main Line Tunnel in 1917, but because of insufficient preparation of the construction technology, the project was abandoned after being halfway

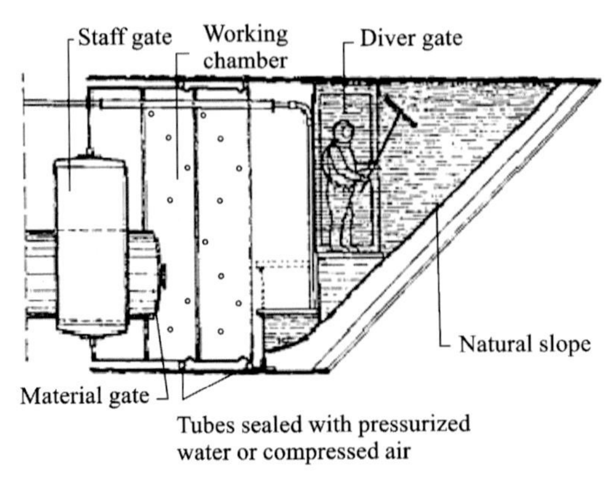

Figure 1.10 The first sealed warehouse liquid-supported excavation face shield machine, as a German patent proposed by Haag in 1896. *Modified from Maidl B., Herrenknecht M., Maidl U., Wehrmeyer G. Mechanised shield tunnelling. 2nd ed. Berlin: Ernst & Sohn; 2012.*

completed. In the construction of the drainage culvert of the old Tanna in 1926, the method also did not achieve the desired results. The first successful project adopting the shield tunnelling method in Japan was the Kanmon Railway Tunnel, which began construction in 1939. The tunnel used a pneumatic method to stabilize the excavation face and excavated it manually. After the Second World War, Japan applied the canopy shield tunnelling method in 1956 and 1957 to the Kanmon Road Tunnel and the Nagatacho Tunnel, respectively, of the Teito Rapid Transit Authority.

In the previous period, the excavation face of the shield tunnelling method was basically open. Even if it was supported by a baffle, it could not adequately support the excavation face. Although the compressed-air method can support the excavation face, it also has visible defects. For example, in the stratum with strong permeability, compressed air is easy to leak, and the excavation face cannot be stabilized well. Therefore to expand the scope of application of the shield tunnelling method, the stability of the excavation face must be improved. In 1961 British engineers Mott Hay, Anderson and John Bartlett first proposed slurry shield machine and applied for a British patent, but the lack of a tunnel project in the United Kingdom to promote this technology caused it to be temporarily shelved. Because of the poor stratum conditions in Japan, stability control of the excavation face was more important. The first slurry shield machine that used a cutterhead to cut soil and used hydraulic slag was successfully manufactured by Mitsubishi Corporation of Japan in 1967. In 1970 Japan Railway Construction

Company used a 7.29 m-diameter slurry shield machine in the Haneda Tunnel project, which increased the maximum diameter of the slurry shield machine, and the construction length reached 1712 m. Subsequently, Wayss and Freytag developed the first slurry shield machine in Germany and applied it to the construction of sewage pipes in Hamburg in 1974.

The EPB shield machine was developed two years after Bartlett proposed the slurry shield machine. In 1963 Sato Kogyo Company Ltd. of Japan developed the EPB shield. The first time that the EPB shield was officially used was in 1974. The shield machine produced by the Japanese manufacturer IHI (Ishikawajima Harima) Co., Ltd. had an outer diameter of 3.72 m. It was used for a 1900 m pipeline tunnel in Tokyo.

1.3.1.3 The development of shield styles

Shield machines were eventually developed having large cross sections, diversified shapes, and intelligence.

In 1985 Wayss Corporation, Freytag Corporation, and Herrenknecht Corporation applied for a combined patent for a hybrid shield machine. Its tunnelling mode could switch between the EPB mode and the compressed-air mode.

In 1986 Japan developed the world's first double-round slurry shield, also known as a double-headed slurry shield or double-joint slurry shield. It was manufactured by Hitachi Shipbuilding Co., Ltd. of Japan for the contractor Kumagaya Group Co., Ltd. This double-round slurry shield was a combination of two shields with a diameter of 7.42 m. It was used in the construction of the Kyobashi double-line tunnel on the newly built Keiyo Line in Japan in 1988. The tunnel length was about 620 m.

In 1992 Japan developed the world's first three-round slurry shield, which was successfully used in the construction of the Business Park station project on the Osaka Metro Line 7.

In 1994 in the Duisburg subway project, the slurry support and EPB support could be switched each other in a shield machine for the first time. In 1997 in Japan, an ultralarge cross-section slurry shield with a diameter of 14.18 m was used on the Eidan No. 7 line of the Tokyo subway system. The shield was also the world's largest-diameter "hugging mother-child slurry shield" at the time. A shield machine could excavate with different sections.

In 2000 Japan adopted the eccentric multiaxis slurry shield machine for the first time in the construction of the Future Port Line 21. In 2001 in the Kyoto subway project in Japan, the world's first double-track subway tunnel began construction with a rectangular shield [8].

In 2001 a 14.2 m slurry-balance shield was used to build the Lefortovo road tunnel in Moscow, Russia. This was the first time that such a large and complex shield system was used in Russia. The system included the world's largest mud separation plant at that time, with a total capacity of 5800 m^3 [9].

In 2003 the fourth tunnel project under the Elbe River in Hamburg, Germany, was successfully completed. The project started in 1997, using a 14.2 m-diameter composite slurry shield with the central leading small cutterhead, a bilge rock crusher, a sonic soft soil survey system, and atmospheric pressure tool change as well as other advanced technology that reflects the international advanced level at that time [10].

In 2005 the north tunnel of the M30 ring in Madrid, Spain, was constructed using the Gemini EPB shield machine (with a diameter of 15.2 m) built by Herrenknecht of Germany. The south tunnel was constructed with an EPB shield machine with a diameter of 15.2 m manufactured by Mitsubishi Heavy Industries of Japan, which had a daily excavation speed of up to 46 m [11].

In 2009 the Orlovsky road tunnel in St. Petersburg, Russia, was planned to be advanced with a 19 m-diameter Herrenknecht shield machine, which was expected to be the world's largest shield machine. However, because of continuous increases in cost estimates, the project was suspended [12].

In 2011 the Sparvo Tunnel in Italy was constructed using an EPB shield machine with a diameter of 15.55 m, the largest section at the time. The high construction risk and great technical difficulty were recognized in the world [13].

In 2013 the Caltanissetta tunnel in Sicily, Italy, was constructed using the super-large-diameter EPB shield produced by NFM. The shield was also one of the super-large EPB shields in Europe at that time, with a diameter of 15.08 m [14].

In 2016 the rainwater discharge pipe of the Tachiaigawa Main Line of Tokyo Sewerage was constructed adopting the spiral H&V shield method for the first time. It overcame the limitation of the tunnel curvature, reduced the cost of ground reinforcement, and made the shield more flexible.

In 2017 Kumagaya Group Co., Ltd. and JIMT of Japan proposed the Sunrise Bit Method of remote controlled-replacement technology for shield machines, which required no manual operation and used shield jacks and ratchets to rotate the tools for change operations.

In April 2017 the SR99 tunnel project in Seattle, Washington, United States, was successfully completed after overcoming difficulties such as ground deformation and seismic design. The tunnel was constructed using the Bertha shield with a diameter of 17.48 m produced in Japan. It was the largest-diameter shield in the world at that time [12].

1.3.2 Overview of development in China

1.3.2.1 Dawn period

Compared with other countries, the shield tunnelling method, especially shield machine production, developed late in China. The first application of the shield tunnelling method in China was in 1953. The Northeast Fuxin Coal Mine used a hand-drilled shield machine to build a 2.6 m-diameter water tunnel [6].

In 1962 the Shanghai Urban Construction Bureau Tunnel Department relied on the Tangqiao Tunnel Project to conduct research on the shield tunnelling method. The project was driven by a hand-digging shield machine with a diameter of 4.16 m (Fig. 1.11). The machine used bolted single-layer reinforced concrete segments as support and dewatering or air pressure to stabilize silty sand and soft clay formations. The tunnel with a length of 68 m was successfully dug, and valuable experience was gained for the follow-up projects.

Figure 1.11 Application of hand-dug shield machine in the Shanghai Tangqiao Tunnel.

In 1966 the Shanghai Tunnel Company began to build the Shanghai Dapu Road Cross-River Highway Tunnel, the first underwater highway tunnel in China. The project used a grid extrusion shield machine with a diameter of 10.22 m, as shown in Fig. 1.12. The shield machine used air pressure to stabilize the excavation face, and the length of excavation reached 1322 m. The tunnel was opened to traffic at the end of 1970.

In 1980 Shanghai first began to use the shield tunnelling method for metro tunnels in China. It used a shield machine with a diameter of 6.412 m for tunnelling a length of 565 m. The shield machine stabilizes the excavation face by pressurized slurry and local air pressure.

In 1986 the China Railway Tunnel Group used the half-section insertion shield machine to the Fuxingmen return line of the Beijing subway, combining the shield tunnelling method and the shallow buried excavation method, as shown in Fig. 1.13.

In 1988 the Cable Tunnel Project, which was built for the Shanghai South Railway Station crossing the Yangzi River, successfully used the first EPB shield machine, with a diameter of 4.35 m, in China. In 1995 Shanghai began to study rectangular shield tunnels and developed a 2.5 × 2.5 m variable grid rectangular shield machine. In 1996 the construction of the south line of the East Yanan Road Tunnel in Shanghai began. The tunnel was 1300 m in length and was excavated by a super-large slurry

Figure 1.12 Grid extrusion shield for the Dapu Road Tunnel.

Figure 1.13 Half-section insertion shield machine.

shield machine with a diameter of 11.22 m. In 2003 the first double-round shield machine in China, constructed by Shanghai Tunnel Engineering Co., Ltd., succeeded in tunnelling on the Shanghai Rail Transit Line 8.

1.3.2.2 Innovation period

The time range of 2003–2008 was the innovation period for China's shield machine manufacture and shield tunnelling technology. With the promotion of the national "863" plan, the independent researches and developments of shield machines officially entered the implementation stage [6].

In August 2002 the national "863" plan project entitled as Φ6.3 m Full-section Tunnel Boring Machine Research and Design was launched. The China Railway Tunnel Group and other units completed the design of the mainframe structure, hydraulic transmission system, electrical system, and back-up system. The project also enabled the research, design, development, and manufacture of shield cutter system and the development and application of shield foam additives and shield sealing grease.

At the end of 2002 the national "863" project entitled as Research and Application of Key Technologies for Shield Tunnelling Machine Cutterhead and Hydraulic Drive System was launched. In July 2004 the cutterhead and hydraulic system developed by the project were successfully applied to the industrial test on Shanghai Metro Line 2, and all the indicators met the project requirements.

In October 2004 China's first EPB shield machine with completely independent intellectual property rights was successfully launched in the interval tunnel of the west extension of Shanghai Metro Line 2.

In July 2005 China began research and development of slurry shield machine. Also under a national "863" plan, the tunnelling system and segment assembly machine of the slurry balance shield was developed.

In 2008 China successfully developed the first long-term composite EPB shield machine, which can provide smooth excavation under different geological conditions such as clay soil, weathered rock, soft and hard uneven strata, and sand and gravel strata. The shield machine was successfully used in construction of Tianjin Metro Line 3.

Also under a national "863" plan, in December 2008, Shanghai Tunnel Engineering Co., Ltd., together with Zhejiang University and China Railway Tunnel Group, successfully developed China's first slurry shield machine with independent intellectual property right and successfully applied it in the Shanghai Dapu Road double-way tunnel.

1.3.2.3 Transition period

Beginning in 2009, China's independent innovation capabilities of shield machine have been significantly improved. The State Key Laboratory of Shield Machine and Tunnelling Technology and the State Key Laboratory of Full-face TBMs have been established, and Chinese shield machine manufacturers have grown rapidly. These include China Railway Engineering Equipment Group Co., Ltd.; China Railway Construction Heavy Industry Group Co., Ltd.; North Heavy Industry Equipment (Shenyang) Co., Ltd.; Shanghai Tunnel Engineering Co., Ltd., Machinery Manufacturing Branch; CCCC Tianhe Machinery Manufacturing Co., Ltd.; Liaoning Sansan Industrial Co., Ltd.; China Shipbuilding Heavy Equipment Co., Ltd.; and other shield machine manufacturers.

In 2013 a tunnel section at the second phase of Line 6 of the Beijing Metro was selected as the test section of construction with an open face shield machine (Fig. 1.14). The open face shield machine has the advantages of a visible excavation face and no need for muck improvement. Despite the relatively large surface settlement at the beginning of the tunnelling process, the open-shield machine test section was successfully completed through continuous ground improvement and optimization of the process.

In December 2013 the world's biggest rectangular shield machine at that time was independently developed and manufactured by China Railway Engineering Equipment Group Co., Ltd. in Zhengzhou, China. The shield machine was 10.12 m wide and 7.27 m high. It was used in the construction of the Zhengzhou Metro tunnel underneath Zhongzhou Avenue.

Figure 1.14 The open-shield machine used in construction of Beijing Metro Line 6.

In May 2014 the first slurry/earth pressure dual-mode shield machine in China was developed in the tunnel section from Huadu Automobile City to Guangzhou North Station on the Line 9 of the Guangzhou Metro. The equipment integrated the functions of an EPB shield machine and a slurry shield machine. The shield machine can quickly switch between the two different driving modes through the control system according to the change of stratum, without disassembling any equipment, to ensure the construction quality and efficiency of the project.

In October 2015 the first ultra-large rectangular shield machine designed, manufactured, and domestically produced by China was successfully developed and was used in the underground connection channel of Shanghai Hongqiao Linkong Park. In comparison to traditional circular shield machines, this shield machine can save about 20% of underground space resources.

In 2018 the first TBM/EPB dual-mode large shield in China was successfully launched on the tunnel section from the Dayuan Station to Taihe Station on the downward part of Guangzhou-Foshan Ring Line in the Pearl River Delta Railway. The shield machine with an excavation diameter of 9.15 m can be used not only for soft ground and extremely upper soft and lower hard formations, but also for long-distance super-hard rock formations.

In November 2019 the first dual-mode shield machine with a screw conveyor for muck/slurry discharging in China was launched in the Shunde section of the westway of Guangzhou Metro Line 7 in Foshan City. It enabled the interchange of the two modes of EPB and slurry supports. When the shield machine digged in the slurry support mode, the screw conveyor was used to discharge the slurry, which effectively avoided the problem of muck stagnation appearing in the traditional slurry shield machine. In the same month the first EPB/TBM dual-mode shield machine with a central screw conveyor in China was used successfully. Both the earth pressure mode and the TBM mode used the central screw conveyor to discharge muck, creatively solving the problems of mud breakout, water gushing, inability to seal the chamber pressure, and other problems that may occur when the EPB mode is converted to the TBM mode or during the TBM digging process. The shield machine was used for the construction of the tunnel section between Buji Station and Shiyaling Station of Shenzhen Metro Line 14.

Overall, at the end of the 20th century and the beginning of the 21st century, driven by the development of rail transit and other civil infrastructure, China has built a large number of tunnels which were constructed with shield machines manufactured by other countries. Through independent innovations over the past 20 years, especially the past 10 years, the shield machine manufacture industry and shield tunnel construction technology have developed rapidly in China.

1.4 Development trend of shield tunnelling method

Because of the development needs of civil infrastructures, besides the continuous development of mechanization, informatization, and intelligence, the shield tunnelling method has the following development trends.

1.4.1 Miniaturization and supersizing

To meet the development needs of tunnel and underground space construction, the trends of shield machine development have been running in two directions of super-large and super-small cross sections [15]. Table 1.1 shows incomplete statistics of large-scale shield tunnel projects in recent years. On April 4, 2017, the world's largest-diameter (17.48 m) super-large EPB shield machine "Bertha" completed the excavation of the Seattle Highway SR99 tunnel in the United States. The Tuen Mun-Chek Lap Kok connection tunnel project in Hong Kong, China, completed in 2020, uses a slurry shield with a diameter of 17.6 m. The diameter of the shield exceeds the diameter of the Bertha shield in the Seattle SR99 project, becoming the shield with the largest diameter in the world at that time. The China's self-developed Chunfeng slurry shield

Table 1.1 Incomplete statistics of large-scale shield tunnel projects in recent years [16,17].

Project name	Shield diameter (m)	Type	Year
Tokyo Bay Road Tunnel, Japan	14. 14	Slurry	1994
Tunnel 4 of the Elbe River, Germany	14. 20	Slurry	1997
Tokyo Metro, Japan	14. 18	Slurry	1998
Green Heart Tunnel, Netherlands	14. 87	Slurry	2000
Lefortovo Highway Tunnel, Russia	14. 20 (reconstruction by the shield machine of the Elbe 4th)	Slurry	2001
Shangzhong Road Tunnel, Shanghai, China	14. 87 (reconstruction by the shield machine of the Dutch Green Heart Tunnel)	Slurry	2004
Silberwald highway tunnel, Russia	14. 20 (reconstruction by the shield machine of the Lefortovo Highway Tunnel)	Slurry	2004
M30 Ring Tunnel, Madrid, Spain	15. 01,15. 20	EPB	2005
Shanghai Yangtze River Tunnel, China	15. 43	Slurry	2006
Jungong Road Tunnel, Shanghai, China	14. 87 (reconstruction by the shield machine of the Dutch Green Heart Tunnel)	Slurry	2006
Shanghai Bund Tunnel, China	14. 27	EPB	2007
Nanjing Yangtze River Tunnel, China	14.93	Slurry	2008
Yingbin 3rd Road Tunnel, Shanghai, China	14. 27 (reconstruction by the shield machine of the Shanghai Bund Tunnel)	EPB	2009
Qianjiang Tunnel, Hangzhou, China	15. 43 (reconstruction by the shield machine of the Shanghai Yangtze River Tunnel)	Slurry	2010
SE-40 Highway Tunnel, Spain	14. 00	EPB	2010
Sparvo Highway Tunnel, Italy	15. 55	EPB	2011

(Continued)

Table 1.1 (Continued)

Project name	Shield diameter (m)	Type	Year
West Changjiang Road Tunnel, Shanghai, China	15. 43 (reconstruction by the shield machine of the Qianjiang Tunnel, Hangzhou, China)	Slurry	2011
Weisan Road Tunnel, Nanjing, China	14. 93	Slurry	2011
Alternative Tunnel for Viaduct in Seattle, Washington, United States	17. 48	EPB	2011
Hongmei Road Tunnel, Shanghai, China	14. 90	Slurry	2012
Waterview Highway Tunnel, Auckland, New Zealand	14. 41	EPB	2013
Caltanissetta Highway Tunnel, Italy	15. 08	EPB	2013
Slender West Lake, China	14. 93 (reconstruction by the shield machine of the Nanjing Yangtze River Tunnel, China)	Slurry	2013
Tuen Mun-Chek Lap Kok Project, Hong Kong, China	17. 63,14. 00	Slurry	2015
Liantang Highway Project, Hong Kong, China	14. 10	EPB	2015
Sanyang Road Tunnel, Wuhan, China	15. 76	Slurry	2015
Hengqin Tunnel, Zhuhai, China	14. 90 (reconstruction by the shield machine of the Hongmei Road Tunnel, Shanghai, China)	Slurry	2016
Santa Lucia Highway Tunnel, Italy	15. 87	EPB	2016
A30 Riverside Expressway, Shanghai, China	15. 43 (reconstruction by the shield machine of the West Changjiang Road Tunnel, Shanghai, China)	Slurry	2016
Beiheng Expressway, Shanghai, China	15. 53	Slurry	2016
Tokyo Outer Ring Tunnel, Japan	16. 10	EPB	2017

(*Continued*)

Table 1.1 (Continued)

Project name	Shield diameter (m)	Type	Year
Chuguang Road Tunnel, Shanghai, China	14. 41 (reconstruction by the shield machine of the Waterview Highway Tunnel, Auckland, New Zealand)	Slurry	2017
Westgate Tunnel, Melbourne, Australia	15. 60	EPB	2017
Suai Passage, Shantou, China	15. 03	Slurry	2017
Heping Avenue Tunnel, Wuhan, China	15. 4	Slurry	2017
Nanjing Heyan Road Tunnel, China	15. 01	Slurry	2017
Crossing the Yellow Tunnel in Jinan, China	15. 74	Slurry	2017
Chunfeng Tunnel, Shenzhen, China	15. 80	Slurry	2018
Zhoujiazui Road Tunnel, Shanghai, China	14. 90	Slurry	2018
Meizizhou Tunnel, Nanjing, China	15. 43	Slurry	2018
Oujiang Tunnel, Wenzhou, China	14. 93	Slurry	2018
Mawan Tunnel, Shenzhen, China	15.5	Slurry	2019
Heao Tunnel, Shenzhen, China	18.1	Slurry	Planned 2024

Source: Tan S., Sun H. Key design technology for cutter replacing of super-large slurry shield under atmospheric condition: case studies of Shantou Gulf tunnel and Shenzhen Chunfeng tunnel. Tunn. Constr. 2019;39(7):1073-1082. Sun H., Feng Y. Statistics on global super large diameter tunnel boring machines. Tunn. Constr. 2020;40(6):925–8.

machine with a diameter of 15.80 m has been put into use in the Shenzhen Chunfeng Road Tunnel Project (Fig. 1.15). In addition, the small shield machine with a diameter of 2.8 m was successfully applied to the water inlet and outlet pipeline project of the main sewage pipe of Nanjing Hongwu Road in 2016, which promoted the development and application of China's small-diameter shield machine. Table 1.2 shows the application of small-diameter shield tunnels in China.

Figure 1.15 The large-diameter slurry shield independently developed for Shenzhen Chunfeng Road Tunnel Project, China.

1.4.2 Diversified forms

To meet the needs of different projects, there are a growing number of cross-sectional forms of shield machine. At present, the shield machines with round, rectangular and horseshoe shape, double or triple-circle shield machines, spherical shield machines, and those with primary-secondary cutterhead have been adopted (Fig. 1.16). In the future, the cross-sectional form of the shield machine will develop in the direction of heteromorphic shape, gradually expanding the scope of application of the shield method.

1.4.3 High level of automation

Shield machines use robot-like technology, computer control, remote control, sensors, laser guidance, advanced geological exploration, and communication technology have been or will be developed and applied. With the rapid development of computer technology, the level of shield tunnelling automation is getting higher and higher, having functions of construction data acquisition, shield posture management, construction data management, equipment management, and real-time remote transmission of tunnelling data. The machine can automatically detect the position and posture of the shield machine and use fuzzy mathematics theory to automatically adjust and control the chamber pressure and automatically realize the segment assembly.

1.4.4 High adaptability

With the development of modern tunnelling technology, the tunnelling technologies for soft ground and hard rock have been merging, greatly

Table 1.2 Some examples of typical minitunnel projects in China.

Number	Project name	Geological condition	Inside diameter (m)	Completion period
1	Shanghai river Crossing water intake tunnel	The soil layer across the river bottom is gray silt clay, gray clay, and dark green subclay. The first two soil layers are soft soils with high compressibility.	2.9	1999
2	Beijing Liangma River sewage tunnel	Light clayey loam–heavy clayey loam; medium-textured loam–heavy clayey loam; light clayey loam–heavy clayey loam	2.7	2000
3	Bahe sewage tunnel	Medium-textured loam–light clayey loam; fine sand–silt; fine sand–medium sand; mostly medium sand	2.7	2001
4	Liangshuihe sewage tunnel	Recently deposited pebbles, boulders, sedimentary fine sand–medium sand layer	2.7	2002
9	Aolin cable Tunnel in Guangzhou	The formation is mainly silty clay, silty sand, and fully weathered mudstone	3.5	2010
10	Beijing Shunyu Road power tunnel	The traversing strata are mainly clay, silt and sand, while the groundwater is abundant	3.0	2012

(*Continued*)

Table 1.2 (Continued)

Number	Project name	Geological condition	Inside diameter (m)	Completion period
11	Shenzhen high pressure gas pipe	Passing through strongly weathered-medium weathered rock, the average compressive strength of the rock is 116.3 MPa	2.2	2014
12	Guangzhou 220 kV Hangyun Transmission and transformation power tunnel	The groundwater is mainly clastic rock fissure water, the aquifer is a strong-medium weathered zone of the Permian rock formation, and the lithology is mainly argillaceous siltstone, siltstone, shale, etc.	3	2015

enhancing the geological adaptability of shield machines. The multimode shield machines with two or more tunnelling modes have become a trend for the shield tunnelling with highly geological adaptability. They adopt different driving modes and different cutterhead layouts to adapt to different strata. Owing to their wide geological adaptability, the multimode shield machines have broad application prospects in China's cities such as Guangzhou, Chengdu, Chongqing, Shenzhen, and Qingdao.

1.5 Book organization

This book systematically introduces engineering survey, design, construction, and maintenance for shield tunnels, including an introduction to shield tunnel engineering, geological survey, shield machine configurations and working principles, shield machine selection, structure types and design of shield tunnel lining, shield advancing and segment assembly, backfill grouting for shield tunnelling, muck conditioning for EPB shield tunnelling and muck recycling,

(A) (B)

Figure 1.16 Special-shape shield machine. (A) Rectangular. (B) Round.

slurry treatment for shield tunnelling and slurry recycling, environmental impact and its control in shield tunnel construction, and shield tunnel structure defects and their treatment.

Chapter 1, Introduction focuses on the basic concepts of shield tunnel engineering and the types of shield machines and summarizes the development histories and technology development trends of shield machines and construction technology.

Chapter 2, Geological Survey and Alignment Design for Shield Tunnel Project introduces the survey content of shield tunnel engineering in detail and emphasizes the difference from other tunnel engineering surveys.

Chapter 3, Shield Machine Configurations and Working Principles introduces the configurations of shield machine, explains the working principles of popular shield machines such as EPB and slurry ones, and introduces the development of several new types of shield machine.

Chapter 4, Shield Machine Selection introduces the principles and main methods of shield machine selection and discusses shield selection techniques commonly in dealing with complex stratum, along with practical cases.

Chapter 5, Structure Type and Design of Shield Tunnel Lining introduces the shield lining structure type and presents in detail the main methods for calculating the surrounding rock pressure of shield tunnel segments and structural internal forces, followed by discussion of the design of shield tunnel structure.

Chapter 6, Launching and Receiving of Shield Machine focuses on stratum reinforcement methods in the shield tunnel ends, the common equipments and safety control technologies for shield launching and receiving, and several new shield launching and receiving technologies.

Chapter 7, Shield Tunnelling and Segment Assembly introduces excavation technology, segment assembly, and construction measurement and monitoring of the EPB and slurry shield machines and introduces the shield attitude control and correction technologies in detail. Some examples are presented to illustrate the shield construction technology under special conditions.

Chapter 8, Backfill Grouting for Shield Tunnelling introduces the materials, equipments, and technologies of the grouting behind segments and discusses the design and construction of the backfill grouting.

Chapter 9, Muck Conditioning for EPB Shield Tunnelling and Muck Recycling introduces the characteristics of EPB shield muck and the reasons for muck conditioning, the types and technical parameters of conditioning agents, the evaluation methods of muck-conditioning effects, and the mechanical behavior of shield tunnelling under muck conditioning.

Chapter 10, Slurry Treatment for Shield Tunnelling and Waste Slurry Recycling mainly discusses slurry treatment, as one of the key technologies during slurry shield tunnelling. In addition to introducing slurry function and performance requirements and slurry parameters, the chapter details slurry treatment equipment and site layout. Finally, to address the increasing importance of environmental protection, it states the recycling of waste slurry as a resource.

Chapter 11, Ground Deformation and Its Effects on the Environment introduces in detail the ground deformation caused by shield tunnel construction and its impact on surrounding buildings and discusses common control standards and control measures. Finally, the control technology of a shield tunnel passing beneath a existing high-speed railway tunnel is introduced as an example.

Chapter 12, Shield Tunnel Structure Defects and Their Treatments introduces the common disease forms and causes of shield tunnel lining structures and investigates the lining structural defects and safety evaluation methods. Finally, the chapter describes the treatment of shield tunnel structural defects.

References

[1] The Japanese Geotechnical Society, translated by Niu Q, Chen F, Xu H. Shield tunnel investigation, design and construction, Beijing: China Architecture Press; 2008.
[2] Zhu Y, Zhu Y, Huang D, et al. Development and application of similar rectangular shield tunnel technology. Mod. Tunn. Technol. 2016;53(S1):1–12.
[3] Maidl B., Herrenknecht M., Maidl U., Wehrmeyer G. Mechanised shield tunnelling. 2nd ed. Berlin: Ernst & Sohn; 2012.

[4] Zhou W. In: Shield tunnelling technology and application, Beijing: China Architecture Press; 2004.
[5] Zhang F, Zhu H, Fu D. Shield tunnel. Beijing: China Communications Press; 2004.
[6] Chen K, Wang J, Shunhui Tan, et al. Shield design and construction. Beijing: China Communications Press; 2019.
[7] Stack B. Handbook of mining and tunnelling machinery. Chichester: Wiley; 1982.
[8] Nakamura H, Kubota T, Furukawa M, Nakao T. Unified construction of running track tunnel and crossover tunnel for subway by rectangular shape double track cross-section shield machine. Tunn. Undergr. Space Technol. 2003;18(2-3):253−62.
[9] Wallis S. Speedy mega TBM in Moscow. Tunnels & Tunnelling International 2002;34(12):24−7.
[10] Falk C. Pre-investigation of the subsoil developments in construction of the 4th Elbe tunnel tube. Tunn. Undergr. Space Technol. 1998;13(2):111−19.
[11] Melis M. Mega'EPBMs lead the way for Madrid's renewal. Tunnels & Tunnelling International; 2006. p. 23−5.
[12] Ledyaev AP, Kavkazsky VN, Ivanes TV, et al. Study in the structural behavior of precast lining of a large diameter multifunctional tunnel performed by means of finite elements analysis with respect to saint-petersburg geological conditions. Civ. Environ. Eng. 2019;15(2):85−91.
[13] Xiao X. Key construction technologies used for the large cross section shield bored tunnel: a case study on sparvo tunnel in Italy. Tunn. Constr. 2013;33(10):866−73.
[14] Capozucca F, Barreca G, Monaco C, et al. Large civil works and complex geological contexts: the importance of the geological model in the building of the Caltanissetta tunnel (central Sicily). Geoing. Ambientale E Mineraria 2013;140(3):21−6.
[15] Chen K, Hong K, Wu X. Shield construction technology. Beijing: China Communications Press; 2016.
[16] Tan S, Sun H. Key design technology for cutter replacing of super-large slurry shield under atmospheric condition: case studies of Shantou Gulf tunnel and Shenzhen Chunfeng tunnel. Tunn. Constr. 2019;39(7):1073−82.
[17] Sun H, Feng Y. Statistics on global super large diameter tunnel boring machines. Tunn. Constr. 2020;40(6):925−8.

Exercises

1. Please briefly describe the basic concepts of the shield tunnelling method and shield tunnel engineering.
2. Please comprehensively analyze the development process of shield tunnelling technology throughout the world.
3. Please briefly summarize the main features of shield tunnelling method.
4. Please give an example to illustrate future development trends in shield tunnelling method.

CHAPTER 2

Geological survey and alignment design for a shield tunnel project

2.1 Purposes of geological survey

Shield tunnelling is a highly mechanized tunnel construction method. Compared with traditional construction methods, the shield tunnelling method is relatively safe and fast. However, the tunnelling efficiency of a shield machine greatly depends on the stratum, hydrology, environment, and other conditions. Except when the shield tunnelling process is shut down for tool changing, it is difficult to change its configuration and other aspects, let alone replacing shield tunnelling with other construction methods, which would increase the cost considerably. Therefore selection of the type of shield machine and its construction parameter settings must take into account the geological conditions of the project, which impose higher requirements for the geological survey of a shield tunnel project.

Accidents may be caused in shield tunnel construction by carelessness or inaccuracy in the geological and environmental survey. For example, during shield crossing through a silt stratum, if the local gravel is not accurately surveyed (and the groundwater level is high) and an extruding shield machine is selected, there are likely to be serious gushing accidents when the machine is passing through the gravel area. In addition, if methane gas, perhaps hidden in the stratum, is not detected during shield crossing through a swamp, failure to take emergency prevention measures and reserve emergency equipment may lead to accidents and loss in shield tunnel construction. Accidents such as pipeline breakage, surface building inclination, and wall cracking may be caused in shield tunnel construction by an inadequate survey of underground pipelines and facilities near the tunnel route before construction. Many engineering accidents in shield tunnel construction have been caused by a careless or inaccurate geological survey, with relevant experiences and lessons too numerous to mention. A meticulous survey of each shield tunnel project is a prerequisite for successful construction of shield tunnels and must be done with care [1].

A geological survey for a shield tunnel project comprises exploration activities carried out to meet the special requirements of shield tunnel design,

Shield Tunnel Engineering
DOI: https://doi.org/10.1016/B978-0-12-823992-6.00002-3

machine type selection, tunnelling parameters setting for the shield boring etc. Necessary fundamental data and related geotechnical parameters are provided for all stages of planning, design, construction, and maintenance management through investigation, analysis, and evaluation of the geological and geographical environmental features and geotechnical engineering conditions of a construction site. Meanwhile, existing geotechnical engineering problems and environmental problems are analyzed and evaluated, and reasonable measures and suggestions are proposed so that economic, reasonable, safe, and reliable design and construction of the shield tunnel project can be achieved.

2.2 Geological survey contents and methods

2.2.1 Geological survey contents

To obtain the necessary geological survey documents and data for safe, fast, and economical construction, the survey activities of a shield tunnel project are generally divided into four stages: planning survey, design survey, construction survey, and follow-up survey after completion. At the design stage, the survey can be divided into the preliminary survey and the detailed survey, and the survey at the construction stage can be divided into the supplementary survey and the construction effect survey. Table 2.1 classifies types and methods of geological survey for a shield tunnel construction project at the stages of planning, design, construction, and maintenance management [2]. As Table 2.1 shows, pertinence and emphasis vary among survey types and methods at different stages.

2.2.1.1 Survey at the feasibility study stage

A geological survey at the overall planning stage, also called a geological survey at the feasibility study stage, has determination of the route plan as a main purpose. The survey at this stage is mainly to determine the route location of a shield tunnel, the location of launching and reception shaft, and the applicability of a shield construction method. The survey results at this stage can also be used as reference data at the later design and construction stages. Survey items generally include site conditions, obstacle distribution and quantity, topography, soil properties, surrounding environment, and previous construction examples.

Site condition survey

A site condition survey includes an investigation of land use and rights; planning of future facilities in adjacent areas; road types and road traffic

Table 2.1 Steps in a geological survey for a shield tunnel project.

Stages		Survey types	Survey methods
Overall planning, feasibility study		Feasibility study survey	Data collection, site reconnaissance and investigation
Design	Preliminary design	Preliminary survey	Site survey, laboratory test, site reconnaissance
	Construction drawing design	Detailed survey	Site survey, laboratory test, site reconnaissance
Construction	Shield tunnelling construction	Construction survey (verification survey)/ supplementary survey	Site survey, laboratory test, site measurement
	Auxiliary measures construction	Construction survey for auxiliary measures (effect survey)	Site survey, laboratory test, site measurement
Operation and maintenance management		Follow-up survey	Site survey, site measurement

conditions; the presence or absence of engineering lands and surrounding environment, including natural environments such as rivers, lakes, and oceans; power supply and water supply conditions; and drainage facilities and equipment status. The tunnel route location and alignment are determined on the basis of survey results, an equipment demand plan is drawn up, and environmental protection measures are determined.

Obstacle survey

An obstacle survey is a survey of original above-ground and underground structures, buried objects, water wells, ancient wells, cultural relics, previous engineering construction records, and so on. If embedded sheet piles or foundation piles of abandoned facilities are found, then measures, for example, manholes for cutting and devices for removing these obstacles, should be arranged on the shield machine in advance.

Topographic and soil surveys

Topographic and soil surveys should give priority to site reconnaissance, investigation, and data collection at the basic planning and feasibility study stages. Survey items mainly include topography, stratum structure, soil properties, groundwater conditions, the presence of anoxic gas and toxic gas in the strata, previous large-scale surface settlement, and the like. It is also necessary to understand and investigate the steady state of the short-term and subsequent long-term settlements caused by the filling loads in land reclamation and the groundwater extraction in construction activities. Such investigation can help to distinguish the influence range of tunnelling induced settlement so that to minimize the impact of the planned shield tunnel construction.

Surrounding environment survey

A surrounding environment survey is done to investigate problems such as noise, vibration, ground displacement, grouting induced groundwater pollution and discharged waste induced environmental pollution. Therefore a survey is required to investigate the effects of such problems and to determine the corresponding control measures.

Engineering case study

Investigations of previous construction cases in the same area or with similar stratum conditions play a very important role in type selection of shield machine and in construction organization. Construction cases in which accidents have occurred are particularly useful in determining the shield machine type, designing the auxiliary construction methods, planning environment protection measures, and other aspects.

2.2.1.2 Survey at the design stage

Preliminary survey

The survey at the preliminary design stage, known as the preliminary survey, aims to explore the regional geological, engineering geological, and hydrogeological conditions along a shield tunnel; analyze and evaluate the applicability of a tunnel route plane, buried depth, structural forms, and construction methods; predict possible risks in geotechnical engineering aspects; provide geotechnical parameters required for preliminary design; and put forward preliminary suggestions for geotechnical treatment in complicated or special strata areas. Related documents and requirements include the following:

1. Collecting related design documents and feasibility study survey reports, analyzing and evaluating the proposed tunnel plan and longitudinal profile of the proposed tunnel route in a topographic map, structural forms and construction methods, and the distribution map of underground facilities along the tunnel line.
2. Preliminarily investigating the geological structure, geotechnical types and distribution, geotechnical physical and mechanical properties, and groundwater conditions; classifying the surrounding strata and determining geotechnical construction type according by dividing the engineering geology sections; evaluating site stability and construction applicability; investigating the types, causes, distribution, scale, and engineering properties of special ground section and adverse geological actions, and analyzing their adverse effects to the projects and predicting their development trends.
3. Preliminarily investigating the surface water level, flow quantity, water quality, and distribution of river and lake sediments, finding out the recharge and discharge relationship between surface water and groundwater; preliminarily ascertaining the groundwater types, recharge, runoff and discharge conditions, highest water level on records, and dynamic change rules of groundwater; and preliminarily evaluating the corrosive effects of water and soil on the construction building materials.
4. Preliminarily analyzing and evaluating possible underground construction and support schemes such as foundation types, vertical shafts, and horizontal passages and groundwater control measures; classifying soil types and site type with regard to the sites with seismic fortification intensity equal to or greater than VI degree; preliminarily evaluating the possibility of seismic liquefaction and subsidence degree; evaluating site stability and construction suitability; analyzing possible engineering problems for project with high environmental risk level and putting forward suggestions for preventive measures.
5. Besides conventional properties of foundation base soil, providing special parameters, such as a permeability coefficient, unconfined compression strength, triaxial unconsolidated undrained shear strength, and consolidated undrained shear strength.

Detailed survey

The survey at the construction drawing design stage, known as the detailed survey, aims to determine the design conditions for the shield machine, lining, and shaft based on soil test results and according to the plane and vertical alignment along the tunnel which are determined in the preliminary survey stage. The detailed survey focuses on surveying various features of special strata on a shield tunnel construction site, identifying technical problems that may occur in the shield tunnel construction process and giving corresponding suggestions. Related data and requirements include the following:

1. Collecting plane and longitudinal profile, load, structural forms and features, construction methods, foundation type and overburden depth, deformation control requirements and other data of the proposed tunnel project including the information of coordinates and terrain.
2. Finding out the distribution of each rock and soil layer; providing physical and mechanical property indexes of each rock and soil layer; calculating the subgrade coefficient, the static lateral pressure coefficient, the thermophysical index, resistivity, and other geotechnical parameters required for construction; investigating the hydraulic connections between surface water and groundwater that affect the project; determining the hydrogeological data such as the groundwater buried conditions, water types, water level, water quality, geotechnical permeability coefficient, and groundwater level change amplitude; analyzing the effects of groundwater on the construction and putting forward suggestions for groundwater control.
3. Determining the types, ages, causes, distribution range, and engineering properties of the strata layers of the site; analyzing and evaluating the stability, uniformity, and bearing capacity of the foundation base; investigating adverse geological actions and special ground conditions; finding out the distribution and features of geological conditions such as saturated sand layers, pebble layers, and boulder layers that may have a negative impact on the tunnel construction; analyzing the potential impact on the project and putting forward suggestions for engineering prevention measures.
4. Carrying out geotechnical engineering analysis and evaluation for auxiliary engineering sections such as entrances and exits, air shafts and passages, construction passages, cross passages and others in accordance with the engineering geological properties, site conditions and surrounding environmental conditions; offering suggestions for engineering inspection of foundation-bearing capacity and stratum reinforcement

effects, and providing suggestions for monitoring the engineering structures, surrounding environment, ground deformation, groundwater level changes, and so on; analyzing and evaluating the impact of engineering precipitation and shield tunnel construction on the surrounding environment of the project and putting forward suggestions for surrounding environmental protection measures.

5. Comprehensively analyzing regional engineering geological and hydrogeological data, and geological hazard data; analyzing and evaluating the stability of the foundation and surrounding rock; predicting possible geotechnical engineering problems and proposing suggestions for engineering treatment; and evaluating the impact of adverse geological actions on the project (predicting the possibility of geological hazards caused by adverse geological actions under the influence of human construction activities based on geological hazard history).

2.2.1.3 Survey at the construction stage

The survey at the construction stage is roughly divided into two steps: a supplementary survey for shield tunnel construction and a construction effect survey of the auxiliary construction method.

Supplementary survey for shield tunnel construction

Before shield tunnel construction starts, a geological supplementary survey should be carried out on the basis of the original detailed geological survey. The supplementary survey focuses on a geological investigation within the range of the shield tunnel route. The ground sections which probably has an impact on the shield tunnel construction mentioned in the detailed survey report should be further investigated in details. Supplementary survey borehole should be carried out beyond a certain range outside the tunnel body to avoid any connection of the shield soil chamber to the surface ground through the survey boreholes to cause pressure loss and leakage. Under special complicated geological conditions, supplementary survey drilling needs to be conducted on the stratum within the tunnel excavation range, the boreholes should be effectively plugged to prevent pressure leakage from soil chamber in the shield tunnelling process.

The methods and key contents of a supplementary survey at construction stage are shown in Table 2.2.

The followings are areas needing attention in a supplementary survey:

1. The quantity and spacing of borehole points and their layout are comprehensively determined according to data and requirements provided

Table 2.2 Lists of methods and key contents of supplementary survey for shield tunnel construction.

No.	Supplementary survey contents	Supplementary survey methods	Key contents
1	Supplementary survey especially for karst stratum	Circular divergent detection with geological borehole	Karst cave diameter, height, and filling pattern
2	Supplementary survey especially for boulder stratum	Geological borehole, ultrasonic detection, and micro motion detection	Boulder size, strength, and distribution
3	Supplementary survey for soft stratum	Geological borehole and coring	Compression coefficient, distribution range, and relation with tunnel location

by professional designer and considering factors such as engineering geological and hydrological conditions, tunnel depth and construction methods. The supplementary survey borehole locations should be selected to avoid the range of the tunnel excavation and to avoid the situation in which a drill pipe falls into the hole in the drilling process, therefore affecting the shield tunnel construction. If the longitudinal slope of the tunnel line is changing, the boreholes can be appropriately deepened to prevent waste of the exploration workload.

2. Adverse geology and special stratum distribution along the tunnel route needs to be investigated in the process of supplementary survey. A detailed understanding of the adverse geological areas will beneficially support the later shield tunnel construction.
3. After completion of the supplementary survey, the adverse geological areas (such as karst, boulder, and soft clay stratum) should be analyzed according to exploration reports, and treatment measures should be determined in advance.

Survey of effects of auxiliary construction measures

To protect the stability of the launching and receiving shaft, the shield machine as it goes into and out of the tunnel entrance and exit, the curved tunnel section and section adjacent to structure, auxiliary construction methods such as a grouting, high-pressure jet grouting deep mixing, and ground freezing are usually adopted to reinforce the stratum. The main purpose of the effect survey of the auxiliary construction measures is to determine the effectiveness of stratum reinforcement. The main survey content includes the final quality of the reinforced ground body, the shape and range of grouting area, groundwater state, and other aspects. Standard penetration, rotary penetration and other survey methods can be adopted because all the above mentioned reinforcement are aims to improve the shear strength of the stratum. For the shield tunnel entrance and exit section, waterproof effect is the most important purpose, therefore electrical sounding exploration, neutron sounding exploration, elastic wave method, ground penetrating radar method, and other methods can be used for surveying.

Although there are many advanced geological prediction methods in the shield boring stage based on seismic waves, geological horizontal drilling, and ground-penetrating radar, however, they are difficult to apply and promote on sites due to the limitation of the shield machine structure and materials and the effect to the tunnelling efficiency.

2.2.1.4 Survey of operation and maintenance stage

A follow-up survey is conducted after the tunnel completed and under operation to continuouly investigate the factors affecting the performance of the tunnel over time. The effects of shield tunnelling induced long-term settlement and the impact to the animals and plants cannot be confirmed at the construction stage, so the follow-up survey must be carried out after tunnel construction to determine those impacts. The contents of follow-up survey vary according to the conditions and causes with the survey frequency slowly reduced. Usually, follow-up surveys need to be performed for quite a long time after completion of the tunnelling construction.

2.2.2 Main survey means

A comprehensive survey method combining engineering geological investigation and mapping, field drilling and sampling, hydrogeological test, engineering geophysical exploration (wave velocity test and resistivity test), in situ testing (static cone penetration test, flat dilatometer test hole, vane shear test, standard penetration test (SPT), and cone dynamic sounding test), and laboratory tests are done in a shield tunnel project. If necessary, load test, pressuremeter test, field direct shear test, geothermal test, natural radioactive logging test, harmful gas test, and other in situ tests also can be performed to obtain the required geotechnical engineering design parameters. Various positioning survey methods such as electromagnetic induction method, magnetic exploration method, ground penetrating radar method, and other nondestructive detection methods for detecting underground structures and buried pipelines also are included.

2.2.2.1 Geological investigation and mapping

Investigation along the route is carried out by using a topographic map on a scale of 1:1000 as a base map to determine adverse geological actions such as topography, stratum lithology, geological structure, and ground fissures. The distribution of surface water bodies and water wells are investigated to determine hydrogeological conditions along the route, and the distribution of underground structures and pipelines is identified. The mapping range usually extends 250 m on each side of the tunnel central line. The investigation mainly conducted by a combination of crossing along the route and fixed-point observation.

2.2.2.2 Engineering exploration and sampling

Engineering geological borehole as an exploration method uses a drilling rig to carry out borehole construction according to a certain design angle and direction. By taking stratum core sample or placing a test instrument into the hole to identify and divide the subsurface stratification. Deep sampling along the borehole can also be performed. Drilling borehole, one of the most extensive exploration methods in an engineering geological survey, can provide geological data for the deep layers. There are many types of drilling equipments, among which alloy drills have the best performance. They can detect ground conditions at very large depths inside the stratum and obtain better core samples, and they have a relatively high core sample recovery ratio. The core samples can be obtained even with small holes. Furthermore, the drilling rig is light, easy to move, and low in cost. There are also other simple drills such as auger drill and percussion drill. With the development of technology, some mature technologies are also being utilized in geological drilling. For example, all conditions in a drilling hole can be shown on a display screen by placing a microcamera in the drilling hole.

The requirements of borehole surveys are different in different survey stages. The borehole distances generally vary with the difference in terrain, stratum structure, and site complexity. The distance between boreholes for general sections should be 30–60 m. Borehole for shield tunnel exploration should be positioned outside a tunnel structure by 3–5 m, outside the tunnel structure by 8–12 m for underwater large-diameter tunnel. The boreholes should be arranged on both sides of the tunnel in a staggered manner, additional boreholes should be positioned near the tunnel entrance, the shield working shaft, cross passages, and positions where construction method will be changed. After completion of the survey, the boreholes should be sealed and remnants in the boreholes should be recorded in detail. The depth of general boreholes should be not less than 2.0 times of the tunnel diameter below the bottom of the tunnel (and not less than 20 m), and the depth of control boreholes is not less than 3.0 times of the tunnel diameter below the bottom of the tunnel (and not less than 30 m). The drilling depth can be appropriately reduced according to the circumstances in case of a strongly weathered or fully weathered rock stratum within the buried depth range of the tunnel structure and should be appropriately deepened in cases of karst or fracture zones.

2.2.2.3 In-situ test

The in-situ test is done to obtain the physical and mechanical parameters of the rock and soil body by directly carrying out the test on the original stratum. The in-situ test does not involve a complicated sampling process and has the characteristics of good reliability, low cost, high distinguishing ability for thin layer, and the like. However, the boundary conditions and the drainage conditions are complicated to be determined in the in-situ test so that a corresponding laboratory tests is required as a supplement to figure out the theoretical basis. The in-situ tests can be classified according to whether there is borehole or not, and the corresponding main operation methods during measurement are shown in Table 2.3.

2.2.2.4 Laboratory test

Laboratory test is an important part of the geotechnical engineering survey. The tester uses an experimental instrument, follows procedures to sample the rock and soil stratum, and conducts tests of various items so that reliable physical, hydraulic, and mechanical parameters can be provided and a reliable basis is provided for engineering design and construction. The main laboratory test types are shown in Table 2.4.

General physical and mechanical properties, compression, and direct shear tests should be carried out on cohesive soil in the shield tunnel cross-section, and the contents of silt particles and clay particles are measured through particle analysis test. The general physical and mechanical property test, direct shear test, and particle analysis test are carried out on undisturbed sand, and particle analysis is carried out on disturbed soil sample, including measurement of the content of clay particles.

Table 2.3 Main in-situ tests and their classifications.

With borehole or not	Operation methods	Measured parameters	Typical tests
Must with boreholes	Groundwater change measured from borehole	Permeability coefficient	In-situ permeation test
	Loading on wall of borehole (intermittent in depth direction)	Deformation coefficient	Horizontal loading test (PMT)
	Penetration or rotation at bottom of borehole (intermittent in depth direction, borehole always as auxiliary means)	Dynamic penetration resistance	Standard penetration test (SPT)
		Blade rotation resistance	Vane shear test (VST)
Without boreholes	Continuous penetration from surface	Dynamic penetration resistance	Dynamic penetration test (DPT)
		Static penetration resistance	Static cone penetration test (CPT)
		Static load settlement and rotary penetration resistance	Swedish soil cone penetration test (CPT)
	Opening blades to draw after penetration	Blade drawing resistance	Drawing resistance test
	Loading in horizontal direction after penetration from surface (intermittent in depth direction)	Deformation coefficient	Dilatometer test

Parameters such as nonuniformity coefficient, curvature coefficient and a representative particle size distribution curve should be provided by particle analysis test.

Second, borehole samples should be selected for the consolidation test, triaxial compression test, static lateral pressure coefficient test, unconfined compression strength test, subgrade coefficient, thermophysical parameter test, and paleosol expansibility test.

In addition, a water sample must be taken from the borehole for water quality analysis to determine the corrosiveness of groundwater on the site,

Table 2.4 Main laboratory tests and parameters.

Laboratory test	Basic geotechnical test	General physical and mechanical property test
	Geotechnical test	Particle analysis of sandy soil
		Collapsibility coefficient
		Compression test
		High-pressure consolidation test
		Quick direct shear test
		Consolidated quick direct shear test
		Saturated quick shear
		Unconsolidated-undrained triaxial test (UU)
		Consolidated-undrained triaxial test (CU)
		Unconfined compression strength
		Natural repose angle of sandy soil
		Content of clay particles
		Quartz content measurement of sandy soil
		Penetration test
		Free expansion rate
		Static lateral pressure coefficient
		Subgrade coefficient
		Organic content
		Thermophysical parameter
	Rock test	Natural rock density
		Uniaxial ultimate compressive strength (natural, saturated and dried)
		Rock elastic modulus and Poisson ratio
		Shear test (c, φ)
	Others	Water quality analysis
		Content analysis of soluble salts

and a certain amount of soil must be taken from the soil layer above the water level for content analysis of soluble salts to find out the corrosiveness of the site soil. Organic matter, the PH value, the content of fulvic acid, and the like parameters in the soft soil should be measured according to engineering requirements.

2.2.2.5 Engineering geophysical exploration

Based on underground physical fields (such as the gravitational field, electric field, and magnetic field) and the impact of differences in physical properties of different geological bodies on the distribution rule of the underground physical fields, the engineering geophysical exploration determines geological distributions and underground structures correlated to the engineering investigation through observation, analysis, and research of these physical fields in combination with related geological data. Engineering geophysical exploration has the advantages of perspectivity, high efficiency, low cost, and can conduct corresponding in-situ physical and mechanical geotechnical tests, gaining more and more attention and development in the engineering geological survey. However, the use of these geophysical exploration methods is conditional and limited, and most of them still have multiple solutions. Therefore a good geological survey should be gained by correctly selecting and adopting geophysical exploration methods and comparing the results with existing geological borehole data. Common engineering geophysical exploration methods are shown in Table 2.5.

The survey methods listed in Table 2.5 are generally applied at all stages of the engineering survey process. All survey methods or means could confirm the results from each other, and all survey results should be comprehensively analyzed. The specific details can refer to relevant textbooks and codes for engineering geotechnical survey. The implementation and operation of the test method at each stage must meet the requirements of the planning, design, construction, operation, and maintenance investigation stages and the requirements of the corresponding result analysis and survey report specified in engineering geological survey code.

Table 2.5 Common engineering geophysical exploration methods.

Method		Detection depth (m)	Accuracy (%)
Electromagnetic induction method (direct or indirect method)		0–3	10–20
Magnetic field exploration method		0–30	20
Acoustic detection method		Several to dozens of meters	20
Resistivity detection method		<100	20
Elastic wave method	Elastic wave velocity measurement	Several to dozens of meters	20
	Reflection wave method	<100	20
	Surface wave method	<100	20
Electromagnetic wave method	Underground radar	≤3	≤5
	Electromagnetic wave CT method	0 to dozens of meters	≤5

2.2.3 Survey results and documentation requirements

In the geological survey report of the shield tunnel project, geotechnical engineering analysis and evaluation should be carried out on the requirements according to survey stages, engineering properties, design schemes, and types of shield machine used. The evaluation should be based on data from engineering geological investigation and mapping, exploration, laboratory test results, and engineering geological and hydrogeological data from the project site so that a reasonable shield tunnel design and construction scheme can be drawn up.

2.2.3.1 Content requirements of survey report

1. The contents of the survey report should meet the documentation requirements stipulated in the relevant standards. The division standard of geological units, engineering geological and hydrogeological regions, and geotechnical stratification should be unified in the report along the tunnel route. The survey results at all stages should be continuous and complete. The survey report should includes text, tables, drawings, and the like, and must be reliable in content and clear in organization.
2. The survey report should include a survey task guideline, an overview of the proposed project, survey requirements and purposes, survey ranges, survey methods and execution standards, finished survey tasks, and so on.
3. The survey report should include geotechnical engineering evaluation and suggestions for engineering measures and must evaluate the impact of adverse geological actions on the project. The detailed survey report should provide the dynamic parameters for seismic design by analyzing the seismic safety assessment report of the proposed tunnel construction site.
4. The survey report should include the other attached tables, drawings and appendixes.

2.2.3.2 Data requirements of the main results

Main parameters of rock and soil body

To understand the engineering geological features of various soil layers in detail, the parameters of the four aspects, shown in Table 2.6, need to be included.

During analyzing the engineering properties of the stratum and in the selection process of shield machine, the following geotechnical parameters can be referenced:

1. Grain size distribution, maximum particle size, d_{50}, d_{10}, ω_L, ω_p, I_p, etc., which can be used to identify what classification the soil belongs to and its basic properties.

Table 2.6 List of main soil parameters.

Parameter nature	Soil parameter
Parameters representing inherent characteristic of soil	1. Grain composition (Gravel content G%, sand content S%, silt content M%, and clay content C%) 2. Maximum soil particle size 3. d_{50} (soil smaller than particle size of d_{50} accounts for 50% of total soil weight) 4. d_{10} (soil smaller than particle size of d_{10} accounts for 10% of total soil weight) 5. Nonuniformity coefficient μ ($\mu = d_{50}/d_{10}$) 6. Liquid limit ω_L 7. Plastic limit ω_p 8. Plasticity index I_p
Parameters representing status of soil	1. Water content ω 2. Saturability S 3. Liquidity index I_L 4. Void ratio e 5. Permeability coefficient K 6. Wet soil bulk density γ_w
Parameters representing strength and deformation property of soil	1. Undrained shear strength S_u 2. Cohesion c 3. Internal friction angle ϕ 4. Standard penetration degree N 5. Sensitivity S_t 6. Compression coefficient a 7. Compression modulus E
Parameters representing groundwater conditions	1. Groundwater level 2. Confined water head 3. Permeability coefficient

2. d_{10}, K, etc., which are important parameters for estimating permeability and cohesiveness of the soil and for predicting the effect of air pressure and precipitation to dry the soil. These parameters are significant for the selection of shield machine type in the water-rich soil layer and of the technical scheme for groundwater control.
3. In the sandy soil layer, a larger void ratio and the permeability coefficient result in a smaller nonuniformity coefficient and a high possibility of soil liquefaction.
4. ω_L, ω_p, I_p, ω, N, etc. are parameters used for analyzing the consistency of the cohesive soil.

5. γ_w, S, c, ϕ, etc. can be used for understanding the soil stability coefficient N_s of an excavation face of the cohesive soil.
6. Soil particle size is an important reference for selecting the way of discharging the soil out of the shield muck chamber.

Hydrogeological data

The groundwater level and type are closely related to the design and construction of the shield tunnel. Hydrogeological conditions and activity characteristics of groundwater must be fully investigated during the survey, and the engineering geological problems such as sand liquefaction, water and sand inrush, and the probability of surface collapse in the saturated sandy soil stratum are analyzed. Therefore the following data should be included in the geological survey of a shield tunnel project.

Distribution and stratigraphy of permeable layers In a complicated stratum, a method combining continuous soil borehole sampling and static cone penetration test is required to determine the thickness and depth of each permeable layer. Then the stratigraphy of an interbedded layer between the cohesive soil layer and the sandy soil layer and the existence of gravels and pebbles should be described in detail. It should be noted that a quicksand phenomenon will occur when a sandy soil layer with a thickness greater than 25 cm, which consequently will affect the stability of the tunnel excavation face. When there are silt thin layers with thicknesses of several centimeters or several millimeters between the cohesive layers, a precipitation method or an air pressure method could be adopted to enhance the shear strength of the cohesive layer of the sandwiched thin sand layer and to improve the stability of the tunnel excavation face.

Lens in complicated strata For complicated aquifer stratum, it is advisable to employ a combination of borehole and geophysical exploration for continuous survey to find out the presence or absence of an ancient river or underground lens which may cause an explosive collapse of the excavation face if the water pressure cannot be reduced by well point precipitation.

Investigation of groundwater level and water pressure distribution When the groundwater level is deep or when the water pressure is being investigated in a second permeable layer, the work of observation well for water level should be very carefully carried out to prevent falsity results.

Distribution of the permeability coefficient The permeability coefficient is strongly related to the soil particle size and the factors such as void ratio, saturation, and shapes and arrangement of soil particles. Therefore when the particle size and the void ratio change significantly in the borehole data, the permeability coefficient will also change significantly. In complicated strata with large changes in soil particle size and permeability coefficient, it is difficult to achieve a better dewatering effect by well-point precipitation. In addition, the flow velocity of the groundwater needs to be confirmed, and when the flow velocity is high, it is necessary to consider whether or not the slurry for protecting the excavation face will be washed away.

Surrounding buildings (structures)

The survey for shield tunnel project should include a detailed investigation of buildings (structures) through which the tunnel will underpass or within the influence scope of tunnel construction, including surface buildings (structures), underground buildings (structures) and civil air defense engineering, subgrade structures, bridge structures, municipal underground pipelines, hydraulic buildings, overhead high-voltage towers (poles), cultural relics, and underground obstacles. Details are listed in Table 2.7.

Adverse geology and special ground

The features, causes, distribution ranges, and development trends of adverse geological actions (such as mined-out areas, karsts, underground rivers, sinkholes, ground settlement, ground fissures, cavities, harmful gases, liquefiable soil, collapsible loess, and artificial fill) should be investigated. The hazards and impacts these factors on the shield tunnel project should be analyzed, and suggestions for a treatment plan should be proposed.

Main drawings

Engineering geological plan The main structures along the route of the tunnel project, terrain objects, line location, types of exploration holes, location, numbers, elevation, stable water levels, adverse geological sections, mileage locations, and the like should be marked on the engineering geological plan. The graphic coordinate system should remain unchanged. The plane scale of the map is generally 1:1000 to 1:5000 and can be adjusted according to actual situations as long as the elements of the drawings can be reflected clearly.

Table 2.7 Types of surrounding buildings (structures) of shield tunnel project.

Building (structure) types	Detailed investigation contents
Surface buildings (structures)	Building floors, height, structural forms, foundation types, foundation buried depth (elevation), and additional pressure at foundation. Main design parameters and construction technologies of the foundation should be contained in buildings (structures) employing composite foundations and pile foundations.
Underground structures and civil air defense engineering	Plane layout of the project, external profile size, elevations of roof and floor, construction methods, structural forms, deformation joints, enclosure structures, antifloating measures, and usage conditions. The protection grade and the locations of entrances and exits should be investigated in civil air defense engineering.
Subgrade structures	Railway (including rail transit) or road grade, pavement materials, pavement width, embankment height, supporting structural forms, and subgrade and foundation forms.
Bridge structures	Bridge types, structure layout, bridge length, bridge width, span, pier base form and bearing capacity, pile foundation or foundation reinforcement design parameters, and operation lifetime.
Municipal underground pipelines	Pipeline types, plane layout, buried depth (or elevation), laying modes, materials, pipe joint length, interface forms, medium types, working pressure, location of working wells and valves, and operation lifetime.
Hydraulic buildings	Project layout, characteristic water level, tunnel flood discharge or diversion standard, reservoir dispatching operation modes, water taking principle of rivers, arrangement of electromechanical equipment and pressure regulation (pressure reduction) gates (valves), types of hydraulic buildings, structural forms, foundation forms, lining conditions, and operation lifetime.
Overhead high-voltage towers (poles)	Line voltage grades, suspension height, corridor width, high-voltage tower (pole) foundation forms, buried depth, and coordinates of intersections of cables and tunnel.

(*Continued*)

Table 2.7 (Continued)

Building (structure) types	Detailed investigation contents
Cultural relics	Cultural relic name, grades, cultural relic protection and control range, structural forms, foundation forms, buried depth, etc. Famous and ancient trees should be included in the cultural relic investigation.
Underground obstacles	Underground cavities, ancient wells, pile foundations left in stratum, anchor rods, and other items affecting shield tunnel construction.

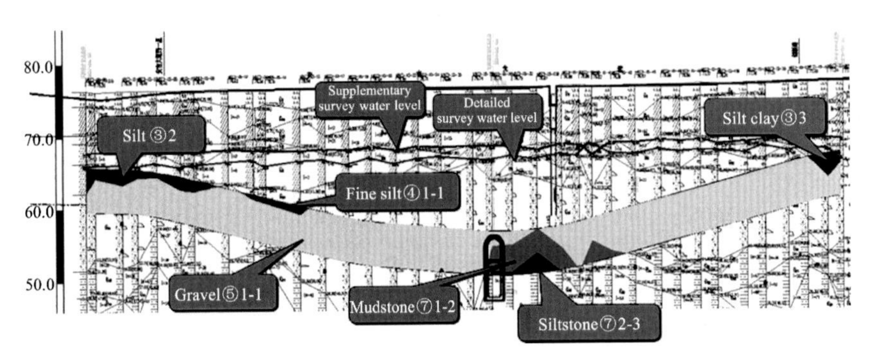

Figure 2.1 Geological longitudinal profile of the running tunnel of a subway [2].

Geological longitudinal profile Drawing the geological longitudinal profile according to a sufficient quantity of reliable geological histogram can maximize understanding of the representative stratum conditions, the engineering soil properties of the difficult sections and various obstacles where the tunnel passing through. The geological longitudinal profile should mark out ground elevation line, exploration hole numbers, orifice elevation, borehole depth and distance, boundary of soil stratification, mileage locations, geological areas, and the like. The longitudinal profile of the shield tunnel generally has the same scale as the plane drawing in a transverse direction while the longitudinal scale is 1:200. A example of geological longitudinal profile of the running tunnel of a subway is shown in Fig. 2.1. In addition, the following key information should be included:

1. The classification of various soil layers along the tunnel route, distribution of the soil layers in the vertical and horizontal directions, engineering characteristics of various types of soil, and the methane content in soil layers.

2. The height of the groundwater where the shield passes through the stratum; the water pressure and soil permeability coefficient when the shield tunnel crosses permeable layer and water-bearing gravel lens, as well as the mobility of the soil under hydrodynamic pressure.
3. The mileage location of various obstacles encountered by shield tunnel excavation, and various underground pipelines and underground buildings (structures) in the vicinity.
4. Engineering properties of an overburden layer when the shield crosses a river.

Borehole geological histogram The name of each stratum, layering depth, elevation, thickness, stratum description, sampling numbers, SPT blow count, and the like, as well as each borehole date and stable water level should be marked in the borehole geological histogram. The vertical scale of the borehole geological histogram is generally between 1:200 to 1:400.

Other drawings

Other accompanying drawings include static cone penetration test result maps, vane shear test comprehensive result maps, flat dilatometer test result maps, wave velocity test result maps, resistivity test result maps, pumping test result maps, river sectional views, soil compression curve charts, geotechnical test result reports, water quality analysis test reports, and rock core photos. These accompanying drawings can be provided upon request, which should clearly reflect the elements of the drawings.

2.3 Evaluation of engineering geological conditions

2.3.1 Contents of the engineering geology evaluation

After completion of the shield tunnel geological survey at each stage, a series of comprehensive analyses and evaluations of the survey results is required. The reasonability, economy, and safety of the construction plan can be determined by reasonably analyzing and evaluating the survey results. In view of differences in survey purposes and means, the analysis and evaluation contents at different survey stages are different.

2.3.1.1 Survey at the feasibility study stage

The engineering geological analysis and evaluation of the survey at the feasibility study stage should include the following:

1. Rough classification of the surrounding rock, approximate analysis of ground stress distribution and hydrogeological conditions, and evaluation

of cavitation conditions and the impact of tunnel construction on the environment.
2. Rough analysis of the influence of adverse geological actions and special ground condition on shield tunnel construction.
3. Suggestions for tunnel route sites, location of launching shaft and receiving shaft.

2.3.1.2 Preliminary survey

The engineering geological analysis and evaluation at the preliminary survey stage should include the followings:

1. Preliminarily determining the classification of surrounding strata and its excavability, determining the physical and mechanical properties of the surrounding rock and geotechnical parameters required for preliminary design, evaluating the stability of the surrounding rock, and putting forward preliminary suggestions for engineering protection measures.
2. Preliminarily evaluating the influence of surface water and groundwater on tunnel construction and carrying out a sectional estimation of tunnel water inflow.
3. Preliminarily evaluating the engineering geological conditions and the ground stability at the tunnel entrance and exit, adjacent deep excavation, the shafts, and the like and putting forward suggestions for engineering protection measures.
4. Analyzing the influence of adverse geological actions and special ground on tunnel construction.
5. Preliminarily evaluating the influence of shield tunnel construction on the urban geological environment and adjacent buildings and other structures.
6. Preliminarily evaluating any underground harmful gas and their influence on the tunnel construction and operation.

2.3.1.3 Detailed survey

The engineering geological analysis and evaluation at the detailed survey stage should include the followings:

1. Segmentally determining the classification of the surrounding rock.
2. Analyzing and evaluating the engineering geological conditions and the ground stability in the shield tunnelling process and putting forward suggestions for treatment measures.

3. Determining any harmful gas distribution within the depth range affected by the tunnel, analyzing and evaluating its possible influence on tunnel design and construction, and putting forward suggestions for treatment measures. When the shield machine needs to cross through adverse geological and special ground regions where may exist boulders, soil caves, karst, radioactive mineral-containing strata, or the like, a engineering geological analysis and evaluation should be carried out to determine the influence on shield tunnel construction and to propose targeted treatment suggestions.
4. Finding out groundwater distribution in the depth range affected by the shield tunnel, analyzing and evaluating its influence on tunnel construction and the influence of groundwater drainage on the environment, segmentally predicting tunnel water inflow and proposing suggestions for treatment measures.
5. Evaluating the applicability of the shield tunnelling method and putting forward suggestions for the selection of shield machine and the monitoring work during construction.
6. Providing the geotechnical elastic resistance coefficient that is required for tunnel design and construction, and providing thermophysical index parameters if necessary.

2.3.2 Influence of common strata on shield tunnel construction

Generally, shield tunnel construction is safest and most feasible in areas where an open-cut method or a sequential excavation tunnelling method is not suitable for construction, in areas with unstable strata, or when the construction time schedule is very tight. However, the selection of shield machine type and construction parameters of shield tunnel construction must be adapted to the geological conditions and environmental conditions, making the engineering investigation particularly important. Geological survey results based on limited-point borehole have some limitations. Therefore it is necessary to provide a thorough analysis of the stability of the strata, particularly when special strata are encountered. The rock and soil types and their distribution characteristics should be analyzed and investigated in combination with causes of regional strata distribution. For the following special ground, the possible geotechnical problems in the shield tunnel construction process should be analyzed and evaluated, and treatment suggestions should be provided [3].

2.3.2.1 Cohesive soil stratum

Cohesive soil is generally an ideal stratum for shield construction. However, the following points must be noted during the survey: A high-plasticity cohesive soil stratum with high clay content could easily lead to clay clogging on the cutterhead; high-sensitivity soft soil is easy to be disturbed and tunnelling in such soft ground easily cause extensive ground deformation and even ground collapse due to overbreaking; and the shielding tunnelling machine easily forward snakelike in soft soil with low strength and hard to control the machine attitude. In addition, if expansive minerals are contained in the cohesive soil stratum, the stratum is likely to expand when meeting with water to cause the cutterhead or the shield body to be stuck in the shield tunnelling process.

Furthermore, as the natural water content larger than the liquid limit, the soft silt clay layer is likely to be disturbed under an external force, with the consequence that its strength is substantially reduced. In such a type of soil layer, not only it is very difficult to maintain the earth pressure balance during shield tunnelling, but also early settlement often occurs before the arrival of the shield, and the settlement lasts and unstable for a long time after the shield passed by. The stratum must be reinforced to prevent this phenomenon. Therefore, it is very important to understand the sensitivity and deformation characteristics of the soil body in advance, since this is a basis for a reasonable reinforcement design.

2.3.2.2 Collapsible sand layer

A collapsible sand layer is characterized by a small nonuniformity coefficient, poor compactness, a large permeability coefficient, and poor stability and is likely to collapse in any case of slight unbalance in earth pressure. Cavities are easily generated in this type of stratum in a short time interval if injected slurry has not been put in place after the shield tail leaves the newly installed lining ring and resulting in collapse in the sand layer or even surface subsidence.

To ensure the stability of the excavation face, the parameters of the slurry mixed or conditioned soil must be designed to match the ground condition when the shield is advancing in this collapsible sandy stratum. Therefore it is extremely important to accurately determine parameters such as particle size, gradation composition, water permeability, and groundwater level of the excavated soil layer. These parameters are important in the design process for preventing water inflow at the joints of a supporting wall and ensuring watertightness of the segment sealing material.

2.3.2.3 Gravelly sand stratum

Gravel and sand stratum

A gravel and sand stratum is low in fine particle content, high in hard mineral content, large in permeability coefficient, and poor in self-stability. Theoratically, a slurry balance shield machine is best for tunnelling in such strata. However, due to considerations of limited site conditions and construction costs in practical engineering, an earth pressure balance shield machine is generally selected for advancing, easily resulting in problems of gushing, rapid wear of the cutterhead, stratum instability, and so on.

Therefore during geological survey of this type of stratum, it is necessary to examine the compactness of the gravel and sand stratum; determine related mechanical parameters; strengthen analysis of particle composition and fraction content; determine the optimal soil conditioning scheme through a field test; conduct mineral content analysis to determine the content of conventional hard minerals such as quartz, feldspar, and hornblende; provide parameters for selection and lifetime prediction of a cutter; judge the probability of sand liquefaction in the stratum by combining groundwater conditions; determine sensitive formation division probably causing water inflow, sand inflow, and tunnel leakage; and provide a reference for synchronous grouting quality control.

High-water pressure gravel layer

A high-water pressure gravel layer mostly contains large pebbles in the lower part of a river. Before advancing through this type of stratum, it is necessary to determine the shapes, sizes, quantity, and harnesses of the pebbles; the groundwater velocity and flow; and other parameters. These parameters are a basis for designing excavation structures (cutter materials, shapes, cut shapes, etc.) of the shield, so a detailed survey must be carried out. When it is difficult to obtain these parameters by a conventional borehole method, methods of large-diameter borehole and trail excavation of deep foundation can be adopted to obtain them [4].

Sandy pebble stratum

A sandy pebble stratum is characterized by almost no fine particles, high hard mineral content, large permeability coefficient, and poor self-stability. The slurry shield machine is the mostly selected type for tunnelling in this ground. The earth pressure balance shield machine can also be selected for tunnelling in such ground condition in view of site conditions and construction cost by giving a detailed design preparation for geological applicability of the shield machine, including structural design of the cutterhead, cutter selection and layout, muck improvement design, and slag discharge system design. Much improvement and synchronous grouting should be completed in the construction process to prevent the cutterhead and screw conveyor from becoming stuck, rapid wear of the cutterhead, and stratum instability. In the construction of the Chengdu and Beijing subway systems, problems of serious wear of the cutterhead and the screw conveyor and cutterhead becoming stuck were encountered.

The lessons of experience have shown that for this type of stratum, the key to a success of shield tunnel construction is to do good engineering geological survey before designing the shield machine. The survey of this type of stratum focuses on investigation of particle size distribution and cementation of pebbles and boulders. It is usually difficult to find larger boulders by borehole, so regional geological conditions, geological origin, and existing geological data should be combined to analyze and judge. The particle size can directly influence the opening ratio of the cutterhead and the size of the muck discharge system. The particle size distribution and water content of the sandy pebble stratum are also important in selection of the slurry concentration (slurry balance shield machine) and mixing ratio setting of a muck modifier (earth pressure balance shield machine).

Composite strata

Composite strata, a new stratigraphic concept, are proposed by the engineering community for the shield tunnelling method. In three-dimensional terms, composite strata are a combination of several strata with greatly different features in geotechnical mechanics, engineering geology, and hydrogeology in the excavation range. Shield tunnel construction in composite strata faces a series of problems, including difficulty in shield machine attitude control, ease of gushing, rapid wear of the cutter (hob eccentric wear), and large ground deformation, even collapse, induced by the geological feature of upper-soft and lower-hard. Therefore, it is important to find out the interfaces between the soft and hard soil layers.

For a rock stratum, it is necessary to determine the rock types, weathering degrees, hard mineral content, integrity coefficient, and rock mass quality. Then, surrounding rock classification should be conducted considering the requirements of shield tunnel construction. These geological features have a direct impact on the design of the shield machine and the setting of shield tunnelling parameters.

Stratum with hypoxia or toxic gas

Before the shield tunnelling passes through a gravel layer with depleted groundwater, or a clay layer containing too much undecomposed organic matter, or any stratum containing hypoxia or toxic gas, water quality should be analyzed first, and then the gas concentration should be measured to determine its content. The shield machine should be equipped with various monitors and alarms to ensure a safe operation.

Other adverse geological strata

For the shield tunnelling method, other adverse geological conditions mainly include expansive stratum, soil caves, karst, fault fracture zones, boulders, and spheroidal weathering bodies. Serious engineering geological problems and even major accidents are easily caused by these adverse geological conditions, so they should be identified by special geological survey, and the possible impact of these problems on shield tunnel construction should be analyzed.

In summary, the geological survey of a shield tunnel project should include analysis, evaluation, and suggestions of engineering geological conditions. The survey should be implemented by a geotechnical investigation unit because an shield machine selection and construction needs professional knowledge to carry out accurate analysis and evaluation of the geotechnical engineering problems. Therefore, reliable geological survey for a shield tunnel project should combine construction requirements for the shield tunnelling method and specific engineering geological conditions, and conduct engineering geological analysis and geotechnical mechanics analysis to provide a survey report with complete documentation, reliable data, correct evaluation, and reasonable suggestions for construction of the shield tunnel.

2.4 Cross-section and alignment design of shield tunnels

2.4.1 Classification of shield tunnels

Generally, shield tunnels can be classified according to section forms and functional purposes. Two different types of classification are shown in Fig. 2.2. The purpose of the shield tunnel takes all line facilities in the underground space as serving objects.

2.4.2 Cross section forms of shield tunnels

The selection of the section forms of shield tunnels depends on factors such as the use requirements of the tunnels, the possibility of construction

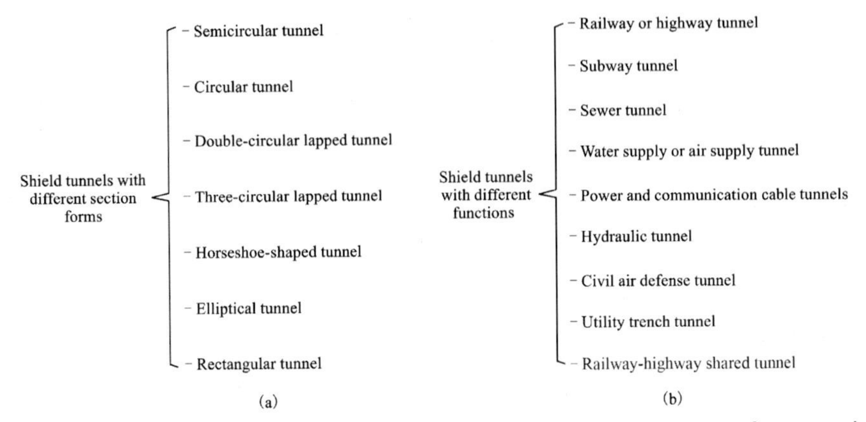

Figure 2.2 Classification of shield tunnels according to (a) section forms and (b) functions.

technologies, stratum properties, and tunnel stress. The clearance size in the tunnel should conform to the requirements of building clearance, curved section widening, functions, and construction technologies. The influence of construction errors, measurement errors, structural deformation, and long term displacement should also be taken into account. In a word, the section design of the shield tunnel should meet the standard regulations of related industry.

2.4.2.1 Circular section

The most commonly used shield tunnel section form is the circular section, which is usually referred to as a standard section. A circular section shield tunnel has the advantages that external pressure in all directions can be equally borne due to the arch action of the circular section, such that stress concentration acting on a lining ring is small (comparing with a noncircular section). The tunnel is durable and safe as the lining ring is less damaged and has a long service lifetime. The circular section shield machine has a simple excavation mechanism, and the excavation system (the cutterhead and a force/torque transfer device) is easy to manufacture with low construction costs. The segments are simple and easy to manufacture and convenient to assembly. Currently, the circular section has problems of a low internal space utilization ratio for some purposes, such as subway tunnels, highway tunnels, and urban utility tunnels. The inner diameter of the circular shield tunnel depends on two factors: sufficient internal space for the use purpose (including maintenance management margin and construction errors) and construction safety, while its outer diameter can be determined by the inner diameter and the lining thickness.

2.4.2.2 Rectangular section

The rectangular section tunnel has the advantage of a high internal space utilization ratio. Compared with the circular section tunnel, soil excavation and discharge can be reduced by about 30% during construction, which is beneficial for cost reduction. The rectangular section occupies a small space in the ground and has a high underground space utilization ratio. The disadvantage of the rectangular section is that external pressure acting on the tunnel lining is large, which is particularly undesirable in construction of a large-size tunnel; Lining segment design and construction are complicated; and the shield machine is complicated to manufacture and high in cost. The rectangular section is an ideal section form for urban subway tunnels, utility tunnels, and the like.

2.4.2.3 Double-circle lapped section

The double-circle lapped section form is mostly used in double-track railways or highways. This section form is characterized by small occupied space and a high space utilization ratio, but the shield machine is complicated to manufacture and expensive, with complicated segment design, assembly, and construction.

2.4.2.4 Three-circle lapped section

The three-circle lapped section form can be designed for subway station construction. With the advantage of a high space utilization ratio, the subway station construction is completely transferred to underground, and the cost of the construction is low. The disadvantages are the complicated design, production, and construction of the shield machine and segments.

The horseshoe-shaped section and the elliptical section have the advantage of a high space utilization ratio but the disadvantage of the high construction cost of the shield machine. However, some manufacturers have invested in their development. For example, the first horseshoe-shaped shield machine in the world is a novel special-shaped section shield machine that was independently developed by China Railway Engineering Equipment Group Co., Ltd. and first successfully used in the Baicheng tunnel of the Haolebaoji-Ji'an Railway.

Table 2.8 lists common section forms and some section forms of shield tunnels that are newly planned or constructed. The section forms of the

Table 2.8 Common forms of shield tunnels.

Items	Section forms	Features	
		Advantages	**Disadvantages**
Circle		1. The circular form is a mechanically stabilized structure, and the lining thickness can be thinner than those of other forms 2. Machine rolling easily to correct	1. Low section utilization ratio and larger occupied dimension in width
Rectangle		1. High section utilization ratio 2. Thickness of an soil overburden layer can be reduced	1. Concentrated stress easily caused at corner, and thus the lining thickness must be increased 2. Complex in structure of tunnelling machine, difficulty in face pressure control 3. Much difficulty in machine rolling correction
Ellipse		1. Small internal force compared with rectangular form, and the lining thickness can be reduced 2. High section utilization ratio and small occupied width compared with circular form	1. Complex in structure of tunnelling machine, difficulty in face pressure control 2. Much difficulty in machine rolling correction
Double-circle		1. Smaller occupied width compared with two circular tunnels excavation 2. Safe tunnelling as compound from circular form section	1. Much difficulty in machine rolling correction
Three-circle		1. Applicability to construction of subway station, etc. 2. Safe tunnelling as compound from circular form section	1. Much difficulty in machine rolling correction

shield tunnels are basically circular. Circular forms are more beneficial for an external load acting on the tunnel from a mechanical point of view. A circular section without conner is more favorable in terms of mechanical driving using panels. Furthermore, fluctuation caused by rotation of the shield machine in a circumferential direction during construction is easily handled with a circular form.

Other section forms besides the circular form can be adopted when the shield tunnel is restricted or limited by adjacent structures. By reducing unnecessary section space, the excavation volume can be substantially reduced, thereby reducing construction cost. Therefore further development of design and construction technologies of special-shaped section shields is expected.

2.4.3 Alignment design of shield tunnel

There are two types of tunnel alignments: plane alignment and vertical alignment (longitudinal slope profile). The design of the vertical alignment needs to consider the tunnel overburden depth. An appropriate alignment can be determined on the basis of use purpose, use conditions, and construction convenience of the tunnel.

2.4.3.1 Plane alignment design

In view of convenience for the shield tunnel construction, the plane alignment of the shield tunnel should be a straight line or curved line with large curvature radius. However, the majority of shield tunnel are underground projects in urban areas, and the alignment arrangement is restricted by stratum condition, surface structure distribution, underground obstacles (structures), and usable land conditions. In other words, the condition that the alignment should be straight line or curved line with a large radius is not easily satisfied. A sharp turn with a small curvature radius will appear in many cases, so the design of the tunnel alignment with small curve radius is the focus of discussion.

When planned in urban areas, shield tunnels, as highly public structures, are mostly built under roads. When the plane alignment is being decided, the influence of shield tunnel construction on road facilities and buried objects under the roads must be taken into account. The tunnel alignment should be as straight as possible, and the curvature radius should be increased as much as possible when a curved line is adopted. However, if the shield tunnel is restricted by layout conditions, obstacles, shaft locations, adjacent buildings (structures), and so on, complicated curve line combinations are common in plane alignment.

The factors influencing shield tunnel construction quality in a curved alignment section are (1) geological features and distribution of the strata through which the shield machine passes; (2) shield machine type and tunnelling parameters (earth pressure balance, slurry balance, driving speed, and advancing control, etc.); (3) shape and structure of the shield machine (diameter, machine length, and presence or absence of an articulated device and other auxiliary equipments) (4) segment width, wedge quantity, and lining ring assembly mode; (5) slope distribution; and (6) adopted auxiliary construction methods (precipitation, underground separating walls, and other measures).

For general curved line tunnel construction, the permissible curvature radius of the shield tunnel depends on the diameter of the shield. For curved alignment with sharp turn and a smaller radius, a rotating shaft or a spherical shield machine must be adopted to achieve the construction. During construction at a position with minimum curvature radius, the use of the articulated device must be carefully planned, and appropriate auxiliary and protective construction methods should be employed, including expanding the excavation space in the stratum or using special shield machines.

Parallel shield tunnels are also commonly used in construction of urban traffic tunnels. The tunnel spacing needs to be determined by evaluating the soil characteristics, the external diameter of the shield, and the type of the shield machine. The spacing between the parallel shield tunnels should generally larger than the tunnel diameter. Reinforcement measures should be taken in case of the spacing is restricted by launching or receiving shaft position, road width, obstacles, and other conditions. In addition, when shield tunnel construction near bridge piers, bridge abutments, buildings, railways, or buried objects, a certain spacing should be left between the tunnel and those structures as far as possible to prevent adverse impacts on these structures, such as bias loading, settlement, and vibration caused by shield tunnel construction. A example plane alignment layout of a sectional parallel shield tunnel of Changsha Metro line 4 in China is shown in Fig. 2.3.

2.4.3.2 Thickness of the overburden soil layer

A minimum overburden thickness is frequently required for the tunnel to ensure normal shield construction and the antifloating requirement of the

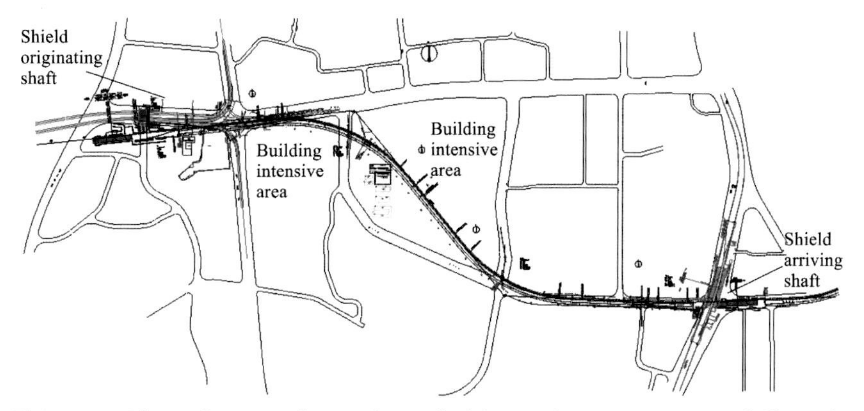

Figure 2.3 Plane alignment layout for a shield tunnel in construction of Changsha Metro line 4, China.

structure. When the overburden thickness is not sufficient, excessive uplift or settlement can easily occur during shield tunnel construction. Furthermore, when the effective weight of the overburden soil and the structure selfweight is not sufficient to resist the buoyancy force of slurry or groundwater, the tunnel structure will locally or completely float upward. The floating of the tunnel will lead to undesirable consequences such as axis deviation structure failure, and poor waterproof effect. Therefore determination of the minimum overburden soil thickness of the shield tunnel is of great significance to ensure the safety of tunnel construction and operation. The following aspects may be considered:

1. Considering the normal driving requirements of the shield machine, the minimum overburden soil thickness is determined according to the principle that the front extrusion force is greater than the static earth pressure but less than the passive earth pressure. In calculating side pressure, water and soil pressure are calculated separately for the sandy stratum but calculated in combination for the sticky stratum.
2. Considering the antifloating requirements of the tunnel structure at the construction stage, the critical overburden soil thickness is determined with the principle that the combined vertical force of the tunnel structure should be zero. The critical overburden soil thickness is multiplied by the corresponding antifloating safety coefficient to obtain the minimum overburden soil thickness. The buoyancy is caused jointly by synchronous grouting slurry and groundwater. The amount of slurry included in calculation is determined by a relationship between slurry solidification and a floating value and it can also be

determined by a method of equivalent slurry gravity in different time durations. Otherwise, the included amount of slurry can be simply and conservatively calculated from the synchronous slurry gravity. At the construction stage, it is not recommended to consider side friction of the overlying soil layer.

3. Considering the antifloating requirements of the tunnel structure at the service stage, the critical overburden soil thickness is determined with the principle that the combined vertical force of the structure should be zero. The critical overburden soil thickness is multiplied by the corresponding antifloating safety coefficient to obtain the minimum overburden soil thickness. The buoyancy is only caused by groundwater. It is recommended that the side friction of the overlying soil layer be considered in the calculation. In calculating side pressure, water and soil pressure are calculated separately for the sandy stratum but calculated in combination for the cohesive stratum.
4. For improving efficiency of the construction work (slag tapping, material transport, and access of operators), difficulty of shaft construction, difficulty of waterproof treatment, reduction of used air pressure and slurry pressure, and maintenance management and operation convenience of the tunnel after completion, the buried depth (i.e., the thickness of the overburden soil layer) of the shield tunnel should be shallow. However, accidents such as stratum subsidence and gushing may easily occur if the buried depth is too shallow. Therefore, the thickness H of the overburden soil layer should be selected under the condition of no adverse effect on the surrounding environment. Usually, H is within 1–1.5D, where D is the external diameter of the shield machine.

The minimum overburden thickness range cannot be generalized and should be evaluated according to specific project conditions. There have been successful cases in which $H < D$, but also cases where subsidence and gushing have still occurred even when $H > 1.5D$. The size of H is related to the soil properties, distribution of surface buildings and underground structures, groundwater level, shield machine types, auxiliary construction methods, construction management measures, and various other factors. Generally, construction management should be strengthened and supplemented by a stratum reinforcement during construction at with small overburden depth, whereas measures of resisting high-water pressure and large earth pressure should be taken during construction at with large overburden depth.

In summary, to determine the minimum overburden soil thickness of the shield tunnel, it is important to consider normal construction of the shield tunnelling and the antifloating safety of the structure. The antifloating calculation of the structure needs to take into account two life stages of tunnel and compare the results when the side friction of the overlying soil layer is included or not included.

2.4.3.3 Vertical alignment design

The vertical alignment of the tunnel should be designed in accordance with the use purpose, including aspects of traffic running stability, passenger comfort, design convenience, water prevention and drainage, and easy for construction, maintenance, and repair. It also depends on the locations of nearby rivers, underground structures, and obstacles. In terms of the use and the construction of the tunnel, the optimal vertical slope along the route is to ensure the water that leaks into the tunnel is naturally discharged to the shaft. For this purpose, a gentle slope where leakage water can be naturally discharged is preferred in principle for normal circumstances, including road tunnels, railway tunnels, power cable tunnels, and communication cable tunnels, preferably no less than 0.2%. However, the slope for sewer tunnels and water supply tunnels must take into account the discharge flow volume, velocity, and other construction aspects. To allow inflowing water to be automatically discharged during construction, the slope can be preferably be increased to a range of 0.2%–0.5%.

For a shield tunnel in a rail transit project, the impact of the locomotive traction mode also needs to be considered in longitudinal slope design, and the traction ability of the traction equipment must meet the requirements the longitudinal slope of the tunnel and the traction coefficient. Meanwhile, the length of the single slope section in the tunnel should not be less than the calculated length of the train. Thus only one slope changing point is incorporated within the running train range and satisfy the requirement for slope difference. As a result, the accumulated influence of additional forces on the slope changing point and frequent changes of the additional forces are avoided, and the running stability of the train is ensured. Furthermore, the length of the straight line between adjacent vertical curves should not be less than 50 m, so that the vertical curves do not overlap each other but are separated by a certain distance, which is beneficial for train operation and track line maintenance.

If the slope is changed to larger than 2%, it will be difficult to transport the excavated soil outside the tunnel, and the material handling efficiency is reduced, and battery car slipping accident may occur. When the slope must be greater than 5% due to condition restrictions, a special transportation mode must be selected, and various safety protection measures must be taken.

In recent years, the buried depth of tunnels has increased in urban areas due to restrictions posed by existing structures and rivers. Sometimes, it is impossible to adjust the vertical alignment in accordance with the use purpose of the tunnels. Therefore, there is a growing trend for the tunnel construction with a large-slope alignment under the existing surface or underground structures. As shown in Fig. 2.4, the maximum longitudinal slope of the shield tunnel in a subway reaches 28‰.

Generally speaking, plane and vertical alignment design parameters of a shield tunnel should conform to the requirements of Table 2.9. A turning shaft can be arranged when the curve radius of the tunnel cannot meet the functional requirements.

2.5 Case study of a supplementary survey

2.5.1 Project overview

The first-stage project of Changsha Rail Transit Line 3 ran from Changsha Pingtang Avenue to Longjiao Road Station, with a total length

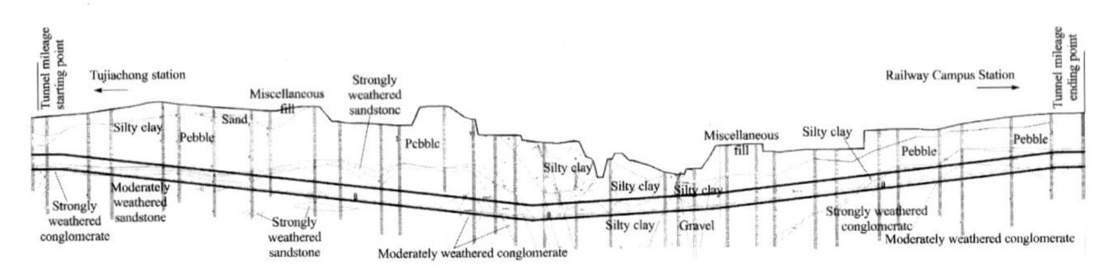

Figure 2.4 Vertical alignment layout of a shield tunnel.

Table 2.9 Plane and vertical alignment design requirements of shield tunnel.

Type of shield tunnel	Minimum curvature radius of plane (D is the external diameter of the tunnel)	Minimum slope	Maximum slope
Subway tunnel	30D or 250 m (choose a larger one)	3‰	35‰
Railway tunnel	40D	3‰	35‰
Highway tunnel	40D	3‰	60‰
Other tunnels	30D	2‰	60‰

of 35.4 km and 25 stations. All stations are underground stations with the rail transit line passing through Changsha from southwest to northeast [5]. The running tunnel between Lingguandu Station and Fubuhe Station has a total length of 2650 m, which is the first slurry shield subway tunnel crossing the Xiangjiang river in Changsha, and is also the first long-distance shield tunnel construction crossing underwater karst cave areas, as shown in Fig. 2.5. In this section tunnel, two Φ6450 slurry balance shield machines were assembled and launched from the west end of Lingguandu Station, starting with the right tunnel first and then the left tunnel, proceeding westward through the air shaft at the east bank, the east branch of Xiangjiang River, Orange Island, the west branch of the Xiangjiang River, and Xiaoxiang Middle Road to arrive at the air shaft at the west bank, moving forward to Fubuhe Station to be received along Tianma Road, and finally were disassembled and lifted out from the shaft at the east end of Fubuhe Station. The shield tunnel was designed with an external diameter of 6.2 m, an inner diameter of 5.5 m, and a segment width of 1.5 m. The karst area has a total length of about 320 m including a 180 m length on Orange Island and a 140 m length on the west branch of the Xiangjiang River (as shown in Fig. 2.5). The construction unit

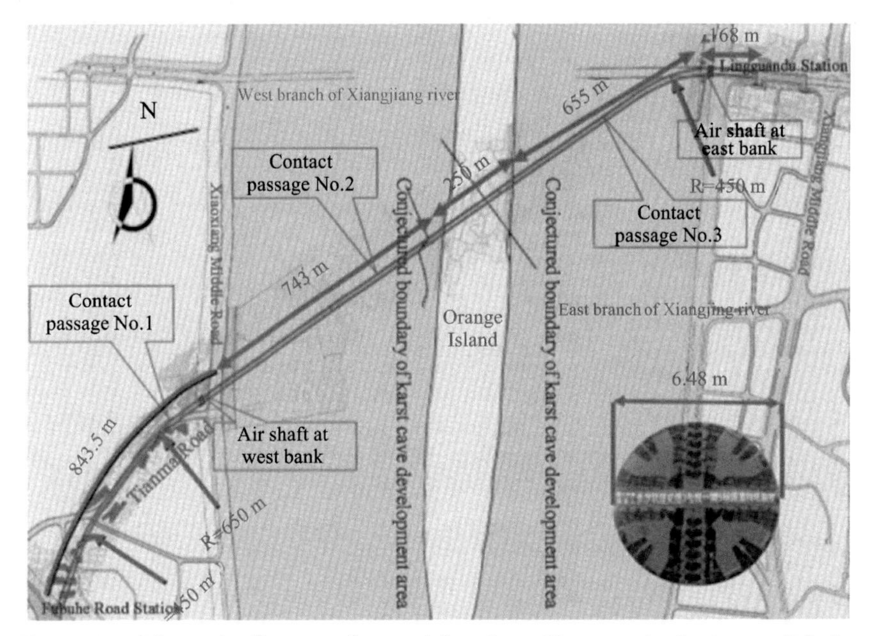

Figure 2.5 Schematic diagram of tunnel line from Lingguandu Station to Fubuhe Station.

planned to carry out a karst cave supplementary survey and a karst cave treatment effect verification survey to provide a basis for the next step of construction (Fig. 2.5).

2.5.2 Purposes of the supplementary survey

According to the requirements of the design department, the adopted Chinese technical standards for supplementary survey mainly include "Code for Geological Survey of Railway Engineering" (TB10012-2001), "Code for Hydrogeological Investigation of Railway Engineering" (TB10049-2004), "Provisional Regulations for High-Speed Railway" in "Geotechnical Classification Standard for Railway Engineering" (TB10077- 2001), "Code for Geological Exploration of Railway Engineering" (TBJ0014—2012), and other related design and technical requirements of the construction unit. In accordance with the above requirements, the main purposes of the supplementary survey were as follows:

1. Determining mechanical properties such as compressive strength of continuous and complete bedrock in the area.
2. Determining the buried depth of the karst caves, cave height, conditions of the rock and soil layer at the top of the cave, and strength of bedrock at the bottom of the cave.
3. Determining the type of the karst cave (closed or opened? connected? multi-karst caves? etc.).
4. Determining water flow conditions in the karst caves and estimating the impact on construction.
5. Determining whether fillers exist in the karst cave or not and estimating the state strength of the fillers.
6. Analyzing and indicating the possibility and range of multilayered karst caves and large karst caves.
7. Submitting treatment schemes and suggestions for various types of karst caves according to the buried depth and size of the karst cave, the fillers, and so on, and putting forward treatment suggestions for a single karst cave in special circumstances.
8. Evaluating the stability of the foundation when a bearing stratum is a tilted stratum and the bedrock surface is uneven, and suggesting treatment measures.
9. Putting forward parameters of a weak interlayer if exists.
10. Requiring numbers and records for each borehole in terms of exploration.

11. Clarifying boring equipment, type of bore bit, and bore diameter.
12. Uniformly adopting absolute elevation and accurately measuring orifice elevation, drilling footage, etc.
13. Carrying out sampling test on weakly weathered bedrock. There is no need to sample and analyze the soil layer above weakly weathered bedrock roof in normal circumstances.
14. Proposing rock layer sampling and testing methods, and providing verification parameters for pile-bearing capacity.

2.5.3 Supplementary survey for karst cave section

To meet the needs of the project, the surveyed section was located in the karst area along the axial direction of the subway tunnel. The distribution of karst caverns along the shield tunnelling alignment was investigated by geological borehole in the geological investigation phase. First, 118 boreholes were drilled with a spacing of 5 m to a depth 10 m greater than the tunnel burial depth at 10 m from the outside boundary of the shield tunnelling alignment. The drilling results revealed a total of 89 karst caverns, and the karst cavern distribution was irregular, with a burial depth range varying from a few meters up to 50 m below the bed river. For safety, the positions, sizes and filling types of the karst caves in this tunnel section were investigated using 42 supplementary boreholes. These supplementary survey boreholes were drilled to 15 m below the tunnel burial depth at 5 m from the outside boundary of the shield tunnelling alignment at a spacing of 2.5 m, and 3 holes were drilled to obtain the soil samples [5]. The basic information of a the borehole column of supplementary survey is shown in Table 2.10.

2.5.4 Engineering geological conditions

2.5.4.1 Geographic and geomorphic conditions

The section of the karst area for supplementary survey is located at the southern end of Orange Island in the center of Xiangjiang River in Changsha. It is one of many alluvial sandbanks along the Xiangjiang River, with Cenozoic and Holocene alluvium (Q^{4al}) from top to bottom as the main stratum. Such stratum distributed along the Xiangjiang River forms the modern floodplain accumulation layer, underlying by an upper Cretaceous (K2) sandy conglomerate. The landform is a sandbank in the river. Now the sandbank is a theme park on the Orange Island, with flat terrain, artificial trees, birds, flowers, and beautiful scenery.

Table 2.10 Borehole information in the supplementary survey.

Construction site name: Karst cave section of Changsha Rail Transit Line 3 Construction date: December 29, 2015, to January 3, 2016 Borehole millage location: CK15 + 850 ~ CK16 + 050, orifice elevation: 36.0638 m Rock core preservation location: stored in warehouse at the construction site					Borehole number:BZ1−3 Borehole coordinates: X = 97123.97, Y = 46625.14 Borehole depth: 45.00 m Groundwater level: 19.70 m			
Age causes	Layer bottom depth (m)	Layer thickness (m)	Layer Bottom elevation (m)	Soil name and feature	Rock core recovery ratio (%)	Sampling depth (m)	Heavy II blow count N(2) /depth(m) SPT blow count N/depth(m)	Basic allowable value of bearing capacity (KPa) Standard value of friction resistance Qik (kPa)
Q_4^{ml}	4.20	4.20	31.863	Miscellaneous fill: brownish yellow, slightly wet, slightly compacted. The composition is clay, and a small amount of construction waste and rough gravel contained.	90			
Q_4^{al}	5.80	1.60	30.263	Silty clay: brownish yellow, soft plastic, damp, about 20% content of silty-fine sand, clay is main composition, and flat and smooth cutting face	90			
	11.00	5.20	25.063	Silt: brownish yellow, medium dense, damp, content of silty-fine sand ≥ 50%, 40%–50% clay, simple composition	85			

	20.40	9.40	15.663	Fine gravel: brownish yellow, medium dense, saturated; gravel content ≥ 50%, grain size mainly between 4–10 mm, mainly in subcircular shape; The parent rock are mainly weakly weathered vein quartz, quartz and sandstone; the void are filled with sand and mud. The grain size distribution is general, and the grain size increasing in greater depth.	80			
K_2	21.60	1.20	14.463	Fully weathered conglomerate (W4): maroon, with argillized cement and gravel, soil-like rock core, and very soft rock.	80			
	28.10 29.60 34.60 35.50 39.00	6.50 1.50 5.00 0.90 3.50	7.963 6.463 1.463 0.563 −2.937	Strongly weathered conglomerate (W3): maroon, gravel structured resided, cement has been softened, rock core fragmented mostly, block diameter mainly between 3–8 cm. The rock body is soft and broken There are dissolution cavities at depth between 28.10–29.60 m and 34.60–35.50 m. There are small amount of filters as endogenous substances at the bottom of cavities.	65			

(Continued)

Table 2.10 (Continued)

	45.00	6.00	−8.937	Weakly weathered conglomerate (W2): light maroon, gravelly structure in massive blocks. about 60% content of gravel, grain size mainly between 10 −40 mm with individual size up to 80 mm, half subcircular and half subangular. Pore type sand mud cementation; The gravel is mainly composed of limestone, sandstone accounts for less than 1/4 of the total; The rock cores are mainly as column section with length of 5−15 cm no longer than 30 cm; The rock body is soft and relatively complete.	85	– 1/39.00−39.20 – 2/40.40−40.60 – 3/43.70−43.90	

2.5.4.2 Stratum rock (soil) characteristics and geological features

Exposed by boreholes, the main rock (soil) layer from top to bottom in the area is composed of Quaternary Holocene artificial fill (Q^{4ml}), and river alluvium (Q^{4al}) and underlying bedrock upper Cretaceous (K2) conglomerate. According to the degree of weathering, the bedrock can be divided into completely weathered layer (W4), strongly weathered layer (W3) and weakly weathered layer (W2), the unexposed slightly weathered layer (W1). The further underlying lower Cretaceous (K1) is not exposed and unknown.

2.5.4.3 Unfavorable geological bodies and adverse geological phenomena

In the process of geological survey, unfavorable geological bodies and adverse geological phenomena were clarified. There were three main items.

Quaternary floodplain alluvium (Q^{4al})

Silty clay, silt, gravel sand, fine gravel, and coarse gravel are contained in this layer from new to old. The buried depth is 3.70–18.00 m with the elevation of the bottom boundary mostly between 17.00 m and 24.00 m. Due to the influence of the long period of abundant groundwater, there is a hidden danger of bottom collapse.

Groundwater

The depth of groundwater level is measured at about 19.00 m with an average elevation of 25.40 m. The groundwater mainly comes from Xiangjiang River, and the groundwater level is obviously affected by seasonal water level of the Xiangjiang River. With a groundwater level elevation up to 30 m, a negative impact definitely can be generated on shield tunnel construction. After the karst cave is filled with water, water inrush accidents may occur in the tunnel construction process.

Karst caves

The statistical results from boreholes of geological survey showed the rate of karst cave encountering along the tunnel section is more than 50%, and the linear karst cave rate was more than 10%. The karst cave development grade in this interval could be classified as level I according to the Chinese technical code for investigation of geotechnical engineering of fossil fuel power plants (DL/T5074-2006, 2007) [6]. The dimension varied from 0.2 m to 22.3 m, and the mean dimension of the cavities was

2.8 m. Among the elevation of these caves, 80 of them were below 6 m (89.9%), and 9 of them were above 6 m (10.1%). For the dimension, 26.5% of them are diameter less than 1 m, 38.8% of them are diameter between 1 m and 3 m, 24.6% of them are diameter between 3 m and 6 m, and 10.1% of them are diameter larger than 6 m. Additionally, most of the caves were filled with groundwater, fine sand or silt; the abundant groundwater had close hydraulic connections with Xiangjiang River water, and the groundwater was slightly basic with pH values from 7.5 to 7.9. Most karst caves were distributed in series and related to each other, and most karst caves contained fillings (silt and sand) and water. In addition, water in the karst caves was interrelated with water in the Xiangjiang River. The surrounding rock in this area is a gently inclined monoclinic stratum with a dip angle less than 10 degrees and inclined to north by east. It was preliminarily judged that there is no obvious correlation between the karst caves so that most of them are independent karst caves or multilayer karst caves. The contact with the outside are mainly through dissolution pores and interlayer cracks. The dissolution characteristics are obviously related to content of the gravel in the stratum and its composition. The dissolution phenomenon is reduced if the cave elevation is below −5 m where is no sign of communication with the outside.

2.5.5 Hydrogeological conditions

The shield tunnel is a section that is approximately 1.4 km long and passes under the Xiangjiang River, and the construction site has a subtropical monsoon climate with high precipitation. Long-term data at the Changsha hydrological observation site on the Xiangjiang River show that the average annual water level is approximately 29.48 m; the highest river water level (39.18 m) occurs from April to July, and the lowest river water level (25.16 m) occurs from November to February. The period from April-July is a high-flow period in which the discharge is at its highest of the year, while the period from November-February is defined as a low-flow period. Typically, the average annual river flow velocity varies approximately between 0.12 m/s and 1.26 m/s. In situ hydrogeological pumping tests show that the groundwater has a close hydraulic connection with the Xiangjiang River water. There are three main types of groundwater within the shield tunnel construction site: pore water in the shallow soft solid layer and the alluvial gravel layer, bedrock fissure water confined in the intensely weathered sandstone and conglomerate, and carbonate karst fissure water confined in limestone. The burial depth of the pore water table is 1~13 m, while the water table of bedrock fissure water is approximately 15~20 m deep, and the water table of carbonate karst fissure water is approximately 21~27 m deep [5]. The hydrogeological conditions are relatively simple but still have a large negative impact on the shield tunnel construction. The water-filling volume of the karst cave cannot be estimated, and water inrush accidents can occur in the construction process.

2.5.6 Testing methods of karst cave treatment result

2.5.6.1 Testing methods

Holes were drilled, of which positions were located by a total station and depths were designed to be 44.40 m, in the construction site in a staggered way within a range of 5 × 6 m. A drilling rig usually used in mining engineering was used for rotary drilling with the protection of the flushing fluid. The drilling hole diameter was 130 mm, and the final hole was 91 mm. A composite sheet and a diamond bit were combined for drilling, and grouting pipes were inserted after completion of the drilling. The quality of drilling and grouting should conforms to the engineering geological design requirements. M15 mortar was adopted for the grouting test with a grouting pressure of 0.6–1.0 Mpa. The whole grouting process was carried out under the supervision of related technicians and completed when slurry returns from the orifice and mortar did not fall down after standing for a period. The grouting of the holes continued until the drilled holes in entire test section had been completed by repeating this procedure.

2.5.6.2 Treatment result verification methods

After completion of grouting, geophysical exploration was used to verify the results of grounding in the karst cave testing section and to check whether or not the whole karst cave was filled with mortar. It was necessary to carry out borehole verification and hydrological observation. The filling effects was checked by sampling the cemented-soil grout, and examining whether or not its compressive strength meet the design requirements. Recommendations can be proposed for the next step construction based on grouting treatment results.

2.5.7 Suggestions for shield tunnel construction

The geological survey concluded that the average elevation of the karst cave is between −8.00 m and 7.05 m, the highest elevation is 14.5 m and the lowest is −2.05 m, while the top elevation of the fully weathered conglomerate is 10.2∼17.80 m. The elevation of the karst cave treatment section was recommended to be between −8.00 m and 17.80 m. To achieve optimal karst cave treatment effect, it was recommended that the whole dissolution zone should be filled with grouts to reach the required strength and all groundwater passages should be plugged. Therefore the arrangement of grouting pipes should be reasonable and all dregs in the hole bottom should be removed. The technical construction party was to be fully competent with construction ability and technical skills and be able cooperate very well. Fillers, additives, grouting pressure and alternative schemes should be selected carefully. In case of severe slurry leakage, appropriate plugging agents and methods should be chosen, and appropriate accelerators should be added. The process was a trial and error correction process. The test data should be collected and recorded such that grouting could be completed sufficiently to achieve the design purpose.

If the optimal effect of karst cave grouting was not achieved in the construction process, it would cause significant negative impacts on shield tunnelling of the Metro construction, even causing accidents such as water inrush, roof collapse, casualties, and heavy economic loss. Therefore, the important role of grouting quality detection is obvious and needs close attention.

Many uncertain events can occur in the grouting process. Dynamic design should be carried out during the construction, and solutions should be adjusted according to different situations to ensure effective grouting and achieve the design purpose.

References

[1] Zhu W, Ju S. Study on risk sources and typical accidents of metro shield construction. Guangzhou: Jinan University Press; 2010.

[2] Ju S, Zhu W. Study on geological investigation methods for shield tunnelling in mixed ground. Mod Tunn Technol 2007;27(6):10−14.

[3] Wang X, Zhang H, Bian Y. On the geological adaptability of shield cutterhead designs. Mod Tunn Technol 2013;50(3):108−14.

[4] Yang Y, Tan Z, Peng B, Li J, Wang G. Study on optimization boring parameters of earth pressure balance shield in water-soaked round gravel strata. China Civ Eng J 2017;50(S1):94−8.

[5] Yang J, Zhang C, Jinyang F, Shuying W, Xuefeng O, Yipeng X. Pre-grouting reinforcement of underwater karst area for shield tunnelling passing through Xiangjiang river in Changsha, China. Tunn Undergr Sp Tech 2020;100:103380. Available from: https://doi.org/10.1016/j.tust.2020.103380.

[6] DL/T5074-2006. Technical code for investigation of geotechnical engineering of fossil fuel power plant. National Development and Reform Commission, Beijing, China 2007.

Exercises

1. What main data should be provided after the geological survey of a shield tunnel project?
2. Please introduce the stages of geological survey for a shield tunnel project? What are the differences between the survey contents at each stage?
3. What are the main geological survey means for a shield tunnel project? What are their functions?
4. What factors should be considered in the design of plane and vertical alignments of a shield tunnel?

CHAPTER 3

Shield machine configurations and working principles

3.1 Composition of the two main types of shield machines

As was mentioned in Chapter 1, Introduction, there are many types of shield machines. After long-term exploration and practice, the most widely used types of shield machines are EPB and slurry shield machines. Here, structural drawings provided by China Railway Construction Heavy Industry Group Co., Ltd. are used as examples to introduce their main components.

As Fig. 3.1 shows, the cutterhead (1) is located at the forefront of EPB shield machine, and its shell temporarily supports the unlined tunnel. The excavation chamber is filled with muck to stabilize the excavation face. The propulsion system (3) provides the thrust for pushing the shield machine forward. The central drive system (6) provides torque to drive the rotation of the cutterhead to cut ground. The pedestrian gate system (4) is used to prevent the excavation chamber from being directly connected to the outside environment and thus resulting in pressure leaking out of the excavation chamber when the staff enters for tool inspection and replacement. The segment erector (7) assembles segments for supporting the surrounding ground under the protection of the tail shield (5). The screw conveyor (8) and belt conveyor (9) are transportation devices

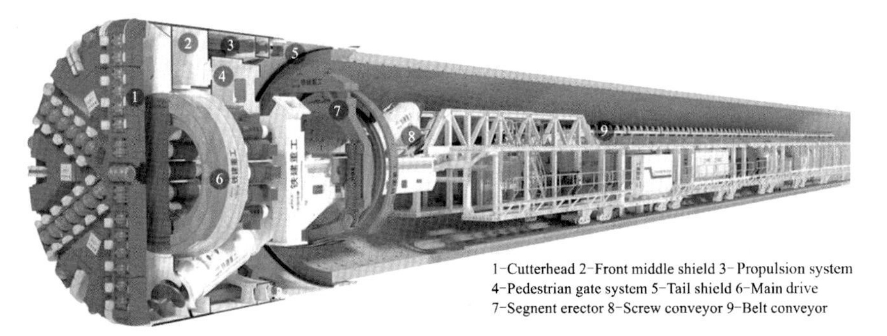

Figure 3.1 The components of an EPB shield machine.

Shield Tunnel Engineering
DOI: https://doi.org/10.1016/B978-0-12-823992-6.00003-5

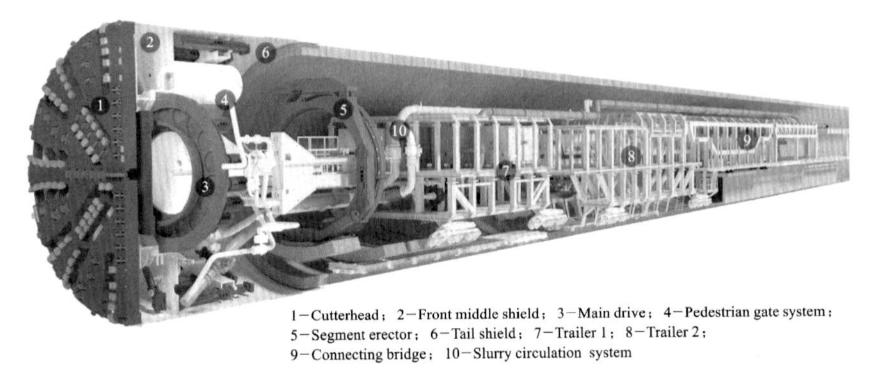

Figure 3.2 The components of a slurry shield machine.

that discharge the muck from the excavation chamber. There is a long space behind the above-mentioned main machine components, which is equipped with the auxiliary backup system for shield tunnelling, including connecting bridges, trolleys, segment cranes, segment conveyors, power distribution boxes, materials tanks for synchronous grouting and muck conditioning, and so on.

As can be seen from Fig. 3.2, the tunnelling system of a slurry shield machine is similar to that of an EPB shield machine, including the cutterhead, shield shell, main drive, propulsion system, and so on. Because of the differences in the medium for balancing excavation face, the slurry shield machine adopts a slurry circulation system (10) to adjust the physical properties of the slurry and provide fresh slurry for the excavation chamber, and then the slurry is discharged with excavated soil through the discharging pipelines. The segment erector constructs the lining support after the excavation. The backup system of the slurry shield machine includes the connecting bridge, power distribution box, slurry circulation system, slurry separation treatment system, slurry preparation system (on the ground surface), and so on.

3.2 Concepts and functions of shield machine components

3.2.1 Basic components

The shield machine basically consists of shield shell, propulsion system, excavation device, soil-retaining device, segment erector, cutterhead driver system, soil-transportation equipment, shield articulation, shield tail sealing system, and so on. These basic parts are described in detail as follows.

3.2.1.1 *Shield shell*

The shield shell is the main component of the shield machine. It is used to protect the workers and mechanical equipments for safe construction. Shield shells were formerly made of cast iron materials but now are made of steel. As shown in Fig. 3.3, the shield shell is composed of the front shield (i.e., the cut ring), the middle shield (i.e., the support ring), and the tail shield. The front shield is located at the forefront of the shield and is the excavation and soil-retaining part. It cuts into the ground first and protects the excavation during construction. The cutterhead and the excavation chamber are located within the scope of the front shield. Generally, the front end of the open shield is equipped with a cutting edge to reduce the disturbance to the ground during tunnelling. The middle shield bears the ground pressure, the jacking force, the frontal resistance produced when the shield head cuts into the soil, and the construction load of the segment erection. A shield jack group is arranged on the beam at the inner wall of the middle shield, and the inner space holds the hydraulic equipment, power equipment, and operation controller of the segment erector. The tail shield is generally formed by the extension of the shield shell and is mainly used to protect the assembly of the segments. The end of the tail shield is equipped with a sealing system to prevent water, soil, and injection materials from entering the shield inside via the gap between the tail shield and the segments.

The outer diameter of the shield shell is equal to the outer diameter of the segment rings plus the shield tail clearance and the thickness of the shield tail shell. The shield tail clearance is at the tail of the shield. To ensure that the segment is installed with enough space and the shield

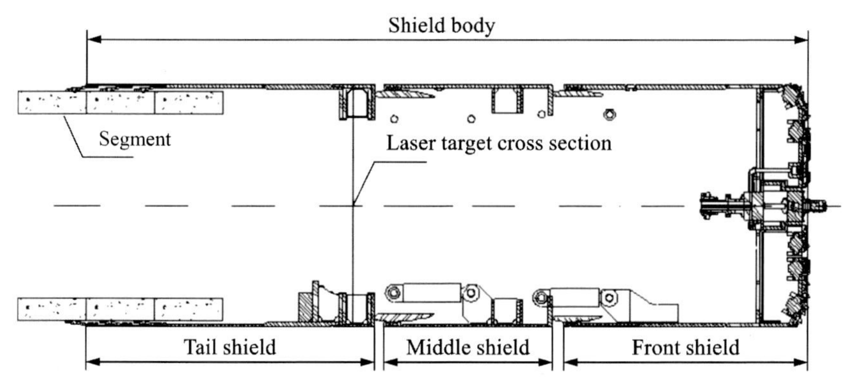

Figure 3.3 The components of the shield shell.

machine can adjust its own attitude at the curved tunnelling line, it is necessary to set the shield tail clearance reasonably. The thickness of the shield tail clearance is affected by (1) the deformation of the shield tail under the action of water and soil pressure and the external pressure when tunnelling along the curved line, (2) the deformation of the segment under the action of water and soil pressures, (3) the inclination of the segment at shield tail when tunnelling along the curved line, and (4) the allowable error of the outer diameter of the segment rings [1].

As is shown in Fig. 3.4, the outer diameter (D_e) of the shield shell can be calculated by the following formula:

$$D_e = D_0 + 2(X + t) \tag{3.1}$$

where D_0 is the outer diameter of the segment ring; X is the shield tail clearance, generally from 25 to 40 mm [1]; and t is the thickness of the shield tail shell, generally from 50 to 100 mm [2].

For shield tunnels with curved alignments, the shield tail clearance should be calculated according to the geometry.

The length of the shield shell is related to factors such as the excavation method, the method of soil discharging, the segment width, the insertion direction of the segment capping block, the number of the shield tail sealing rings, and whether or not the middle shield and the tail shield are hinged. Its length is calculated as follows:

$$L = L_C + L_G + L_T \tag{3.2}$$

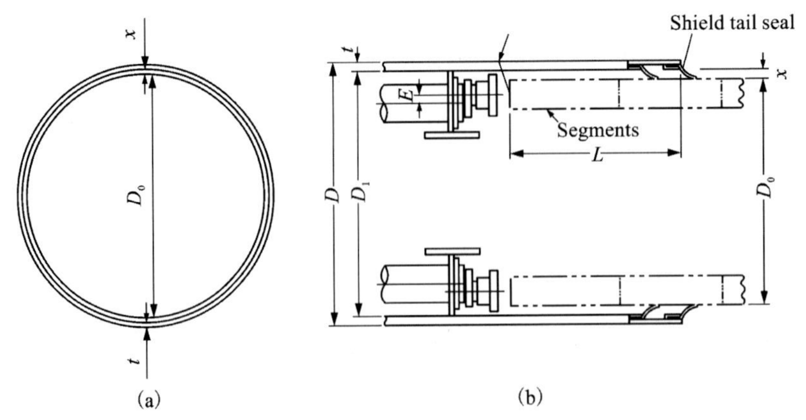

Figure 3.4 The outer diameter of the shield shell. (a) Front view. (b) Side view.

where L_C is the length of the front shield, L_G is the length of the middle shield, and L_T is the length of the tail shield.

For hand-dug and semimechanical shields, the length of the front shield is determined according to factors such as the penetration depth that the front shield bores into the ground, the maximum travel distance of the jack cylinders, and the length of the excavation workspace. For closed-chamber shield machines, the length of the front shield depends on whether the cutterhead is equipped within the scope of the front shield. When the cutterhead is equipped in the front shield, the length of the front shield does not need to include the length of the cutterhead and is mainly determined by the capacity (length) of the closed chamber and others. When the cutterhead is equipped in front of the front shield, the length of the front shield needs to accommodate the lengths of the cutterhead, the closed chamber, and others.

The length of the middle shield depends on the specifications of the shield propulsion jacks, muck discharging device (screw conveyor, belt conveyor), and segment erection system, and its length should satisfy the length of the jack cylinders at their maximum travel state.

The length of the tail shield is determined by the length of the jack support (L_D), the width of the segment (B), the margin of the assembled segment (C_F), and the rear margin (C_R) including the tail sealing materials. It is calculated as follow:

$$L_T = L_D + B + C_F + C_R \tag{3.3}$$

The ratio of the length of shield shell to the diameter (L/D_e) is defined as the sensitivity (ξ) of shield machine. The lower the sensitivity, the more convenient the operation of the shield machine. Zhang et al. [2] proposed the sensitivity values for different diameters of shield machines. For large-diameter shields with a D_e larger than 6 m, the ξ is recommended to be 0.7–0.8 (generally 0.75); for medium-diameter shields, the ξ is recommended to be 0.8–1.2 (generally 1.0); for small-diameter shields, the ξ is recommended to be 1.2–1.5 (generally 1.5).

3.2.1.2 Propulsion system

The propulsion system is a mechanism that pushes the shield machine forward in the ground, and it is a key system of the shield machine. Its main component is a group of propulsion jacks arranged on the inner ring beam of the shield shell. In addition to the simple axial force generated by the shield jack group, 5%–8% of the lateral load is possibly generated by the tilting of the jack pad. As the shield advances, the angle between the shield axial direction and the end of the segment changes. To be adaptive this angle change, the jack cylinders are usually not rigidly connected to the shield body but are installed on the shield body utilizing elastic materials such as hard rubber.

The main factors that affect the design of the shield jack stroke include the segment width, the idling stroke margin, the jacking stroke margin, and the insertion direction of the segment capping block. In determining the jacking stroke margin, the inclination between the shield and the segment must be considered. For shields with an outer diameter of 7 m or less, a margin of about 150 mm is sufficient. A shield machine with a larger diameter requires a larger margin. In using a shaft-insert type of segment capping block, there must be sufficient space for inserting this type of segment block. The space size varies depending on the insertion angle of the segment capping block.

To monitor and control the attitude of the propulsion system, a guidance system is required. The guidance system is capable of grasping and analyzing various parameters during shield tunnelling process and has the functions of tunnel design axis management, space position detection, attitude detection, graphic display, measurement base point verification, and communication with the host control system. The guidance system is an indispensable system for guiding the smooth shield tunnelling. The guidance system is composed of theodolite, electron laser system (ELS) target, rear-view prism, computer, and so on, and it can update the attitude information of the shield machine in real time and can control the shield machine within an allowable tolerance range of the designed tunnel alignment when turning. The main reference point of the guidance system is a laser beam emitted from a laser theodolite, which is installed on the segments behind the shield [3].

3.2.1.3 Excavation tool

Hand-dug shield machines are used for manual excavation by tools such as sharp picks, cross picks, or spades. The semimechanical shield machine uses single-bucket excavators, backhoes, and outrigger rock drills for partial excavation. Sometimes it is necessary to use a spade to excavate the hard-to-reach parts of the excavator. Except for hand-dug and semimechanical shield machines, all other shield machines use a cutterhead for excavation. This cutterhead is a rotating or oscillating excavation tool, which is composed of a cutterhead board for stabilizing the excavation face, lot of cutters for digging the ground, a rotating or oscillating drive mechanism, and a bearing mechanism.

Cutterhead

The cutterhead is welded to steel structures and is normally at the forefront of the shield machine. It not only excavates the ground but also supports the excavation face to promote the stability of excavation face. As is shown in Fig. 3.5, the cutterhead can be located inside the notched ring of the shield head or be protruded from the notched ring, or it can be flush with the notched ring. The cutterhead that protrudes the notched ring has the widest applications [1].

As is shown in the front views in Fig. 3.6, the cutterheads for shield machines may be spoke cutterhead, panel cutterhead, or composite cutterhead. The spoke cutterhead is composed of spokes, cutters arranged on the spokes, and openings. It is used mostly for (open) mechanical shield machines and EPB shield machines. Its opening ratio is relatively large, about 60%–95%. Owing to the small contact area between the spokes

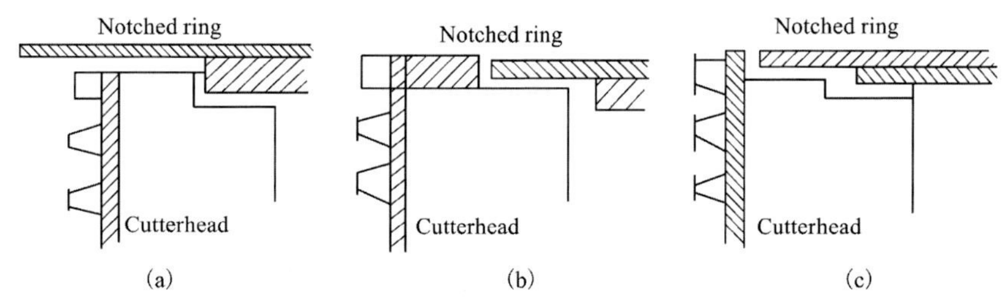

Figure 3.5 Excavating cutterheads. (a) Inside the notched ring of the shield head. (b) Protruding from the notched ring. (c) Flush with the notched ring.

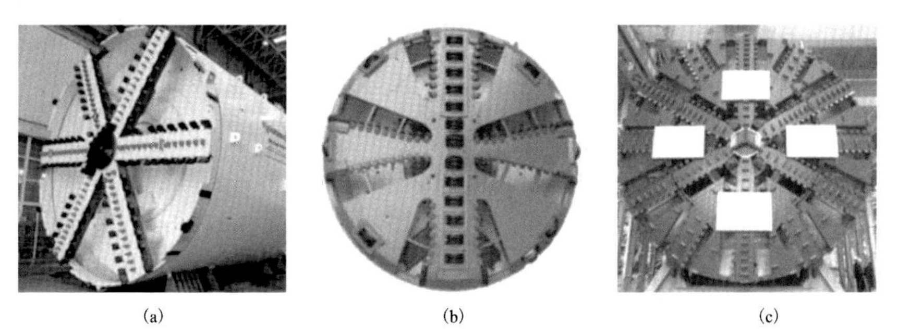

Figure 3.6 The three types of cutterheads. (a) Spoke cutterhead. (b) Panel cutterhead. (c) Composite cutterhead.

and the stratum, water spraying and mud spraying are easy to occur in a weak stratum with high groundwater pressure. The panel cutterhead is composed of panels, cutters, and openings. It can be a panel cutterhead with a fixed opening or a panel cutterhead with an adjustable opening, and it can be used for EPB shield machines and slurry shield machines. Its opening ratio is relatively small, about 30% [4]. The composite cutterhead has both characteristics of the panel cutterhead and the spoke cutterhead, and its opening ratio is normally between 35% and 50%.

There are three types of support methods for the cutterhead: central support, intermediate support, and peripheral support (Fig. 3.7). Table 3.1 compares the performance of the three different cutterhead support methods.

Cutter

The cutters are arranged on the cutterhead. By using different cutters, rock-breaking by rolling, rock-breaking by cutting, and even overexcavation can be performed.

1. Disc cutters (Fig. 3.8): The disc cutters for rolling rock-breaking can be divided into toothed ones and smooth ones. Due to the different tooth shapes, toothed disc cutter can be divided into spherical-tooth disc cutters and wedge tooth disc cutters. The toothed disc cutters are mostly used in soft rock formations, and the disk hobs are mostly used in hard rock formations.

In hard stratum (hard rock, boulder, hard clayey soil), it is necessary to configure the disc cutters to break the hard stratum. Under the influence of

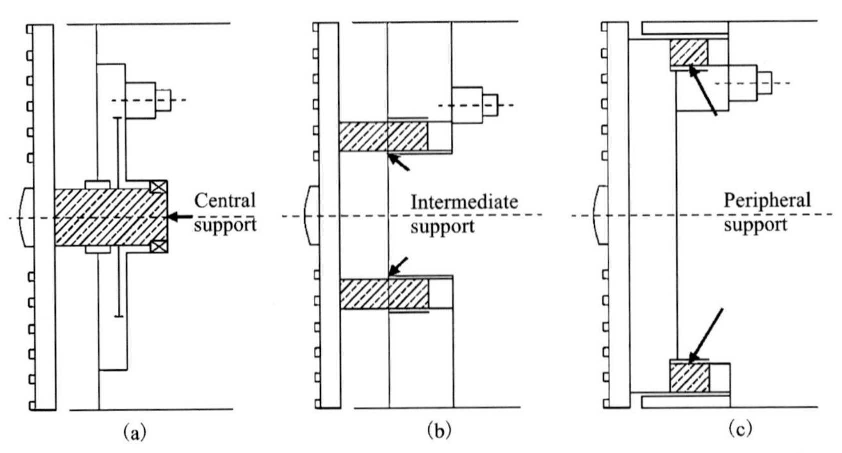

Figure 3.7 The support methods of cutterheads. (a) Central support. (b) Intermediate support. (c) Peripheral support.

Table 3.1 Performance comparison of the cutterheads with different support methods [2].

Performance	Central support	Intermediate support	Peripheral support
Torque of screw conveyor and drives	The screw conveyor is equipped in the lower part of the excavation chamber with small impeller and small torque	The screw conveyor is mounted in the middle of excavation chamber	The screw conveyor is equipped in the periphery of the excavation chamber with large impeller and large torque
Diameter of screw conveyor	Small	Large	Large
Mechanical torque consumption	Low consumption and high efficiency	Low consumption and high efficiency	Large consumption and low efficiency
Position of mixing blade	The blade is equipped inside the cutterhead	The blade is mounted on the spoke	The mixing is done by the inner soil bucket
Condition of soil adhesion	Low	Middle	High
Ability to dig hard soil	Normal	Good	Good
Applicable shield diameter	Middle, small	Middle, large	Large
Soil sand sealing effect	Short sealing and good durability	Middle	Long sealing and low durability
Chamber space	Small	Middle	Large
Suitability for long-distance tunnelling	Strong	Middle	Weak
Production difficulty	Little	Little	Big
Swing ability during shield advancement	Large	Middle	Small

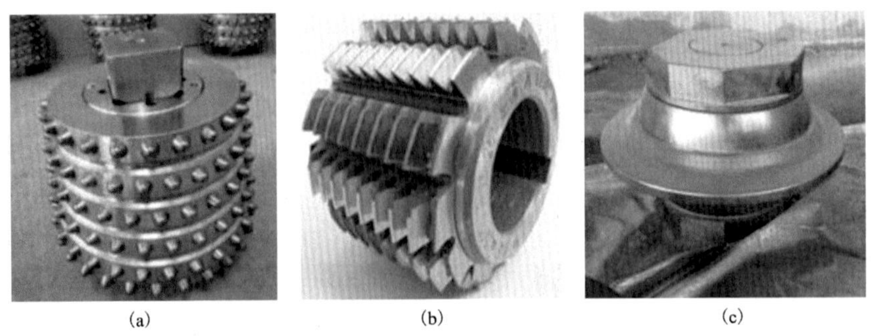

(a) (b) (c)

Figure 3.8 Hobs. (a) Spherical-tooth disc cutter. (b) Wedge-tooth disc cutter. (c) Smooth disc cutter.

huge thrust and rotational torque, the disc cutter performs the functions of pressing, rolling, splitting, and grinding the rock to get the comprehensive effect of fracturing, expansion, shearing, and grinding and achieve the purpose of breaking the hard layer. The breaking the complete rock via disc cutters is achieved as follows [5]:

a. The cutter edge cuts into the rock mass under the huge thrust pressure, forming cutting marks. With the cutterhead rotation and the cutter rolling, this part of the rock is first broken into powder, accumulating on the top of the blade and forming a powder core area.

b. The cutter edge splits into the rock mass on both sides, causing many micro-cracks in the weak part of the rock mass.

c. As the depth spiltting into the rock increases, the rock dust continues to fill in the microcracks. Stress concentration occurs at the ends of the microcracks, and the microcracks gradually expand and become apparent cracks.

d. Rock fractures are formed when the apparent cracks intersect with the micro-cracks generated by adjacent cutters or when the apparent cracks extend to the rock surface.

e. When the fractured body falls into the bucket from the excavation face, new fractured bodies and rock dusts are generated due to the collision of the fractured body and the cutterhead and each other. But in the soft layer, the disc cutters generate axial pressure (in the direction of tunnel tunnelling) and cutting shear pressure (in the tangential direction of cutterhead) in the soil around the excavation face as the cutterhead rotates, continuously cutting soil from the excavation face [6].

Because of the geological applicability and the size of the cutter seat, the disc cutters are more suitable for composite or panel cutterheads. The disc cutters are generally equipped on the spokes of the cutterhead, and few ones can be equipped on the cutterhead panel if there are special requirements for the construction conditions. The layout area, shown in Fig. 3.9, can be divided into central area, frontal area, transition area, and marginal area. The cutter layout in the cutterhead is as follows.

a. The disc cutters arranged in the central area are usually called as central cutters, and their main arranging forms are either "—" or "†" type. The "—" type is mainly divided into eight-edged and ten-edged types, and the "†" type is mainly divided into eight-edged and twelve-edged types. When the cutter layout and the number of blades have been determined, the position of each disc cutter is determined accordingly.

b. The disc cutters arranged in the front area are called as front ones, which are the most important cutters on the cutterhead and are arranged in the largest number compared to other areas. The front disc cutters are mainly arranged on the spoke or panel of the cutterhead, and the cutter spacing is normally a constant of *S*, which is generally determined by geological parameters. In the consideration of the force balance of the cutterhead, the arrangement method generally adopts the Archimedes spiral form and the cutters are grouped in a symmetry form.

c. The transition area refers to the area between the front one and the marginal one. The transition area and the front area are basically the same in disc cutter height from the cutterhead board; but based on the consideration of equal wear, the cutter spacing in the transition area is generally slightly smaller than in the front, and the number of disc cutters is generally not more than 2 in the transition area.

d. The disc cutters arranged in the marginal area are called the edge ones, and the geometry of the cutter tip envelope is mostly arched.

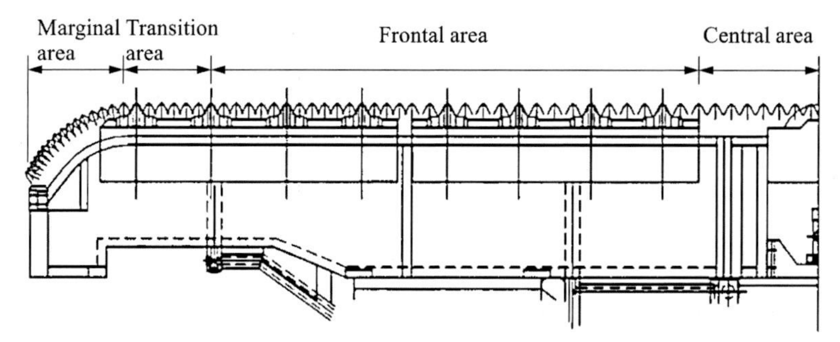

Figure 3.9 Schematic diagram of the layout of the disc cutters on the cutterhead.

However, since the cutting radius in the marginal area is larger than that of other areas, the cutting trajectory line is longer, and the wear is more serious. The cutter spacing S should generally not be greater than the cutter spacing in the transition zone, and as the radial radius increases, the cutter spacing should gradually become smaller.

2. Scrapers (Fig. 3.10): This type of cutters contain blades, which are used to cut the soft soil layer and scrape the formed muck into the excavation chamber.
3. Advance cutters (Fig. 3.11): The advance cutters are arranged ahead of other cutters, cutting the ground in advance to prevent other cutters from first cutting hard rocks such as pebbles and boulders and to protect them. The advance cutters are divided into three types: shell cutters, rippers, and tooth cutters. Japanese shield machines usually use shell cutters, German Herrenknecht shield machines more often use

Figure 3.10 Scraper.

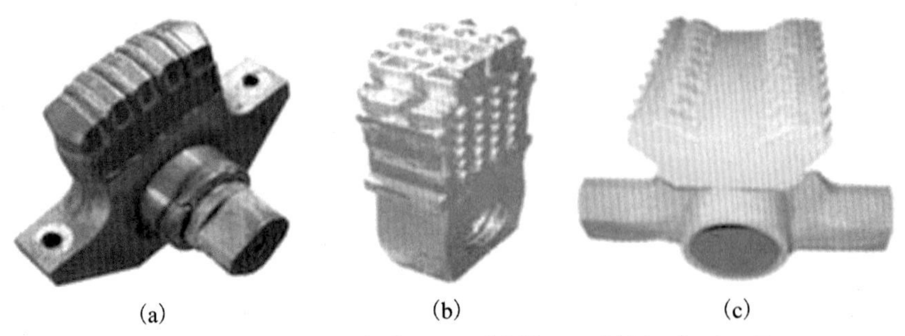

(a) (b) (c)

Figure 3.11 Advance cutters. (a) Shell cutter. (b) Ripper. (c) Toothed cutter.

toothed cutters, and Canadian Lowett and French NFM shield machines more often use rippers.

4. Edge Scrapers (Fig. 3.12): The edge scrapers, which can also be called as spatulas, are arranged on the outer circumference of the cutterhead. They are used to remove the excavated muck from the edge to avoid muck accumulation, to ensure the excavation diameter of the cutterhead and to avoid indirect wear on the outer edge of the cutterhead.
5. Profiling cutters (Fig. 3.13): The profiling cutters are arranged on the outer edge of the cutterhead, operated by hydraulic cylinders, and controlled by programming. The operator of shield machine can control the overexcavation depth of the profiling cutters and the position of overexcavation. As shown in Fig. 3.13b, when it is decided to expand the left side of tunnel, the profiling cutters can be extended when turned to the left side.

Figure 3.12 Edge scrapers.

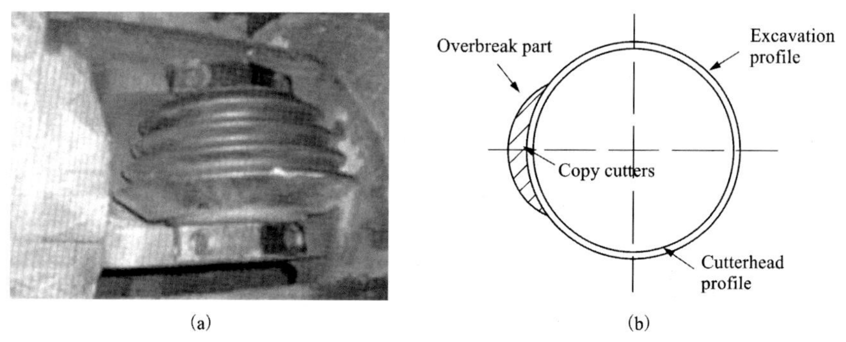

Figure 3.13 Profiling cutter. (a) Photograph of a profiling cutter. (b) Overexcavation on the left side.

3.2.1.4 Pedestrian compartment and compressed air adjusting system

The pedestrian compartment allows workers to enter the excavation chamber for inspection and maintenance of the cutterhead, cutter inspection and replacement, and other operations, as shown in Fig. 3.14. The pedestrian compartment is classified into single chamber and double chambers. The single-chamber compartment has small structure and occupies small radial space. The double-chamber compartment can separate people and materials, with large capacity and high efficiency. Therefore the double-chamber compartment is more reasonable and more popular than single-chamber one in the market.

If the shield stops in a stratum with poor self-stability, the excavation chamber must be pressurized to stabilize the excavation face. In this case, to carry out inspection and maintenance operations, it is necessary to pressurize and depressurize the pedestrian compartment before people enter and exit the excavation chamber.

The compressed air adjusting system installed in the shield body is used to adjust the support pressure at the excavation face and adjust the air pressure in the pedestrian compartment. This compressed air adjusting system includes an air compressor, a pressure regulator, a pressure sensor, an air tank, a control valve, and a breathable air filter [3].

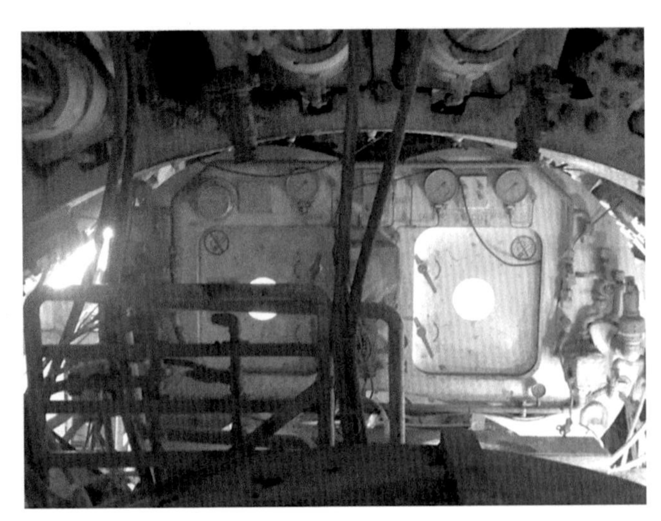

Figure 3.14 Pedestrian compartment.

3.2.1.5 Soil-retaining device

To stabilize the excavation face in weak formations and to avoid large deformation of the ground, it is necessary to support the excavation face properly. In addition to using the cutterhead to support the excavation face, the EPB and slurry shield machines respectively rely on the muck and slurry in the closed excavation chamber to support the excavation face. Fig. 3.15 shows how to balance the excavation face of the slurry shield. There are earth pressure and water pressure on the left side of the excavation face and chamber pressure on the right side. It is necessary to establish a pressure balance between the left and right sides to stabilize the ground effectively and avoid large ground deformation.

The hand-dug shield machine relies on the movable front eaves and the sub-chamber baffles to retain the soil and to avoid the collapse of upper soil at the excavation face (see Fig. 3.16). The movable front eaves can be adjusted horizontally to penetrate the ground deeply and can also be adjusted to support the excavation face in the vertical direction. However, because of the destruction of the initial ground stress, the front eaves and the sub-chamber baffles cannot effectively control the ground settlement and so it is often necessary to prereinforce the ground. Semimechanical shield machines are similar to hand-dug shield machines and often require prereinforcement of weak ground to control its deformation. The difference

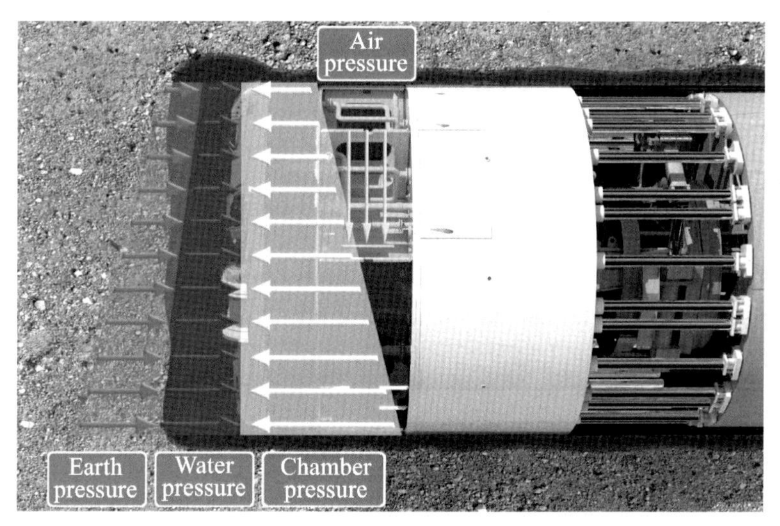

Figure 3.15 Schematic diagram of excavation face stability for slurry shield machine as an example.

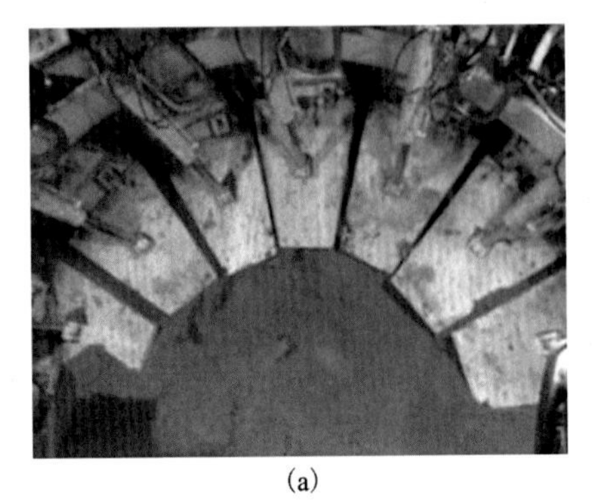

(a)

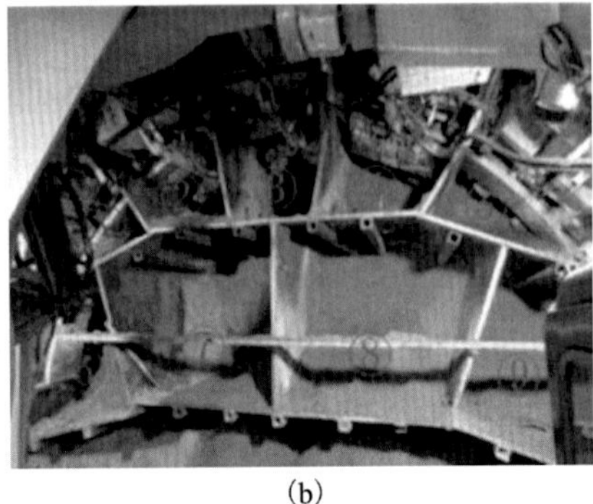

(b)

Figure 3.16 The retaining structures of the hand-dug shield machine. (a) Front eaves. (b) Sub-chamber baffle.

is that to facilitate the excavation of weak ground with a single-bucket excavator and a backhoe, the semimechanical shield machine does not require a separate baffle. The mechanical shield machine uses a cutterhead for excavation, and the cutterhead also plays a key role in stabilizing ground.

3.2.1.6 Segment erector

The segment erector can achieve segment clamping, rotation, up-and-down movement, front-and-rear movement, and lifting and antiswaying of segment blocks (see Fig. 3.17). The segment clamping device clamps the segment blocks by inserting a pin into the segment lifting hole. The pin should have enough strength to avoid failure accidents caused by the segment block falling. The rotating device is controlled by a hydraulic motor and sends segment blocks to their preset position according to the design requirements. After the segment blocks have been sent to the preset position, the up-and-down movement device uses a jack to transport the clamped segment blocks to the preset position in the radial direction. The front-and-rear movement device is mainly set for the capping blocks of segments. When the capping blocks are inserted axially, they must be moved along the axial direction of the shield. For radially inserted capping blocks of segments, the front-and-rear movement is required to be about 200 mm to meet the space requirements of segment assembly. When the capping blocks are in place, it is often necessary to slightly lift the adjacent blocks on both sides to provide enough space for the capping blocks to be inserted. The lifting device uses a hydraulic jack to extend and retract the piston rod, which is generally installed on the shield body or the segment erector. The segment antiswaying device is set to prevent the

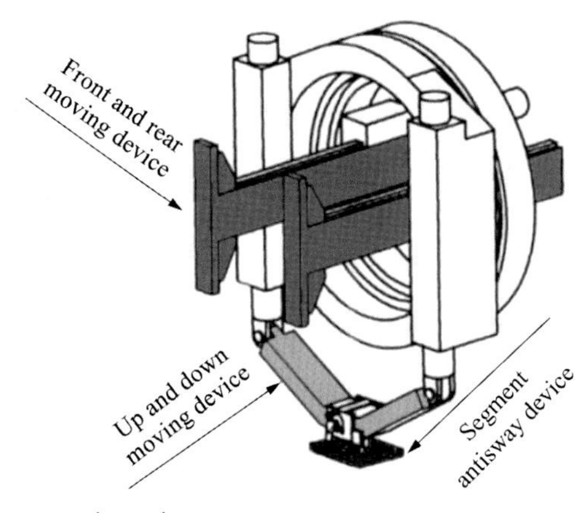

Figure 3.17 The segment erector.

segment blocks from swaying. In the process of segment block rotation and positioning, due to their considerable weight, they are easy to sway and are likely to collide with the segment erector or other structures and break themselves and the segment erector [1].

3.2.1.7 Cutterhead driver

Since the cutterhead undertakes the tasks of boring the ground at the excavation face and mixing the muck in the excavation chamber, the working performance of its driver determines the tunnelling efficiency of shield machine. The cutterhead driver provides the necessary torque for the cutterhead rotation. The driver drives the gear or pin-lock mechanism behind the cutterhead to rotate the cutterhead through either an electric motor or a hydraulic motor with a reducer. To obtain a large torque, a hydraulic cylinder can also be used to drive the cutterhead to rotate. Electric motors convert electrical energy into mechanical energy and have the advantages of high energy efficiency, low noise, and low heat generation and are thus commonly used. To prevent the cutterhead torque from being significantly increased when encountering pebbles or having clogging on the cutterhead, which may damage the motor, the clutch must be used to stop the cutterhead rotation when the cutterhead torque exceeds the warning value. Compared with electric motors, hydraulic motors convert hydraulic energy into mechanical energy, being easy to adapt to overload and easy to adjust rotation speed. Therefore hydraulic motors are recommended in formations where hard particles such as pebbles that greatly affect loads are prone to occur. Another advantage of the hydraulic motor is that the axial length is short. This feature can shorten the overall length of the shield machine and can be applied to the construction of small-radius curved shield tunnels.

3.2.1.8 Soil-discharging system

1. For hand-dug and semimechanical shield machines, the muck is directly discharged through screw conveyors or belt conveyors, usually with artificial assistance or the aid of excavator to send the muck to the inlet of the screw conveyor or the belt conveyor. The muck discharging system of (open) mechanical shield machine is relatively complicated. As shown in Fig. 3.18, the slag is discharged through buckets, sliding guides, hoppers, and belt conveyors.
2. The slurry shield machine uses a slurry circulation system to discharge the slurry with excavated soil in the excavation chamber and send the treated (fresh) slurry (as shown in Fig. 3.19) to the chamber for requirement of filter cake forming at the excavation face. The excavated soil is transported from the slurry tank to the slurry separation station through the slurry pump.

The slurry circulation system consists of slurry feeding pumps, slurry discharging pumps, relaying pumps, slurry feeding pipelines, discharging pipelines, measuring sensors (flow rate, density), extension pipelines, and slurry storage tanks. In addition, to avoid clogging of the suction ports of the slurry pumps, it is necessary to install slurry stirring rod at the suction

Figure 3.18 Soil discharging system of a mechanical shield machine.

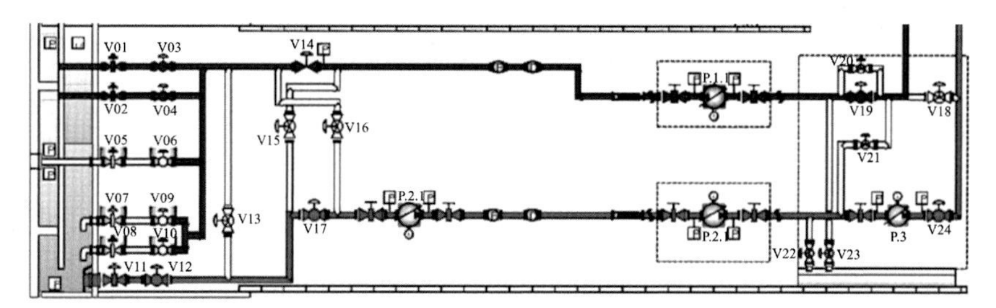

Figure 3.19 Schematic diagram of a slurry circulation system.

port of the excavation chamber. The main working processes of the slurry circulation system are as follows: The feeding pump and relaying pump transport the fresh slurry prepared in the slurry pool to the slurry storage tank through the slurry pipelines. Then discharging pump and relaying pump discharge the slurry that carries the excavated soil and transport it to the slurry treatment system on the ground for separation.

The slurry separation system is used to separate the water and soil in the discharged slurry containing excavated soil. The system is set on the ground and consists of two parts including slurry separation station and slurry preparation equipment. The slurry separation station separates and treats soil particles with different sizes and is mainly composed of vibrating screens, cyclones, slurry storage tanks, adjustment tanks, and slurry pumps. The (fresh) slurry preparation equipment is used to mix the slurry and consists of sedimentation tanks, slurry tanks, and slurry production system.

3. For an EPB shield machine, the muck in the excavation chamber is usually discharged by a screw conveyor, and the discharged muck is then transported to the muck carriage by the belt conveyor. The screw conveyor is composed of a cylinder, a driver, spiral blades, and slag discharge gates. There are mainly two type of screw conveyors, including with a shaft and without a shaft, as shown in Fig. 3.20. The screw conveyor with a shaft directly rotates the central shaft with blades, which has a better antispewing capability, but the grain size of the excavated muck that can be discharged is small. The screw conveyor without a shaft directly rotates the blades, which can discharge muck with large grain size, but the antispewing capability is poor. The screw conveyor continuously discharges the muck out of the excavation chamber. The muck forms a sealed soil plug under good muck conditioning during the discharging process in the screw conveyor to prevent the water in the soil from being dispersed and to maintain the

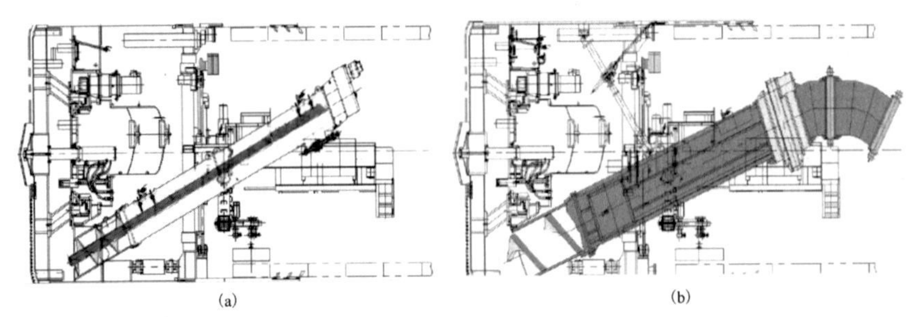

Figure 3.20 Types of screw conveyor. (a) With a shaft. (b) Without a shaft.

stability of the earth pressure in the excavation chamber. The shield machine driver compares the earth pressure value in the excavation chamber with the expected earth pressure value and adjusts the soil discharging rate accordingly. In this way, the muck pressure in the excavation chamber is controlled to achieve a continuous dynamic pressure equilibrium to ensure the stability of the excavation face and the smooth shield tunnelling.

The belt conveyor adopts either an electric driver or a hydraulic driver and is composed of belt conveyor brackets, front passive wheels, rear driving (active) wheels, upper and lower supporting wheels, belts, belt tensioning device, mud scraping device, and drive motor with a reducer. The conveyor is arranged on the rear supporting bridge and trailer. There are multiple emergency stop switches on the belt conveyor or an emergency stop pulling device throughout the entire process to ensure the safety of muck discharging. Curve adjustment and antideviation devices are installed on the belt conveyor to adapt to the possible curve alignments during shield tunnelling [3].

3.2.1.9 Shield Articulation

The shield articulation is an added component, and the shield machine was not equipped with an articulation at first. When the shield machine passes through the curved tunnel, the segment forms an angle with the shield shell tail. Restricted by the angle limitation, the radius of the tunnel curve cannot be too small. When the sensitivity of the shield machine (the ratio of the diameter of the shield to its length) is about 1, the minimum construction radius generally requires 300 m for normal metro tunnel with a diameter of about 6 m [7]. When the shield machine is used for a small-radius tunnel, due to the limited angle between the axis of the segment and the axis of the shield, the segment assembly is complicated. In the curved tunnel construction the shield machine relies on the jack cylinders to distribute different pressures to achieve the shield machine turning, probably causing the fragmentation risk of the segments. The gap between the shield shell and the segment ring cannot be too small, because it affects the turning of the segment rings. On the other hand, this gap should not be too large, as a large gap brings difficulties to sealing around the shield tail. For a shield machine with a articulation system, the shield head can be pushed in the directions according to the tunnel curve requirements, and the shield tail remains unchanged in attitude. Such a strategy can avoid the generation of the angle between the shield tail shell and the segment ring and can maintain the suitable attitude of the segment ring and the shield. Curved tunnelling relies on the articulation jacks connecting two shield shell parts. As long as the gap between the shield tail shell and the segment ring meets the requirements of segment assembly, there is no need to turn the shield, and the gap can be slightly reduced for easy sealing the shield tail.

Shield articulation systems are classified as positive articulation and passive articulation. As shown in Fig. 3.21a, the propulsion jack cylinders are fixed to the tail shield for the positive articulation, and thus the thrusts of the propulsion jacks act on the tail shield and then are transmitted to the middle shield through the articulation jacks. The bending angle of the two shield parts of the shield machine is adjusted by the active expansion and contraction of the articulation jacks to realize the turning of the shield machine. The active articulation not only facilitates the turning of the shield machine in curved tunnel sections, but also can be used to correct the attitude of the shield machine [8–10]. As is shown in Fig. 3.21b, the

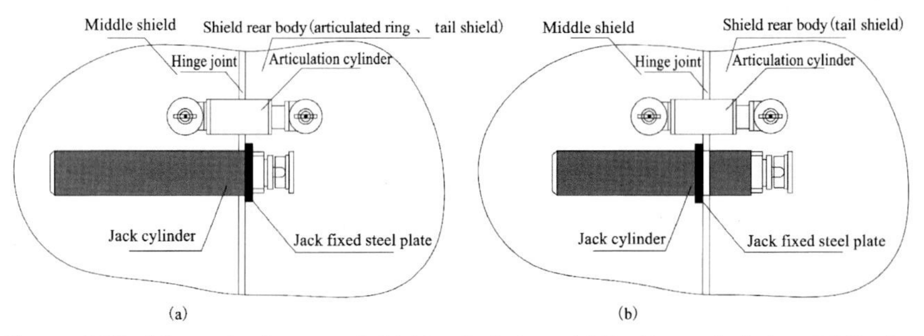

Figure 3.21 Schematic diagram of shield articulations. (a) Positive articulation. (b) Passive articulation. *Modified from Zhang T, Xu P. Comparison of active articulation and passive articulation of shield machine. In: Proceedings of China Shield Technology Symposium, 2011.*

passive articulation relies on external forces to make the articulattion jack expand and contract so that a certain angle may occur between the middle shield and the tail shield of the shield machine. The propulsion jack cylinders of the shield machine are fixed on the middle shield, and the thrusts of the propulsion jacks directly act on the middle shield of the shield machine. Table 3.2 shows a comparison between the positive articulation and the passive articulation.

Table 3.2 Comparison of the active articulation and passive articulation [10].

Articulation type	Working method	Features
Positive articulation	The stroke difference of the articulated cylinder is adjusted according to the radius of the tunnel curve. The thrust of the propulsion jack acts on the tail shield of the shield machine, and the middle shield and the front shield are pushed through the articulation jacks.	Precise controls of the stroke of the articulation jack and the angle between the middle shield and the tail shield can be achieved. The operation is relatively complicated.
Passive articulation	The hydraulic pressures of the propulsion jack groups are adjusted according to the radius of the tunnel curve and the stroke difference of the articulation cylinders. The thrusts of the propulsion jacks act directly on the middle shield, and the middle shield drags the tail shield through the articulation jacks.	The operation is relatively simple when turning, and the turning can be achieved by adjusting the zoned hydraulic pressures of propulsion jacks. It cannot directly control each articulation stroke and requires external forces to change the angle between the middle shield and the tail shield. Correction that is too fast may cause the positioning gap to be too small and the segments may be broken at the shield tail.

3.2.1.10 Sealing system at the shield tail

Since there is a gap between the shield tail and the segment ring, a sealing system is required to prevent groundwater and grouting slurry from penetrating into the shield body. Multiple metal sealing brushes (as shown in Fig. 3.22) are often mounted at the shield tail. To avoid leakage effectively, it is necessary to fill the cavities between any two adjacent rings of sealing brushes with high-viscosity oily or resinous materials. As the shield machine advances, the filled sealing material is consumed gradually and needs to be replenished in a timely manner. The sealing material can be injected from the supply pipelines of sealing material on the shield tail shell. The feeding pipelines pass through the shield tail shell, and injection pumps are required to inject the sealing material into the sealed cavities at the shield tail. The consumption of sealing material along each segment ring can be calculated according to the following formula:

$$Q = (1.5 \sim 2.0)\pi DBt \tag{3.4}$$

where D is the outer diameter of the segment (in meters), B is the width of the segment (in meters), and t is the thickness of gap between the segment and the shield tail shell (in meters).

The sealing performance of the shield tail has a great impact on the safe tunnelling of the shield, especially when the shield passes through unfavorable geological conditions. For example, when the shield machine excavates through the coarse sand layer with high groundwater pressure on Foshan Metro Line 2, China, due to improper operation during the replacement of the sealing brushes, the sealing system at the shield tail failed and the external water and sand suddenly flowed into the tunnel, causing ground collapse as well as casualties and economic losses.

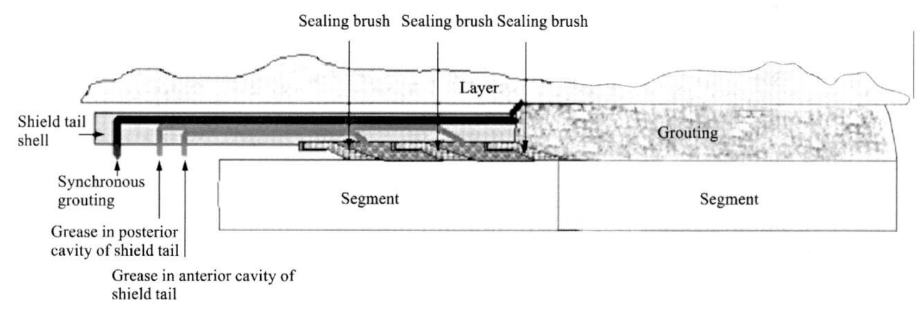

Figure 3.22 The sealing system at the shield tail.

3.2.1.11 Synchronous grouting system

The excavation diameter of the cutterhead is larger than the diameter of the segment ring at the shield tail, resulting in a gap between the excavated ground and the segment ring. When the shield tail escapes from the assembled segment ring and the gap cannot be replenished with grout in time, the stratum is prone to subsidence. To effectively control the ground settlement during shield tunnelling, as shown in Fig. 3.22, the cement slurry can be injected synchronously at the shield tail. The grouting capacity of the synchronous grouting should meet the needs to fill the annular gap created at the highest tunnelling speed. Owing to the influence factors such as the location and quantity of the injection points, the grout fluidity, and the geological conditions, the synchronous grouting system often cannot sufficiently fill the gap between the ground and the segment ring, resulting in that it is necessary to perform a secondary grouting process in time to fill the remaining gap to minimize settlement as much as possible [3]. More details will be described in Chapter 8, Backfill Grouting for Shield Tunneling.

3.2.1.12 Muck conditioning system

The EPB shield machine needs to be equipped with a muck conditioning system to inject additives such as foam, bentonite, or polymer into the excavation face, excavation chamber, and screw conveyor. The muck conditioning system uses the stirring rods at the cutterhead back and bulkhead front and the blades in the screw conveyor to mix the additives and muck. The purpose of muck conditioning is to make the excavated muck have suitable consistency, low water permeability, and low friction resistance for achieving stabilization of the earth pressure in shield chamber and avoiding the risk of soil clogging and water/muck spewing. It shows that the selection of the muck conditioning system is essential. The muck conditioning systems are normally equipped with a foam injection system, a bentonite injection system, and a polymer injection system. The details are described in Chapter 9, Muck Conditioning for EPB Shield and Muck Recycling.

The basic parts of shield machine has been described above. Some other detailed components such as electrical equipment and cutter wear detection devices are not explained here. The interested readers can refer to relevant books.

3.2.2 Backup systems

Different shield machines have different backup systems. For a metro shield with a tunnel diameter of about 6 m, the backup system mainly includes connecting bridges, trolleys, segment cranes, segment conveyors, belt conveyors, and other auxiliary equipments, and it is responsible for muck transportation, segment transportation, mortar transportation, and so on.

3.2.2.1 Backup trolley

The configuration of backup trolleys should consider (1) the tunnel section, tunnel curve, and tunnel slope; (2) the convenience of equipment installation, maintenance, and management; (3) the ways of transporting out the muck and transporting segments; and (4) construction sites and safety facilities such as safety passages, handrails, and railings.

In a curved tunnel section with a small radius or steep slope, enough space between the trolleys and the segment rings must be ensured when the trolleys are in motion, and the trolleys must be prevented from overturning and derailing.

3.2.2.2 Hydraulic power unit

The shield machine is equipped with hydraulic power units, which are used mainly for the hydraulic motor drive of screw conveyor, the gate drive of screw conveyor, the motor system of the segment erector, the profiling cutter system, and so on, as shown in Fig. 3.23a. The control room, muck conditioning system, power unit, and hydraulic station related to the shield are assembled on the rear supporting trolley and arranged in a form that is convenient for operation and inspection, as shown in Fig. 3.23b.

(a)

(b)

Figure 3.23 Layout of the backup system of the shield machine. (a) Hydraulic power unit. (b) The control room.

3.2.2.3 *Connecting bridge*

The connecting bridge is a connection arranged between the main shield machine and the rear working platform. It is used to connect the hose rack and cable rack. The connecting bridge and the lower part of the working platform at the rear are equipped with a transport device for conveying materials such as segments. The layout of the connecting bridge is shown in Fig. 3.24, and its connection with the rear supporting trolley is shown in Fig. 3.25.

3.2.2.4 *Segment crane*

The segment crane includes two electric hoists and a driving device, and its function is to lift the segment blocks from the segment cart to the segment transport trolleys. The trolleys are placed at the lower part of the connecting bridge, which can temporarily store the segment blocks transported by the segment crane. The transport trolley can also transfer the segment blocks to the area that the segment erector can reach and move forward with the shield machine under the action of the traction trolleys.

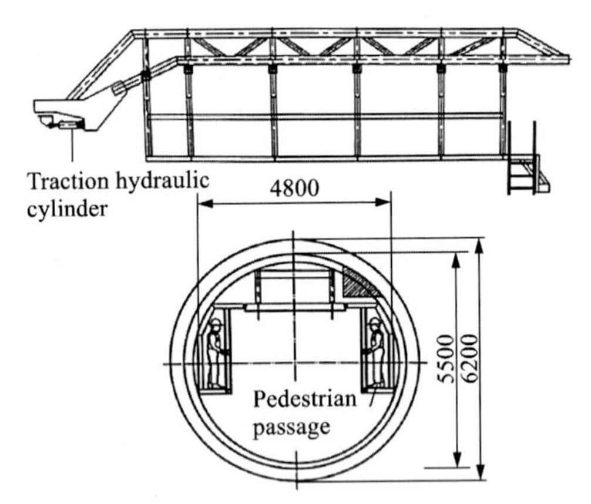

Figure 3.24 Schematic diagram of the connecting bridge (unit: millimeter).

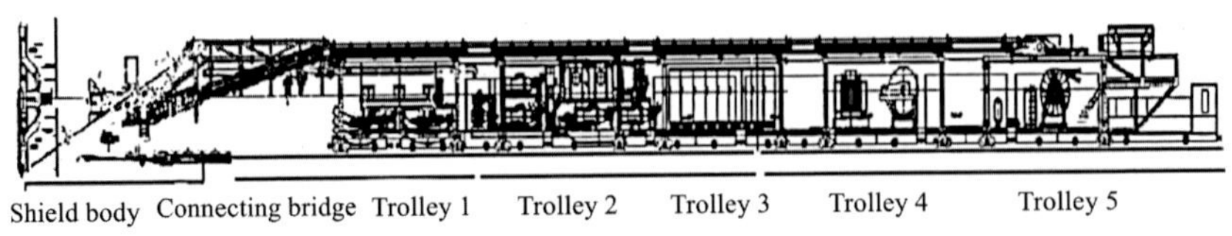

Figure 3.25 Location diagram of the backup trolleys and connecting bridge.

3.2.2.5 Segment and material delivering system

The segments and materials are transported by a single-beam hoist and a second double-beam hoist. The lower part of the second double-beam hoist meets the space requirements of storing a ring of segments. The transport scheme is shown in Fig. 3.26.

3.2.2.6 Remote monitoring system

The remote monitoring system of shield machine includes an automatic measurement system and an monitoring information management system. It may have the functions of real-time monitoring of the construction site, remote recording of monitoring data, real-time and historical data/curve display, parameter abnormal alarm, tunnelling data report generation, tunnelling operation and measurement data access and query functions, data feedback function, and settlement prediction. The functions can be divided into three levels [11].

The first level is the data measurement and acquisition system. The function is transmitting the output signals of shield tunnelling parameters (excavation speed, jack thrust, cutterhead rotation speed, screw conveyor rotation speed, excavation chamber pressure, and so on), forces and deformations of various supporting structures to the corresponding data acquisition system after preprocessing.

The second level is the data processing and transmission system, which transmits the digital signals collected by the data acquisition equipment to the upper processing system in the operation room of the shield machine. The shield machine operating room is located in the rear trolley, and it contains industrial controlling computers and on-site display screens. The industrial controlling computer are controlled by field technicians to display measurement data and to set shield tunnelling parameters. As shown in Fig. 3.27, the display screens are used for the setting and monitoring of tunnelling parameters for the propulsion jack cylinders, articulation jack

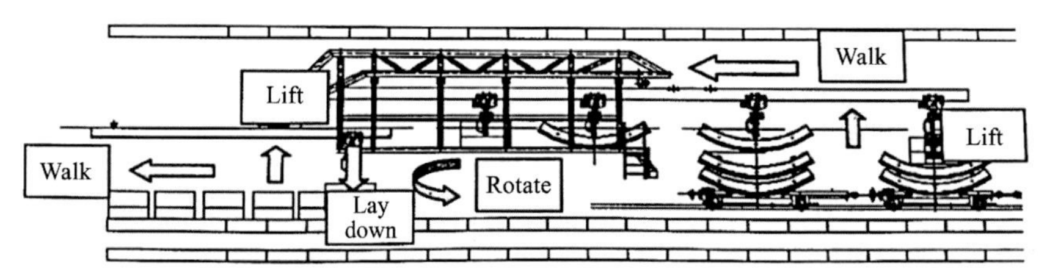

Figure 3.26 Schematic diagram of the transportation of segments.

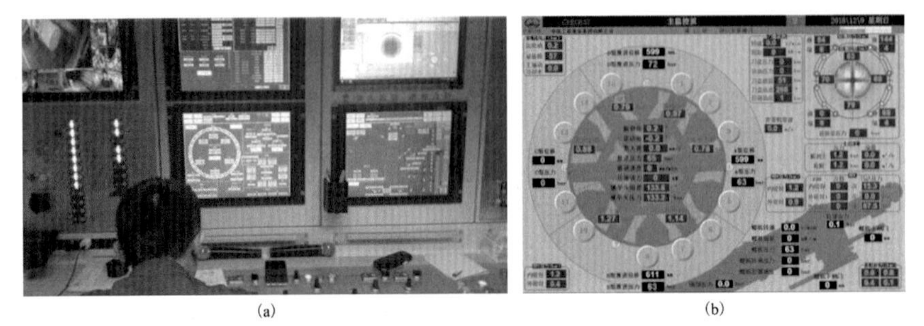

(a) (b)

Figure 3.27 Shield machine operating room. (a) Industrial controlling computer and operation desk. (b) Main monitoring screen of an industrial controlling computer.

Figure 3.28 Remote monitoring center on the ground surface.

cylinders, backfill grouting system, the foam generation system, and so on. Each interface in the screens can be switched by function keys.

The third level is a remote monitoring center located on the ground surface, as shown in Fig. 3.28. This monitoring center is convenient for ground personnel to monitor the operation of the shield machine remotely and in-time fault diagnose, and it provides a network platform for data processing and data storage during tunnelling.

3.3 Working principles of main shield machines

3.3.1 EPB shield machine

The cutterhead of an EPB shield machine cuts the ground, destroying the in-situ stress balance in the initial ground. It is necessary to provide the excavation face with sufficient supporting pressure to balance the earth pressure and water pressure (as shown in Fig. 3.29) and then stabilize the excavation face. After the excavated ground form muck and enter the excavation chamber, the muck is then discharged through the screw conveyor. Owing to the dynamic relationship between the volume of soil entering and exiting the excavation chamber, the supporting pressure at the excavation face is affected. When the amount of muck discharged is greater than the amount of muck inflow, the supporting pressure at the excavation face is less than the initial earth stress, and the excavation face tends to deform toward the excavation chamber. Otherwise, when the amount of muck discharged is less than the amount of muck inflow, the muck fills the excavation chamber and is compressed, and the excavation face tends to deform in the direction of shield tunnelling. To ensure the smooth discharge of the muck, it is necessary to inject conditioning materials into the excavation face and the excavation chamber. The conditioning materials can improve the workability of the muck and reduce its friction and permeability for reducing the torques of the cutterhead and screw conveyor, and avoiding soil clogging in clayey ground and water/muck spewing in sandy ground with high groundwater.

Determination of the ideal muck discharging volume is difficult to achieve directly because the original stratum forms loose muck after excavation, and

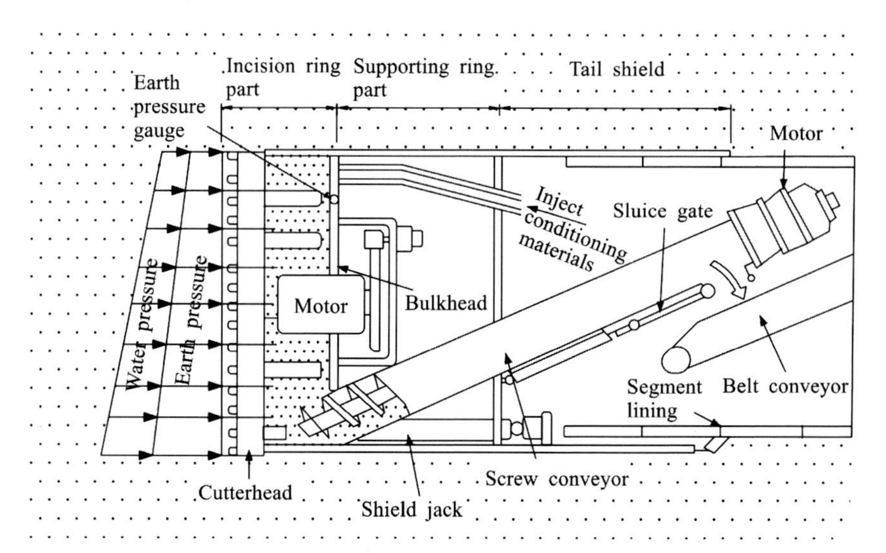

Figure 3.29 Schematic diagram of the pressure balance at the excavation face of an EPB shield machine.

the discharging volume is affected by the dynamics of the looseness coefficient of the muck (the ratio of muck volume to the original ground volume), the injection of conditioning agents, and the muck stirring in the excavation chamber. The stability of the excavation face often needs to be achieved by controlling the pressure of the excavation chamber. Owing to the rotation of the cutterhead, it is often impossible to directly and accurately measure the earth pressure at the excavation face, and it needs to be indirectly controlled by monitoring the pressure on the bulkhead of the excavation chamber. Because of the influences of the cutterhead and the muck, there is a pressure difference between the bulkhead and the excavation face, inducing that it is not easy task to obtain the planned supporting pressure at the excavation face.

3.3.2 Slurry shield machine

Slurry shield machines are mainly used in formations with high water pressure and strong permeability. Similar to the EPB shield machine, it is necessary to rely on the slurry pressure in the excavation chamber to balance the earth pressure and water pressure at the excavation face [6] to maintain the stability of the excavation face, as shown in Fig. 3.30. The excavated soil enter the slurry tank and form a high-concentration slurry

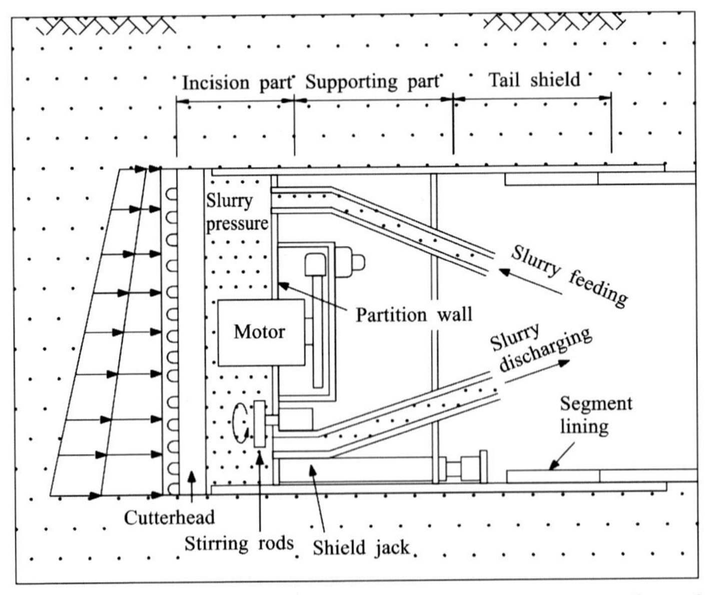

Figure 3.30 Schematic diagram of pressure balance at the excavation face of a slurry shield machine.

containing after being mixed. The slurry is discharged to the slurry separation system on the ground surface by the slurry pumps, and after slurry treatment and sand precipitation the fresh slurry is repressurized to the slurry tank. In this way, excavation, dumping, and advancing are completed in a continuous cycle. Compared with the muck of the EPB shield machine, the slurry is liquid, and the pressure of the slurry tank is easy to control.

To effectively stabilize the excavation face and prevent the infiltration of groundwater into the excavation chamber, the formation of mud film on the excavation face is the key to the slurry shield tunnelling. When the slurry pressure is greater than the groundwater pressure in the chamber, the muddy water penetrates into the soil according to Darcy's law, forming suspended particles with a particular proportion to the soil pores, and these particles are captured and accumulated on the excavation face to form a mud film. The thickness of the mud film continues to increase over time, and the resistance to water penetration gradually increases, so the slurry pressure in the excavation chamber can effectively support the external earth pressure and water pressure in the ground.

3.4 Special shield machines

With the development of the shield tunnelling method and the construction requirement of some special tunnel projects, a large number of new special shield machines have begun to emerge. These emerging shield machines not only solve some construction problems that are difficult to be solved with conventional technology, but also greatly improve the efficiency, accuracy, and safety of shield tunnelling. Special shield machines include free-section shield machine, radial-expanding shield machine, spherical shield machine, multicircle shield machine, H&V (horizontal variation and vertical variation) shield machine, variable-section shield machine, and eccentric multiaxis shield machine [12].

3.4.1 Free-section shield machine

The excavation section of the free-section shield machine has an unconventional shape, which is realized by setting multiple planetary cutterheads around the central cutterhead. The planetary cutterheads revolves around the main cutterhead while rotating along its rotation shaft, and the orbits of the planetary cutterhead rotations are determined by the corner angle of the entire excavation face. Therefore through the design of the orbits of the planetary cutterheads, the excavation face can be selected to be rectangular, elliptical, horseshoe, oval, or another cross-sectional form. This shield is especially suitable for small and medium-sized tunnel projects that need to shuttle through the ground with existing pipelines and in which the underground space is limited.

In 2018 the world's first large-section horseshoe shield machine (see Fig. 3.31) was successfully used in the Baicheng Tunnel of the Haolebaoji-Ji'an Railway. The excavation diamater of the shield machine is 10.95 m high and 11.9 m wide. It uses nine small cutterheads to form a horseshoe-shaped section with each small cutterhead rotating freely.

3.4.2 Radial-expanding shield machine

The radial-expanding shield machine is used to expand the diameter of a part of the original shield tunnel. During construction the first step is to remove parts of the original lining and to excavate parts of the surrounding rock in sequence and then build a space capable of holding an enlarged shield as its starting station. Owing to the removal of the original lining, the pressure distribution around the tunnel will change, so reinforcement measures are required. After the lining has been removed and the radial-expanding shield machine is assembled and tested in the space, the shield machine is ready to advance.

Figure 3.31 Horseshoe shield machine.

3.4.3 Spherical shield machine

The spherical shield machine is normally used for the continuous tunnelling at right angles of tunnel route. It is mainly used for construction fields in which it is difficult to guarantee sufficient spaces for the construction of the deep launching shafts of shield machine or when turning at right angles is required. The significant feature of the spherical shield machine is that it can turn from a vertical shaft to a horizontal shaft, or rotate 90 degrees for continuous excavation [13].

Spherical shield machine can be classified into the vertical-horizontal type and the horizontal-horizontal type [14]. The vertical-horizontal spherical shield machine is equipped with ring-shaped overexcavation tools in the outside of the shield cutterhead. It is capable of excavating two kinds of underground spaces with different functions and cross-sections via one excavation machine. The structure and construction process of such a spherical shield are shown in Fig. 3.32.

The horizontal-horizontal spherical shield machine (Fig. 3.33) is a ring shield that first completes the horizontal tunnel construction in one direction and then rotates the sphere horizontally to carry out the construction of another horizontal tunnel, which can meet the 90° turning requirements of the shield machine.

3.4.4 Multicycle shield machine

The multicycle shield machine (Fig. 3.34) is used for the MF (multicircular face) shield tunnelling method [14]. Through various combinations of

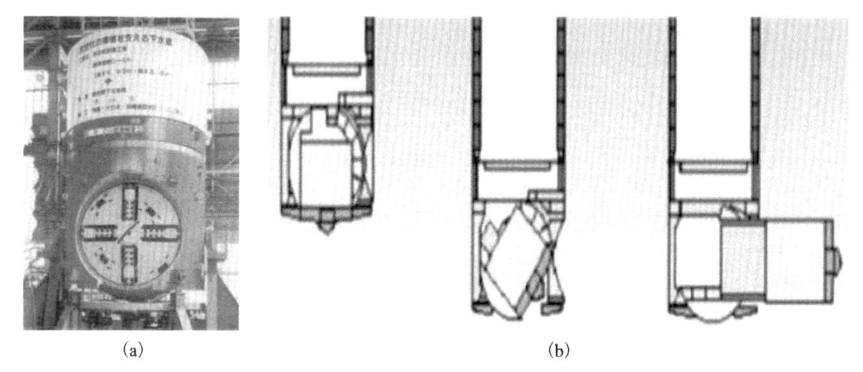

Figure 3.32 Vertical-horizontal shield machine and construction technology. (a) The spherical shield. (b) Construction process of such a spherical shield. *From Li B., Song M., Zheng B. Discussion on sphere shield. Sci Technol Wind 2009;14:234. Chen K., Hong K., Wu X. Shield Constr Technol. Beijing: China Communications Press; 2016.*

Figure 3.33 Horizontal-horizontal shield machine. (a) Cutterhead in the forward direction. (b) Cutterhead rotates 90°. *From Chen K., Hong K., Wu X. Shield Constr Technol. Beijing: China Communications Press; 2016.*

Figure 3.34 Multi-circle shield machine. (a) Double-circle. (b) Three-cirlce. *From Chen K., Hong K., Wu X. Shield Constr Technol. Beijing: China Communications Press; 2016.*

circles, tunnels of various cross sections can be constructed [13]. Multicircle shields are used mostly to construct metro stations and underground parking stations.

3.4.5 Horizontal and vertical shield machine

The H&V shield machine (Fig. 3.35) uses a special hinge mechanism to combine two circular shields to construct tunnels with spiral curves [14]. It can be separated from the double-circular shield to construct a single-circular tunnel. It is also possible to twist two adjacent tunnels from vertical to horizontal, or from horizontal to vertical, into a spiral shape [15]. In 2016 the H&V shield construction method was applied for the first time in the rainwater drainage pipe project of the Tachiaigawa Main Line in Tokyo, Japan. In the construction, two connected articulated shield machines were used to drive two superadjacent tunnels synchronously.

Figure 3.35 H&V shield machine. *From Chen K., Hong K., Wu X. Shield Constr Technol. Beijing: China Communications Press; 2016.*

Figure 3.36 Variable-section shield machine. *From Chen K., Hong K., Wu X. Shield Constr Technol. Beijing: China Communications Press; 2016.*

The outer diameter of the shield machine is 5.58 m with an interval of 9 cm.

3.4.6 Variable-section shield machine

The variable-section shield machine (Fig. 3.36) combines the main cutters and the profiling cutters on the cutterhead. The main cutters are used to excavate the central part of the circular section, and the profiling cutters are used to excavate the surrounding parts [16] [14]. According to the tunnel design, the strokes of the hydraulic jacks for profiling cutters are adjusted by the automatic control system to perform overexcavation. By

Figure 3.37 Eccentric multiaxis shield machine. *From Chen K., Hong K., Wu X. Shield Constr Technol. Beijing: China Communications Press; 2016.*

adjusting the amplitudes of the profiling cutters, it is possible to construct sections with any cross-sectional shapes.

3.4.7 Eccentric multiaxis shield machine

The eccentric multiaxis shield machine (Fig. 3.37) adopts multiple main shafts, a set of crankshafts mounting with cutter frame are fixed perpendicular to the main shaft direction [14]. Both the main shaft and the cutter frame make a circular motion in the same plane, and the excavated section is close to the shape of the cutter frame [17]. The cutter frames can be designed to be rectangular, round, oval, and so on, according to the requirements of the tunnel section shape. The Yokohama Line MM21 and TRTA (Teito Rapid Transit Authority) Metro Line 11 respectively used the eccentric multiaxis shield machines with diameters of 7.15 and 9.60 m for tunnel construction.

References

[1] The Japanese Geotechnical Society. translated by Niu Q, Chen F, Xu H Shield method investigation, design and construction. Beijing: China Construction Industry Press; 2008.
[2] Zhang F, Zhu H, Fu D. Shield tunnel (fine). Beijing: China Communications Press; 2004.
[3] Chen K, Wang J, Tan S, et al. Shield design and construction. Beijing: China Communications Press; 2019.
[4] Bernhard M, Martin H, Ulrick M, et al. Mechanised shield tunnelling. 2nd ed. Berlin: Ernst & Sohn; 2012.

[5] Peng L, Wang W, Zhang Y. Tunnel engineering. Wuhan: Wuhan University Press; 2014.
[6] Yuan D, Shen X, Liu X, et al. Study on stability of excavation face of slurry shield. J Highw Transp 2017;30(08):24–37.
[7] Xu H. Application of articulation and profiling knife of articulated shield machine. Undergr Eng Tunn 2003;000(002):41–4.
[8] Han X, Li P. Application of the articulated system in shield machine. Hydraulics Pneumatics & Seals 2011;10:29–32.
[9] Liu P, Gao F, Guo W, et al. Research on the controllability of shield hinged device. J Shanghai Jiaotong Univ 2009;1:106–9.
[10] Zhang T, Xu P. Comparison of active articulation and passive articulation of shield machine. Municipal Engineering Technology 2011;29(S2):216–17.
[11] Xiang Y, Zhao Y. Research on remote monitoring and feedback analysis system of tunnel shield construction. China Foreign Highw 2009;29(01):167–70.
[12] Hou X. Analysis of shield technology. Sci Technol Inf (Sci Teach Res) 2007;18:103 + 114.
[13] Li B, Song M, Zheng B. Discussion on sphere shield. Sci Technol Wind 2009;14:234.
[14] Chen K, Hong K, Wu X. Shield Constr Technol. Beijing: China Communications Press; 2016.
[15] Hu X. Research on calculation method of internal force of double-circle shield tunnel lining structure. Shanghai: Tongji University; 2007.
[16] Wang J. Development trend and application of tunnel shield construction technology. Chin Foreign Architecture 2018;04:151–2.
[17] Chen D, Yuan D, Zhang M. Development and application of shield technology. Mod Urban Rail Transit 2005;05:25–9.

Exercises

1. According to the tunnelling principles, what types of shield machines there are?
2. What are the basic configurations of shield machines?
3. What are the differences between the positive articulation and the passive articulation for shield machine?
4. Briefly describe the balance principles at the excavation faces of the EPB shield and the slurry shield.
5. Briefly describe the classification of shield cutters and their application conditions.
6. Combining a literature review, briefly introduce several special shield machines and their working principles.

CHAPTER 4

Shield machine selection

The proper selection of the shield machine is crucial for the success of shield tunnel construction. Since the French engineer Marc Isambard Brunel applied for the patent for a shield machine in 1818, the shield tunnelling method has experienced more than two centuries of development. Different types of shield machines have gradually emerged, and even the earth pressure balance (EPB) shield, which was a relatively late development, has undergone more than 40 years of development since its birth in 1974. However, accidents have occurred during shield tunnelling resulting from improper selection of the shield machine. Common accidents include cutter wear, clogging, water spewing, large ground deformation, and excavation face failure. Therefore it is important to carefully and rationally select the type and configuration of shield machine before using it.

4.1 Selection principles and methods of shield machines

4.1.1 Selection principles

Shield machine selection involves the reasonable selections of the type and configurations of shield machine to be used according to the tunnel size, length, overburden depth, ground conditions, surrounding environment, construction period, and other factors. Proper shield machine selection is a prerequisite for safe, environmentally friendly, high-quality, economical, and rapid construction. The shield machine selection should minimize auxiliary construction measures.

Shield machine selection involves the determination of shield machine and its configurations, including opening ratio, stiffness and strength of the cutterhead, cutters, advancing system, hydraulic system, and so on. For projects with a complex construction environments and geological conditions, various factors and criteria must be considered so that the chosen shield machine is targeted and adaptable, so as to make full use of the shield equipment functions to avoid and overcome various risks during shield tunnel construction. Shield machine selection plays a very important role in tunnel

Shield Tunnel Engineering
DOI: https://doi.org/10.1016/B978-0-12-823992-6.00004-7

construction, duration, quality, and cost control [1]. It should be tailored according to the specific characteristics of engineering geology, hydrogeology, landforms, above-ground buildings, and underground pipelines and other structures. The core of shield machine selection lies not only in the equipment itself, but also in how suitable the machine is for the geological conditions and adjacent environments along the tunnel route.

The overall principle of shield machine selection is the combination of safety, technology, and economy. The first principle is to ensure the safety during the tunnel construction. For this reason, attention should be paid to the geological conditions (its stratum, strength, coefficient of permeability, fine particle content, and gradation) and groundwater conditions. At the same time, it is necessary to fully clarify the functions required by site conditions, the environment surrounding the shield shaft, above-ground and underground structures on the construction line, and other factors. These must be considered together with technology and economy to choose a suitable shield machine. If the wrong machine is chosen, many auxiliary measures will have to be taken to avoid engineering accidents and to ensure tunnelling efficiency.

The following principles should be followed in selecting a shield machine and determining its configurations [2,3]:

1. The selected shield machine should have strong adaptability to engineering geology and hydrogeology and should meet the requirements of construction safety.
2. Safety, technological advancement, and economic rationality should be unified, and in the case of safety and reliability, technological advancement and economic rationality should be considered.
3. Conditions of tunnel diameter, length, buried depth, construction site, surrounding environments, and so on should be taken into account.
4. Requirements of quality, construction period, engineering cost, and environmental protections should be satisfied.
5. The capacity of the backup system should be matched with the main machine to meet the requirements of production capacity and shield advancing speed.
6. Reputation, achievement, and technical service of the shield manufacturer should also be fully considered during the selection of machine brand so as to avoid ineffective maintenance of the shield machine during its use.

With these principles in mind, the shield type, configuration, and main technical parameters are researched and analyzed to ensure the safety and reliability of shield tunnel construction and to select the most suitable shield machine for specific tunnel project.

4.1.2 Selection methods

Proper shield machine selection requires lots of engineering experiences and even researches for special conditions. Relevant scholars have put forward the triangle theory for shield machine selection after taking into account various factors that affect shield tunnelling performance [3]. The core of the theory can be stated simply as follows: The stability of the excavation surface is considered to be the center; the engineering geology and hydrogeology are analyzed as the basic points; the soil particle size, coefficient of permeability, and groundwater pressure are researched as the principles with comprehensive consideration of the actual project; the selected shield meets the overall goals of a stable excavation face, efficient cutting, and smooth muck discharging. These can be summarized as "one center, two basic points, three principles, and three goals," as shown in Fig. 4.1. With the popular application of shield machines in tunnel excavation, the machine durability (high reliability and long life of shield key components) as the fourth goal of shield machine selection has received

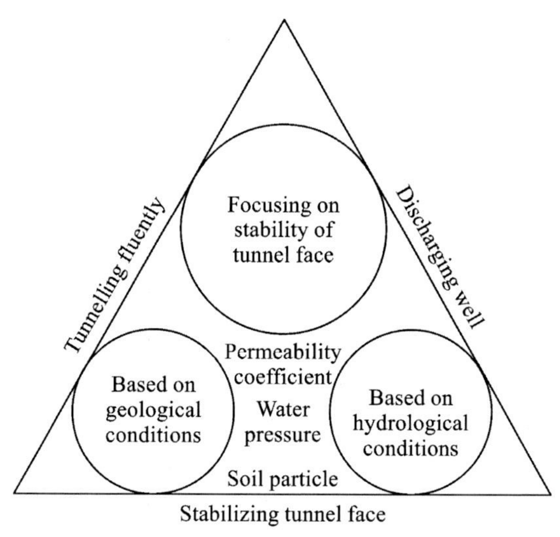

Figure 4.1 Schematic diagram of the triangle theory for shield machine selection. *Translated from Hong K. Key technology of shield tunnelling. China Communications Press; 2018.*

increasing attention to adapt to long-distance tunnel construction in complicated geological and environmental conditions.

Specifically, the shield machine selection should be carried out with consideration of the following points [3]:

1. The ground layers through and beneath which the tunnel passes should be researched comprehensively. The physical and mechanical properties of the rock and soil in the layers greatly affect the stability of the tunnel excavation face, and they are the primary factors in determining the shield machine type and configurations.
2. The groundwater conditions strongly affect the stability of the excavation face, and high groundwater pressure in the strong-permeability stratum endangers the stability of the excavation face and control of ground deformation.
3. The shape of the shield machine depends on the designed section of the tunnel. Although the current rectangular, horseshoe-shaped, and other special-shaped shield machines have been developed, the most common cross-section type is still a circular shield machine, which is less difficult than other shields in terms of equipment manufacturing and tunnel construction.
4. The surrounding environment is also a key factor affecting the selection and configuration of the shield machine. The existence of important surrounding structures determines the control importance of ground settlement and stability. However, there is a big difference in the ground stability and settlement with closed shield machines (including EPB and slurry shield machines) and open shield machines (hand-dug, semimechanized, and mechanized shield machines). Among them, the open shield machine normally needs to be equipped with auxiliary measures such as jet grouting to prereinforce the ground for controlling the ground stability and settlement.

As shown in Table 4.1, the "Tunnel Standard Specification (Shield Tunnelling Method) and Compilation Instructions" from the Japanese Civil Engineering Society provides the adaptability relationship between shield machines and ground [4].

4.2 Selection of shield machine types

Shield machine selection should first consider whether the selected type of shield machine can effectively ensure the stability of the excavation face and control the ground settlement, based on factors such as the

Table 4.1 Adaptabilities of shield machines to strata.

Soil			Shield											
			Open						Closed					
			Hand-dug		Semimechanical		Mechanical		EPB				Slurry	
									Earth pressure		Earth pressure with soil conditioning			
Geological classification	Soil	N value	Appli-cability	Precautions	Appli-cability	Precautions	Appli-cability	Precautions	Appli-cability	Precautions	Appli-cability	Precautions	Appli-cability	Precautions
Alluvial clay	Humus	0	×		×		×		×		△	Foundation deformation	△	Foundation deformation
	Muddy clay	0–2	△	Foundation deformation	×		×		○		○		○	
	Sandy mud and sandy clay	0–5	△	Foundation deformation	×		×		○		○		○	
		5–10	△	Foundation deformation	△	Foundation deformation	△	Foundation deformation	○		○		○	
Proluvial clay	Silt and clay	10–20	○		○		△	Soil clogging	△	Soil clogging	○		○	
	Sandy silt and sandy clay	15–25	○		○	—	○		△	Soil clogging	○		○	
		≥ 25	△	Excavation machine	○		○		△	Soil clogging	○		○	
Soft rock	Stiff clay and mudstone	≥ 50	×	—	△	Groundwater pressure	△	Groundwater pressure	△	Soil clogging	△	Cutter tool wear	△	Cutter wear
Sandy soil	Muddy clay-sand mixture	10–15	△	Groundwater pressure	△	Groundwater pressure	△	Groundwater pressure	○		○		○	
	Loose sand	10–30	△	Groundwater pressure	×		△	Groundwater pressure	△	Fine content	○		○	
	Dense sand	≥ 30	△	Groundwater pressure	△	Groundwater pressure	△	Groundwater pressure	△	Fine content	○		○	

(Continued)

Table 4.1 (Continued)

Soil			Shield											
			Open						Closed					
			Hand-dug		Semimechanical		Mechanical		EPB				Slurry	
									Earth pressure		Earth pressure with soil conditioning			
Geological classification	Soil	N value	Appli-cability	Precautions	Appli-cability	Precautions	Appli-cability	Precautions	Appli-cability	Precautions	Appli-cability	Precautions	Appli-cability	Precautions
Sandy gravel and pebbles	Loose sandy gravel	10–40	$\triangle$	Groundwater pressure	$\triangle$	Groundwater pressure	$\triangle$	Groundwater pressure	$\triangle$	Fine content	○		○	
	Consolidated sandy gravel	≥ 40	$\triangle$	Groundwater pressure	$\triangle$	Groundwater pressure	$\triangle$	Wear of tools and panels, groundwater pressure	$\triangle$	Groundwater pressure	○		○	
	Sandy Gravel containing pebbles	—	$\triangle$	Safety of excavation operations, groundwater pressure	$\triangle$	Groundwater pressure, overexcavation	$\triangle$	Wear of tools and panels, groundwater pressure	$\triangle$	Screw conveyor standard, tool panel wear, tool standard	○		$\triangle$	Cutter standard, mud feed countermeasures
	Boulders and pebbles	—	$\triangle$	Breaking of gravel, groundwater pressure	$\triangle$	Groundwater pressure, overexcavation	$\times$		$\triangle$	Tool standard, screw conveyor standard	$\triangle$	Tool standard, screw conveyor standard	$\triangle$	Gravel crushing, Slurry-feeding countermeasures

Notes: 1. Symbols are explained as follows: ○—Applicable soil conditions in principle; $\triangle$—Auxiliary construction methods and auxiliary mechanisms should be studied during application; $\times$—Soil conditions not applicable in principle. 2. Most open shields are combined with a pneumatic construction method, but whether it is applicable should be fully studied. 3. Extrusion shields have certain restrictions in the application of alluvial clay. In addition, it also tracks the deformation of foundation, which has not been used recently and was deleted from the object. 4. The N value is the standard penetration value of each type of soil. 5. Precautions give only the most important items in the foundation and form of $\triangle$, and other precautions are omitted.

Translated from The Japanese Geotechnical Society. Translated by Niu Q, Chen F, Xu H. Investigation, design and construction of shield tunnelling. Beijing: China Architecture Press; 2008.

geological and hydrological conditions and surrounding environments, and then consider the limiting factors, such as construction period and cost. Finally, combined with the available auxiliary construction measures, the appropriate type of shield machine and its configurations are chosen.

4.2.1 Selection of shield machine types

4.2.1.1 Permeability coefficient of ground

The permeability coefficient is an evaluation index of ground permeability. The larger the pores in the soil, the looser the soil mass and the stronger the soil permeability. By contrast, the denser the soil, the lower the permeability. Generally, the greater the coefficient of permeability, the higher the water content of the stratum under the groundwater table.

The permeability coefficient of the stratum is a very important factor for the selection of shield machine type. If the stratum is mainly composed of water-rich sand and gravel layers, a slurry shield machine is recommended to be selected. For other strata, considering the machine cost, an EPB shield machine is recommended to be chosen.

As shown in Fig. 4.2, the basic criteria can be followed in selecting the shield machine type [5]: When the permeability coefficient of the

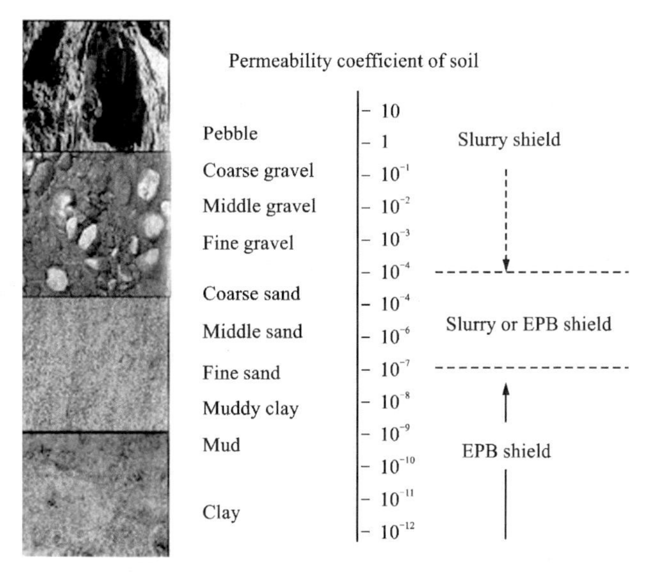

Figure 4.2 Comparison of the adaptability of shield machines in strata with different ranges of permeability coefficients. *Translated from Lv S. Shield type selection and comprehensive construction technology research. Shijiazhuang: Shijiazhuang Railway University; 2017.*

stratum is less than 10^{-7} m/s, an EPB shield machine is recommended. When the permeability coefficient is 10^{-7}–10^{-4} m/s, either an EPB or a slurry shield machine can be used. When the permeability coefficient is greater than 10^{-4} m/s, a slurry shield machine is recommended.

4.2.1.2 Grain size distribution

The EPB shield machine is mainly suitable for tunnel construction in cohesive soils, such as silt, silty clay, and clay. Generally speaking, the muck with a large amount of fine particles easily forms an impermeable plastic-fluid body, which can easily fill the shield chamber so that earth pressure can be easily established in the soil chamber to balance the excavation face. By contrast, in a stratum with a lot of coarse particles, such as sand pebbles, the muck cannot be evenly distributed in the soil chamber, so it is not suitable to establish the EPB mode. In this case, a slurry shield is more suitable. Fig. 4.3 shows the selection criteria of shield machine type considering the effect of grain size distribution [6]. The soils in the range of clay and silt are the applicable grain size gradations for EPB shields; the soils in the range of gravel and coarse sand are the suitable grain size gradations for slurry shields. For coarse and fine sand areas, slurry shield machines can be used; or after muck has been conditioned, EPB shields can be used.

As can be seen from Fig. 4.3, the tolerable particle size for an EPB shield is less than 0.2 mm without soil conditioning, and the particle size

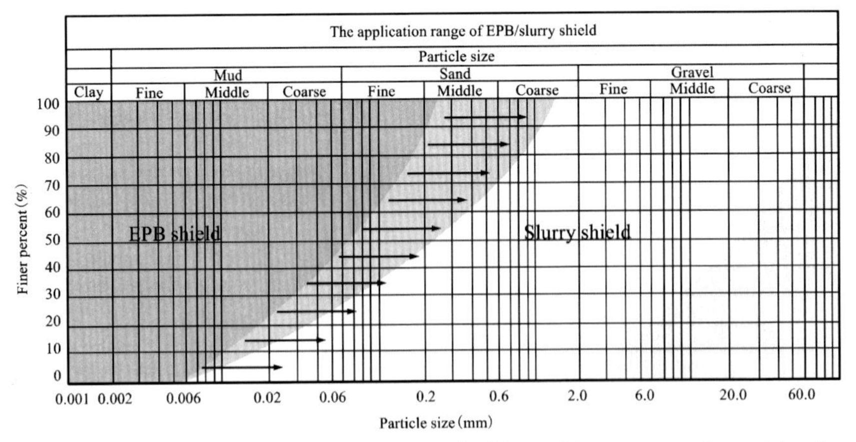

Figure 4.3 Curve of relationship between shield machine type and particle size. *Translated from Spagnoli G. Electro-chemo-mechanical manipulations of clays regarding the cloggingduring EPB-tunnel driving. Germany: RWTH Aachen University; 2011.*

range can extended up to about 1.5 mm at most. By contrast, the applicable particle size of the slurry shield ranges from 0.01 to 80 mm.

It is worth noting that the scope of application of EPB shield and slurry shield is not unchangeable. Some strata that are not originally suitable for an EPB shield can become amenable to EPB shield tunnelling through reasonable muck conditioning, which will be introduced in details in Chapter 9.

4.2.1.3 Groundwater pressure

When the groundwater level is high, water spewing is likely to occur at the outlet of the screw conveyor, causing the earth pressure in the excavation chamber to drop and leading to excavation face collapse and large ground settlement. Generally speaking, when the water pressure is greater than 0.3 MPa, a slurry shield is suitable. If an EPB shield is used, it will be difficult for the screw conveyor to form an effective soil plug, and water spewing will be likely to occur at the outlet of the screw conveyor, causing the earth pressure in the soil chamber to drop and the excavation face to collapse. When the water pressure is greater than 0.3 MPa, if the EPB shield must be used for geological reasons, the length of the screw conveyor needs to be increased, or a secondary screw conveyor or a pressure-maintaining pump should be used. Although the pressure-maintaining pump has been used in a small number of projects, it has not been effectively promoted because it will greatly affect the shield tunnelling efficiency. Increasing the length of the screw conveyor or adding a secondary screw conveyor requires that the shield machine has enough space, and the space in the commonly used machine is relatively limited. The increase in screw conveyor length or the addition of secondary screw conveyor will bring space problems to construction.

4.2.1.4 Selection considering tunnel section and environmental factors

For tunnel sections, the slurry shields should be used when sections are large and extra-large, and the EPB shields are recommended when sections are medium-size and small. However, large-diameter EPB shields have constantly been setting new records. From an environmental point of view, the slurry shields have poor environmental performance, but the EPB shields have strong environmental performance.

The research by Japanese scholars [7] shows that except for the use of slurry shields under certain environmental conditions, such as when the tunnel crosses the river, lake, or sea or when the diameter of the excavation face is greater than 10 m, the EPB shields are always recommend for tunnel construction. This is because the construction cost of using a slurry shield is high, and the construction area is required to be large, affecting traffic and city appearance. By contrast, the EPB shields have the advantages of a small occupied area, a small impact on traffic, and low equipment cost. However, this type of shield has a large excavation torque and causes a large stratum settlement, so the tunnel face generally cannot be too large. However, with the development of shield equipment and construction technology, the EPB shield machines with a diameter of more than 10 m continue to emerge. As was stated in Chapter 1, Introduction, the largest diameter of an EPB shield machine has reached about 17.5 m.

4.2.2 Application extension of EPB and slurry shield machines

The EPB shields and the slurry shields are the two most widely used types of shield machines. They have their own advantages and disadvantages, as shown in Table 4.2. However, their application range can be extended through reasonable auxiliary measures.

4.2.2.1 EPB shield machine

The traditional application stratum of the EPB shield is a low-permeability one that contains clay, silt, or a amount of fine sand. In a soil layer with high groundwater pressure, in order to prevent groundwater from flowing into the excavation chamber, the coefficient of permeability of the muck in the chamber is required to be at least less than 10^{-5} m/s. Furthermore, the muck must be stirred to achieve a better workability state. For example, the Shanghai silty clay can reach a good plastic-flow state by being stirred in the shield chamber. However, more and more EPB shields are used to tunnel in strong-permeability ground. To extend the application strata of the EPB shield, the muck in the excavation chamber must be conditioned. The graph in Fig. 4.4 suggests that different soil-conditioning schemes should be adopted according to the soil characteristics during the EPB shield tunnelling [8]. It can be seen that the EPB shield is most suitable for use in clay and silt layers. Adding water and foam can reduce the soil plasticity and prevent the soil from adhering to the cutters. As the particle size of the soil transitions from the sand to the round gravel layer, polymer as well as foam should be added to the soil, while the hydrostatic pressure should not be too large. As the particle

Table 4.2 Comparison between an earth pressure balance shield and a slurry shield.

Items	EPB shield machine		Slurry shield machine	
	Brief descriptions	**Evaluation**	**Brief descriptions**	**Evaluation**
Stabilizing excavation surface	Earth pressure in the chamber is maintained to stabilize the excavation face	Good	Pressurized slurry keeps the excavation face stable.	Excellent
Adaptability to geological conditions	Special measures for soil conditioning are required in high-permeability strata such as sandy soil.	Good	Strong adaptability	Excellent
Resistance against water and soil pressures on the excavation face	The soil and water pressures are balanced with chamber pressure normally with soil conditioning.	Good	Mud film is formed on the excavation face to reduce the ground permeability.	Excellent
Control of ground surface settlement	The ground surface settlement is controlled by maintaining chamber pressure, controlling advancing speed, and keeping the balance between the muck-entering and exit the chamber.	Good	The ground surface settlement is limited to controlling mud quality, chamber pressure, and advancing speed and keeping the dynamic balance of mud feeding.	Excellent

(Continued)

Table 4.2 (Continued)

Items	EPB shield machine		Slurry shield machine	
	Brief descriptions	**Evaluation**	**Brief descriptions**	**Evaluation**
Soil discharging in shield	A muck truck is towed by a locomotive for transportation, and the muck is lifted by a door crane, which is slow in efficiency.	Good	Slurry is discharged efficiently with pumps.	Excellent
Muck/slurry treatment	It can be dumped directly.	Simple	Slurry separation treatment is normally required before being dumped.	Complex
Construction site	Smaller construction site	Good	A large site is required for slurry treatment.	Poor
Equipment cost	Muck carriages are required.	Medium	The machine is much more expensive. A slurry separation station is required to be set up.	High
Engineering cost	The shield tunnelling does not require slurry treatment equipment and station; it only requires configuration of an additive injection system.	Low	The slurry shield requires slurry production and transportation and slurry separation system.	High

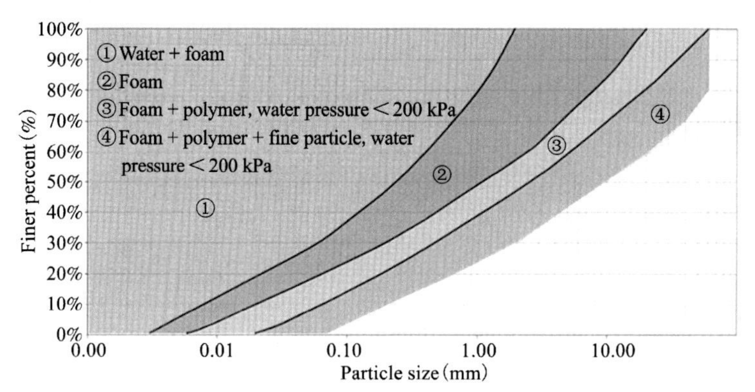

Figure 4.4 Application range of EPB shield machine after muck conditioning. *Translated from Li C. Research on calculation model of key parameters of EPB shield in sandy gravel stratum. China University of Mining and Technology (Beijing); 2013.*

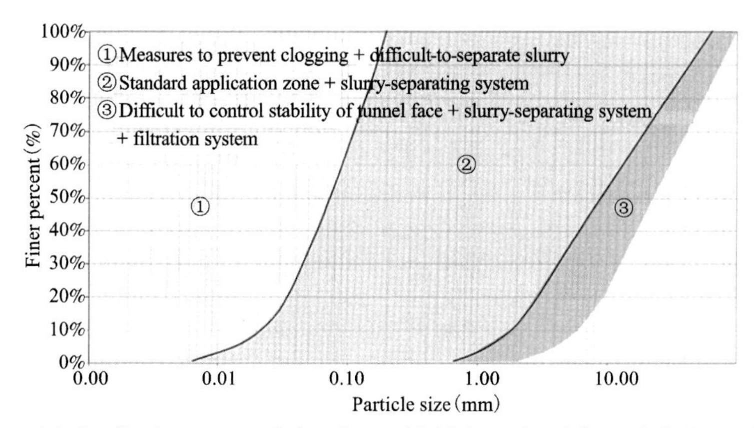

Figure 4.5 Application range of the slurry shield. *Translated from Li C. Research on calculation model of key parameters of EPB shield in sandy gravel stratum. China University of Mining and Technology (Beijing); 2013.*

size of soil increases, the soil transitions from round sandy gravel to gravel. In muck conditioning, appropriate amounts of fixed fillers such as bentonite slurry should be added along with foam and polymer.

4.2.2.2 Slurry shield machine

The slurry shield can form a filter cake on the excavation face to reduce the ground permeability, so it can be used in non-cohesive soil layers such as water-rich gravel and pebbles [8] (Fig. 4.5). If the ground permeability is too large, it is easy to delay the development rate of the mud film. When the permeability coefficient $k \geq 10^{-3}$ m/s, the slurry will diffuse to the external soil through the soil pores and cannot form a mud film on the excavation face. Therefore it is necessary to add fillers, fine particles, or chemical additives to the slurry to improve its rheological properties or to reduce the ground permeability through treatment methods such as pregrouting reinforcement. Both of these methods expand the application range of slurry shields. However, when there are too many fine particles in the slurry, the shear strength of the mud film will be reduced. In addition, the slurry separation system cannot technically separate the fine particles from the bentonite, which increases the slurry treatment cost.

4.3 Selections of shield machine configurations

4.3.1 Cutterhead selection

The cutterhead has the functions of cutting soil, stabilizing excavation face, and mixing muck. Therefore during the tunnelling process, the working conditions of the cutterhead are very harsh, and the forces that the cutterhead bears are complicated. The cutterhead structure is related to the excavation efficiency and service life of the shield. Its type and configurations depend mainly on the geological and hydrogeological conditions. Different cutterhead types should be adopted for different strata, but there is a lack of complete theoretical basis and enough test data in the design of shield machine configurations, considering their adaptablity to geological and hydrogeological conditions. Currently, the cutterhead selection mainly relies on engineering experience, as discussed in the following subsections.

4.3.1.1 Cutterhead type and opening ratio

In actual construction, which cutterhead type is used in specific project should be determined according to the influence factors such as shield machine type and ground conditions. The slurry shield generally uses a panel cutterhead or composite cutterhead, while the shield can use a panel cutterhead, a spoke cutterhead, or a composite cutterhead, depending on the geological conditions.

The cutterhead opening ratio is a key factor in cutterhead selection and directly affects the efficiency of muck entering the excavation chamber. The opening position, shape, size, and other parameters must be determined according to geological conditions. The following considerations are recommended:

1. Owing to the small radius of rotation at the cutterhead center, shield tunnelling in clayey and even sandy strata (due to the conditioning

using bentonite) can easily clog the cutterhead center, affecting the normal shield tunnelling. Therefore the local opening ratio in the center of the cutterhead should be designed as large as possible.

2. Slurry shields are mostly used in highly permeable and easily failed strata, and their opening ratio is relatively small, generally 10%–30%, while EPB shields have a wide range of opening ratios according to geological conditions.

The opening ratio of the cutterhead with a panel-type is small, and the largest diameter of muck particles entering the soil chamber can be restricted by the opening size of the cutterhead. The panel cutterhead is used mostly in weak and inhomogeneous strata. Owing to the heavy load of the cutterhead in a hard rock stratum, in order to prevent the cutterhead from being worn, the panel cutterhead is also often used in the stratum. Affected by the cutterhead panel, the earth pressure at the excavation face is quite different from the earth pressure in the soil chamber measured on the bulkhead, making earth pressure management difficult. For a clayey stratum, when soil conditioning is not appropriate, muck cannot enter the soil chamber smoothly and sticks easily to the steel board, resulting in clogging, which leads to low tunnelling efficiency, heavy load, and short service life.

The spoke cutterhead has a large opening ratio, which can ensure the smooth flow of muck and help to avoid clogging. It also has a small working load and a long service life. Owing to the limited blocking capability of the spoke cutterhead, the excavated rock and soil are relatively large, so the screw conveyor might not discharge muck smoothly. Therefore the spoke cutterhead is highly adaptable to solely soft strata such as sand. As for the strata containing large particles such as boulders, there is an increased risk of screw conveyor blockage when a spoke cutterhead is configured.

The composite cutterhead has the characteristics of both the panel one and the spoke one. The scrapers are equipped on both sides of the spokes and panels, and the disc cutters are installed in the front of the wide panels. Therefore the composite cutterhead can adapt to tunnelling in various complex geological conditions and is gradually becoming the first choice of cutterhead.

4.3.1.2 Driver system type of the cutterhead

The driver system of the cutterhead should have the following working functions: its power and torque can be output to facilitate monitoring of the shield tunnelling states; the impact resistance from hard rock and boulder-containing strata is as high as possible so as to prevent its components from being damaged; the rotation speed of the cutterhead is continuously adjustable and can be rotated forward and backward to deal with tunnelling in abnormal strata; the driver system of the cutterhead is one of the most power-consuming equipment in shield machine, so reducing the working power as much as possible can save energy to a great extent; and the driver system should have high reliability and good working performance. At present, to ensure these functions, the driving mode of the cutterhead is mostly designed to be hydraulic. However, with the continuous development of variable-frequency motor technology, variable-frequency motor driver systems are gradually being adopted for shield machines. A comparison of the two driving modes is shown in Table 4.3.

4.3.1.3 Support form of the cutterhead

As was mentioned in Chapter 3, Shield Machine Configurations and Working Principles, the support of the cutterhead includes three forms: central support, intermediate support, and peripheral support. The support selection of the cutterhead should be related to the opening ratio and geological conditions. The cutterhead with a central support has a simple structure and is mostly used for shields with medium and small diameters. When this support type of cutterhead rotates and cuts the soil, the soil

Table 4.3 Comparison between a variable-frequency motor drive and a hydraulic drive.

Feature		Variable-frequency motor driver	Hydraulic driver
Volume	Driver part	Big	Small
	Subsidiary part	Medium-sized	Big
Transmission efficiency		High	Low
Maintenance cost		High	Normal
Speed control performance		Good	Good
Stratum adaptability		Good	Big
Temperature in the shield tunnel		Little heat and low temperature	Much heat and high temperature
Overload capacity		Strong	Strong
Equipment cost		High	Medium

Modified from Song Y. Research on the selection and design theory of cutter wheel for shield machine. Southwest Jiaotong University; 2009.

chamber has a large space, which makes the muck easily stirred and smoothly flow. There is little possibility that clayey muck adheres to the cutterhead and soil chamber, and it is not easy to cause clogging. In addition, the pressure in the soil chamber is relatively stable, but it is difficult to handle large rocks and boulders, owing to the narrow space of the shield machine using this cutterhead. The cutterhead with a intermediate support has good structural balance and is used mainly for shields with large and medium diameters. When it is used for machines with small diameters, the treatment of large particles should be considered to prevent soil blocking. Because of the existence of the intermediate support structure, the soil chamber is divided into two areas. The fluidity of the muck in these two areas is quite different, and it is difficult to ensure effective stirring of muck in the chamber. The muck in the peripheral area is easily mixed, while the mixing of the muck in central area is relatively poor, resulting in problems such as soil clogging and poor muck discharging, an increase in the cutterhead torque, and difficulty in controlling earth pressure in the chamber. The cutterhead with a peripheral support is generally used for shields with small diameters. Owing to the large space in the soil chamber, handling of large particles is easier. When it is used in clayey soil, the focus will be on how to avoid the clay clogging.

4.3.1.4 Rated torque and speed of the cutterhead

When tunnelling in complex strata, it is important for the safety of tunnel construction to maintain the dynamic stability of shield tunnelling parameters. The rated rotation speed and torque of the cutterhead should be adapted to other parameters, such as shield thrust, opening ratio, and soil chamber pressure. The angular velocity of the cutterhead rotation is related to the linear speed of cutters. The linear speed should also be considered for cutting ground and mixing muck. Engineering experiences have shown that the linear speed of cutters in a rock stratum should generally not exceed 2.5 m/s, while the linear speed in a sandy or soft stratum is lower, generally around 0.3–0.5 m/s [9]. The rated torque of the cutter wheel can be determined by theoretical calculation or experience.

The excavation diameter of the cutterhead depends mainly on the diameter of the designed tunnel. During the construction process, owing to the influence of the radius of turning around and the slope of the excavation line, a profiling cutter needs to be installed at the edge of the cutterhead for overcutting ground when turning around or driving upward or downward. However, considering that cutters experience a certain amount of wearing abrasion when cutting in hard stratum, it is necessary, in selecting the cutter type to use, to take into account the reduction of the shield diameter resulting from the cutter wearing abrasion. To avoid the risk that the shield will become locked in a rock stratum, the cutter-head and shield body should be designed specially. For example, in a composite stratum, the diameter of shield from the front to the tail can be gradually reduced by 10 mm to control the locking risk of shield.

4.3.2 Cutter selection

The cutter selection should take into consideration the following aspects: (1) the cutter adaptability to stratum, (2) the cutter height, (3) the cutter spacing, (4) the mounting method of the cutter holders, (5) the cutter layout, and (6) the layout of the profiling cutter. For a specific tunnel project, a detailed study of the cutter configuration is required.

4.3.2.1 *Selection and layout of the disc cutters*

For rock strata and soft strata containing obstacles that include boulders, piles, and others and cannot be pretreated, the disc cutters are required to be configured for the shield cutterhead. The disc cutters have different forms depending on the number of cutter rings, such as single-edge, double-edge, and multiedge disc cutters. The double-edge disc cutter has a low capability for breaking rock, and it normally adapts to rocks with a compressive strength of less than 8 MPa. In addition, it has a small starting torque, and so it is most suitable for soft strata. The single-edge disc cutter is currently the most widely used one, because engineering practice has proved that it has a strong capability for breaking rock.

There are generally two mounting methods of the disc cutters, including front installation and rear installation, as shown in Fig. 4.6. For the front installation, the disc cutters need to be moved to the cutterhead

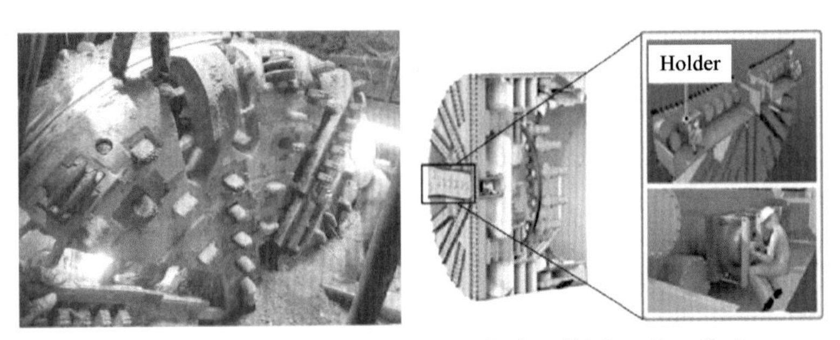

Figure 4.6 Disc cutter installation. (a) Front installation. (b) Rear installation.

front, normally through a shaft when the cutters are being changed. If the cutters are heavy, it will be difficult to change them. Therefore rear installation has become a welcome method of changing cutters.

Disc cutters are prone to wear during shield tunnelling. The setting of the disc cutters should take into account the following aspects [10]:

1. When the disc cutters are configured, other types of cutters should be supplemented. According to the practical research in Guangzhou and Shenzhen, China, when tunnelling in composite strata, especially soft strata, the disc cutters are recommended to be 3-4 cm higher than other cutters and their diameters are recommended to be 17 inches.
2. For the strata that are soft on the top and hard on the bottom at the excavation face and other strata that need to be excavated by an EPB shield, the quality of the bearing seal should be stressed to ensure that the disc cutters will not be worn out due to the failure of the bearing seal under high earth pressure.
3. The assembly torque (also called as the launching torque) of the disc cutters should be reasonably configured. When it is necessary to use the EPB mode to tunnel in composite strata, the assembly torque of the disc cutters can be set at 3–8 kg · m. However, for hard rock whose full face is stable (that is, tunnelling without the need of establishing soil pressure), the assembly torque can be increased to a maximum of 25 kg · m.
4. When tunnelling in hard rock and uneven strata, the ring material of the cutters that are installed in the cutterhead side should be specially strengthened, and the number of side disc cutters should be appropriately increased to reduce the distance between cutters. The maximum compressive stress to which the bearings of the disc cutters can adapt also needs to be increased to meet the needs of overexcavation.

4.3.2.2 Selection and layout of the scrapers

When the scrapers are used in a soft stratum, their designed capacity of breaking rock is normally less than 20 MPa. They can also be used to scrape rock debris in hard stratum. From a geometric point of view, the scrapers can be laid on the cutterhead following a concentric circle and an Archimedes spiral. To achieve the uniform distribution of stress on the cutterhead and the full-face cutting, the Archimedes spiral is mostly used for the layout of scrapers. The Archimedes spiral is a trajectory generated by a point leaving a fixed point with a constant speed while rotating around the fixed point with a constant angular velocity. According to the

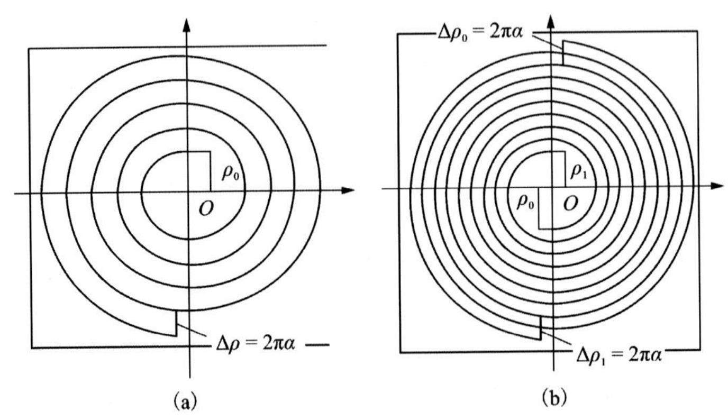

Figure 4.7 Schematic diagram of the trajectories caused by scrapers arranged in the form of Archimedes spirals. (a) Single spiral. (b) Double spirals. ρ_0, ρ_1: Initial value of polar axis (mm); α: Constant coefficient.

definition of the Archimedes spiral, when the scrapers are arranged on the cutterhead in the form of an Archimedes spiral, any area of the excavation face can be cut by the scrapers, avoiding the appearance of the cutting blind zone caused as the concentric circle arrangement mode [11]. Depending on the number of Archimedes spirals, the trajectories are divided into two forms: single spiral and double spirals (Fig. 4.7). To achieve the clockwise and anticlockwise cutting, the scrapers need to be distributed symmetrically on both sides of the spoke and panel of cutterhead. In addition, there is a slight difference in radial layout between scrapers and disc cutters. In the excavation of a composite stratum, the disc cutters first excavate ground, and the scrapers are used mainly to scrape debris. Since the maximum allowable wearing abrasion of the 17-inch disk cutter is usually 25 mm, the heights of the scrapers should be 25 mm lower than those of disk cutters.

4.3.2.3 Selection and layout of the special cutters

1. Advance cutters: The advance cutters are mainly configured to work synergistically with other cutters in design. During shield tunnelling, the advance cutters cut ground before other cutters, for breaking the soil into pieces, loosening the ground, reducing the cutting resistance of other cutters, and so creating good cutting conditions for other cutters. According to their functions, their cross sections are generally smaller than those of other cutters and so easily cut the ground. The use of advance cutters can significantly increase the fluidity of the

excavated soil, greatly reduce the torque of other cutters, promote their cutting efficiency, and reduce their wearing abrasion. For example, during construction of the section from Haizhu Square Station to Shiergong Station on Guangzhou Metro Line 2, China, except for four overexcavation disc cutters that were retained, all of the disc cutters were replaced with advance cutters for shield tunnelling in the mudstone ground. When these advance cutters were 110 mm in height ahead of the cutterhead panel, the advancing speed was increased; when they were 150 mm in height ahead of the cutterhead panel, the tunnelling advancing speed was further increased. [12]

2. Shell cutters: The shell cutters (Fig. 4.8) can also be used as advance cutters. When the shield machine excavates in sandy and gravel strata, especially with large-diameter particles, if the disc cutters are used, large ground deformation will occur by the extrusion of disc cutters, greatly reducing their cutting effect and even sometimes losing the cutting and crushing capability. However, the use of shell cutters can solve the problem. The shell cutters are always installed on the front face of the cutterhead especially for cutting sand and gravel.
3. Profiling cutters: A shield machine is often designed with two profiling cutters (one of them is spare), which are installed at the radial ends of the cutterhead. According to the stratum difference, the profiling cutters include two types, overexcavation scraper and overexcavation disc cutter, as shown in Fig. 4.9. During construction, according to the

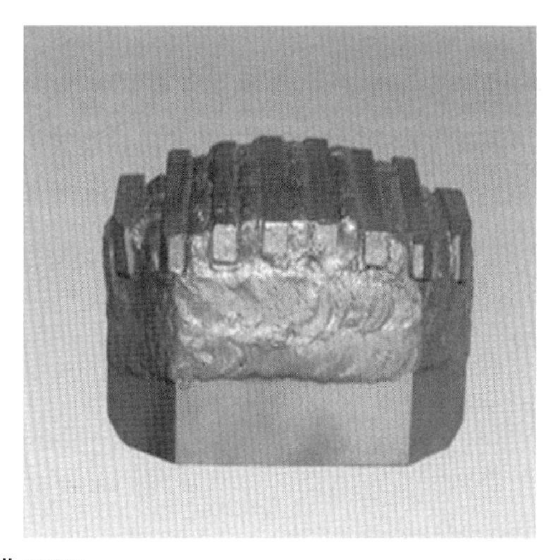

Figure 4.8 A shell cutter.

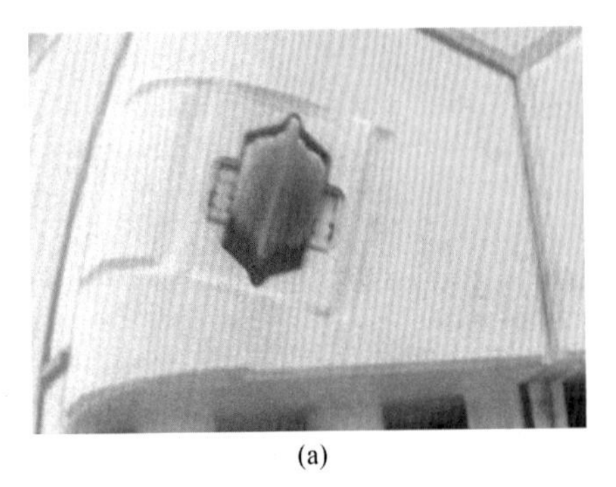

(a)

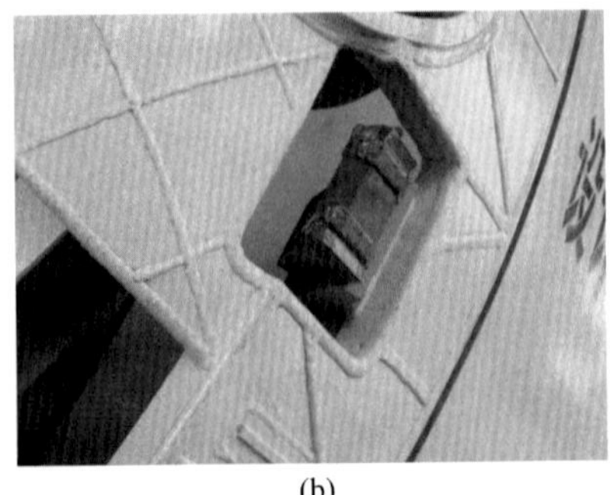

(b)

Figure 4.9 Layout of profiling cutters. (a) Overexcavation disc cutter. (b) Overexcavation scraper.

requirement of overexcavation thickness, the profiling cutters can be extended and retracted radially from the cutterhead edge to achieve overexcavation. The maximum extension of the profiling cutters generally ranges from 80 to 130 mm. During shield turning in a curved tunnel or correcting for shield attitude, the required space is created by changing the form of profiling cutters, ensuring that the shield can achieve curved excavation, smooth turning, and attitude correcting. The configuration of the profiling cutters needs to take the following issues into account: (1) The overexcavation range of the cutters can be set to overexcavate ground at any position of the circumferential area. For example, in the shield tunnel section from Datansha South Station to Zhongshan Ba Station on Guangzhou Metro Line 5, China, there was a sharp turning section, and the overexcavation was carried out at the inner side of the tunnel curve during shield tunnelling [13]. (2) For a sharp curved tunnel section that is relatively long, in order to reduce the wearing abrasion of profiling cutters, they can be used in cooperation with folding the shield via articulation and selecting reasonable thrusts for jack cylinders to achieve a sharp turn.

4. Fishtail cutters: Fishtail cutters are normally configured on the center of the cutterhead to improve the cutting and stirring capability of other cutters. Two techniques can be applied to the configuration of the fishtail cutters. One technique is to let the shield cut soil in two steps: The fishtail cutters cut the ground in a small circular section (about 1.5 m in diameter) at the center of excavation face first, and then the full face is excavated with other cutters. In other words, the fishtail cutters cut ground ahead of the other cutters. The other technique is to design the fishtail cutters in a cone shape (Fig. 4.10) so

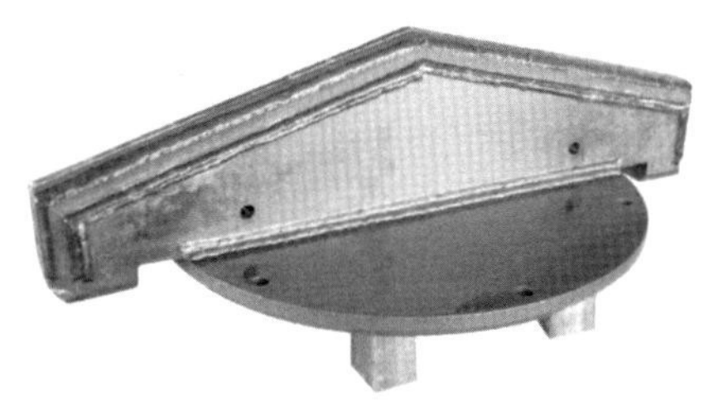

Figure 4.10 Fishtail cutter.

that the ground cut by the fishtail cutters can roll the excavated soil (similar to plowing) besides tangential and radial movement. This can solve the cutting problem of the central part of the excavation face, improve the fluidity of the excavated soil, and increase the overall tunnelling efficiency of the shield.

When tunnelling in a high-strength stratum, more attention should be paid to the reasonable protection of the fishtail cutters. For example, when the Section 3 of Beijing Metro Line 9, China, was tunneled in a stratum with large boulders, the blades of the fishtail cutter were impacted and fractured, and the cutter seat was severely worn. After that, the following protective measures were taken: (1) On the four spokes that are different from the layout direction of the fishtail cutter, four advance cutters were added to protect the fishtail cutter. (2) Two strengthened advance cutters were fixed beside the fishtail cutter to protect the ejecting outlet of the soil conditioner.

5. Tooth cutters: When tunnelling in soft rock, because the ground cannot put enough resistance on the disc cutters, it will cause them to fail to roll and lead to eccentric wearing abrasion of their rings, resulting in loss of their rock-breaking ability. In this case the disc cutter holder can be used to install tooth cutters instead (Fig. 4.11) for rock breaking. The trajectories of the tooth cutters with symmetrical edges are exactly the same as those of the disc cutters, so the cutterhead can break rock well when it rotates both clockwise and anticlockwise. For example, during the construction of shield tunnel in the hard rock of Shenzhen Metro, China, the configured cutters included disc cutters, tooth cutters, and scrapers, among which the disc cutters and tooth cutters had the same cutter seats. The disc cutters were 35 mm higher than other cutters to protect tooth cutters and scrapers when tunnelling in hard rock. It is worth mentioning that the disc cutters should be used instead of tooth cutters [14] when tunnelling in a hard rock stratum (such as granite).

Figure 4.11 Tooth cutter.

Table 4.4 The relationship among the outer diameter of shield, the outer diameter of screw conveyor, and the maximum particle diameter of the muck discharged.

Outer diameter of shield	Outer diameter of screw conveyor	Maximum particle diameter of the muck discharged	
		With axle	Without axle
2.0–2.5 m	300 mm	Φ105 × 230 L	Φ200 × 300 L
2.5–3.0 m	350 mm	Φ125 × 250 L	Φ250 × 340 L
3.0–3.5 m	400 mm	Φ145 × 280 L	Φ270 × 375 L
3.5–4.5 m	500 mm	Φ180 × 305 L	Φ340 × 400 L
4.5–6.0 m	650 mm	Φ250 × 405 L	Φ435 × 650 L
>6 m	700 mm	Φ280 × 415 L	Φ470 × 700 L
	1000 mm	Φ425 × 750 L	Φ650 × 1000 L

Translated from The Japanese Geotechnical Society. Translated by Niu Q, Chen F, Xu H. Investigation, design and construction of shield tunnelling. Beijing: China Architecture Press; 2008. Xie J, Tang H, Xie X. Disc cutter with pressure inside the cavity of cutter. CN202441384U, 2012.

4.3.3 Selection of muck discharging systems

4.3.3.1 EPB shield machine

The EPB shield uses screw conveyor as the muck discharging system. Table 4.4 shows the relationships among the outer diameter of shield, the outer diameter of screw conveyor, and the maximum particle diameter of the muck discharged [4], and it can be referenced for configuration of the screw conveyor.

For the screw conveyor with an axle, the axle should extend into the excavation chamber for enhancing the soil discharging efficiency of the screw conveyor.

When the groundwater level is high in high-permeability ground, it is difficult for the muck in a screw conveyor to form a soil plug, resulting in water spewing. To avoid this phenomenon, the following options are available:

1. The outlet of the screw conveyor is configured with a double-gate system. The two gates are opened alternately to reduce the spewing pressure.
2. The injection ports of the solution of bentonite and high-molecular polymer are reserved for muck conditioning. When necessary, the solution can be injected into the excavation chamber and the screw conveyor to condition the muck and avoid the water spewing.
3. The screw conveyor is equipped with a pressure holding pump, which can be connected with a muck discharging pipeline to avoid water spewing if necessary. However, the discharging efficiency is greatly reduced with the configuration of the pressure holding pump, so construction workers do not want to use it in many projects.
4. One more option is to increase the length of screw conveyor or to use the secondary screw conveyor (if there is enough space in the shield machine) to reduce the pressure of muck discharge outlet. Moreover, the free water in the muck can be drained through the drainage pipe at the bottom of the secondary screw conveyor, further reducing the risk of water spewing.

4.3.3.2 Slurry shield machine

The slurry shield uses a slurry circulation system to enable the input of fresh slurry and the output of waste slurry. The configurations of the slurry circulation system are as follows:

1. Slurry-feeding pump: The top-mounted slurry pump is usually installed on the ground surface.
2. Slurry-discharging pump: Usually, this slurry pump with adjustable discharging rate is selected and installed on the trolley behind the main machine of shield.
3. Slurry-feeding pipe: To reduce pressure loss, the diameter of the slurry-feeding pipe is usually 50 mm larger than the diameter of slurry-discharging pipe. However, the diameter of the slurry-feeding pipe can be set to be the same as that of slurry-discharging pipe at locations close to the locations such as trolleys, valves, flexible pipe, and so on.
4. Slurry-discharging pipe: The diameter of the slurry-discharging pipe depends on factors such as the maximum diameter of soil particles in the transported slurry, the sedimentation baseline flow rate of soil particles, the shield advancing speed, and the outer diameter of the shield.

Table 4.5 Selection criteria for the diameter of the slurry-feeding and slurry-discharging pipes.

Outer diameter of slurry shield (m)	Diameter of slurry-discharging pipe (cm)	Diameter of slurry-feeding pipe (cm)
<2	5–10	5–10
2–4	7.5–15	10–20
4–6	10–15	15–20
6–8	10–20	20–25
8–10	20–25	25–30
10–12	25–30	30–35

Modified from The Japanese Geotechnical Society. Translated by Niu Q, Chen F, Xu H. Investigation, design and construction of shield tunnelling. Beijing: China Architecture Press; 2008.

In gravel strata, the diameter of the slurry-discharging pipe should not be less than 200 mm. The Japanese Geotechnical Society [4] proposed selection criteria for the diameter of the slurry-feeding pipe and slurry-discharging pipe, as shown in Table 4.5. The table does not consider the effect of the particle size in the slurry.

4.3.4 Selection of other main configurations

Other main configurations of shield machine include the muck-conditioning system, synchronous grouting system, segment assembly system, backup system, shield attitude control system, and data acquisition and monitoring system. This section mainly describes the selection of the muck-conditioning system, synchronous grouting system, segment assembly system, and backup system.

4.3.4.1 Muck-conditioning system

As described in Chapter 3, Shield Machine Configurations and Working Principles, muck conditioning system of EPB shield are commonly equipped with a foam injection system, a bentonite injection system, and a polymer injection system, depending on geological conditions.

Foam injection system

The foam injection system can be set on the control panel. It has the following operation modes:

1. Manual control. It is arbitrary to operate the solution or air flow of each line of conditioning agent and to debug the optimal conditioning parameters.

2. Semiautomatic control. The required flow of foam can be injected according to the supporting pressure in excavation chamber.
3. Automatic control. The volume of foam injection can be automatically adjusted according to parameters such as the advancing speed and the pressure in excavation chamber.

The foam injection system is configured in one of two forms: a single pump for each pipeline or a single pump for multiple pipelines.

With a single pump for each pipeline, a nozzle for the conditioning agent on the cutterhead directly corresponds to a single pump. When the resistances among different foam pipelines are different, foam can still be injected according to its set flow, reducing the problem of nozzle blockage. At the same time, two materials are fully mixed in the mixing box by mechanical stirring and then are delivered via pump, so manual operation can be omitted for the single pump of each pipeline.

With a single pump for multiple pipelines, the foam solution and water are mixed in the pipeline, and the flow of the mixed liquid is adjusted mainly through regulating valves. However, some impurities may cause the regulating valves to become stuck, resulting in the valve core being unable to be effectively adjusted. Moreover, this affects the use of this system and causes waste of foam materials. The regulating valves are controlled by switches, which will also lead to low control accuracy. When pressure increases, it is also necessary to manually adjust the pressure valves to provide the injection pressure of the system.

Bentonite injection system

For EPB shield tunnelling in sandy ground under high groundwater pressure, muck may be conditioned with bentonite slurry. The bentonite is stirred with water and left there for several hours for swelling to obtain excellent slurry. There are two ways of stirring bentonite slurry: air-blown stirring and mechanical stirring. Air-blown stirring can quickly disperse the bentonite particles, while mechanical stirring can disperse the particles only through field flow. Extrusion pumps (Fig. 4.12) and screw pumps are mostly used as delivery pumps for the bentonite slurry. Owing to high pressure and large flow fluctuations, plunger pumps are not often used.

The bentonite injection system can also be used as a system for injecting other conditioners. For example, when a stratum with cohesion is encountered, it is necessary to inject dispersant solution to avoid soil clogging, and when a water-rich stratum is encountered, it is necessary to add superabsorbent resin to prevent from water spewing.

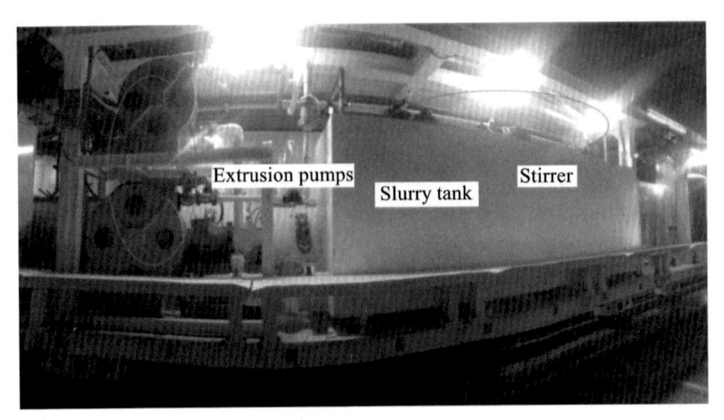

Figure 4.12 Extrusion pump of a bentonite injection system.

4.3.4.2 Synchronous grouting system

Selection of the grout delivery type

There are four types of grout delivery for synchronous grouting:

1. Direct pressure feeding: Grout on the ground surface is delivered directly to the injection port for synchronous grouting. This type is suitable for small-diameter shield tunnel with a short tunnelling distance.
2. Relay delivery: Grout on the ground surface is delivered to relay equipment placed on the rear trolley and is then injected by the grouting pump installed on the trolley. This type is suitable for large-diameter shield tunnel with a long tunnelling distance.
3. In-tunnel transportation: Grout on the ground surface is hydraulically injected into the mortar tank on the horizontal train, transported in the tunnel, and then injected by the grouting pump on the trolley. Because the delivery pipeline is not long, this type of synchronous grouting can be used without worrying about grout blockage in the pipeline.
4. In-tunnel grout mixing: Various grout materials are transported to the mixing equipment on the trolley and are then mixed and injected.

Currently, the relay delivery and in-tunnel transportation types are commonly used. The application of direct pressure feeding and in-tunnel grout mixing has become obsolete.

Layout of the grouting pipe

According to the relationship between the position of the injection port of the grouting pipe and the shield shell body, the layout of the grouting pipe can be classified into external and internal ones, as shown in Fig. 4.13.

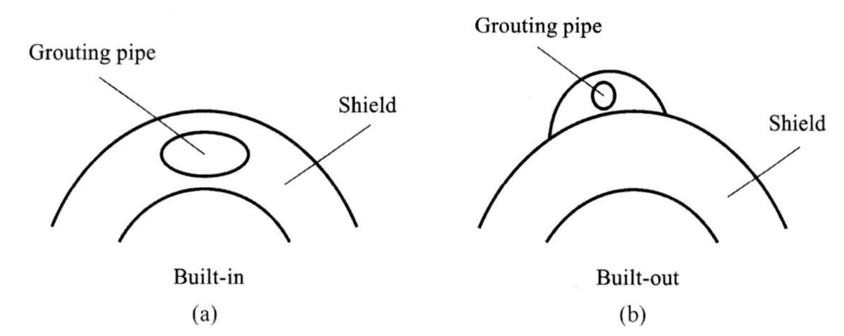

Figure 4.13 Layout of grouting pipe. (a) Internal grouting pipe. (b) External grouting pipe.

The external grouting pipe is placed on the outer end of shield, and a protective sleeve is installed outside the pipe injection port. It has a simple structure and does not need to increase the diameter of shield. Since this type of grouting pipe is easily damaged by friction, it is only suitable for soft soil strata.

The internal grouting pipe means to slot the inner side of shell plate to place the grouting pipe. The internal grouting pipe is installed in the shield shell and can adapt to all kinds of strata. However, the diameter of shield should be increased in the design, that is, the gap between the outer diameter of the shield tail and that of the segment should be increased, resulting in increase in the grouting amount and construction cost.

4.3.4.3 Segment assembly system

Selection of the segment assembly machine

The segment assembly machine is driven by a hydraulic system, and there are three types: ring type, hollow axle type, and rack and pinion type.

The segment assembly machines with the hollow axle type (Fig. 4.14a) and the rack and pinion type are easier to stably control during axial translation. Their braking is convenient and safe, but it is unstable during axial rotation and prone to produce shaking due to the long travel distance and complicated forces. To provide stable axial rotation, a high-rigidity structure is required for the axial rotation.

The ring-type assembly machine (Fig. 4.14b) has a smaller axial distance, lower rigidity requirements, more stable movement during axial rotation, and more stable axial positioning during braking but high requirements for the cantilever beams and general assembly accuracy. The machine features a hollow circular rotation, which can supply more working space even during the machine driving. Meanwhile, the muck transportation operation is

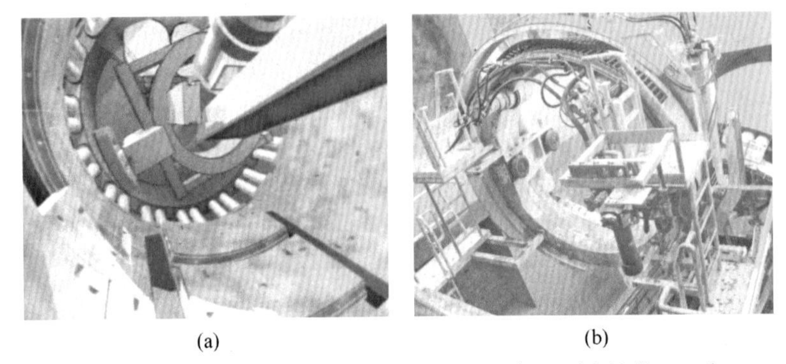

(a) (b)

Figure 4.14 The form of the segment assembly machine. (a) Hollow axle type. (b) Ring type.

not affected as usual and has the characteristics of high positioning accuracy and a relatively simple structure. It is installed mainly for medium and large tunnel sections and is currently the main structural form for segment assembly.

Selection of the segment clamper

There are two types of segment clampers, including mechanical grabbing and vacuum suction, as shown in Fig. 4.15. The mechanical grabbing with a grabbing head is suitable for small segments with low mass. For the segments used in large-diameter tunnels, the clamper generally adopts vacuum suction to ensure that the segments can be clamped safely and quickly. Many countries have been adopting vacuum devices to clamp segments. However, owing to the relatively high risk in using vacuum suction, some countries do not recommend its use to clamp segments, in consideration of personnel safety.

4.3.4.4 Backup system

The shield machine and its backup system form the construction line of the tunnel, and any problems caused by experiments in the line influence construction efficiency. While the selection of shield machine and its configuration is of primary importance, the configurations of the backup system should also be stressed. Taking the amount of soil excavation in each segment and one-time muck discharged as an example, the selection of the backup system behind the shield machine, such as muck carriage, segment carriage, battery tractor, and gantry crane, should consider both the normal tunnelling of the shield and maximizing the satisfaction of their own requirements.

(a) (b)

Figure 4.15 Segment assembly machine. (a) Mechanical grabbing type. (b) Vacuum suction type.

Selection of the muck carriage

The muck carriage is a direct carrier for muck transportation. The following points should be considered in its design: (1) The muck carriage must have enough rigidity and strength to ensure that it will not be deformed or broken during its lifting using the gantry crane, and it cannot be so heavy that it will cause overweight of the gantry crane. (2) The muck bucket and chassis of the muck carriage should be independent. When a gantry crane is used, only the muck bucket is lifted and the chassis is not lifted at the same time. (3) The muck volume transported by all muck carriages must be enough to load the amount of soil excavated in each segment ring at one time. (4) The width of the muck carriage should not exceed the inner diameter of the shield, and space should be left for workers and others in the tunnel. (5) The length of the muck carriage should be considered together with the mortar carriage and segment carriage, and the length of the entire train should not be beyond the reaching area of the horizontal belt conveyor for muck delivery. (6) A pair of lifting holders and an eccentric turning rod should be configured on the two sides of the muck bucket to ensure that the muck carriage is smoothly turned over when the gantry crane lifts the truck bucket and dump the muck into the pool on the ground surface.

Selection of the segment carriage

The segment carriage is a carrier that transports segments from the shaft to the segment assembly location. Attention should be paid to the following points when configuring it: (1) The internal boundary size of shield and the design size of segment should be considered simultaneously. The width of the segment carriage cannot be more than that of a segment. (2) The upper part of the segment carriage should be equipped with a special rubber pad or wood block to ensure its flexible contact with the segment carriage when segments are placed, and the position of the rubber pad or wood block should be suitable for the segment weight to prevent from being damaged during transportation. (3) For the arched structure of a segment, the top surface of the segment carriage is recommended to be designed as an arched structure to best ensure that the segment bottom is in full contact with the top surface of the segment carriage. (4) The turning radius of the segment carriage normally should not be lower than 25 m. (5) The driving system of the carriage should be equipped with a buffer device to achieve reliable braking.

Selection of the battery tractor

The battery tractor delivers the battery to supply power for the entire train transporting segment, mortar, and muck from the shield tunnel to the launching shaft. Its proper selection is directly related to the normal transportation in the tunnel. (1) The battery tractor must be capable of running at full load for all configured transport vehicles. (2) It can start safely on a ramp larger than that of the constructed tunnel and tow the entire train to run normally. (3) It should be equipped with a frequency converter to adapt to different working conditions. (4) It is equipped with storage batteries, and it is recommended that one-time charging ensure the transportation with a total distance of at least 10 km. (5) The battery tractor, muck carriages, and segment carriages are a unified system, and the brakes, gauges, and connection methods of the entire train should be arranged as a whole.

Selection of the gantry crane

The gantry crane is a lifting system that vertically transports the excavated muck in the muck bucket to the storage pool on the ground surface. (1) It must have the capacity to lift the total weight of the muck bucket and the full load of muck. (2) Its span should be designed according to the conditions at the construction site to determine whether to adopt a cantilever structure, but the net height is generally designed to be higher than 8 m to ensure a normal height of 3 m for turning the muck bucket. (3) Its drive vehicles generally adopt a frequency converter, which avoids the traveling speed to exceed the allowable one and ensures the stable traveling of large vehicles. (4) The lifting pole has a self-balancing device. It is detachable and can be easily be picked up and hung. (5) It should have subsidiary configurations such as sound and light alarms, lightning protection, moisture proof, manual rail clamping device, lifting height limit device, lifting weight limit device, large and small cars walking limit devices, lifting height indicator, lifting weight indicator, lightning rod, anemometer, and buzzer.

4.4 New technologies in the configuration of shield cutters

4.4.1 Technologies for the cutterhead configurations

The success or failure of the entire shield tunnel project depends greatly on the cutterhead configuration, as a complicated mechanical design process. Taking the configuration of disc cutters (as the main rock-breaking tool) as an example, there are a variety of design theories and configuration methods. Based on the parametric profile analysis theory, Bian [14] proposed a systematic assessment method for disc cutter performance under multigeological cutting conditions and a matching weight selection strategy, which can provide an optimal solution using a genetic algorithm. In addition, in view of the complexity and ambiguity of the geological adaptability of the shield machine, Bian [14] established a fuzzy adaptive relationship between the cutting abilities of the cutters and the two geological parameters including compressive strength and integrity of rock strata, and their adaptive membership functions were determined. A cutter selection scheme considering geological adaptation based on fuzzy mathematics was proposed for the composite cutterhead EPB shield tunnelling, as shown in Fig. 4.16. The TOPSIS in the Fig. 4.16 is an abbreviation of technique for order preference by similarity to an ideal solution.

4.4.1.1 Digital design technology

Owing to the lack of theoretical guidance and professional auxiliary design tools, most current cutters are designed according to empirical methods, whose design efficiency is low, and design quality cannot be guaranteed. More important, the cutterhead is designed for particular geological conditions, so the design experience cannot be effectively accumulated and used for reference. To this end, some scholars have developed a series of digital design platforms for the configuration of shield cutters with different functions. For example, the Colorado School of Mines in the United States established a tunnelling machine cutter layout model using a two-dimensional polar coordinate system based on testing results for the linear cutting machine over years. This model not only can

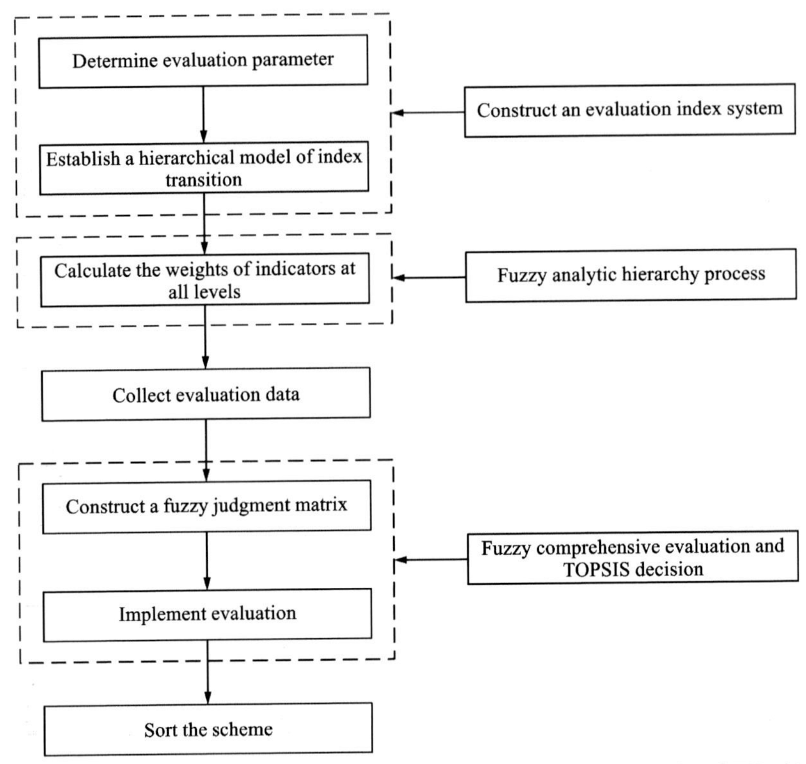

Figure 4.16 Selection flowchart of the cutters for the composite cutterhead EPB shield tunnelling. *Translated from Bian Z. Research on performance evaluation method of cutter wheel of composite EPB shield machine. Changsha: Central South University; 2013.*

optimize the layout of cutters, but also can predict tunnelling performance and tool life and cost. The State Key Laboratory of Shield and Tunnelling Technology of China independently developed digital software for cutter configuration. This software can perform parametric modeling and optimization analysis for various types of cutters. In addition, on the basis of design theory and performance evaluation method of shield cutterheads, Central South University in China developed a computer-aided design system for cutter configuration for EPB shield machines using platforms including Visual Basic and SolidWorks. The developed modules of the key parameter calculation and performance evaluation of a cutterhead enable the design system to have the functions of three-dimensional modeling, key parameter calculation, and performance evaluation of the cutterhead.

4.4.1.2 Digital management technology

The essence of cutter management is to track, manage, and optimize the information of cutter health and to build scientific planning for cutter change and maintenance. The establishment of a closed-loop digital management system is an indispensable core technology for shield tunnel construction. This system includes some functions, including data description and storage of cutters, positioning recognition, state monitoring, and scheduling optimization. It is of great significance for the effective use of cutters, reduction of engineering costs, reduction of construction risks, and promotion of construction efficiency.

4.4.2 New structures and materials for shield cutters

4.4.2.1 New structures for disc cutters

For achieving good performance and extending the adaptability of disc cutters in complicated conditions, some new structures for disc cutters were proposed. A new hub structure was developed to effectively block the permeation of slurry and avoid the failure of its sealing. New disc cutters with internal cavity pressure were proposed to prevent mud infiltration (see Fig. 4.17). New disc cutters with internal and external pressure balance devices were developed for the slurry shield in high water

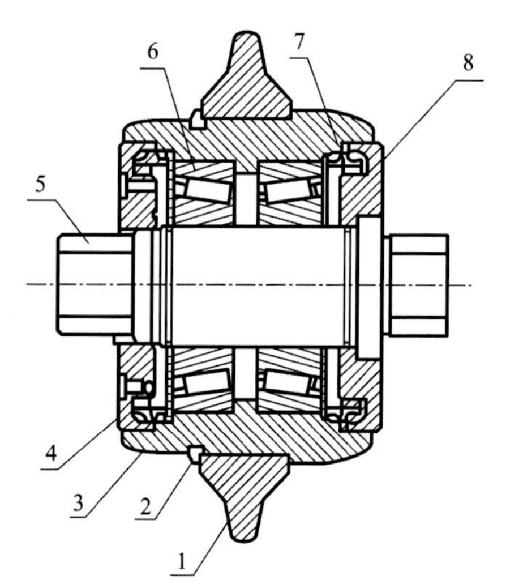

Figure 4.17 Schematic diagram of the partial structure of disc cutter with internal cavity pressure. 1: Cutter ring; 2: Protection ring; 3: Cutter body; 4: Upper cover; 5: Cutter axle; 6: Bearing; 7: Floating seal; 8: Lower cover [15]. *From Xie J, Tang H, Xie X. Disk cutter with pressure inside the cavity of cutter. CN202441384U,2012-09-19.*

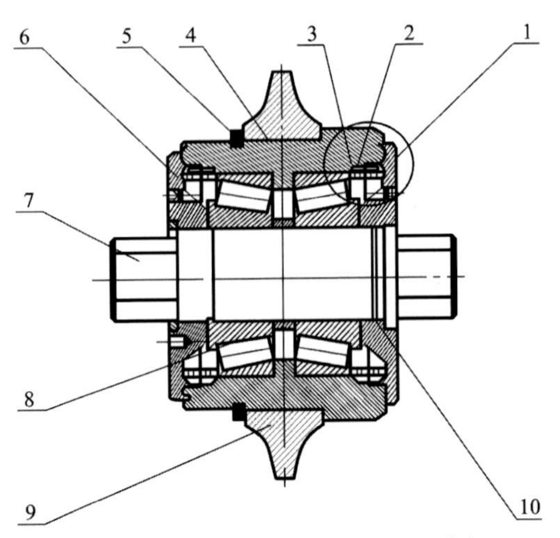

Figure 4.18 Schematic diagram of the partial structure of disc cutter for high water pressure with internal and external pressure balance devices. 1: Right cover; 2: Seal ring; 3: Oil seal matrix made from alloy cast iron; 4: Hob; 5: Protecting ring; 6: Left cover; 7: Cutter axle; 8: Bearing protecting ring; 9: Cutter ring; 10: Seal ring [16]. *From Liu M, Gan W, Zhao J. A shield disk cutter suitable for high water pressure and high earth pressure environment. CN202483584U,2012-10-10.*

pressure (≥0.5 MPa) (see Fig. 4.18). In addition, the development of disc cutters has been showing a trend toward two extremes. On one hand, miniature disc cutters have been studying for meeting the needs of tunnelling in hard rock. On the other hand, large-scale (20 inches or more) multiblade disc cutters have been researching to have high rock-breaking capability and high abradability.

4.4.2.2 New materials for cutters

The use of high-strength steel for cutters has become widespread in recent years. It has good abradability and can promote the long-term cutting performance of cutters. In Japan, South Korea, and other countries the steel material SKD11 is commonly used to make cutter rings. This material is produced with high abradability and toughness that can extend the service life of disc cutters. A new cutter ring made of ledeburite die steel material has been developed as a basic organizational structure, which has high hardness and low brittleness. In terms of mechanical properties and cost effectiveness, ledeburite die steel is an ideal forging material for making cutter rings. In addition, material production technology is expected to make breakthroughs in new types of cemented carbide cutter rings, coarse-grained particle carbide cutter rings, new abradable surfacing materials, and more. Cemented carbide cutter rings with good hardenability, hardening capacity, and tempering stability are being developed for tunnelling in full-face rock ground with an undrained compression strength higher than 120 MPa and composite strata. The coarse-grained particle carbide cutter rings have good abradability, thermal conductivity, and fatigue shock resistance and low thermal expansion coefficient. The new abradable surfacing materials have good structure and toughness, few cracks in the welding layer, and good welding performance.

4.4.3 New detection technologies of cutter states

There are two directions in the development of the detection technologies of cutter working states. On one hand, real-time acquisition of cutter wear information has been developing for cutter state detection. New feasible wear detection technologies include cutter wear monitoring technology based on ultrasonic sensors, disc cutter wear monitoring technology based on eddy current sensors, and cutter visualization monitoring technology. On the other hand, multipurpose sensors have also been developed to monitor parameters such as the rotation and force of disc cutters. They can be helpful for achieving long-distance monitoring of disc cutters to better serve the design and development of cutters and control of tunnelling parameters.

4.5 Examples of shield machine selection in common strata

The success or failure of shield tunnel construction depends greatly on whether the selection of a shield machine and its configurations are reasonable or not. A reasonable shield machine selection can effectively promote shield tunnelling efficiency and reduce construction costs, especially in a complex stratum. As two examples, the difficulties and countermeasures are discussed for shield tunnel construction using an EPB shield and a slurry shield, as the two mostly common shield machines for tunnelling in weak ground. One example is the shield machine selection in composite strata (a soil layer underlaid by a rock layer); and the other is the shield machine selection for tunnelling under a river.

4.5.1 Shield machine selection for tunnelling in composite strata: EPB shield

4.5.1.1 Project overview

The Xin-Guang section on the Line 5 of Nanning Metro was located in Xixiangtang District, Nanning, China, as shown in Fig. 4.19. The section

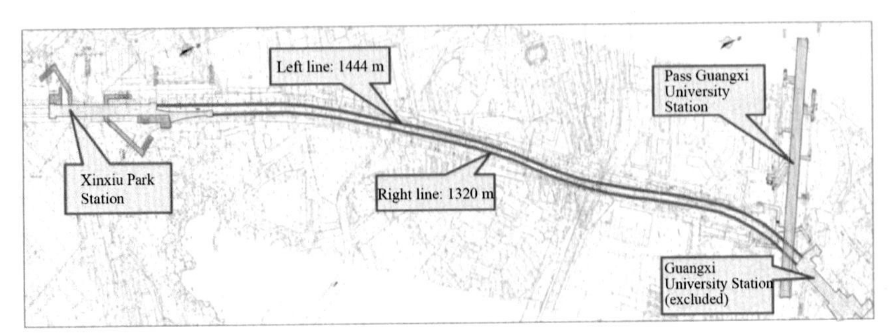

Figure 4.19 Section layout of the Xin-Guang section.

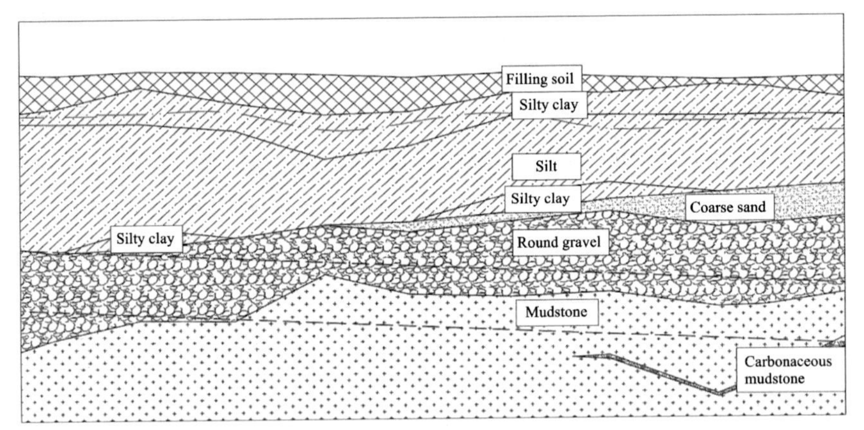

Figure 4.20 Geological profile of the left tunnel in the Xin-Guang section.

mileages of the left and right tunnels were ZDK20 + 588.134 to ZDK22 + 031.821 and YDK20 + 712.134 to YDK22 + 031.821, respectively, and their total lengths were 1443.687 and 1319.687 m, respectively. The EPB shield tunnelling method was adopted for both of the tunnels.

This tunnel section mainly passed through a stratum with a round gravel layer underlaid by a silty mudstone layer. The geological profile of the left tunnel section is shown in Fig. 4.20. The types of groundwater in this section are mainly perched water, pore water, and crack water. Groundwater mainly occurs in the Quaternary sand layer and round gravel layer. As shown in the engineering investigation, the buried depth of the groundwater level in this section is 3.50–11.30 m, and its average is 5.84 m. According to the results of the water pumping tests and local engineering experience, the permeability coefficient is about 0.01–0.5 m/d for the fine-grained soils and about 20–70 m/d for the coarse-grained soil.

4.5.1.2 Difficulty analysis in shield machine selection

Wearing abrasion of cutters in long-distance shield tunnelling

The tunnels are located mostly in the gravel layer, which has a low cohesion and a high quartz content. The cutters and screw conveyor are easily worn in the layer, so it is important to reduce wearing abrasion of cutters during the long-distance shield tunnelling.

Shield tunnelling in mudstone strata

The mudstone has a high content of clay and silt particles. During excavation of the silty mudstone strata by the shield machine, there may be stagnant discharge of muck in the chamber, leading to muck accumulation and resulting in clogging on the cutterhead and in the excavation chamber and screw conveyor. Moreover, since the wearing abrasion of cutters is greatly increased, resulting in that the cutters become abnormal and their service lifetime is shortened.

Opening of shield chamber for cutter maintenance and change

The section passed underneath Guangxi University Station in the Line 1 of the Nanning Metro after the left and right lines are tunneled for 1407 m (i.e., 937 segment rings) and 1283 m (i.e., 855 segment rings). Before passing underneath, the section tunnel is located in the composite strata of a round gravel layer underlaid by a mudstone layer. There were no good positions in which to open the excavation chamber for changing cutters during the whole tunnel section. Therefore it was difficult to check or change cutters before crossing the Guangxi University Station.

4.5.1.3 Target selection of shield machine configurations

Main parameters of the shield machines

The tunnels adopted EPB shield machines with a type of CTE6250 produced by China Railway Engineering Equipment Group Co., Ltd. The length of the entire machine (main machine + backup system) was about 80 m in total for each one, and the total length of each main machine was 8388 mm. The total weight (main machine + backup system) was about 500 t. The excavation diameter of the shield machines was designed to be 6280 mm. The cutterhead rotation rate was designed to be 0–3.7 rpm. The maximum advancing rate and thrust was designed to about 80 mm/min and 3991 t, respectively. The tunnel segments were designed to be 6000 mm/5400 mm–1500 mm/36 degrees (external diameter/internal diameter – segment width/central angle between two adjacent bolts in the segment ring). For each of the shield machines, the minimum horizontal turning radius was 250 m, and the maximum longitudinal climbing slope was ± 5%.

Each shield machine incorporated a hydraulic driver mode, and its shield body was configured with a passive articulation. The screw conveyor for each of the shield machines was equipped with a screw shaft, having a maximum muck discharging capacity of 335 m^3/h. The segment erector system was configured with a central rotary type and a VMT guiding system.

Cutterhead configuration

To pass through the composite stratum of a round gravel layer underlaid by a silty mudstone layer and the four diaphragm walls of the Guangxi University Station, the shield cutterhead was arranged for each of the shield machines as follows.

1. Cutterhead design

   A composite cutterhead was adopted, as shown in Fig. 4.21. It consisted of six main cutter beams, six corbels, six corbel support

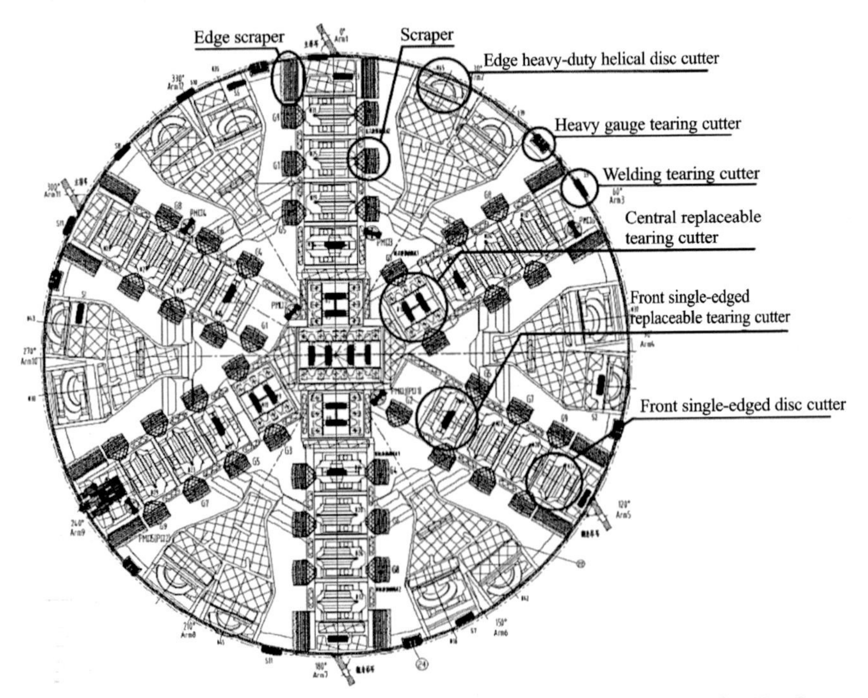

Figure 4.21 The layout of cutters for China Railway R81# and R82# in the Xin-Guang Section of Nanning Railway Line 5.

beams, and one outer ring beam. The overall opening rate of the cutterhead was 34%, and the central opening rate was 38% (larger than the opening rate on the peripheral area of the cutterhead) to control soil clogging in the cutterhead center.

2. Anticlogging design
   a. The height of the support beam in the axial direction of the cutterhead was lower than that of main cutter beam, the central area of cutterhead, and the peripheral area.
   b. The cutterhead openings were connected with each other. This facilitated the flow of muck from the central area to the peripheral area and so improved the flow of muck on the cutterhead in the radial direction.
   c. The cutterhead and its supporting system should have a certain muck-stirring capacity. As shown in Fig. 4.22, the muck can be stirred by the relative movement between the cutterhead (rotatable) and the pressure-bearing bulkhead (fixed). Therefore a stirring rod was added to the bulkhead to enhance the stirring effect.
   d. The main bearing was configured with an outer diameter of 3061 mm. With this main bearing, the cutterhead flange was relatively large, and the central space in the excavation chamber was relatively large to reduce the soil clogging potential in the center.
   e. The pipeline connection structure in the cutterhead was designed in the form of an "L" beam. To stir and flush muck in the central soil chamber, high-pressure flushing ports were installed in the

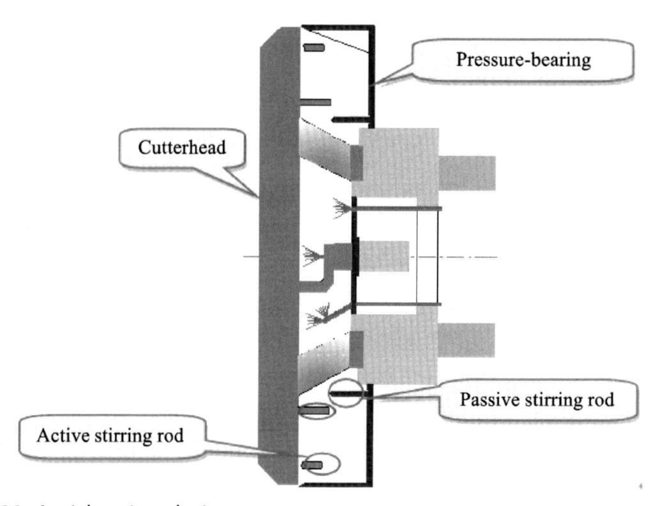

Figure 4.22 Anticlogging design.

center of the beam, and passive stirring rods and high-pressure water flushing ports were configured on the bulkhead. This could effectively prevent soil clogging in the center of excavation chamber.

f. Reinforced scrapers were designed to be 135 mm high. To ensure the cutting efficiency of the cutters and so that the muck entered the soil chamber smoothly, it was necessary to reduce the number of scrapers and increase the distance between adjacent scrapers on the same beam. This could facilitate the flow of muck through the openings among the cutter beams and prevent muck accumulation on the beam surface, which would result in soil clogging.

g. The front panel of the cutterhead was uniformly equipped with six foam injection ports, each of which was configured with single pipeline and single pump. Two of the ports were shared for injecting bentonite slurry, in case the EPB shield passed through the gravel under groundwater.

h. Four active stirring rods were placed on the back of the cutterhead., and two passive stirring rods were placed on the bulkhead. This design could be helpful to stir the muck in soil chamber, improve its fluidity, and prevent muck from accumulating in the chamber to form clogging.

3. Cutter configuration

The cutterhead was equipped with six central replaceable tearing cutters (height: 175 mm), six front single-edged disc cutters (height: 175 mm), 16 front single-edged disc cutters (height: 175 mm), 12 edge heavy-duty disc cutters (height: 175 mm) with oblique teeth, 40 front scrapers (height: 135 mm), 24 edge scrapers (height: 135 mm), 12 welding tearing cutters (height: 150 mm), and six heavy tearing cutters (extension: 50 mm) for the cutterhead edge protection, as shown in Fig. 4.23.

4. Wearing abrasion detection

The cutterhead was additionally equipped with four hydraulic wearing abrasion detection devices to detect wearing abrasion of the cutters over time through alarms once after the hydraulic oil leaks.

5. Abradability design

To promote abradability of the cutterhead, the front surface of the cutterhead panel was welded with 6.4 + 6.4 mm composite steel plate, and the outer surface of outer ring beam was welded with 12.5 + 12.5 mm composite steel plate.

6. Heightening design of peripheral disc cutters

   Because of the long-distance tunnel construction, the peripheral disc cutters were heightened by 10 mm so that the excavation diameter of the cutterhead could be enlarged to adapt to the long-distance shield tunnelling.

7. Foam and bentonite nozzles at the cutterhead

   All nozzles used for injecting conditioning agent at the front of the cutterhead panel were mounted from the panel back to protect the nozzles. They could be replaced in excavation chamber if they were completely blocked and could not be cleaned. The nozzles could be drawn out from the back of cutterhead panel for repairing or replacement, as shown in Fig. 4.23.

Other configurations

1. Front shield

   The diameter of the front shield was designed to be 6250 mm, and the notch of the front shield was welded with a 5-mm thick wear-resistant layer to increase the abradability of the notch. To improve the muck fluidity, four stirring rods were arranged on the bulkhead of the excavation chamber, and the surfaces of the stirring rods were welded with highly wear-resistant welding materials to increase their antiabradability.

   The bulkhead was equipped with air, water, and electrical connections, all of which were convenient for use when changing cutters in the excavation chamber.

2. Middle shield

   There were two seal rings between the middle shield and the tail shield. One was a rubber seal, and the other was an emergency airbag

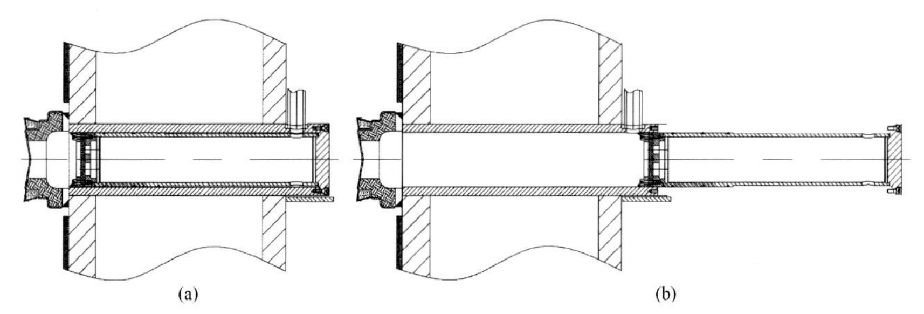

Figure 4.23 Schematic diagram of nozzle replacement. (a) The nozzle was installed in the cutterhead panel. (b) The nozzle was taken out from the panel back.

seal. Under normal circumstances the rubber seal ring worked. The emergency airbag seal ring was for use when water gushed or the rubber seal ring needed to be replaced. The compression of the rubber seal ring could be adjusted by some bolts of the adjusting blocks, thereby adjusting the seal gap between the middle shield and the tail shield.

3. Tail shield

The shield tail consisted of a hinged sealing ring and a shell. The diameter of shell was 6230 mm, and its thickness was 50 mm, and the length of the shield tail was 3890 mm.

All grouting and grease pipelines were embedded in the shell of the tail shield. Each grouting pipe had two small observation windows, which were beneficial for protection, cleaning, and maintenance of pipelines. There were 10 grouting pipes in total, six of which were reserved. There were 12 grease pipes, every six of which were arranged in one group and led to one of the two grease chambers between two closed rings of tail brushes.

The tail brush sealing system was composed of three rings of sealing brushes welded on the shell, which can prevent grouting material and water from leaking into the shield body.

The shield tail clearance was design to be 30 mm thick, which satisfied the requirements of the segment assembly and shield attitude adjustment. There was a ring of grouting-stop boards at the shield tail, which were made of abradable steel plates. The grouting-stop boards not only prevent mortar from filling the gap between the shield shell and the excavated ground, but also prevent slurry leaking out of the shield chamber from reaching the shield tail gap to influence the grouting effect.

4. Screw conveyor

The screw conveyor with a shaft was adopted and had an inner diameter of 800 mm, where the largest particle size can pass through would be as large as 300 × 560 mm. The first screw shaft had five groups of drivers, and it could supply a maximum torque of 210 kN•m and a maximum rotation rate of 25 rpm. The abradability design adopted a wear-resistant layer welded on the front part of the shaft. The front barrel of the screw conveyor had a replaceable bushing, which could be replaced when it was severely worn. The bottom of the front section of the screw conveyor barrel was equipped with a replaceable wear-resistant window, which could be replaced when severe wearing abrasion happened, as shown in Fig. 4.24. To prevent water spewing, the axial distance between every two adjacent blade joints was reduced from 630 mm (used conventionally) to 550 mm. It was equipped with double muck discharge gates, which could be helpful to prevent from water spewing, as shown in Fig. 4.25.

5. Propulsion and articulation system

Single and double cylinders were uniformly distributed along the circular direction of the shield, with a total of 30 cylinders divided into four groups. There were seven, seven, eight, and eight cylinders

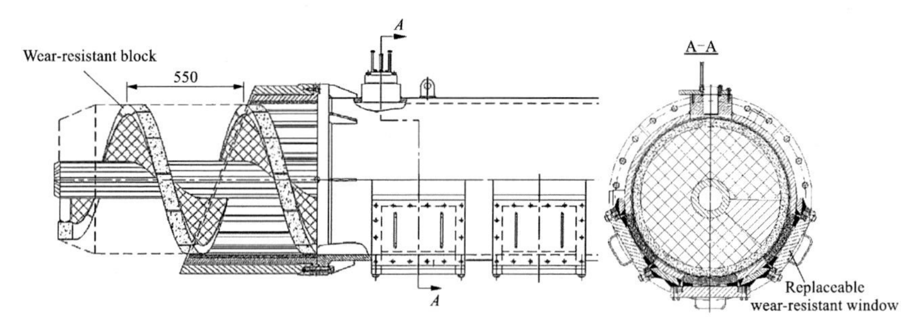

Figure 4.24 Antiabradability design.

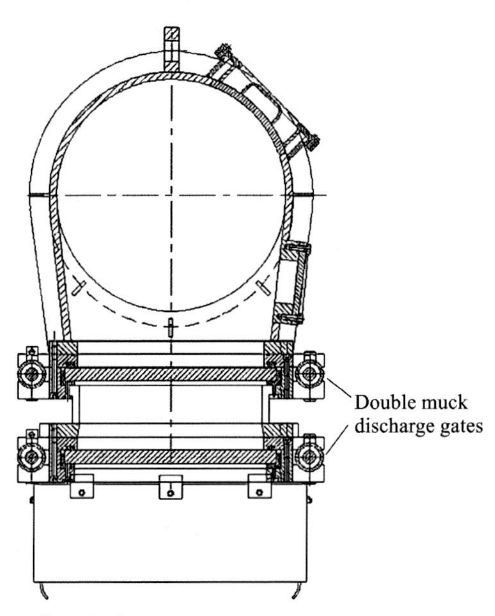

Figure 4.25 Antispewing design.

in the top, bottom, left, and right groups, respectively. Shield attitude could be corrected and adjusted by controlling different advancing speeds of each group of cylinders. Each group of cylinders was equipped with a built-in stroke sensor. According to these four groups of evenly distributed cylinders with sensors, the shield tunnelling attitude could be determined. The distance per shield advancing cycle satisfied the requirements of assembling segments with a segment width of 1500 mm.

The EPB shield machine adopted a passive articulation system, which consisted of 18 cylinders (divided into four groups) in the circular direction of shield and had a maximum articulated tension of 1200 t. When tunnelling in a curved section, the shield tail could automatically adjust its attitude along with segments. Four groups of the articulated cylinders at different positions were equipped with built-in displacement sensors to monitor the stroke of the articulated cylinders at the different positions along the circular direction. The articulation system was applicable to the shield tunnelling along the curved line with a radius of 250 m at a minimum.

6. Grouting system

The EPB shield was equipped with two grouting pumps, each of which had two outlets. The grouting system used four grouting pipes, which were connected to the pump outlets. To achieve the function of the automatic control of the grouting pressure, a pressure sensor was installed at each injection end of the pipe to detect the grouting pressure.

The operation of a synchronous grouting system at the shield tail had both a manual mode and an automatic mode. The pipe was equipped with both water and air cleaning devices, which were installed under the equipment bridge. When needed, the grouting pipes can be flushed forward and backward. The installation location of grouting pumps were reserved for the secondary grouting at the back of segments and advanced grouting.

7. Foam and polymer injection system

Foam solution was pumped to the foam generators through four pumps. The foam solution was mixed with air to generate foam in the foam generators, and the foam was then injected into four foam nozzles on the cutterhead, soil chamber, and screw machine. The concentration of foam solution was adjusted through a frequency conversion system for soil conditioning in different geological conditions. The foam injection volume of each channel was determined by a frequency conversion system, and the added volume was adjusted according to the muck condition and pressure requirements in the excavation chamber. The foam generation system was equipped with manual, semiautomatic, and automatic control modes, which could be selected depending on different requirements.

8. Bentonite injection system

   The bentonite injection system was configured with a tank (with a volume of 6 m^3) for storing bentonite slurry on a trailer. The bentonite slurry could be injected into the excavation face, soil chamber, and screw conveyor where muck needs to be conditioned through two extrusion pumps. A single pump was attached to each bentonite slurry pipe at the cutterhead to ensure the injection pressure. Each pipe was equipped with a pressure sensor and a flow meter to detect the pressure and flow of bentonite slurry in each channel. The opening and closing of each channel were achieved by a pneumatic ball valve, whose button was set in the main control room. Additionally, to fill the gap between the shield shell and the excavated ground, a bentonite injection system was used. There were six injection pipes for the middle shield. The bentonite slurry in each pipe could be injected separately.

4.5.2 Shield machine selection for tunnelling under a river: slurry shield

4.5.2.1 Project overview

The left and right tunnels in the Wangjiang Road Tunnel Project in Hangzhou, China, were 1837 and 1830 m long, respectively. They were constructed by using two slurry shields. Construction started from the south working shaft, went northward and crossed the Qiantang River, and reached the north shield shaft. The segments were designed to be 11.3 m in outer diameter, 10.3 m in inner diameter, and 2.0 m in width.

The tunnels mainly passed through the mucky silty clay and round gravel, and sometimes silt sand, mucky silty clay, sandy silt, and silty clay. The geological profile of the left tunnel section is shown in Fig. 4.26. Generally, the soil particles of the shield-excavated soil layers became coarser from the top to the bottom, and the characteristics of the soil layers were quite different. The diameters of the excavated gravel particles were generally 0.5–3.0 cm, and the largest ones were more than 10 cm. Their shapes were mainly subcircular, and some were subangular. The length of tunnel section with the round gravel layer exceeding 20 cm in thickness was about 540 m. There was a riprap in the excavation area of

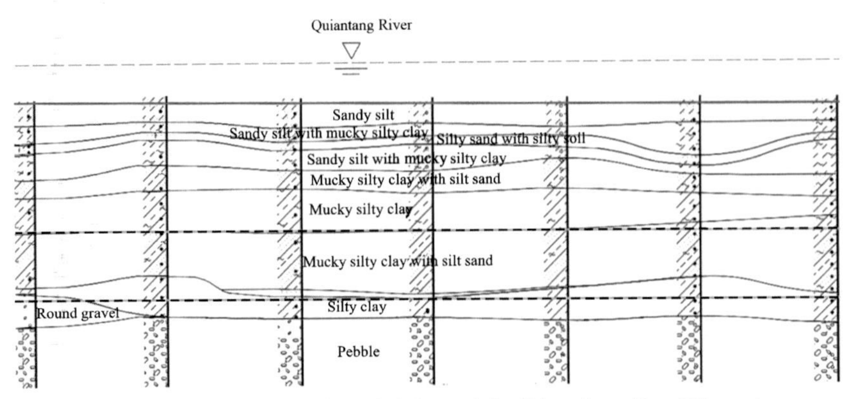

Figure 4.26 Geological profile of the left line of the Wangjiang Road Tunnel.

the right tunnel to the north of the Qiantang River, and the length of the riprap area was about 30 m in the longitudinal direction of the tunnel. In the middle area of the river there was a biogas layer, whose general pressure was 0.05–0.2 MPa. The geological investigation around YK2 + 650 to the south of the river revealed that there were large hidden ponds, which were artificially backfilled. In addition, some small hidden ponds were scattered in other area. Generally, the soil in the backfilled ponds was extremely weak.

The groundwater in the site was mainly divided into three types: Quaternary loose rock pore water, Quaternary loose rock pore confined water, and bedrock fissure water. The pore phreatic water was supplied by the vertical infiltration of atmospheric precipitation and the infiltration of surface water. The water body in the middle section of the Qiantang River was closely related to the phreatic hydraulic water, and the river water constituted the water source of phreatic aquifer. The groundwater level changed significantly with seasonal climate and the water level of the Qiantang River. According to regional data, the dynamic range of the groundwater level difference was generally about 1–2 m. During the geological investigation the buried depth of the phreatic level was 1.30–4.50 m, and its elevation was 3.44–5.72 m with an average of 4.80 m. The fissure water mainly came from the lateral seepage of groundwater and the infiltration of the upper confined aquifer. It flowed slowly and drained downstream. Combined with the analysis of the characteristics of the site landform as a plain area, the water volume in the bedrock fissure was weak, so the fissure water was of little significance to this project during shield tunnelling.

4.5.2.2 Difficulty analysis in shield machine selection

1. Adaptability requirements for silty clay and round gravel strata

   The strata passed by shield tunnel were mainly mucky silty clay and round gravel. The silty clay was highly cohesive and easily blocked the slurry pipelines, and it was difficult to discharge the slurry with the excavated clay during the slurry shield tunnelling. The round gravel strata had strong water permeability, and it would be easy to produce quicksand under the disturbance of shield construction. The type of the shield and its configurations should be selected according to the tunnel stratum characteristics.

   The slurry shield tunnelling should have the following functions: (1) high abradability of the cutters and slurry pipelines and automatic monitoring of their wearing abrasion; (2) reasonable cutterhead design, including proper opening ratio and opening position of the cutterhead; (3) good sealing performance of the shield under high water pressure and high adaptability of the synchronous grouting system to the high water pressure; (4) being able to enter the excavation chamber under pressure; (5) being equipped with a flushing system on each of the slurry suction ports at the cutterhead center and the front panel to ensure smooth muck discharge; (6) being equipped with a pressure filter for each of slurry separation system to achieve nonpollution discharge of slurry.

2. Adaptability to high water pressure

   Owing to the high water pressure in shield tunnelling under the river, it was easy to cause sudden water inrush and quicksand and ground collapse during the tunnel construction. The maximum water pressure at the top of tunnel was 0.45 MPa, which had high sealing requirements for the main bearing seal, shield tail sealing, slurry pump, and tunnel water drainage. Therefore the shield design incorporated the following capabilities: (1) High sealing performance of the shield was required for the safe shield advancing under high water pressure conditions; (2) The shield tail had strong water-blocking performance and could effectively prevent a sudden water inrush.

3. Shield tunnelling through riprap or other isolated anomalous ground

   According to the geological investigation, there was riprap or other isolated anomalous bodies in sections from YK1 + 055 to YK1 + 070.8 and from YK1 + 162.5 to YK1 + 174.4 of the right tunnel. It was difficult to reduce the impact of riprap or other isolated abnormal bodies on shield construction. The shield machines should have the following capabilities: (1) The cutterhead should be configured to ensure that the cutters can break the riprap with large particle size. (2) Large-size stones or obstacles should be crushed for effectively preventing discharging pipes from being blocked. (3) There should be a means of detecting wearing abrasion of the cutters.

4. Shield construction in a biogas-bearing stratum and its influence from tides

   According to the geological investigation, there was shallow biogas in the middle and southern sections of the tunnels under the river. No biogas overflow was found in the north and south land sections. The biogas overflowed sporadically from the exploration points with a short-time blowout phenomenon in some cases near the main channel of the River, and the gas pressure was generally 0.05–0.17 MPa. There was a potential problem that the biogas would leak into the shield tunnel. It was important to ensure the safety of construction personnel and equipment when the shield passed through the biogas section. The shield tunnel construction were configurated with (1) a gas detection system, (2) a strong sealing performance of shield to prevent the leakage of biogas into the tunnel, and (3) the ventilation and inspection requirements.

4.5.2.3 Shield machine selection

Shield machine selection

A slurry shield machine can adapt to various strata, such as silty clay, silt sand, pebble, and sandy gravel. The slurry penetrates the excavation face to effectively resist water apressures, and the occurrence of water and sand spewing can be avoided. The excavation muck is transported through slurry pipeline, so the discharging efficiency of muck is high. Thus, the slurry shield machine was most suitable for the geological and hydrological conditions of the Wangjiang Road Tunnel. The shield machine for each of the tunnels had a cutterhead diameter of 11.64 m and a front shield diameter of 11.61 m.

Cutterhead configurations

The panel cutterhead was used for each tunnel, as shown in Fig. 4.27. The surface and the opening of the cutterhead were welded with a wear-resistant layer, and the outer ring was welded with a wear-resistant plate. To maintain good mechanical support for the excavation face, the opening ratio of the cutterhead was designed to be as low as 33%. The cutterhead center was equipped with a flushing system for flushing muck into the excavation chamber to prevent the center from being clogged. The

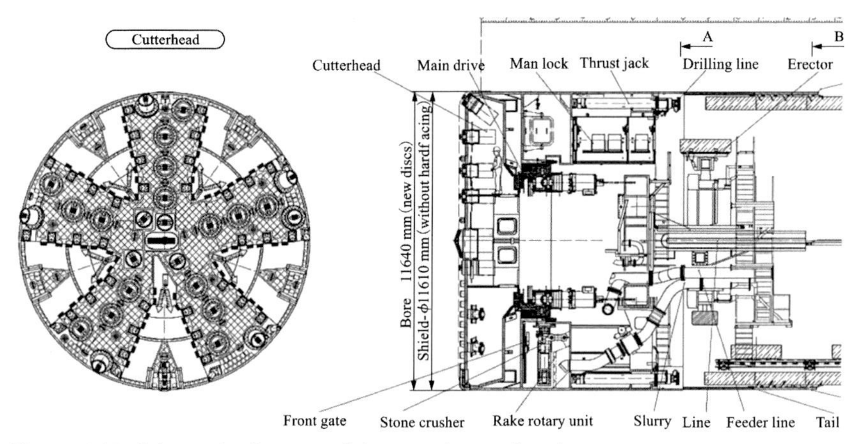

Figure 4.27 Schematic diagram of the panel cutterhead.

cutterhead openings were designed as wedge-shaped structures (the opening gradually becomes larger along the muck-entering direction), which were conducive to the flow of muck.

The cutterhead was mounted on the inner gear ring of the main bearing through flanges and was driven by a frequency conversion motor. The cutterhead was designed to rotate in both directions, and its rotation rate could be steplessly adjusted. For safety and convenience in cutter changing, all cutters were back-mounted and could be replaced on the back of the cutterhead.

Cutter configuration

1. Advance cutters

   The cutterhead was equipped with 20 replaceable advance cutters and 31 fixed welded advance cutters, all of which were 175 mm higher than the cutterhead board and 25 mm higher than the scrapers. The advance cutters cut strata such as sandy gravel and riprap or other isolated anomalous ground, damaging their integrity and facilitating the excavation of the scrapers and protecting the scrapers. The cutting width of each advance cutter was 60 mm, which was relatively narrow and allowed them to have high cutting efficiency in the difficult stratum. The advance cutters were designed to be adaptable for the two-direction rotation of the cutterhead.

2. Front scrapers

   There were 118 front scrapers in total installed on both sides of the cutterhead panels. They were 150 mm higher than the cutterhead boards. The front scrapers were used to cut the soft soil layers and scrape the cut soil into the excavation chamber. Their cutting widths were 100 mm, and the blades made from double-layer carbide-tungsten alloy were used to promote abradability of the front scrapers. The allowable wearing abrasion thickness of the double-layer blades was as large as 64 mm (2 × 32 mm). After the blade of the first layer was worn, the second layer could continue to play a role. The backs of the front scrapers were welded with double-row carbon-tungsten alloy stud teeth. Three front scrapers in different areas at the cutterhead were equipped with wearing abrasion detection devices to monitor the wearing abrasion of the scrapers over time and ensure that they would be able to work normally.

3. Edge scrapers

   There were 10 edge scrapers placed on the outer ring of the cutterhead. The edge scrapers were welded with single-layer tungsten-carbide teeth and double-row carbon-tungsten alloy cylindrical teeth to ensure high abradability of the edge scrapers and a long work life for the 1837 m long tunnel excavation.

Other configurations

1. Shield shell

   The structure of the shield shell was designed to withstand 10 bar static water pressure plus 7.5 bar dynamic fluid pressure. The shield shell was made of welded steel plates, and all parts of the shield shell were assembled with bolts.

2. Main driver system for cutterhead

   The main driving system included the main bearing, a variable-frequency motor, a gear box, and a frequency control cabin installed on the rear supporting trailer. As shown in Fig. 4.28, the main bearing has two sets of sealing systems, including internal and external ones. The external sealing system incorporated four lip-shape seals and one labyrinth-shape seal. The sealing system had the functions of automatic lubrication, automatic sealing, and automatic detection of the working conditions of the seals. The internal sealing system incorporated two lip-shape seals and an optimized design scheme. Unlike the traditional design, which could directly bear the earth pressure in the excavation chamber, the optimized design scheme incorporated the internal sealing in the main bearing body, touching the central cavity of the cutterhead, greatly improving the internal sealing effect.

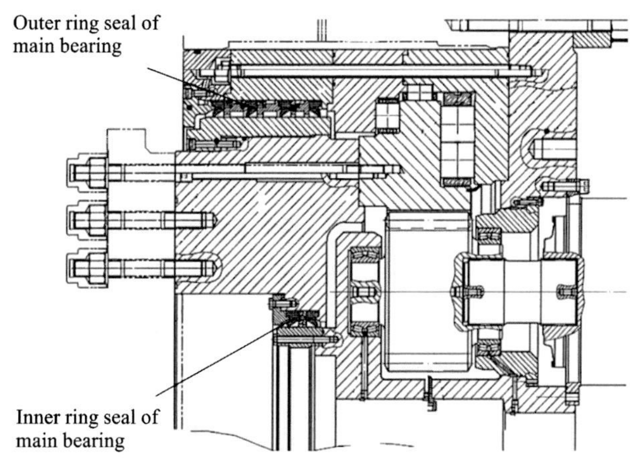

Figure 4.28 Schematic diagram of the main bearing seal.

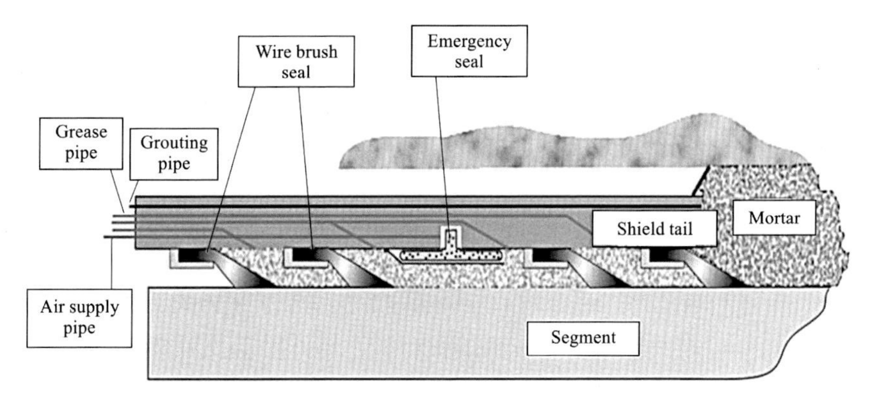

Figure 4.29 Schematic diagram of the shield tail seal.

3. Shield tail seal

To avoid groundwater, grout, and biogas leaking in the shield tail, four rings of wire brushes were equipped there (Fig. 4.29). Additionally, an expansion emergency seal was configured at the shield tail. If the first two rings of wire brushes failed, the emergency seal would be expanded by inflation to completely seal the gap between the outer side of segment and the inner side of shield tail, thereby preventing groundwater, grout and biogas from leaking into the tunnel and allowing the first two rings of failed wire brushes to be safely replaced in the tunnel.

4. Propulsion system

The propulsion system, which included the propulsion cylinders and the corresponding hydraulic pump stations, provided the propulsion power for the shield machine. Since the diameter of the shield was as large as 11.6 m in this project, the propulsion cylinders were divided into six groups along the circumferential direction. Each group of cylinders had an independent pressure adjustment. To make the shield machine excavate in the correct direction, the operator could individually adjust the pressure and stroke of the six groups of the cylinders to correct and adjust the shield advancing direction. The six groups of propulsion cylinders were equipped with stroke measurement detectors, and the cylinder advancing speed could be adjusted in the main control room through the control buttons.

5. Rock breaker

Pebbles or rocks might be encountered during the construction of the shield tunnels, and the maximum particle size allowed by slurry pump was 200 mm in diameter, so larger rock particles had to be crushed. For each tunnel, a hydraulically driven jaw crusher was installed in front of the slurry pipeline at the bottom of the slurry chamber (see Fig. 4.30) to crush large rock particles. The jaw crusher was driven by heavy-duty oil cylinders to the breaking plates, and the crusher teeth could be replaced and lubricated automatically. A strong steel filtering net with holes of 180 × 180 mm was installed at the entry port of the slurry discharging pipeline to ensure that no larger rock particles entered the slurry discharging pipeline. To avoid pipeline blockage due to the clay clogging, two water flushing pipes were installed at the slurry discharging outlets. A safety gate was set on the bulkhead bottom of the slurry chamber. After the gate was closed, the crusher could be inspected and repaired under atmospheric pressure.

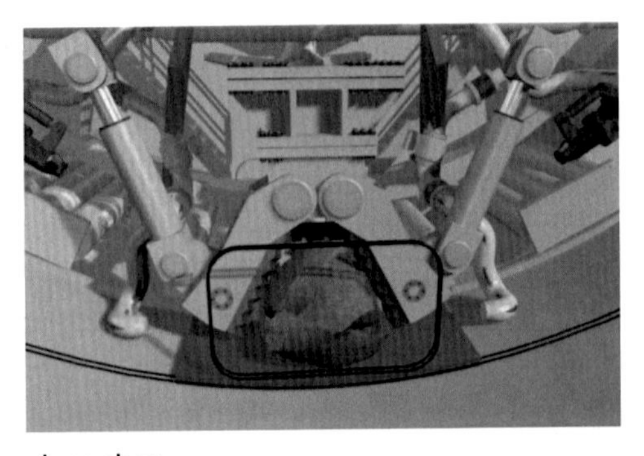

Figure 4.30 The rock crusher.

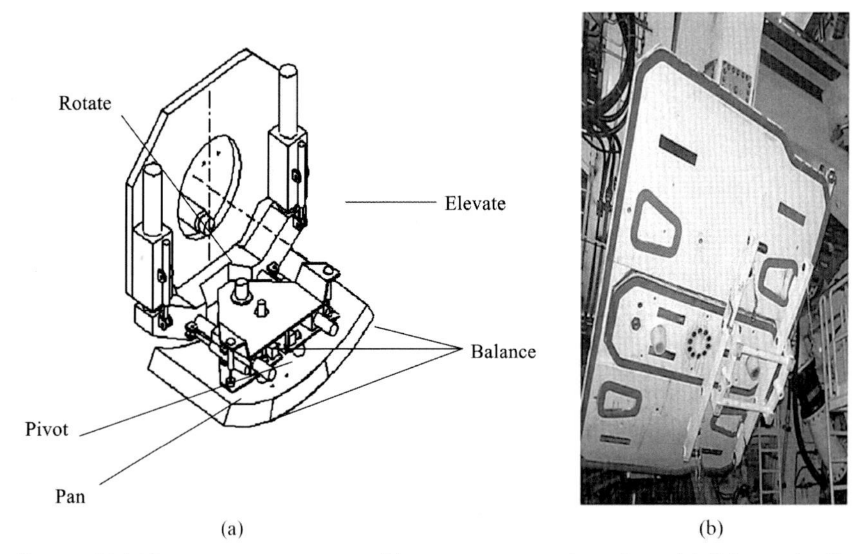

Figure 4.31 The segment erector with a vacuum suction plate. (a) Schematic diagram. (b) Photograph of vacuum suction plate.

6. Segment erector

The segment assembly system (see Fig. 4.31) was equipped with a vacuum suction plate. Its rotation part was driven by a hydraulic gear motor and a pinion and rotated on a ring gear fixed on the stator of the segment assembly system. All movement functions of the segment erector were hydraulically driven and wirelessly controlled.

7. Synchronous grouting system

According to engineering experience from similar projects and the geological conditions of this project, a single-liquid grouting material was adopted. The backup system trailer was equipped with a mortar tank and three grouting pumps. During the shield tunnelling process, the mortar in the mortar tank was pressed into the gap between the shield tail and the segment by the grouting pump. The grouting pump was a double-piston one, as shown in Fig. 4.32.

8. Gas detection device

If the slurry shield encountered biogas, except for the situation that the biogas could be drained with slurry, a large amount of biogas might accumulate at the top of the slurry chamber and might probably cause the top of soil chamber to lose stability without slurry support. In addition, when there was an abnormal situation that required personnel to enter the chamber in pressure, the biogas in the excavation

Figure 4.32 The double-piston grouting pump.

chamber would threaten their lives. Therefore the shield machine of this project required a gas detection device and a special exhaust device in the excavation chamber. The exhaust device could safely and harmlessly drain any biogas that accumulated at the top of the slurry chamber. In addition, the middle shield and trolleys needed to be equipped with automatic gas detection and alarm devices. During the passing of the biogas stratum, a hand-held gas detector was also required to conduct an all-round inspection of the shield tunnel every 2 hours, and tunnel ventilation had to be ensured continuously.

References

[1] Xiao M. Development of design technology of large diameter shield tunnel in China. Railw Stand Des 2008;(8):84–7.
[2] Li J. Analysis on the selection of Nanning metro shield. Eng Technol Res 2019;4 (04):123–4.
[3] Hong K. Key technology of shield tunnelling. China Communications Press; 2018.
[4] The Japanese Geotechnical Society. translated by Niu Q, Chen F, Xu H Investigation, design and construction of shield tunnelling. Beijing: China Architecture Press; 2008.
[5] Lv S. Shield type selection and comprehensive construction technology research. Shijiazhuang: Shijiazhuang Railway University; 2017.
[6] Spagnoli G. Electro-chemo-mechanical manipulations of clays regarding the cloggingduring EPB-tunnel driving. Germany: RWTH Aachen University; 2011.
[7] Tah JHM, Carr V. Towards a framework for project risk knowledge management in the construction supply chain. Adv Eng Softw 2001;32:835–46.
[8] Li C. Research on calculation model of key parameters of EPB shield in sandy gravel stratum. Beijing: China University of Mining and Technology; 2013.
[9] Song Y. Research on the selection and design theory of cutter wheel for shield machine. Southwest Jiaotong University; 2009.
[10] Liu J. Construction technology and experience of shield tunnel in Shenzhen metro. Tunn Constr 2012;32(01):72–87.
[11] Lin L. Research on the opening characteristics of cutter wheel of the EPB shield machine and the method of cutter layout. Changsha: Central South University; 2013.

[12] Ding Z, Zhang Z, Bai Y. Shield type selection and shield improvement application in Guangzhou metro tunnel construction. Chin J Rock Mech Eng 2002;12:1820–3.
[13] Liu X, Tang X, Li S, Lai W. Shield tunnelling and segment selection for the sharp turning section of Guangzhou Metro Line 5. Constr Technol 2010;39(09):47–9.
[14] Bian Z. Research on performance evaluation method of cutter wheel of composite EPB shield machine. Changsha: Central South University; 2013.
[15] Xie J, Tang H, Xie X. Disc cutter with pressure inside the cavity of cutter. CN202441384U, 2012.
[16] Liu M, Gan W, Zhao J. A shield disc cutter suitable for high water pressure and high earth pressure environment. CN202483584U, 2012.

Exercises

1. What principles should shield machine selection follow?
2. What are the respective characteristics and adaptability of EPB and slurry shield machines?
3. What types do shield cutters have? Please explain the requirements for their layouts on the cutterhead.
4. What types do shield cutterhead have? Please state their applications.
5. What are the components of shield muck/slurry discharging system and their characteristics and applicable conditions?
6. Please collect the shield machine information and write a report for an existing project in special geological conditions such as a sandy gravel stratum or a bedrock protruding stratum, and then propose any suggestions for the optimization of shield machine configuration.

CHAPTER 5

Structure type and design of shield tunnel lining

5.1 Types and materials of shield tunnel lining

The cylindrical lining of a shield tunnel is a permanent structure. This structure bears the surrounding earth pressure, ensure the internal space, meet the demands of applicability and durability, and must be appropriate for the tunnel construction conditions (jack thrust and backfill grouting pressure). Generally, shield tunnel lining can be categorized as single-layer lining and double-layer lining. Single-layer lining segments are assembled in the shield tail at one time, act on supporting the surrounding rock and bearing shield thrust during construction, and become permanent structures after enclosed as a ring. The double-layer lining comprise primary lining and secondary lining. The primary lining have the same structure as single-layer lining, and the secondary lining is usually used for improving structural stiffness, strengthening the waterproof and rustproof capabilities of segments and achieving internal decoration, and acting as antivibration measures.

Single-layer segment lining structure are generally adopted in shield tunnelling because their mechanical properties and durability can meet the construction and operation requirements of the tunnels. Moreover, single-layer segment lining are simple to construct, have a short project implementation period, are economical, and are controllable in terms of their waterproof effect. A section of the single-layer lining of a shield tunnel with a certain length (usually 1−1.5 m) along the axial direction (longitudinal direction) is called a ring, and the ring is partitioned into n arc blocks (i.e., segments) along the circumferential direction. To increase the construction speed of the shield tunnel, the segments are generally prefabricated parts made in factories in advance and transported to the site to be assembled into the rings during tunnel construction.

5.1.1 Structure types

There are three types of shield tunnel lining: prefabricated segmental lining, double layer lining, and extruded concrete integral linings. The

Shield Tunnel Engineering
DOI: https://doi.org/10.1016/B978-0-12-823992-6.00005-9

prefabricated lining is formed by assembling segments, which were prefabricated at factories, at the shield tail. The double layer lining, referring to a layer of integral concrete or reinforced concrete lining made inside the assembled segmental lining, prevents tunnel seepage and lining corrosion, corrects tunnel construction errors, reduces noise and vibration, and serves as internal decoration. The extruded concrete lining (ECL) adopts a set of lining construction equipment to place reinforced concrete integral lining at the shield tail as the shield advances. It is called ECL because it has an extrusion effect on shield jack thrust after pouring the concrete [1].

5.1.1.1 Fabricated segmental lining structure

Fabricated segmental lining is also called as assembling segment linings. With the development of shield machines and the demands of engineering construction, cast-in-place concrete technology cannot meet the construction requirements of shield tunnelling, so fabricated segmental linings was invented. As a result, good segment quality can be ensured during shield construction, and the construction speed is increased.

The fabricated segment lining as a main assembly component in shield construction, plays a role in resisting earth pressure, groundwater pressure, and some special loads. Fabricated segments are generally made of high-strength impermeable concrete to ensure reliable bearing capacity and waterproof performance. They are formed mainly by hermetically pouring concrete into finished segment molds.

In metro engineering, segment manufacturing costs account for a large proportion of total investment, about 45%. The structure diagram of the fabricated segmental lining is shown in Fig. 5.1.

The main functions of the fabricated segmental lining can be summarized as follows:

1. Safe enough to bear the load acting on the tunnel
2. Having functions applicable to the purpose of the tunnel
3. Having a structure type applicable to the tunnel construction conditions.

Since the segment lining supports an external load by means of main beams, panels, longitudinal ribs, joint plates, and so on, it needs to be considered separately in design. In addition, the design of the segments of the tunnel at any sharp turn as well as the seismic impact on the shield tunnel also need to be considered.

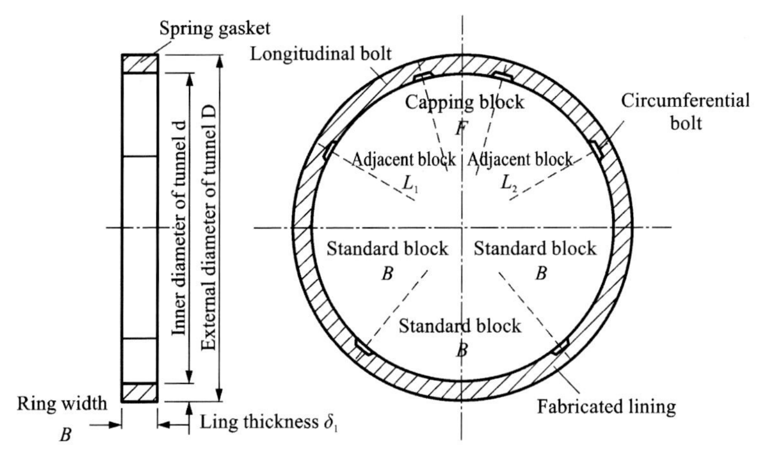

Figure 5.1 Single-layer fabricated lining ring.

5.1.1.2 Segment layer plus cast-in-place layer lining structure

When it is difficult for shield tunnelling to achieve the purpose of the tunnel only by using a primary segment lining, the design function demand can be met by pouring a concrete secondary lining inside the segment ring [2]. Although the purposes of the secondary layer lining vary with the application of the tunnel and the intentions of the management department, in most cases it is adopted in accordance with the requirement to meet the necessary functions of the tunnel.

For example, the purpose of the secondary layer lining used in the Shiziyang Tunnel of the Guangzhou-Shenzhen-Hong Kong Passenger Dedicated Line crossing the Pearl River is to prevent fire and collision. Because the tunnel was designed to be a high-speed railway tunnel, the construction of the secondary lining can also strengthen the integral stiffness of the segment lining of the weak ground section at the tunnel entrance and reduce the vibration effect brought by the train load. The main structure of the tunnel adopts the primary lining segment, and the secondary layer lining is mostly used for reinforcing segments, preventing corrosion, seepage, and shaking, correcting central line deviation; achieving a smooth inner surface; and tunnel interior decoration, shown in Fig. 5.2 [3].

The functions of the secondary lining adopted in shield tunnelling are mainly divided into two categories, shown in Table 5.1.

With the increasing application of secondary lining, the time of when to place the secondary lining has been of wide concern. The construction

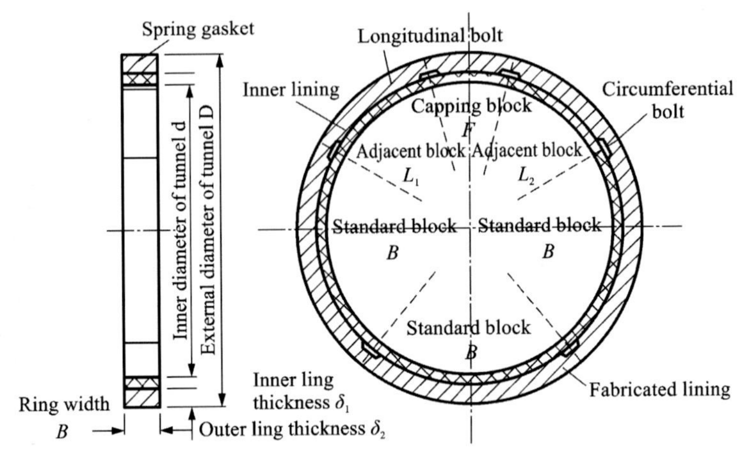

Figure 5.2 Structure of a double lining ring.

Table 5.1 Functions of secondary lining adopted in shield tunnel.

Load bearing type	Functions	Explanation
Without consideration of load bearing	Segment reinforcement	Preventing segment deformation and aging in the long term
	Segment corrosion prevention	
	Water prevention and stop	Isolating internal environment from external environment of segment lining
	Smooth inner surface of tunnel	Reducing roughness and decoration
	Serpentine correction of tunnel	Usage requirement
	Tunnel vibration reduction	Railway tunnel shall be considered
	Collision and fire prevention	Special consideration for important tunnels
With consideration of load bearing	Inner pressure bearing	Water, oil delivery tunnel, etc.
	Load from accessory structure of tunnel	Bifurcation section for water delivery tunnel; Cross passages for traffic tunnel
	Later additional load sharing	Tunnel with inner pressure
	Local load sharing	-
	Axial stiffness improvement of tunnel lining	Preventing differential settlement and resisting earthquake

time is closely related to the internal force and deformation of its own structure. The later the construction time, the more stable the segment lining stress tends to be and the smaller the surrounding rock pressure from the segments that is applied on the secondary lining structure, which is very beneficial for structure stress of the secondary lining.

In view of the impact of newly built adjacent structures (tunnels, buildings, and fill loads) after completion of the tunnel, as well as section reinforcement or differential settlement treatment at the tunnel bifurcation or dynamics during earthquakes, it is sometimes necessary to consider the secondary lining as the main structure in the tunnel design. Moreover, the integral stiffness of the tunnel or the uncertain problems in the working mechanism of the composite structure should be appropriately considered and fully demonstrated.

5.1.1.3 Extruded concrete lining structures

ECL is a construction method by which a framework mechanism is equipped at the shield tail, a plug plate is arranged between a shield shell and the framework, and during shield machine advancing, the lining structure is extruded and the fresh concrete is continuously poured with a certain pressure to form the new lining section, as shown in Fig. 5.3.

An ECL can be plain concrete or reinforced concrete. The most widely used is steel fiber–reinforced concrete (SFRC). ECL is formed at one time, having a smooth inner surface and no gap behind the lining, so there is no need for grouting, and it is particularly effective for controlling ground movement. However, ECL needs more construction equipment, including frame molds for concrete formation, systems for assembling and disassembling the frame molds, and concrete distribution systems composed of

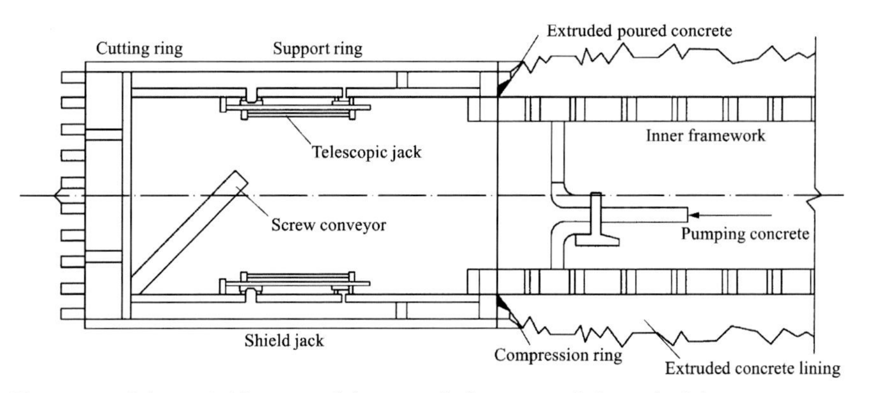

Figure 5.3 Schematic diagram of the extruded concrete lining principle.

concrete preparation trucks, pumps, valves, and pipes. Furthermore, the processes of concrete preparation, distribution, and bar erection are relatively complicated, and it is still difficult to meet the requirements in soil layers with larger permeability For these reasons, ECL is not widely used [4].

5.1.2 Materials of fabricated segment lining

The materials for segment production are mainly cast iron, steel, concrete, and the like. Except for special demands, reinforced concrete is generally used as the material of lining segments, and the concrete grade should not be less than C50. In addition, segments made of composite materials have increasingly been shown to have important advantages in tunnel water prevention, reasonable stress performance, economy, and other features.

5.1.2.1 Ductile iron

Ductile iron is processed into cast-iron segments by metal-cutting machinery after being molded and formed. The cast-iron segments have the advantages of corrosion resistance; ductility; good waterproof performance; being light in weight; having high strength, high precision, accurate appearance, and fast installation speed; and being convenient to carry and easy to be made into thin-walled structures. However, the use of the cast-iron segments consumes metal, requires a large amount of machining, and causes high construction costs. Furthermore, it is not suitable for bearing impact loads, so it has been seldom used in actual engineering.

5.1.2.2 Steel

Steel segments are mainly produced by processing section steel or steel plate welding (Fig. 5.4) and have the advantages of high strength, good ductility, and transportation and installation convenience. The precision of the segments fabricated in this way is slightly weaker than that of cast-iron materials. Because of their small stiffness, the segments are easily deformed under construction stress during construction and are used only in some special situations (e.g., temporary linings at cross passage ports of parallel tunnels) due to poor corrosion resistance.

5.1.2.3 Reinforced concrete

Reinforced concrete segments, the most commonly used segment type, are easier to process and manufacture, have good strength, are low in construction cost, and are resistant to corrosion (Fig. 5.5). With a steel mold to fabricate, the segment precision can reach ± 0.5 mm. However, in

(a) (b)

Figure 5.4 (a) Steel segment. (b) The transition area of steel and concrete segments.

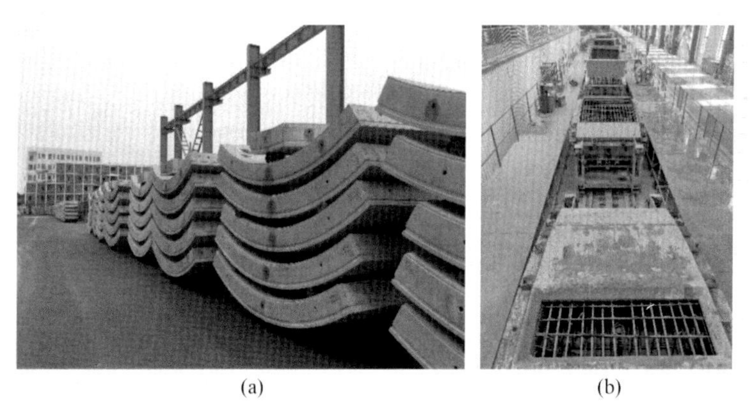

(a) (b)

Figure 5.5 Reinforced concrete segments. (a) Segment fabrication. (b) Segment stacking.

engineering practice, some shortcomings of reinforced concrete segments are gradually exposed. Reinforced concrete is heavier (can be up to 5.6T) as a result of larger thickness. During transportation and installation, the edges are easy to damage. In particular, box segments are very easily broken under the action of shield jacks. In addition, when assembled into rings, large concentrated stresses are often generated locally in the segments during bolt tightening, causing the segments to be cracked due to insufficient segment production precision and uneven end faces.

Therefore in producing segments, it is necessary to select high-strength concrete with good impermeability, strictly control the water-to-cement ratio, reasonably control the steel content and the thickness of the reinforced concrete protection layer, improve the production precision, improve the assembly quality during splicing, and adopt a staggered assembly method. In addition to conduct a comprehensive treatment in segment production processes, joint waterproof materials, measures, and so on, the segments also must have excellent impermeability.

5.1.2.4 Steel fiber reinforced concrete (SFRC)

The toughness of SFRC is much higher than that of plain concrete. Owing to the crack-resistant effect of steel fiber in concrete, the anticracking ability of concrete materials is effectively improved. The SFRC has a better softening property and better fatigue resistance than plain concrete. Furthermore, the durability of SFRC under the action of various physical factors is generally improved to different extents. The chemical resistance is also improved [5].

5.1.2.5 Composite materials

Steel shells made of steel segments are used as basic structures and then these structures are spaced with longitudinal ribs to form a simple composite segment structure (Fig. 5.6) after being filled with concrete. Compared with the steel segments, shield segments made of such materials are lighter in weight than reinforced concrete, easier to produce, have higher stiffness than the steel segments, and consume less metal than the steel segments, but have poor corrosion resistance and are complicated to process. Therefore, composite segment structure is a compromise solution (Table 5.2).

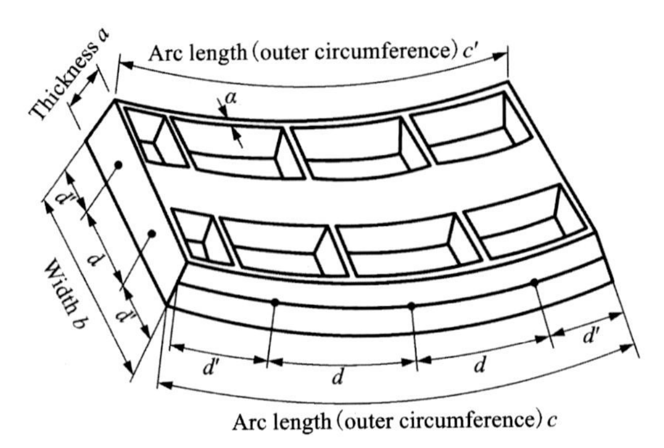

Figure 5.6 Composite segments.

Table 5.2 Segment types and advantages/disadvantages.

Segment type	Section shape	Joint mode	Features	Disadvantages
Reinforced concrete segment	Flat	Straight bolt and curved bolt	1. Low cost, most widely used 2. Good durability 3. Capable of building practical and barrier-free linings	1. Larger thickness requiring a large tunnel excavation face 2. Heavy weight, manual assistance needed during transportation and assembly, and easy to damage
Steel fiber reinforced concrete segment	Flat	Straight bolt, curved bolt, plug (or pin) and hinge joint	1. Good in tensile, bending resistance, and physical and mechanical properties, high toughness 2. Not easy to be damaged 3. Good durability and high anticracking ability	1. Complicated to produce and high steel consumption 2. Heavy weight, manual assistance needed during transportation and assembly
Cast-iron segment	Box	Straight bolt	1. High strength, good durability, and high production precision 2. Light weight, requiring smaller tunnel excavation face compared with concrete segment 3. Capable of selecting special structures for local positions bearing special loads	1. High cost 2. Difficult to weld 3. Easy to generate brittle fracture

(Continued)

Table 5.2 (Continued)

Segment type	Section shape	Joint mode	Features	Disadvantages
Steel segment	Box	Straight bolt	1. Light weight and easy to assemble and transport 2. Capable of randomly installing reinforcement materials and easy to process 3. Mostly used in small- and medium-sized shield tunnels	1. Easy to deform 2. Poor corrosion resistance 3. Complicated in processing
Composite segment	Flat	Straight bolt	1. Effective composite structure of concrete and steel plate, and smaller thickness than reinforced concrete segment	1. Poor corrosion resistance of the steel plate 2. Complicated joint structure

5.2 Segment types and features

5.2.1 Classification of segments

Segments vary according to used materials, section shapes, and joint modes. The segments can be classified in several ways [5,6].

5.2.1.1 Classification according to materials

Currently, materials for producing segments are mainly cast iron, steel, reinforced concrete, composite materials of steel and reinforced concrete, and the like. The segments made of the composite materials have increasingly shown incomparable advantages, playing important roles in tunnel water prevention, reasonable stress, economy, and so on.

5.2.1.2 Classification according to shapes

According to different construction conditions and design methods, reinforced concrete segments have different types. The segments can be roughly divided into box segment and flat segment on the basis of differences in the handhole formation sizes of the segments.

5.2.1.3 Classification according to connection modes

Segments

Joints of segments are connected by bolts. The integrity of the lining is enhanced, and the bearing capacity is high. Segments can be used for tunnels with various diameters in unstable ground strata, but the construction cost and the material cost are increased, and the workload is large.

Blocks

Blocks are assembled without a connection between them, and the rings are stabilized by the constraints provided by the surrounding strata. This assembly form is applicable to stable ground condition with a low water content. With a block assembly form, the construction speed is fast, thereby reducing the construction cost and the material cost. However, there are also some problems; for example, waterproofing and mud-proofing between joints are not well solved, causing the block ring to lose stability.

5.2.1.4 Classification according to alignment purposes

Common rings and wedge rings can be divided in accordance with the alignment purposes of the segments, as shown in Fig. 5.7. The common rings are used for normal splicing of straight sections, and the wedge rings are mainly used for ring serpentine correction and curve transition.

5.2.2 Features of segments with different section shapes

5.2.2.1 Box segments

Box segments are segments with ribbed structures due to large handholes, as shown in Fig. 5.8. Because the handholes have large cavities, the segment look like boxes composed of ribs and panels. The handholes not only facilitate penetration and tightening of bolts, but also save a lot of concrete material and reduce the weight of a single segment. Box segments are usually used in large-diameter tunnels, but they are easily cracked under the action of shield jacks if designed improperly.

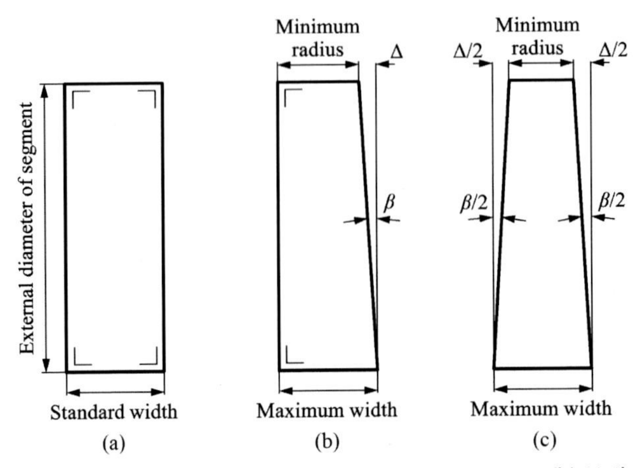

Figure 5.7 Plane projection of segment ring. (a) Common ring. (b) Unilateral wedge ring. (c) Double wedge ring.

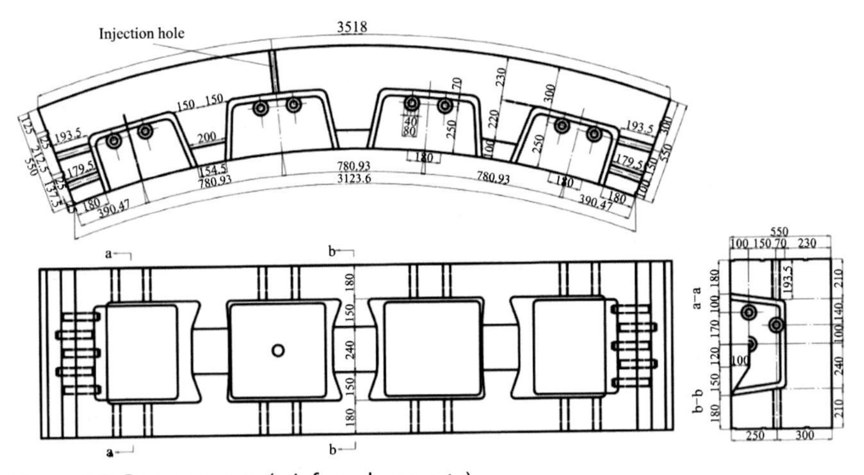

Figure 5.8 Box segments (reinforced concrete).

5.2.2.2 Flat segments

Flat segments refer to segments with curved plate structures due to small handholes, as shown in Fig. 5.9. Since only a few segment concrete section are weakened, the flat segment requires a smaller thickness than the box segment under the same load. In addition, because of its simple shape, the process of steel mold production, steel bar erection, and demold are all convenient. This type of segment has high resistance to the jacking force of the shield jack and also is low in tunnel ventilation resistance during normal operation.

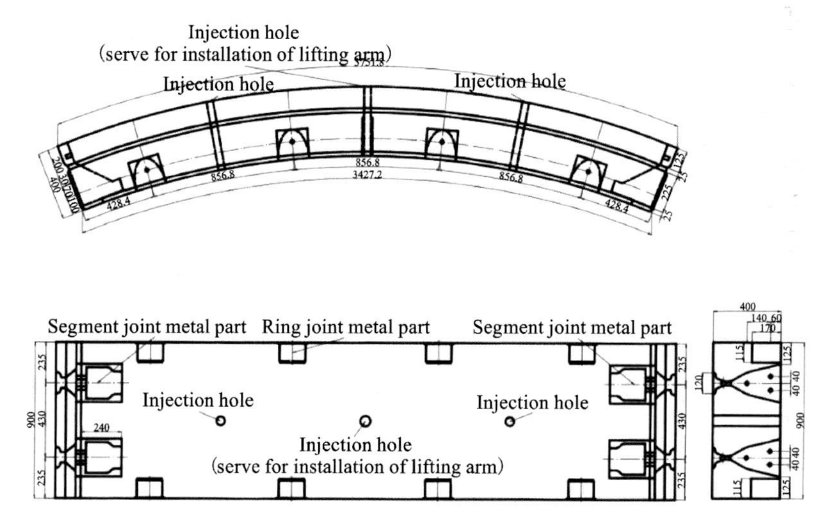

Figure 5.9 Flat segments (reinforced concrete).

5.2.2.3 No bolt segments

Flat segments without bolts are usually called blocks. The fabricated reinforced concrete block lining is a structure type commonly used in circular tunnels. Generally speaking, blocks can be assembled into stable and rigid lining rings without bolt connection, because the design shape can make protruding and recessed parts of the blocks of two adjacent rings coincide with each other and can be clamped against the contact faces of the block slopes. When the slopes are in close contact, the compression shear force at the end of each block in each ring can be transferred to the central part of each block in the adjacent ring. Compared with the segment, the block has the advantages that the reinforcement ratio can be reduced under the same load due to the larger section. In some cases, there are also plain concrete and stone blocks without steel bars.

The blocks are generally applicable to stable ground condition with low water content. Because of the partition requirements of tunnel linings, rings (more than three pieces) assembled from the blocks become an unstable multihinged circular structure. After the lining structure is deformed within a limit range, the rings are stabilized because of the constraint of surrounding strata on lining rings. However, the waterproof and mud-proof problem of joints between the blocks and between adjacent rings must be satisfactorily solved, or sharp deformation of the ring will occur making the ring lose stability and causing engineering accidents.

There are many types of blocks, which mainly depends on surrounding geology and construction conditions. The blocks can be divided into two

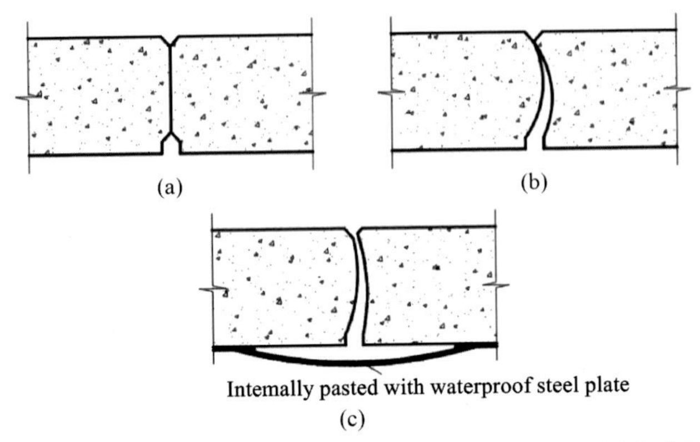

Figure 5.10 Ball joints of blocks. (a) Plane joints. (b) Ball joints. (c) Ball joints pasted with waterproof steel plate.

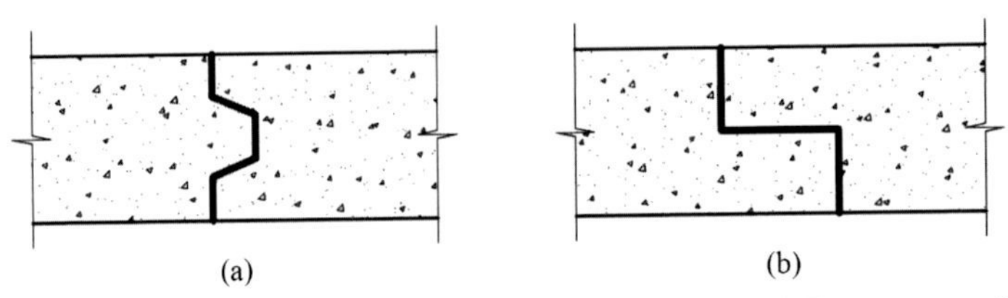

Figure 5.11 Tongue-and-groove joints of block structures. (a) Tongue-type joint. (b) Groove-type joint.

types, ball type and tongue-and-groove type, in accordance with connecting modes of the blocks, as shown in Figs. 5.10 and 5.11, respectively.

5.2.3 Structure characteristics of segments

5.2.3.1 Standard blocks, adjacent blocks, and capping blocks

The ring is usually composed of three types of segments, namely, several (numbered x) standard blocks (also called A-type blocks) that are axially and equally partitioned, one capping block (also called K-type blocks) for capping, and two adjacent blocks (also called B-type blocks) at the two sides of the capping block, as shown in Fig. 5.12.

There are two types of capping blocks for assembly: radial insertion and axial insertion (Fig. 5.13). With the characteristic of a certain taper in the radial direction, the radial insertion-type capping block is inserted along the radial direction from the inner side of the tunnel. With the characteristic of a taper in the axial direction, the axial insertion-type capping block is inserted along the axial direction of the tunnel. Under normal circumstances

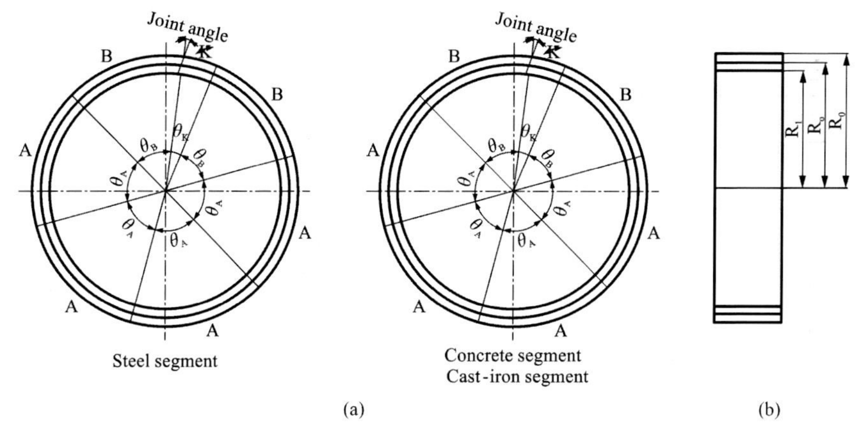

Figure 5.12 Ring composition (a) Cross section. (b) Side face.

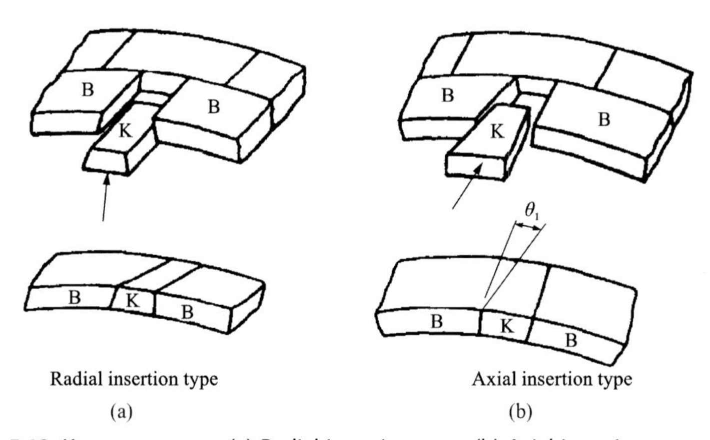

Figure 5.13 K-type segment. (a) Radial insertion type. (b) Axial insertion type.

the circumferential length of the capping block should be smaller than the lengths of the standard block and the adjacent block.

5.2.3.2 Ring partition number

The partition number n (i.e., the segment number) of the ring is equal to $(x + 2 + 1)$, where x is related to the external diameter R_0 of the ring. The larger the R_0, the greater the x. In addition, the partition number should not be too high in terms of composition, production, and assembly speed of the ring during shield tunnelling, while more partition numbers are better in terms of transportation and convenience. For railway tunnels, the x can be 3~8 and generally be 3~5; for waterways and power and communication cable tunnels, the x is 2~4.

5.2.3.3 Segment width and height

The segment width is the segment size measured along the axial direction (longitudinal direction) of the tunnel, as shown in Fig. 5.14. In terms of transportation and assembly convenience, curved construction, and shield tail length, a small segment width is better, but in terms of reducing the segment manufacturing cost, reducing joint numbers, and increasing the construction speed, a large segment width is better. Comprehensive consideration should be given to the section size of the tunnels in combination with the actual construction experience, economy, and construction conditions. The segment width varies with the section size of the tunnel, generally within a range of 300~1500 mm. For concrete segments, the width is mostly 900–1500 mm.

For box segments, the segment height is about the sidewall height of the segment, and for flat segments, it is called segment thickness (as shown in Fig. 5.13). The selection of the ratio of the segment thickness to the external diameter of the ring depends on various factors, such as soil conditions, overlying strata thickness, construction loads, application of tunnels, and segment construction conditions. In terms of current application, the segment thickness is generally about 4% of the external diameter of the segment. For large-diameter segments, especially box segments, it is mostly about 5.5%.

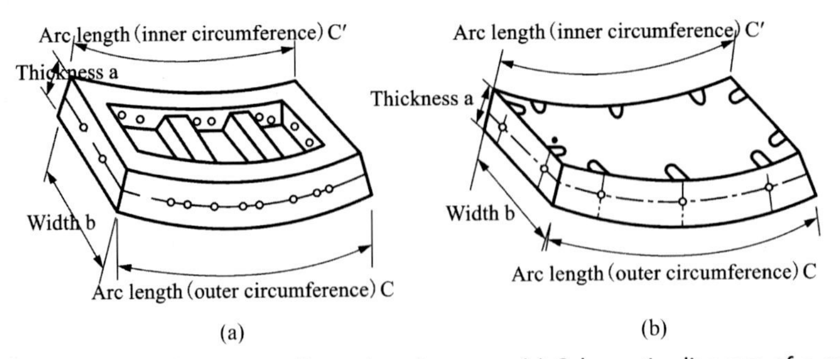

Figure 5.14 Typical segment dimension diagrams. (a) Schematic diagram of a reinforced concrete box segment. (b) Schematic diagram of a reinforced concrete flat segment.

5.2.3.4 Deformation joint

For single layer circular tunnel linings, factors such as differential settlement and earthquake impact probably cause longitudinal deformation of the tunnel which may lead to cracks of the lining. For double layer lining, owing to the difference between internal temperature and external temperature of the lining, and differences in ages, elastic modulus, and residual shrinkage, the postpoured concrete cannot contract freely but is affected by eccentric tension, which often leads to cracks. Therefore, in saturated water-bearing weak ground, deformation joints need to be arranged at certain distances along the longitudinal direction of the tunnel lining structure, especially at the junction of the tunnel and the shaft, where deformation joint is preferred due to the large differences in stiffness. Common deformation joints are shown in Fig. 5.15.

From theoretical analysis and engineering practice, the structure of a deformation joint should meet the following three main requirements:

1. The joints can adapt to a certain range of linear deformation and angular deformation;
2. The joints are capable of releasing longitudinal bending stress, but during construction and operation stage, they can transfer the shearing force during thereby controlling the differential shear deformation;
3. The joints are capable of avoiding the entry of water before and after deformation.

5.3 Segments assembly and waterproofing

5.3.1 Segment connection forms

Segment joints are mainly connected in two directions: circumferential joints for connecting the adjacent segmental rings and longitudinal joints for connecting adjacent segments. When segments are used to form lining

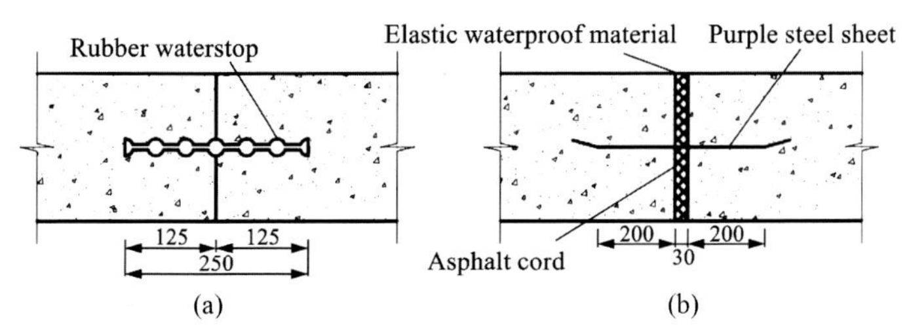

Figure 5.15 Type of deformation joint. (a) Deformation joint with rubber waterstop. (b) Deformation joint with asphalt cord waterproofing and purple steel sheet.

rings in shield tunnels, the inner surface is generally smooth so that it is beneficial for backfill grouting. At present, the bolt connection is the standard segment joint connection with holes reserved in the concrete segments, but the exposed bolts and joints will be rusted. Therefore it is better to use fewer bolts for segment assembly [7].

Early segment joints were mostly rigid, and it was believed that the more rigid the connection was, the safer the structure was. Through long-term experimentation, practice, and research, this traditional concept gradually gave way to flexible structure. The segment connection form has also experienced a transition from rigid connection to flexible connection.

5.3.1.1 Longitudinal connection

Bolted joint

1. Straight bolt. The straight bolt has good bending stiffness under the condition of reaching a certain bolt pretightening force, has good bending stiffness and engineering application effect, is simple to produce and saves materials, but also needs to undergo bolt head treatment. The existing method is to add fast-hardening concrete and a plastic sealing cover. The straight bolt is widely used because of the economic and reasonable construction cost and installation convenience. Longitudinal and circumferential joints between segments are connected by straight bolts, a single row of bolts being arranged at positions distant from inner side of the lining by $h/3$ (h refers to segment thickness), as shown in Fig. 5.16. It is necessary to consider the matching between the bolts and segment ribs; that is, the bolts should first enter plastic states, and at the same time, various construction impacts should be considered to avoid using too small bolts.

   Straight bolt connection passes through the segment where has been strongly reinforced with steel end ribs and steel box, as shown in

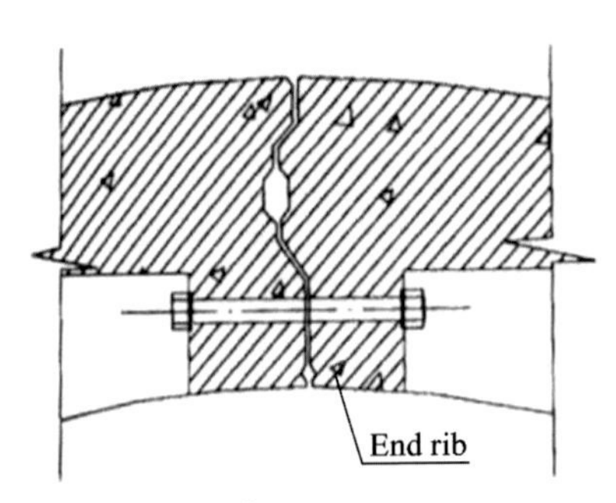

Figure 5.16 Straight bolt connection form.

Fig. 5.16. Such connection form has been used in the Shanghai subway. Although this connection can shorten the bolt length and reduce steel consumption, the steel consumption of the end rib plate is high. In addition, precision is often not guaranteed during steel box embedment, so reinforced concrete end ribs are now usually used.

2. Bent bolt. Bent bolt connection is mostly used for flat segments. Originally, bent bolt was arranged for strengthening the stiffness of joints. As bent bolt requires small bolt handholes and is low in section weakening, thereby the structure have large stiffness in bearing positive and negative bending moments, as shown in Fig. 5.17. Compared with the straight bolt, the bent bolt connection is high in construction cost, and the joints are easily deformed. Moreover, if the bent bolts and segment steel molds cannot be strictly processed in accordance with the designed radian and precision, it will be very difficult to conduct bolt perforation during construction, especially during staggered assembly. The assembly work will take a lot of time and labor to conduct bolt perforation. Experiments show that the bent bolt joint is more easily deformed than the straight bolt joint. Furthermore, engineering practices show that the bent bolt is not convenient for construction and consumes more materials. Therefore, bent bolt has been gradually replaced by the straight bolt. However, the bent bolt is more commonly used in situations with high bolt joint flexibility and large calculated bending force.

Nonbolted joint

Nonbolted joint means that, in addition to the shape change of the block, there are no other attachment joints to strengthen the connection. Generally, there are ball forms and tongue-and-groove forms. The tongue-and-groove joint has the advantages of simple installation, fast construction speed, and low construction cost (Fig. 5.18). The protruding tongue-and-groove block

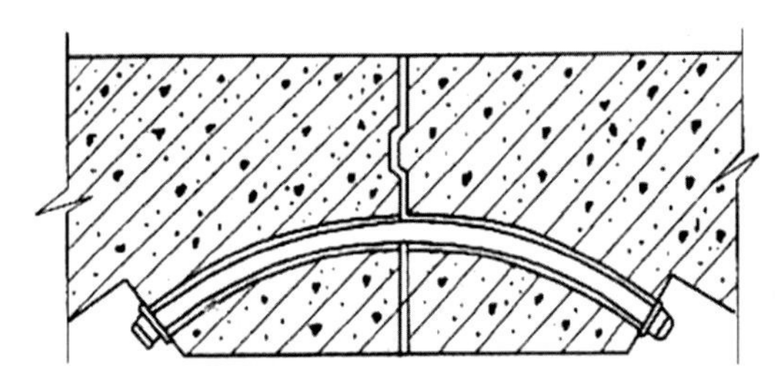

Figure 5.17 Bent bolt connection form.

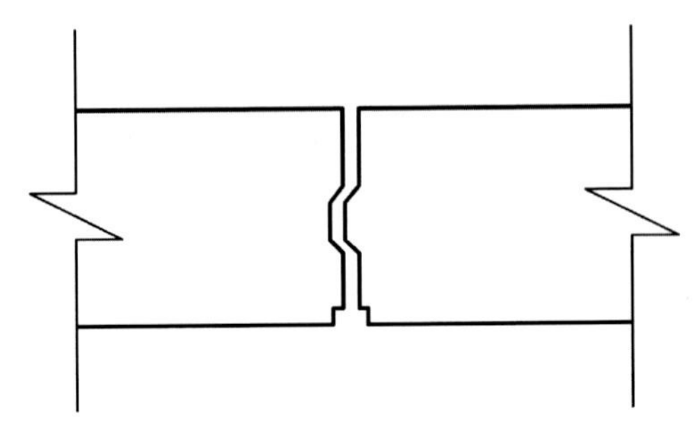

Figure 5.18 Schematic diagram of a tongue-and-groove joint.

can provide high shear capacity for tunnels with longitudinal differential settlement. However, because of its small bending stiffness, tongue-and-groove joint cannot resist to the bending moment caused by an external load and must rely on the resistance of the surrounding rock to achieve its own stress balance. Therefore when there is a bending moment caused by differential settlement in the radial direction, the tongue-and-groove joint will be greatly expanded, which is not beneficial for the integrity and waterproofing of the segment.

Pin connection

There are many forms of pin connection. As shown in Fig. 5.19 and Fig. 5.20, some are arranged in the circumferential direction, and some are radially inserted. Some of the used connectors are pre-embedded with components, and some are installed during assembly. Their structural functions are to strengthen component connection, prevent two sides of the joint from moving relative to each other, and bear the shearing force on the joint, so they are sometimes called shear pins. The segments that use pin connection are simple in shape and consistent in section strength, the inner wall of the formed tunnel is smooth and flat and is easy to clean, and no additional lining is needed unless special demands. Compared with bolt connection, the pin connection has the advantages of high shear resistance, large bending stiffness, connection convenience, no handhole, and savings of labor and time. It can be said that pin connection is a good form using less material and simpler process to achieve a very good connection effect. However, in insertion and installation, a gap is often formed between wedge blocks, which is not beneficial for waterproofing. Therefore this type of joint is often used in water supply and drainage engineering, and it is often used with a thin layer of secondary lining.

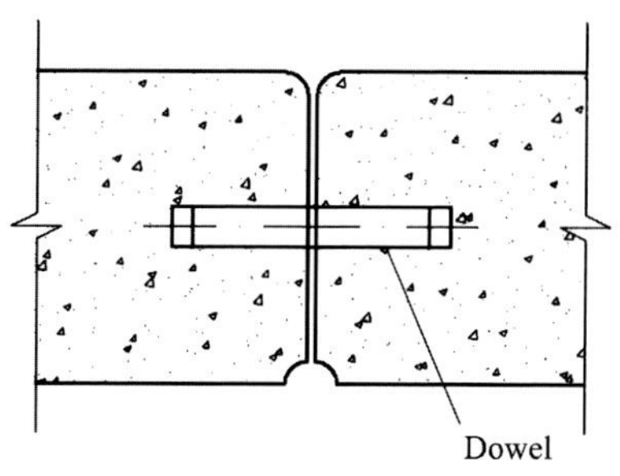

Figure 5.19 Dowel plug.

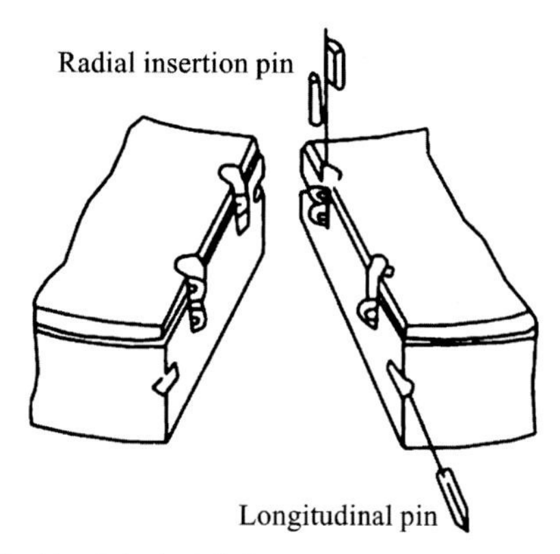

Figure 5.20 Longitudinal/radial pin joint.

5.3.1.2 Circumferential connection

During shield advancing in the uneven strata, the degree of influence and disturbance to the strata is also different along the longitudinal length of the tunnel because of complicated and changeable construction processes. Fabricated tunnel linings built in the strata will cause longitudinal deformation of the tunnel, and as a result of various other factors, the sealing conditions of fabricated tunnel lining joints may not be good, which will cause water leakage and mud leakage at the bottom of the tunnel, resulting in longitudinal differential settlement of the tunnel and mutual rubbing of circumferential faces. In addition, as the tunnel passes through buildings or existing tunnels, the longitudinal deformation of the tunnels will be caused as a result of large eccentric load caused by the jacking force of the jack and instantaneous moving of uniform loads during shield advancing. Therefore, for the design of tunnel lining ring joint structures, to meet satisfactory engineering quality requirements, it is necessary to take effective circumferential connections.

Tongue-and-groove connection for circumferential joints

Tongues and grooves of different geometric shapes are arranged along the circumferential face of the segmental lining ring so that protruding and recessed parts of adjacent segmental ring can be coincided and connected with each other and can be clamped with each other by means of tongue-and-groove contact of the segments (blocks). When the segments (blocks) are produced with accurate sizes and close contact can be ensured, then the effect of transferring a longitudinal force between rings can be achieved. To improve waterproofing, a circle of groove pasted with a rubber tape sealing pad is designed on the ring ribs and longitudinal ribs of the segment. The depth of the groove and the thickness of the rubber tape, and the rebound size of the adhesive tape after compression must conform to the expected waterproof requirement after deformation. Such tongue-and-groove connection of joints is applicable to ring connection structures of the fabricated tunnel segment (block) linings in various geological ground condition.

Tongue-and-groove/bolted connection of circumferential joints

To better strengthen the circumferential connection between lining rings, a certain number of longitudinal bolts are arranged along the longitudinal axis of the tunnel on the basis of the circumferential tongue-and-groove connection with protruding and recessed parts. The longitudinal bolts arranged on ring joints further improve the ability of the tunnel lining structure to resist the longitudinal differential settlement. Positions of bolt holes are shown in Fig. 5.21.

Smooth plane connection of circumferential joints

For circular tunnel linings in stable and impermeable stratum, circumferential connection joints of the blocks are often connected by smooth

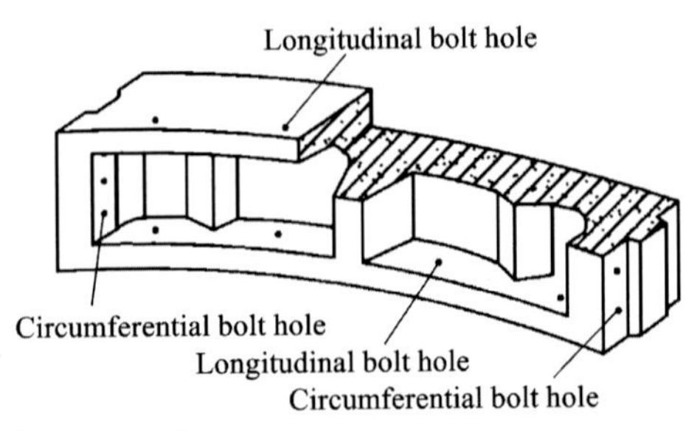

Figure 5.21 Bolt hole position layout.

planes. Since the blocks are wedge blocks along the radial direction of the tunnel, the blocks are extruded and stressed under the action of external pressure, just like the built stone arch bridge, and the lining structure is still stable, safe, and reliable.

5.3.1.3 Novel special joint for segment

Second-generation One-Pass segment

The second-generation One-Pass segment is obtained through further improving the One-Pass segment originally developed by Japanese Obayashi for single-layer lining segment (Fig. 5.22) [8]. Such segment does not require bolt tightening and is applicable to rapid construction. Its joint between blocks uses a horizontal pin joint, and its joint between rings uses a Push-press fastening joint. The horizontal pin joint is composed of a group of C-shaped pieces and H-shaped pieces with supports. The H-shaped pieces are arranged on one of the C-shaped pieces, connected with another one of the C-shaped piece after sliding, and then fastened by the counterforce of the supports. The Push-press fastening joint is a pin connection joint using a wedge, in which a bolt pin on the male joint side is connected to the female joint side to complete fastening.

Sliding-pin quick joint segment

The sliding-pin quick joint segment is a type of segment that was developed by Maeda Corporation in Japan [8] (see Fig. 5.23). It is labor-saving and automatic during splicing and has a completely smooth inner arc surface. Its joint between blocks uses a sliding-pin joint, and its joint between

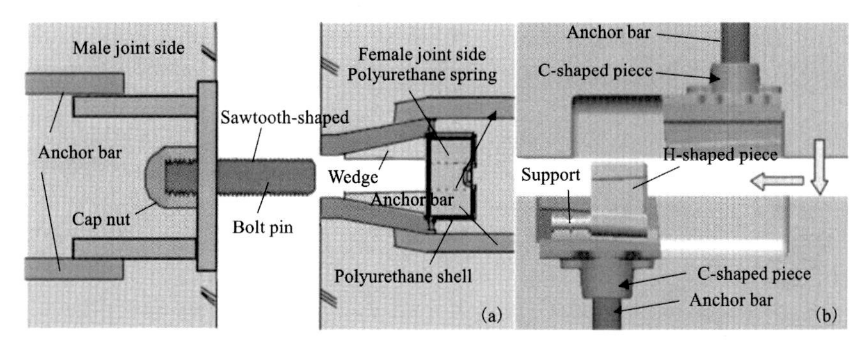

Figure 5.22 Schematic diagram of second-generation One-Pass segment joints. (a) Horizontal pin joint. (b) Push-press fastening joint. *From Jin S.T.E.C. International vision: big PK of segment quick connector in Japan[EB/OL]; 2019, p. 3–23. <https://mp.weixin.qq.com/s/5m41wGQfKbdGumzh1Kf6eQ>.*

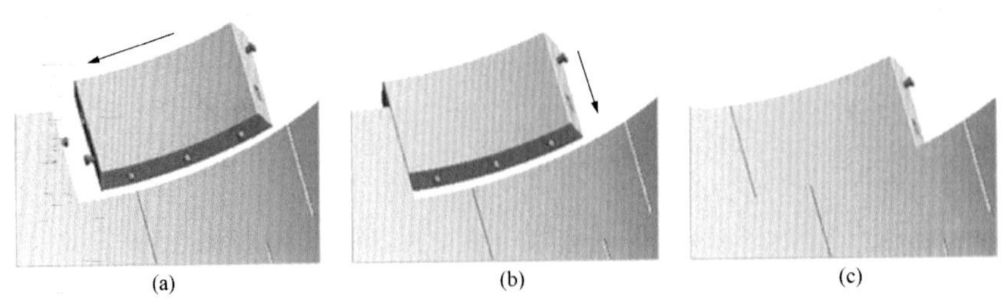

Figure 5.23 Schematic diagram of sliding-pin joint segments. (a) Align splicing faces of segments. (b) Axially insert. (c) Complete fastening. *From Jin S.T.E.C. International vision: big PK of segment quick connector in Japan[EB/OL]; 2019, p. 3–23. <https://mp.weixin.qq.com/s/5m41wGQfKbdGumzh1Kf6eQ>.*

rings uses a sun quick joint. A wedge and polyurethane rubber serving as a counterforce piece are arranged in a C-shaped piece of the joint between the blocks, which not only can be used for generating a joint fastening force, but also can adjust and control the fastening force by changing the hardness of the polyurethane rubber.

The sliding-pin sun quick joint is a joint that can be fastened only by push, without a retightening operation. In addition, the joint can expand and contract to a certain extent through elastic deformation of a steel bar, can follow expansion of the seam between rings during an earthquake, and has favorable seismic performance. Since the first construction and application of the sliding-pin quick joint segment in 2003, it has been widely used in many tunnels for utility, railway, road, sewer, and other projects. The external diameters of the segments range from 2550 to 13,700 mm.

Slide locking joint

The slide locking joint is a joint suitable for fast construction without bolt tightening, that was developed by Maeda Corporation as well (Fig. 5.24) [8]. This joint does not require bolt tightening and meets the high stiffness requirements. The joint structure comprises a male joint side composed of bolts and elastic members and a female joint side with a bolt sliding groove. Furthermore, slide locking joints are divided into single-bolt types and multibolt types, which can be applied in the range from small and medium diameter to large diameter.

The segment using the slide locking joint can be fastened through axial sliding. Currently, slide locking joints have been used in underground railways, utility tunnels, and other projects. Compared with

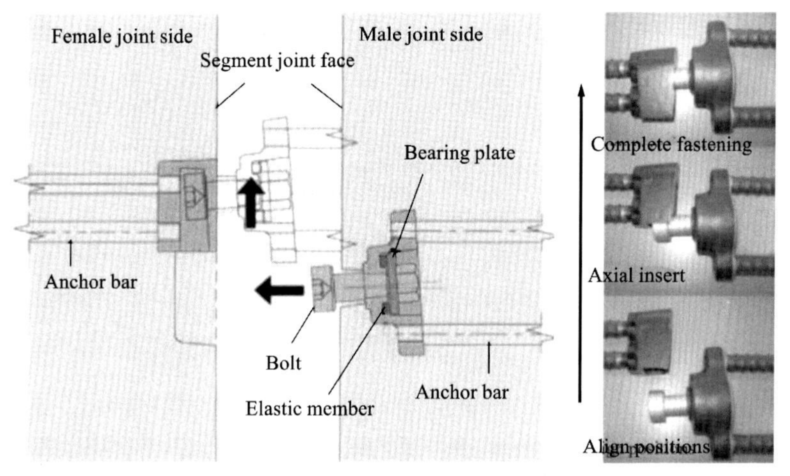

Figure 5.24 Details of the slide locking joints and its sliding direction. *From Jin S.T.E.C. International vision: big PK of segment quick connector in Japan[EB/OL]; 2019, p. 3–23. <https://mp.weixin.qq.com/s/5m41wGQfKbdGumzh1Kf6eQ>.*

a conventional bolt joint, the slide locking joint has the advantages of short assembly time of about 5 minutes per block, and high roundness and fewer cracks on rings.

5.3.1.4 Worm wheel joint

The worm wheel (WW) segment is a novel segment that was developed by Kajima Construction in Japan and uses a worm wheel joint to solve the problem that the joint structure of a conventional segment with a smooth inner arc face cannot be tightened (see Fig. 5.25).

The WW joint has the characteristic that a tightening force can be introduced to the joint through a transmission mechanism after the segment has been assembled, and the tightening force can be adjusted. A C-shaped piece is arranged at the female joint side, and the male joint side is fastened through pulling a bolt at the male joint side by rotating a WW nut (see Fig. 5.26) [8].

5.3.2 Segment assembly modes

The fabricated segment lining structure of the shield tunnel is formed by assembling a plurality of arc segments into rings and connecting the

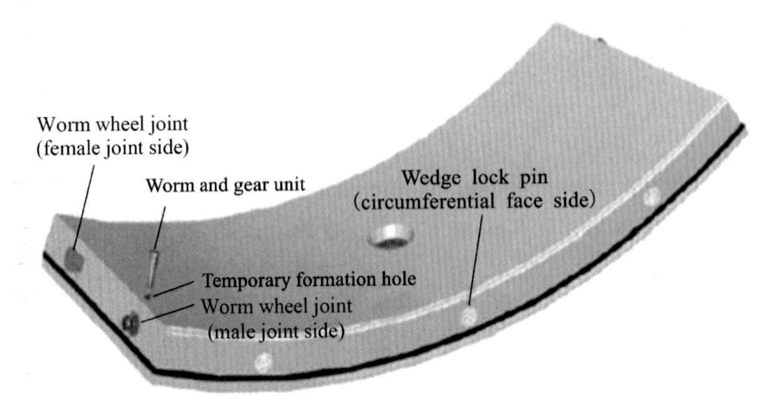

Figure 5.25 Schematic diagram of a worm wheel joint segment. *WLP*, wedge lock pin. *From Jin S.T.E.C. International vision: big PK of segment quick connector in Japan [EB/OL]; 2019, p. 3–23.* <*https://mp.weixin.qq.com/s/5m41wGQfKbdGumzh1Kf6eQ*>.

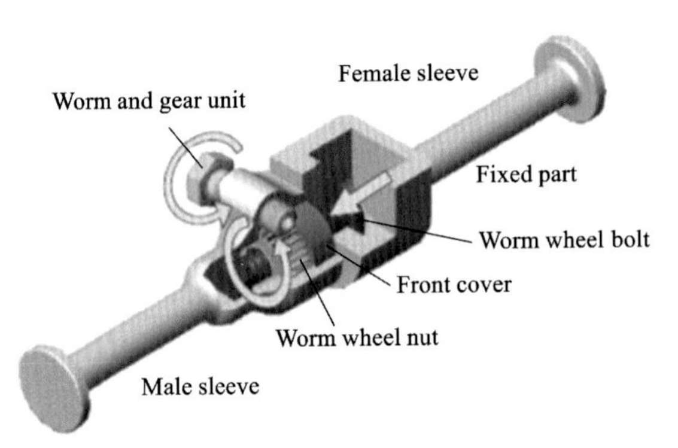

Figure 5.26 Schematic diagram of a worm wheel joint. *From Jin S.T.E.C. International vision: big PK of segment quick connector in Japan[EB/OL]; 2019, p. 3–23.* <*https://mp.weixin.qq.com/s/5m41wGQfKbdGumzh1Kf6eQ*>.

rings one by one. The connection between the segments and between the rings was described earlier in this chapter. There are two modes of assembling the segments: straight-joint assembly and staggered-joint assembly [7]. In straight-joint assembly, the longitudinal joints of all rings are in a straight line; in staggered-joint assembly, the longitudinal joints of two adjacent rings are staggered. Both of the assembly modes are shown in Fig. 5.27.

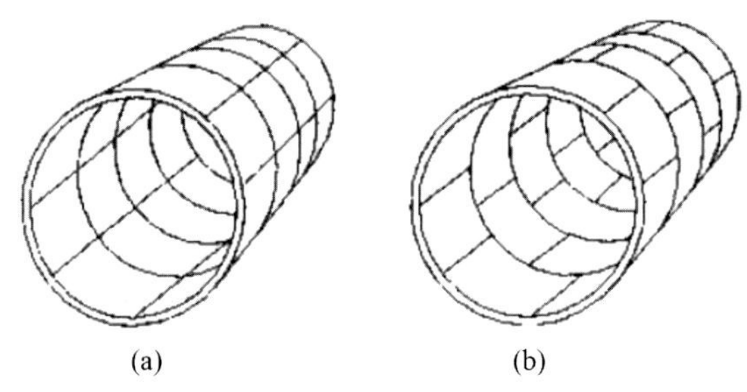

Figure 5.27 Schematic diagram of a joint assembly. (a) Straight-joint assembly. (b) Staggered-joint assembly.

5.3.2.1 Straight-joint assembly

Straight-joint assembly is closely related to the flexible lining design theory. The flexible lining design theory is based on the idea that a tunnel lining is not a separate structure subjected to a definite load. The lining design is not only a structure problem, but also a stratum and structure interaction problem. The lining acts as a thin film for redistributing the load to the surrounding strata rather than an arch ring of supporting the load from the stratum. Therefore by properly adjusting the relative stiffness between the tunnel lining and the surrounding strata, the ground deformation can be adjusted, and the mutual effect between the lining and the strata is changed, thereby being beneficial for lining force. Generally, there are three ways to achieve a flexible lining structure:

1. Reducing the stiffness of the fabricated joint.
2. Increasing the number of the joints.
3. Reducing the lining thickness.

In the straight-joint assembly mode, there is no mutual transmission of shear force and bending moment between adjacent rings. The deformation of the longitudinal joint is free from constraint of adjacent segment bodies, but only relies on bolt connection of the joint. Therefore, the lining structure can obtain better flexibility by the straight-joint assembly mode. The straight-joint assembly is applicable to formations with good ground condition, which can fully mobilize the strata resistance and make the design of the lining structure more reasonable as well as ensuring that the lining meets the application requirement.

In China, the shield tunnel that has mostly adopting straight-joint assembly is the Shanghai subway shield tunnel, which is related to the fact that the shield tunnels in Shanghai have been used to employing such assembly mode for a long time. Rich engineering experience has been accumulated, and segment production precision and construction precision are also very high.

5.3.2.2 Staggered-joint assembly

Staggered-joint assembly is mainly realized by the following two modes: (1) Adjacent rings are staggered with each other, and rectangular segment blocks are placed in this staggered-joint mode. The longitudinal joints of the tunnel are not in the same straight line because of staggering of the ring joints of two adjacent segments, so the stiffness of the whole tunnel lining is improved. (2) Nonrectangular segment blocks are used. Since the circumferential joint face of the nonrectangular segment block is not parallel to the axis of the tunnel, the longitudinal joints of the tunnel lining formed by assembling this type of segments are not definitely in a straight line, thereby forming staggered joints.

The advantages of staggered-joint assembly are that the stiffness distribution of the ring joint can be uniform, the longitudinal stiffness of the segment lining is improved, and the deformation of the joint and the whole structure is reduced. In the staggered-joint assembly mode, there are only three joints intersections between circumferential and longitudinal ring joints. Compared with straight-joint assembly, the ring and longitudinal joints cross to each other, which makes it easy to conduct waterproof treatment on the joints. In the staggered-joint assembly mode, the joint deformation is slight which is also conducive to waterproofing.

Because of its advantages, the staggered-joint assembly mode is frequently used in shield tunnels (such as river-crossing tunnels and undersea tunnels) that have high waterproofing requirements or shield tunnels in soft soil areas, where it can achieve higher space stiffness, controlling the deformation of the tunnel lining structure and ensuring normal application of the tunnel.

5.3.3 Combination forms of segment lining ring

Three segment lining ring combination forms are commonly used for shield tunnels in China, as shown in Table 5.3, and each method can fit the plane alignment curve and for correcting deviation. The first two forms are often used in metro shield tunnels, and the third is often used in large-diameter shield tunnels and in some urban metro lines [9].

5.3.4 Segment lining waterproofing

During tunnel construction in water-bearing formations, the fabricated segment required for shield tunnel construction should not only meet the requirements of structural strength and stiffness, but also meet the requirements of tunnel lining waterproofing. Otherwise, there will be problems of structure damage,

Table 5.3 Segment lining ring combination forms.

Methods	Features
Standard lining ring, left-turn lining ring and right-turn lining ring	Except construction of alignment rectification, the standard lining ring is mostly used in straight section; the curved section can adopt the standard lining ring and a left-turn lining ring and right-turn lining ring combination to fit the alignment. The combination method is simple to construct and operate.
Left-turn lining ring and right-turn lining ring combination	The line is fit by the left-turn lining ring and right-turn lining ring combination. Since each ring is wedge-shaped, construction operation is relatively troublesome during assembly.
Universal wedge segment (universal segment)	A type of wedge segment is used for fitting straight line section, curved section and conducting construction rectification. When the segment is arranged, the lining ring needs to be twisted at various angles, the capping block is sometimes located in the lower half part of the tunnel, and segment assembly is relatively complicated. It is used in some of urban metro tunnels in China.

equipment corrosion, lighting reduction, poor traffic safety, and unattractive appearance due to water leakage.

5.3.4.1 Segment lining waterproofing design principle

The waterproofing design of shield tunnel segment rings should follow the principle of prevention first, blocking second, multiple measures sealing at joints, and comprehensive treatment. The segments are made of high-precision steel molds, taking self-waterproofing segment structure as a basis with the focus on joint waterproofing to ensure integral waterproofing of the tunnel.

5.3.4.2 Segment lining waterproofing standards and waterproofing measures

The waterproofing grade and the corresponding waterproofing measures are determined in accordance with the functions of tunnels and related specifications, as shown in Table 5.4.

Table 5.4 Waterproof grades and corresponding waterproofing measures.

Waterproof grade	Waterproof concrete	High-precision segment	Joint waterproofing				Outer coating of segment	Exposed metal piece corrosion protection	Cathode protection	Secondary layer lining
			Elastic gasket	Caulking	Sealant injection	Screw hole sealing ring				
I	Mandatory	Mandatory	Mandatory	Mandatory	Optional	Mandatory	Optional	Mandatory	Mandatory	Suggested
II	Mandatory	Mandatory	Mandatory	Mandatory	Optional	Mandatory	Optional	Mandatory	Mandatory	Optional
III	Mandatory	Mandatory	Mandatory	Mandatory	—	Suggested	Optional	Mandatory	Mandatory	—
IV	Optional	Optional	Mandatory	Optional	—	—	—	Mandatory	Mandatory	—

From GB50157-2013, Code for design of metro. Beijing, China: China Architecture & Building Press; 2013.

5.3.4.3 Segment self-waterproofing

Reasonably choosing concrete grades and impermeability

When the tunnel lining segment works under groundwater pressure in a water-bearing formation, the lining structure is required to have certain impermeability to prevent groundwater infiltration. The higher the concrete strength, the higher the tensile strength, and the better the anticracking ability, which is beneficial for segment waterproofing. However, simply improving the impermeability grade and strength of concrete in the segment will lead to an increase in unit cement consumption, an increase in hydration heat, and an increase in shrinkage, thereby causing cracks. Therefore, the concrete grade, impermeability and additives for the segment must be reasonably chosen.

Improving the production precision of the segments

For waterproofing of fabricated reinforced concrete segments, the use of high-precision steel molds to improve the production precision of the segments is an important link, according to the practices of tunnel construction worldwide. If the production precision does not meet the requirement, coupled with cumulative errors of lining assembly, it will lead to larger initial gaps in the lining joint. At the same time, inadequate production precision of the segment can easily cause lining breakage and collapse during shield advancing and bright water leakage. In the past, the application and development of reinforced concrete segments in water-bearing formations were often restricted, and the main reason was that the production precision of the segment was insufficient, thereby causing tunnel leakage.

Outer waterproof coatings of segments

Reinforced concrete segments buried in the ground will cause concrete structure to be damaged and steel bars to be corroded as a result of groundwater enriched with sulfate radicals and chloride ions. These harmful substances permeating the pores of the lining by chemical reaction can cause corrosion. Generally, in areas with large buried depth or obvious corrosiveness, the segment must be coated with an enhanced waterproof and corrosion-resistant outer coating.

5.3.4.4 Segment joint waterproofing

Segment joint waterproofing includes elastic gasket waterproofing between segments, caulking waterproofing between adjacent segments inside the tunnel, and grouting into joints when necessary. Elastic gasket waterproofing is the most important and reliable measures and is the focus of joint waterproofing. It is important to consider the impact of segment production precision on joint waterproofing. The joint width is usually required not to be larger than 1.5 cm.

Elastic gasket waterproofing

1. Circumferential joint gasket. This gasket must have enough bearing capacity and elastic restoring force, withstand and uniformly distribute the jacking force of the shield jack, and prevent segment from being broken. The gasket still maintains favorable elastic deformation performance under the reciprocating action of the jacking force of the jack.
2. Longitudinal joint gasket. This gasket has lower bearing capacity than the circumferential joint gasket, can fill and align initial gaps in the longitudinal joints of the segments, and has a certain buffer and suppression effect on local concentrated stress.

Caulking waterproofing and plugging

In caulking waterproofing, caulking materials are arranged in caulking grooves inside the segments to form a secondary defense line of joint waterproofing. The shape of the caulking groove should take into account the need for no filler falling and running at vault caulking, so it is usually designed to be narrow in the mouth and wide in the middle. The caulking material should have the characteristics of good watertightness, corrosion resistance, telescopic restorability, short hardening time, small shrinkage, and construction convenience. The materials that meet these requirements include epoxy, polysulfide rubber, and urea resin materials.

Several main caulking sealing waterproof design structures are shown in Fig. 5.28.

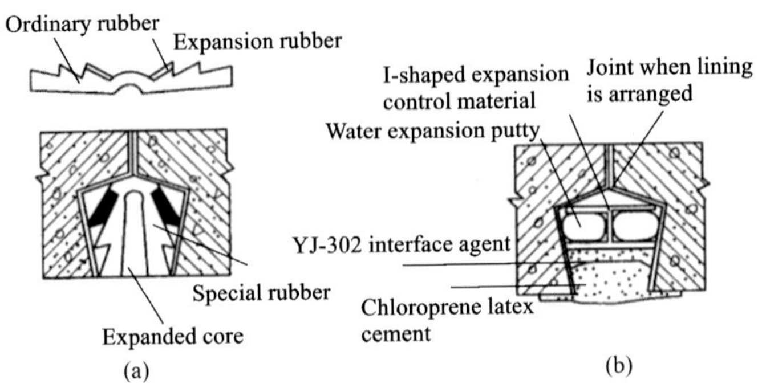

Figure 5.28 Caulking sealing waterproof design structure views. (a) Type I. (b) Type II.

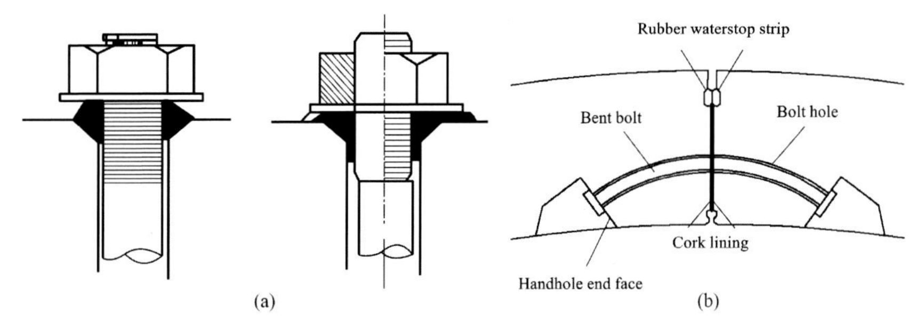

Figure 5.29 Joint bolt hole waterproofing. (a) Front view and profile view. (b) Waterproofing at joint.

Miscellaneous

Other additional measures can be used in segment joint waterproofing, such as grouting at joints, plugging at bolt holes and grouting holes, and plugging of cracks on segment surfaces, which can be adopted depending on cases.

5.3.4.5 Bolt hole waterproofing

The sealing and waterproofing of bolt holes is another important part of segment waterproofing. The segment joint is equipped with waterproof gaskets outside the bolt hole. If the gasket has a good waterstop effect, no leakage occurs at the bolt hole. However, in parts with gasket failure or poor segment assembly precision, water may leak from the segment joint and the bolt hole. Therefore special waterproofing treatment is necessary for the bolt hole when necessary.

The most widely used method is to make the bolt hole in one side of a cavity rib into a taper shape. An annular sealing ring made of synthesis resin, natural rubber, and polyethylene is padded between a bolt washer and a cavity rib face. When the bolt is tightened, the sealing ring is extruded and deformed to run into the screw hole and fill in between the bolt and the pore wall to achieve a waterstop effect, as shown in Fig. 5.29.

5.4 Load calculation for shield tunnel

5.4.1 Load hypothesis

Various loads to be considered in the design of shield tunnel linings should be hypothesized on the basis of different conditions and design methods, and these loads should be combined according to the application purpose of the tunnel to calculate the internal forces of the lining sections [10].

The loads can be divided into the three classifications such as primary loads, additional loads and special loads, as shown in Table 5.5.

The primary load is a basic feature that must be considered during tunnel design. Fig. 5.30 shows the schematic diagram of the primary load distribution. The additional load is an imposed load during construction or after completion and is a load considered according to tunnel applications, construction conditions, and surrounding environments. The special

Table 5.5 Load classifications.

Primary loads	Vertical and horizontal earth pressure Water pressure Self-weight of the lining Overlying surface load Subgrade reaction
Additional loads	Internal load, e.g., water pressure in intake tunnel, etc. Construction load Earthquake effect
Special loads	Effects of parallel configured tunnel Effects of adjacent construction Others

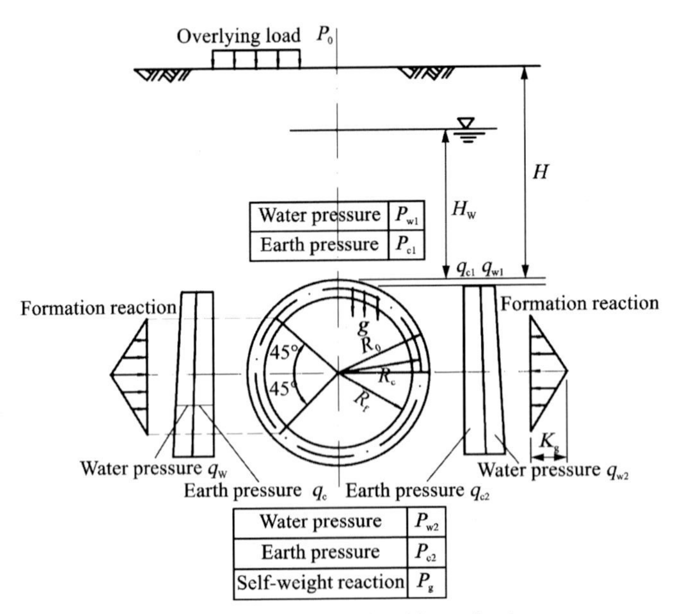

Figure 5.30 A schematic diagram of primary load hypothesis.

load is a load considering ground conditions and special tunnel applications. The above load components is usually regarded as a static load in the design. Construction loads such as the thrust force of the jack, back grouting pressure, and earthquake effects all belong to instantaneous loads. The primary load hypothesis is shown in Fig. 5.30.

5.4.2 Load calculation methods

5.4.2.1 Vertical and horizontal earth pressure

In considering the earth pressure acting on tunnels, the vertical and horizontal earth pressure is used for determining design load and is unrelated to tunnel deformation. In addition, the earth pressure at the bottom of the tunnel is regarded as a subgrade reaction. There are two methods for calculating the earth pressure; one is to consider water pressure as a part of earth pressure, and the other is to calculate water pressure and earth pressure separately. Generally, the former is applicable to cohesive soil, and the latter is applicable to sandy soil. But for hard clay and consolidated silt with good self-stability, water pressure and earth pressure are suggested to be separated. When water pressure and earth pressure are calculated integrally, the wet unit weight is used for strata above the groundwater level, and the saturated unit weight is used for strata under the groundwater level. When water pressure and earth pressure are calculated separately, the wet unit weight is used for strata above the groundwater level, and the buoyant unit weight is used for strata under the groundwater level.

Vertical earth pressure

The vertical earth pressure is regarded as a uniformly distributed load acting on the top of the lining, and its value should be determined on the basis of overburden thickness of the tunnel, the section shape of the tunnel, the external diameter, and the surrounding rock conditions.

1. Full overburden pressure theory

   In considering the earth pressure acting on the tunnel for a long time, the arching effect of soil cannot be obtained if the overburden thickness is smaller than the external diameter of the tunnel, so the total overburden pressure is adopted:

$$P_{el} = P_0 + \sum \gamma_i H_i + \sum \gamma_j H_j \tag{5.1}$$

   where P_{el} is the vertical earth pressure at the vault of the lining; P_0 is the additional load; γ_i is the unit weight of the ith stratum above the groundwater level; H_i is the thickness of the ith stratum above the groundwater level; γ_j is the unit weight of the jth stratum under the groundwater level; H_j is the thickness of the jth stratum under the groundwater level; and H is the overburden thickness with $H = \sum H_i + \sum H_j$.

Full overburden pressure theory, without consideration of stress transfer between soil bodies, it is applicable to tunnels in soft stratum with shallow covered depth, but it is not applicable to stratum with harder soil or buried deeply.

2. Terzaghi theory on loosening earth pressure

When the overburden thickness is greater than the external diameter of the tunnel, it is more likely to generate an arching effect above the tunnel, and the loosening earth pressure in the ground arch may be considered in design calculation (Figs. 5.31 and 5.32). In sandy soil, loosening earth pressure is often used when the overburden thickness is larger than $(1 \sim 2)$ D (D is the external diameter of the segment ring). In clay the loosening earth pressure is often used in hard clay when the overburden thickness is larger than $(1 \sim 2)$ D. For moderately consolidated clay and soft clay, it is common to regard the full overburden gravity above the tunnel as the earth pressure.

The Terzaghi formula is usually used for calculating the loosening earth pressure. Generally speaking, when the vertical earth pressure adopts the loosening earth pressure, an additional lower limit of the earth pressure is set in condensation of loads during construction and changes after

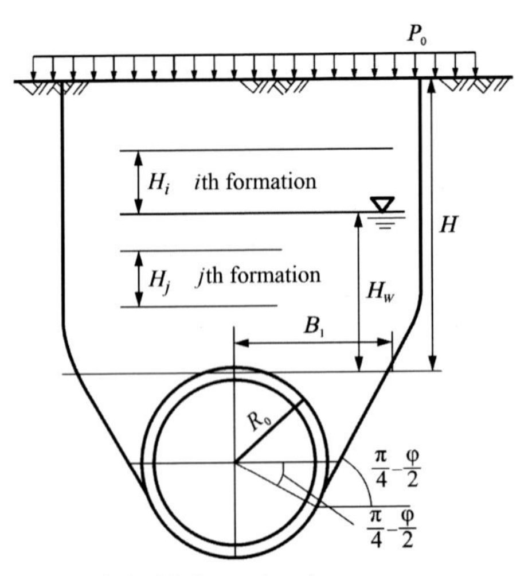

Figure 5.31 Calculation model of full overburden pressure.

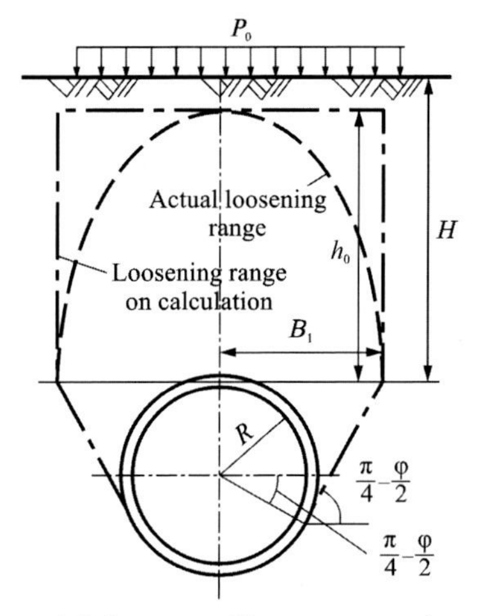

Figure 5.32 Calculation model diagram of loosening earth pressure.

completion of the tunnel. Although the lower limit of the vertical earth pressure varies with the usage purpose of the tunnel, it is generally regarded as the earth pressure value equivalent to twice the overburden pressure with thickness of tunnel external diameter. When the stratum is in multiple layers, calculation is conducted by taking the dominant layer in the strata as a basis and assuming the ground as a single layer, or the loosening earth pressure is calculated in a multilayer state:

1. $P_0 > \gamma H$

$$h_0 = \frac{B_1\left(1 - c/B_1\gamma\right)}{K_0\tan\varphi}\left(1 - e^{-K_0\tan\varphi\cdot H/B_1}\right) + \frac{P_0}{\gamma}e^{-K_0\tan\varphi\cdot H/B_1} \tag{5.2}$$

$$\sigma_v = \frac{B_1\left(\gamma - c/B_1\right)}{K_0\tan\varphi}\left(1 - e^{-K_0\tan\varphi\cdot H/B_1}\right) + P_0e^{-K_0\tan\varphi\cdot H/B_1} \tag{5.3}$$

$$B_1 = R_0\cot\left(\frac{\pi/4 + \varphi/2}{2}\right) \tag{5.4}$$

where h_0 is the height of loosening earth in the ground arch; B_1 is the half width of the bearing ground arch; σ_v is the Terzaghi vertical earth pressure; K_θ is the ratio between horizontal earth pressure and vertical earth pressure (lateral earth pressure coefficient); φ_0 is the internal friction angle; P_0 is the overlaying surface load; γ is the unit weight of earth; c is the cohesion of the earth; φ is the internal friction angle of the earth; R_0 is the external diameter of the lining ring; and H is the overburden thickness.

2. $P_0 < \gamma H$

$$h_0 = \frac{B_1\left(1 - c/B_1\gamma\right)}{K_0\tan\varphi}\left(1 - e^{-K_0\tan\varphi\cdot H/B_1}\right) \tag{5.5}$$

$$\sigma_v = \frac{B_1\left(\gamma - c/B_1\right)}{K_0\tan\varphi}\left(1 - e^{-K_0\tan\varphi\cdot H/B_1}\right) \tag{5.6}$$

The concept of vertical soil strips in the Terzaghi formula is obtained through experiments. However, the breakage of the overlying rock-soil mass is no longer along the vertical side face of the whole rock-soil pillar as the buried depth is increased.

3. Platts theory of earth pressure

The Platts theory of earth pressure holds that a parabolic balanced arch (Platts pressure arch) will be formed above when a tunnel is excavated in a loose medium, and the pressure acting on the tunnel structure is the gravity of the loose rock-soil body in a natural arch:

$$\sigma_v = \gamma h = \gamma\frac{B}{f} \tag{5.7}$$

$$B = R(1 + 1/\sin(45°\text{-}\varphi/2)) \cdot \tan(45°\text{-}\varphi/2) \tag{5.8}$$

where B is the half width of the bearing arch; f is the Platts coefficient; φ is internal friction angle of the rock-soil body; R is the radius of the structure; and h is the height of the bearing arch.

The Platts formula considers the influence of factors such as span of bearing arch above the tunnel ($2B$), the heightof the tunnel ($2R$), and the internal friction angle of the rock-soil body (φ); assumes that the balanced arch can be naturally formed above the tunnel vault; and is simple and clear in physical concept. However, the Platts formula has limitations as a result of no consideration of the influence of the buried depth of the tunnel and the cohesion of surrounding stratum.

4. Bill Bowman formula

Bill Bowman thinks that the slip fracture face can be substituted by two straight lines at an angle of $(45° + \varphi/2)$ with the horizontal line when a shallow tunnel is excavated in loose rock-soil body, and a holding effect of the two side soil bodies is also taken into account. The Bill Bowman formula is

$$q_b = \gamma H\left[1 - \frac{H\tan\varphi\tan^2\left(45^\circ - \frac{\varphi}{2}\right)}{2B_1} - \frac{c\left(1 - 2\tan\varphi\tan\left(45^\circ - \frac{\varphi}{2}\right)\right.}{B_1\gamma}\right] \tag{5.9}$$

where q_b is the surrounding ground pressure obtained by calculation of the Bill Bowman formula.

Compared with the Terzaghi formula, the Bill Bowman formula is more applicable to shallow tunnels with poor surrounding ground conditions. If the surrounding ground conditions have not been properly selected, the ground pressure may even be negative, which is inconsistent with engineering practice.

5. Recommended formulas in the Chinese *Code for Design of Railway Tunnel*

The formulas in Chinese *Code for Design of Railway Tunnel* [11] are based on a large number of case study statistics of tunnel collapse. The vertical pressure above the tunnel is

$$q = \gamma h_q = 0.45 \times 2^{s-1} \times \gamma w \tag{5.10}$$

where h_q is the equivalent load height; s is the surrounding stratum grade; γ is the unit weight of surrounding stratum; w is the width influence coefficient, with $w = 1 + \mathrm{I} \times (B - 5)$, where i is an ratio coefficient of the surrounding stratum pressure when B is increased or decreased by 1 m. Taking the vertical uniformly distributed pressure of the surrounding stratum when B is equal to 5 m as a reference; i is equal to 0.2 when B is smaller than 5 m; and i is equal to 0.1 when B is larger than 5 m.

The calculation formula for surrounding earth pressure of shallow tunnels in the Chinese *Code for Design of Railway Tunnel* takes into account the influence of the internal friction angle (φ) of the surrounding stratum and the surrounding stratum grade classification (s). Although the calculation formula for surrounding stratum pressure of deep tunnels does not directly consider the influence of the internal friction angle and the cohesion (c) of the surrounding stratum, these two factors are reflected in the surrounding stratum classification. Furthermore, the Chinese *Code for Railway Tunnel* is based on 1046 collapse samples of more than 400 railway tunnels in China, the collapse sample data of the tunnel itself reflects the influence of geological conditions and engineering factors on the surrounding earth pressure. Thus the influence factors considered by the calculation methods given in the Chinese *Code for Design of Railway Tunnel* are relatively comprehensive.

Horizontal earth pressure

The horizontal earth pressure is regarded as a distributed load acting on two sides of the lining horizontally from the vault to the bottom of the tunnel. Its value is calculated on the basis of a vertical earth pressure coefficient and a lateral earth pressure coefficient.

When it is difficult to obtain the subgrade reaction, the static earth pressure coefficient under the construction condition can be considered as the lateral earth pressure coefficient. When the subgrade reaction can be obtained, an active earth pressure coefficient is used as the lateral earth pressure coefficient, or the above static earth pressure coefficient is taken as a basis with appropriate reduction. The value of the lateral earth pressure coefficient to be adopted in the design calculation should be between the values of the static lateral earth pressure coefficient and the active lateral earth pressure coefficient. Generally speaking, the lateral earth pressure coefficient can be determined according to a relationship with a subgrade reaction coefficient within a range shown in Table 5.6.

Table 5.6 Lateral earth pressure coefficient (λ) and subgrade reaction coefficient (k).

Water-earth pressure calculation	Types of soil	λ	k (MN/m^3)
Separate calculation of water pressure and earth pressure	Dense sandy soil	0.35–0.45	30–50
	Medium-dense sandy soil	0.45–0.55	10–30
	Loose/slightly dense sandy soil	0.50–0.60	0–10
	Consolidated clay	0.35–0.45	30–50
	Hard/hard plastic clay	0.45–0.55	10–30
	Plastic clay	0.50–0.65	0–10
Combined calculation of water pressure and earth pressure	Plastic clay	0.55–0.65	5–10
	Soft plastic clay	0.65–0.75	0–5
	Fluid plastic clay	0.70–0.85	0

Formulas (5.11)–(5.13) give a method for determining the horizontal earth pressure:

1. $H_w \geq 0$

$$\begin{aligned} P_{e1} &= P_0 + \sum \gamma(H - H_w) + \sum \gamma' H_w \ (\textit{If}\ H < 2D) \\ P_{e1} &= \sum \gamma(h_0 - H_w) + \sum \gamma' H_w \ (\textit{If}\ H \geq 2D \textit{ and } h_0 > H_w) \\ P_{e1} &= \sum \gamma h_0 \ \ (\textit{If}\ H \geq 2D \textit{ and } h_0 < H_w) \\ q_{e1} &= \lambda(p_{e1} + \gamma' t/2) \\ q_{e2} &= \lambda\left[p_{e1} + \gamma'(t/2 + 2R_c)\right] \end{aligned} \tag{5.11}$$

2. $0 > H_w \geq -2R_c$

$$\begin{aligned} P_{e1} &= P_0 + \sum \gamma H (\text{If}\ \ H \leq 2D) \\ P_{e1} &= \sum \gamma h_0 (\text{If}\ \ H \geq 2D) \\ q_{e1} &= \lambda(p_{e1} + \gamma t/2) \\ q_{e2} &= \lambda[p_{e1} + \gamma(-H_w) + \gamma'(t/2 + 2R_c + H_w)] \end{aligned} \tag{5.12}$$

3. $H_w < -2R_c$

$$\begin{aligned} P_{e1} &= P_0 + \sum \gamma H (\text{If}\ \ H < 2D) \\ P_{e1} &= \sum \gamma h_0 (\text{If}\ \ H \geq 2D) \\ q_{e1} &= \lambda(p_{e1} + \gamma t/2) \\ q_{e2} &= \lambda[p_{e1} + \gamma(t/2 + 2R_c)] \end{aligned} \tag{5.13}$$

where P_{e1} is the vertical earth pressure at the top of the lining; P_0 is the overburden load; H_w is the height from vault to groundwater level; γ' is the buoyant unit weight of soil; h_0 is the height of the bearing arch; q_{e1} is the horizontal earth pressure at the top of the lining; q_{e2} is the horizontal earth pressure at the bottom of the lining; H is the overburden thickness; D is the external diameter of the lining ring; R_c is the centroidal radius of the lining segment; t is the thickness of the lining segment; and λ is the lateral earth pressure coefficient.

5.4.2.2 Water pressure

In normal cases the water pressure acting on the lining is hydrostatic pressure (Fig. 5.33). However, to simplify calculation, the water pressure can be uniformly distributed vertical water pressure along the tunnel width whose value is equal to the hydrostatic pressure at the vault and bottom of the tunnel, respectively, and the horizontal pressure linearly changing from the vault to the bottom of the tunnel.

The gravity of water due to tunnel excavation acts as buoyancy on the lining. If the resulting force of the vertical earth pressure at the vault and

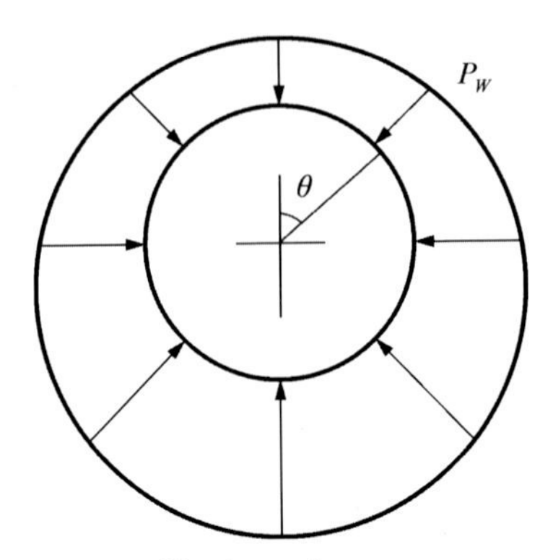

Figure 5.33 Schematic diagram of hydrostatic pressure.

the lining self-weight (static load) is larger than the buoyancy, their difference will be the vertical earth pressure (subgrade reaction) acting on the bottom of the tunnel. When the sum of the vertical load (water pressure excluded) acting on the top of the lining and the self-weight of the lining is smaller than the buoyancy, the earth pressure generated by resistance in the strata at the top of the lining resists the buoyancy. Such a phenomenon may occur in a strata that has a small overburden thickness and high groundwater level and is prone to liquefaction. If it is difficult to generate resistance equivalent to the buoyancy at the top, the tunnel will float up, so measures such as a secondary lining must be taken to increase the gravity of the tunnel or loading on the ground surface.

Formulas (5.14)–(5.20) represent methods of calculating water pressure, buoyancy, and verticaluniformly distributed earth pressure at the bottom of the tunnel. If hydrostatic pressure is adopted, the water pressure at each point on the segment lining is

$$P_w = \gamma_w[(H_w + t/2) + R_c(1 - \cos\theta)] \tag{5.14}$$

where γ_w is the unit weight of water; H_w is the height from vault to groundwater level; t is the thickness of the lining segment; and R_c is the centroidal radius of the segment lining.

A combination of vertical uniformly distributed pressure and horizontal pressure which is linearly increased with depth is adopted:

$$\begin{aligned} P_{w1} &= \gamma_w H_w \\ P_{w2} &= \gamma_w[H_w + (t + 2R_c)] = \gamma_w(H_w + 2R_0) \end{aligned} \tag{5.15}$$

where P_{w1} is vertical water pressure at the top of the lining; H_w is height from vault to groundwater level; and P_{w2} is vertical water pressure at the bottom of the lining.

If hydrostatic pressure is adopted, the buoyancy is

$$F_w = \gamma_w \pi R_c^2 \tag{5.16}$$

If the combination of vertical uniformly distributed water pressure and horizontal pressure linearly changing with depth is adopted, the buoyancy is

$$F_w = 2R_c(P_{w2} - P_{w1}) = 4\gamma_w R_0 R_c \tag{5.17}$$

The uniformly distributed earth pressure at the bottom of the tunnel is

$$P_{e2} = P_{e1} + \pi g - F_w / 2R_c \tag{5.18}$$

where P_{e1} is vertical earth pressure at the top of the lining; P_{e2} is vertical earth pressure at the bottom of the lining; and g is calculated according to Eq. (5.19).

5.4.2.3 Self-weight of segment

The self-weight of the segment lining is a vertical load acting on a centroidal line of the cross section of the tunnel and is calculated according to Eq. (5.19):

$$g = W / (2\pi R_c)$$

$$g = \gamma_c \times t \text{ (\textit{If the section is rectangular})} \tag{5.19}$$

where γ_c is unit weight of segment reinforced concrete; and W is longitudinal gravity per linear meter of lining.

5.4.2.4 Ground overload

Overload reference values are

$$P_0 = 10kN/m^2 \cdot \text{(Highway vehicle load)}$$

$$P_0 = 10kN/m^2 \cdot \text{(Railway vehicle load)}$$

$$P_0 = 10kN/m^2 \cdot \text{(Gravity of building)}$$

5.4.2.5 Subgrade reaction

The range, size, and direction of subgrade reaction must be determined for calculating the internal force of a lining member. There are usually two types of subgrade reaction, which are determined in combination of design calculation methods; one is the reaction independent of subgrade displacement, and the other is the reaction dependent on subgrade displacement. The former is used as the reaction balanced with a given load and is presupposed to be uniformly distributed, and the latter is considered to be related to the displacement in the lining subgrade and is proportional to the displacement of the stratum. The proportional factor is defined as a subgrade reaction coefficient, and the value of the proportional factor depends on the toughness of the surrounding strata and the dimension of the lining.

The subgrade reaction is the product of the subgrade reaction coefficient and the displacement of the segment lining and is determined by the toughness of the surrounding rock and the stiffness of the lining. The stiffness of the segment lining depends on the stiffness of the segment and numbers and types of joints.

In the common calculation method, the subgrade reaction, having nothing to do with the subgrade displacement, in the vertical direction is a uniformly distributed reaction force balanced with the vertical load. On the other hand, the subgrade reaction, acting on the side face of the tunnel in the horizontal direction, is generated with deformation of the lining in the surrounding strata direction. Therefore within the range of a central angle ± 45 degrees, the horizontal subgrade reaction is distributed in a triangular shape and at the tunnel horizontal diameter level is the maximum value point. In the calculation, the subgrade reaction is proportional to the tunnel horizontal deformation towards the surrounding strata, as shown in Fig. 5.34:

$$q = k\delta \tag{5.20}$$

where k is subgrade reaction coefficient, determined according to geological conditions and in consideration of a relation with the lateral earth pressure coefficient λ; and δ is the displacement value.

Aside from the common calculation methods, another method to determine the subgrade reaction is in a way that considers the interaction between the segment ring and the subgrade through a subgrade spring model. This method considers the subgrade reaction to be a reaction force generated when the segment is deformed towards the subgrade side.

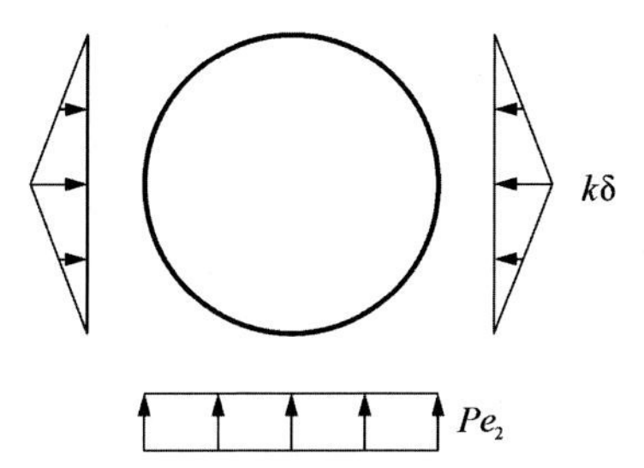

Figure 5.34 Subgrade reaction calculation model.

Different from conventional calculation methods, the subgrade reaction is regarded as a reaction force generated by the displacement of the ring toward the subgrade side when the subgrade spring model is selected to describe the interaction between the segment ring and the subgrade. The circumferential spring subgrade model (Fig. 5.35a) is used in some countries in Europe and United States, while the partial spring subgrade model (Fig. 5.35b) is used in Japan mostly. The later is also widely used in China. From the perspective of application examples, most examples consider only radial springs to be effective, and there are also some examples considering tangential springs for design. However, in most cases the ground spring coefficient is determined with reference to the subgrade reaction coefficient of the conventional calculation methods.

5.4.2.6 Internal load

The internal load is a load acting on the inner side of the lining after the tunnel has been completed. The load varies with the application purpose of the tunnel.

For the internal load of railway vehicles acting on the bottom of the lining (except in extremely soft ground), it can be considered that the load is directly supported by the ground around the lining because of hardening of the back grounding material, which has a little effect on the lining. However, the load concentrated inside the tunnel, such as the support of the base plate and a hanging load in the tunnel, will affect the strength and deformation of the segment lining. Therefore the load should be set according to the actual situation and should be considered during segment design. In terms of the tunnel bearing internal water pressure, including the secondary lining, an appropriate structure model must be selected, and the water pressure and earth

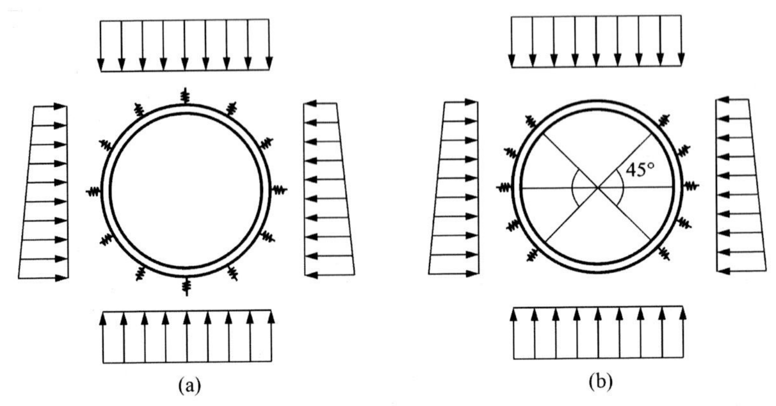

Figure 5.35 Schematic diagram of spring subgrade model. (a) Circumferential spring subgrade model. (b) Partial spring subgrade model.

pressure that act on the segment must be carefully selected considering load history and structure stress history to ensure the safety of the tunnel structure.

5.4.2.7 Load during construction

The construction load is an collective term of temporary loads acting on the segment ring from the beginning of segment assembly to the hardening of tail gap grouting, including the thrust of the jack, the pressure of synchronous grouting, static pore water pressure, and operation load during segment assembly. The construction load varies with the ground conditions and construction methods, and it is difficult to accurately calculate its numerical value. However, in the segment design, comprehensive, reasonable and full consideration should be given to the construction load.

5.4.2.8 Impact of earthquakes

The impact of earthquakes on underground structures can generally be considered according to the following methods. Generally, the linear unit weight (weight per unit length) of the tunnel is smaller than that of tunnel excavated soil. Therefore when the seismic force acts, the inertial force acting on the surrounding stratum is larger than the inertial force acting on the tunnel. However, test results found that when the overburden thickness of the tunnel reaches a certain level, it could be considered that the tunnel and the stratum basically produce same vibration. Therefore when the tunnel is located in the a homogeneous stratum and the overburden thickness is large enough, it can be considered that an earthquake will have less impact on the tunnel. Compared with other tunnels, the shield tunnel has reduced stiffness due to joints, thus shied tunnel has a better performance in following the stratum displacement. Under the following conditions, the earthquake has a great impact on the tunnel and cannot be ignored:

1. Changes of lining structures at position of underground joints and shaft connection parts (changes in types of segments, presence or absence of secondary lining, etc.)
2. In a weak stratum
3. Abrupt changes in ground conditions such as geology, overburden thickness, and bed depth
4. Sharply curved alignment
5. In loose saturated sandy layers and cases probably generating liquefaction.

If a detailed and thorough engineering geological survey is carried out at the tunnel planning stage, it is also a method to choose a correct tunnel alignment as a aseismic measure. On the other hand, it can also be considered to make enlarged tunnel section to give space for repair after an earthquake damage.

5.4.2.9 Impact of adjacent construction

In recent years there have been more examples of shield tunnels or structures constructed nearby the completed tunnel. Since there is new tunnel construction nearby, the soil around the previous tunnel is disturbed, and the load acting on the lining is changed. When the soil is disturbed heavily and the load is changed significantly, there will be a great impact on the tunnel lining. If such a situation was anticipated when the tunnel was designed, its impact should be taken into account in the design. If it was not considered, the lining should be reinforced and protected according to actual damage condition of the lining, and the ground should also be reinforced.

5.5 Internal force calculation of segment lining

5.5.1 Introduction of internal force calculation methods

In the development of underground structure calculation theory, the design calculation method of the shield tunnel lining structure has gradually developed. At present, there are mainly two methods, load-structure calculation mode and stratum-structure calculation mode.

The load-structure calculation method takes the lining support structure as the main bearing structure, and the surrounding ground restricts the deformation of the support structure. This method has the advantages of clear stress distribution, simple calculation and easy safety evaluation. This method is still the shield tunnel lining structure design method mainly used in various countries. On the basis of different mechanical treatment methods of the segment joint, the structure model of the segment ring, during internal force calculation of the structure, can be roughly divided into three categories: a method of assuming that the segment ring is a ring with uniform bending stiffness, a method of assuming that the segment ring is a multihinged ring, and a method of assuming that the segment ring is a ring with a rotary spring and evaluating the splicing effect of the staggered-joint by a shear spring. The common load-structure calculation methods include the routine and modified routine calculation method, multihinged ring calculation method, elastic-hinge ring calculation method, beam-spring model calculation method, etc. These methods should be selected according to the actual conditions of the project [12].

The stratum-structure model calculation method regards the tunnel support structure and the stratum as a tunnel support system to jointly bear loads, and the support structure restricts the deformation of the surrounding strata into the tunnel. This method is a way to reflect the principle of the tunnel support structure and is more consistent with the current tunnel design idea. In theory, the stress and displacement states of the surrounding strata and the support structure can be accurately calculated. However, because of uncertainty of initial ground stress field and variability and measurement complexity of various physical and mechanical parameters of rock-soil body and the lining materials, it is very difficult to conduct simulation analysis calculation and the result variability is large. Generally, such method is used for qualitative analysis and complicated working conditions. The stratum-structure model calculation methods are commonly included in numerical simulation based on ABAQUS, ADINA, FLAC, ANSYS, and other software.

5.5.2 Routine calculation method and its modified form

In the routine calculation method, the segment is evaluated by taking the segment ring with the same stiffness as the segment by ignoring the presence of the joints. Owing to excessively high evaluation of the stiffness of the segment ring, excessive large section internal force is obtained for a hard ground with a subgrade reaction, while smaller section internal force is obtained for soft ground without a subgrade reaction. Sometimes the design results are prone to cause danger. The constructed tunnel calculated in this way is expected to be deformed greatly, and so full attention must be paid.

The modified routine calculation method, the same as the routine calculation method, also takes the segment ring as a uniform stiffness ring. The difference is that this method adopts effective rate (η) of the bending stiffness and increase-decrease coefficient (ζ) of the bending moment to modify the stiffness of the segment ring in the calculation method. That means the stiffness reduction of the segment ring caused by the joint and the splicing effect caused by staggered assembly is considered by these two coefficient in the calculation. Therefore, this method is closer to actual practice than the routine calculation method.

The modified routine calculation method is a relatively mature method. The impact of the vertical ground reaction is not considered in the calculation model, the vertical subgrade reaction is calculated as the uniformly distributed load according to the vertical load borne by the tunnel on the basis of the load balance condition, and the horizontal ground reaction is hypothesized to be a horizontal triangular load distributed at the middle of the tunnel in a range of 90 degrees, as shown in Fig. 5.34. Whether the horizontal subgrade reaction generated by ground deformation due to self-weight of the segment is considered or not is determined on the basis of the grouting condition and its parameters behind the segment ring. This method can calculate the section internal forces of the segment ring under different soil conditions. The section internal force calculation formula is shown in Table 5.7 (when calculating the left half ring, θ in the table is θ_{left} in Fig. 5.36; when calculating the right half ring, θ in the table is θ_{right} in Fig. 5.36). It should be noted that the content in the Fig. 5.36 is a calculation diagram during separated calculation of water pressure and earth pressure, while in combined calculation, the water pressure and the earth pressure are need to be combined together. During combined calculation of the water pressure and the earth pressure, the natural unit weight of soil is used for strata above the groundwater level, and the saturated unit weight of soil is used for strata under the groundwater level.

Table 5.7 Segment internal force calculation formulas of routine method and modified routine calculation method.

Load	Bending moment (*M*)	Axial force (*N*)	Shear force (*Q*)
Vertical load $(p_{e1}+p_{w1})$	If $0\le\theta\le\pi$ $\frac{1}{4}(1-2\sin^2\theta)(p_{e1}+p_{w1})R_c^2$	If $0\le\theta\le\pi$ $(p_{e1}+p_{w1})R_c\sin^2\theta$	If $0\le\theta\le\pi$ $-(q_{e1}+q_{w1})R_c\sin\theta\cos\theta$
Horizontal load $(q_{e1}+q_{w1})$	If $0\le\theta\le\pi$ $\frac{1}{4}(1-2\cos^2\theta)(q_{e1}+q_{w1})R_c^2$	If $0\le\theta\le\pi$ $(q_{e1}+q_{w1})R_c\cos^2\theta$	If $0\le\theta\le\pi(q_{e1}+q_{w1})R_c\sin\theta\cos\theta$
Horizontal triangular load $(q_{e2}+q_{w2}-q_{e1}-q_{w1})$	If $0\le\theta\le\pi$ $\frac{1}{48}(6-3\cos\theta-12\cos^2\theta+4\cos^3\theta)$ $(q_{e2}+q_{w2}-q_{e1}-q_{w1})R_c^2$	If $0\le\theta\le\pi$ $\frac{1}{16}(\cos\theta+8\cos^2\theta-4\cos^3\theta)$ $(q_{e2}+q_{w2}-q_{e1}-q_{w1})R_c$	If $0\le\theta\le\pi$ $\frac{1}{16}(\sin\theta+8\sin\theta\cos\theta-4\sin\theta\cos^2\theta)$ $(q_{e2}+q_{w2}-q_{e1}-q_{w1})R_c$
Subgrade reaction $(q_r=k\delta)$	If $0\le\theta<\pi/4$ $(0.2346-0.3536\cos\theta)k\delta R_c^2$ If $\pi/4\le\theta\le\pi/2$ $(-0.3487+0.5\sin^2\theta+0.2357\cos^2\theta)k\delta R_c^2$	If $0\le\theta<\pi/4$ $0.3536\cos\theta k\delta R_c$ If $\pi/4\le\theta\le\pi/2$ $(-0.7071\cos\theta+\cos^2\theta+0.7071\sin^2\theta\cos\theta)k\delta R_c$	If $0\le\theta<\pi/4$ $0.3536\sin\theta k\delta R_c$ If $\pi/4\le\theta\le\pi/2$ $(\sin\theta\cos\theta-0.7071\cos\theta\sin\theta)k\delta R_c$
Self-weight $(P_{g1}=\pi g_1)$	If $0\le\theta<\pi/2$ $\left(\frac{3\pi}{8}-\theta\sin\theta-\frac{5}{6}\cos\theta\right)g_1R_c^2$ If $\pi/2\le\theta\le\pi$ $\left[-\frac{\pi}{8}+(\pi-\theta)\sin\theta-\frac{5}{6}\cos\theta-\frac{1}{2}\pi\sin^2\theta\right]g_1R_c^2$	If $0\le\theta<\pi/2$ $\left(\theta\sin\theta-\frac{1}{6}\cos\theta\right)g_1R_c$ If $\pi/2\le\theta\le\pi$ $\left(-\pi\sin\theta+\theta\sin\theta+\pi\sin^2\theta-\frac{1}{6}\cos\theta\right)g_1R_c$	If $0\le\theta<\pi/2$ $-\left(\theta\sin\theta+\frac{1}{6}\sin\theta\right)g_1R_c$ If $\pi/2\le\theta\le\pi$ $\left[-(\pi-\theta)\cos\theta+\pi\sin\theta+\pi\sin\theta\cos\theta-\frac{1}{6}\sin\theta\right]g_1R_c$

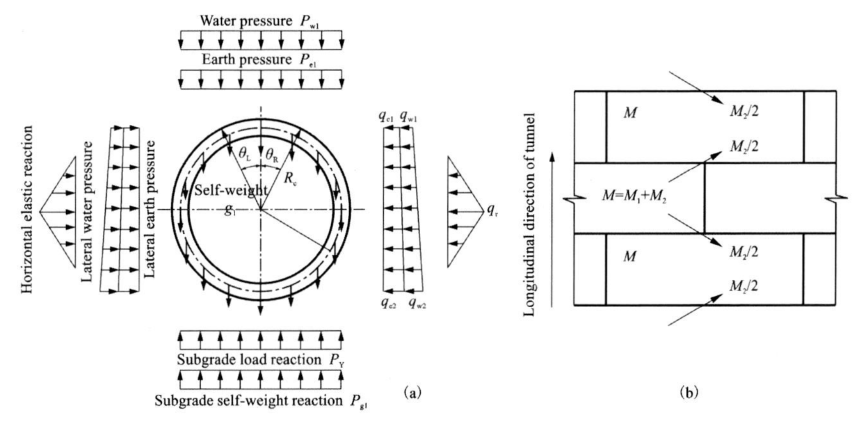

Figure 5.36 Calculation model of modified routine calculation method. (a) Cross-sectional load distribution. (b) Longitudinal load distribution.

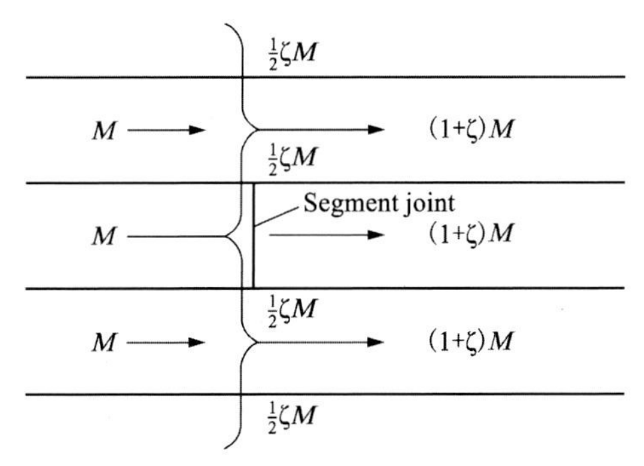

Figure 5.37 Bending moment transfer caused by joint (two rings are a group of staggered-joint assembly).

The consideration of the presence of the segment joint in the model reduces the integral stiffness of the segment ring by an average amount. The reduction coefficient is η ($\eta \leq 1$), that is, the segment ring has a equivalent stiffness ηEI, as shown in Fig. 5.37.

Further considering the influence of staggered-joint assembly of segments, the bending moment is calculated by multiplying an scaling coefficient ξ ($\xi \leq 1$) based on the internal force calculated from the ring with equivalent stiffness being ηEI. Then the bending moment of the main section of the segment lining is $(1 + \xi)M$, and the bending moment of the segment joint is $(1 - \xi)M$. According to the results of a large number of segment joints loading tests in several countries, the parameter η is approximately 0.6–0.8, and ξ is approximately 0.2–0.3.

This model is a homogeneous ring model if specifying $\eta = 1$ and $\xi = 0$. Therefore this model is actually a modification of the homogenous ring model. The present situation of the modified routine calculation method for the calculation of η and ξ is limited to qualitative evaluation, which is calculated on the basis of experience or experiment results. Furthermore, the section internal force generated at the joint cannot be directly determined by this method.

5.5.3 Multihinged ring calculation method

The multihinged ring calculation method is based on a design guidance of taking the segment lining ring as a hinged ring for evaluation where the segment joint is a hinge (Fig. 5.38). The principle of a typical multihinged ring internal force calculation method is that the multihinged lining ring is deformed under the action of active and passive earth pressure, the ring gradually becomes from an unstable structure to a stable structure, and the hinge is not abruptly changed during the ring deformation process. The calculation hypothesis as follows:

1. When the segment of lining ring rotates, the segment or block is treated as a rigid body.
2. The subgrade reaction at the periphery of the lining ring is uniformly distributed, the calculation of the subgrade reaction meets the requirement for the stability of the lining ring, and the direction of the subgrade reaction all faces the center of the circle.

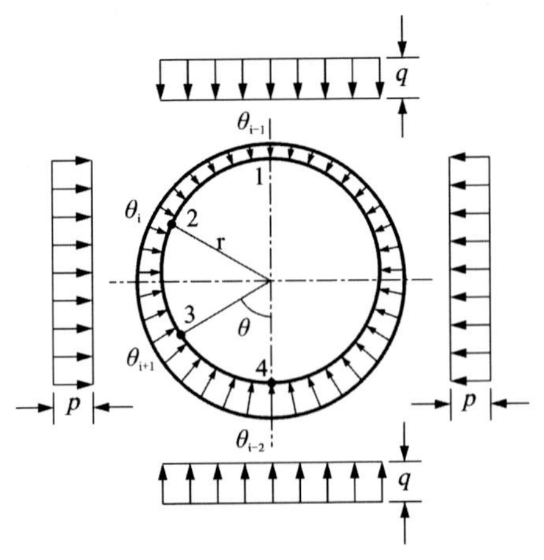

Figure 5.38 Multihinged ring calculation model.

3. The friction between the ring and the stratum medium is not taken into account in the calculation.
4. The relationship between the subgrade reaction and the deformation is calculated in accordance with Winkler formula.

Taking the beam 1–2 as an example for calculation (Fig. 5.39), $\theta_{i-1} = 0°$, $\theta_i = 60°$, the ground reaction at each section of the lining is

$$q_{\alpha_i} = q_{i-1} + \frac{(q_i - q_{i-1})\alpha_i}{\theta_i - \theta_{i-1}} \tag{5.21}$$

Obtained from $\sum X = 0$: $H_1 = H_2 + 0.5pr + 0.327q_2r$
Obtained from $\sum Y = 0$: $V_2 = 0.866qr + 0.388q_2r$
Obtained from $\sum M_2 = 0$: $H_1 = (0.75qr + 0.25p + 0.34q_2)r$

Where X, Y, M_2 means the horizontal, vertical, and in plane rotation direction; H_i and V_i is the internal force of joint i at the horizontal and vertical direction respectively. Similarly, through statically determinated solution on beam 2–3 and beam 3–4, the internal forces H_1, V_1, H_2, V_2, H_3, V_3, H_4, and V_4 at each joint can be obtained, and the internal force of an arbitrary section (taking the 1–2 beam as an example) is

$$H_{\alpha_i} = H_2 + pr(1 - \cos\alpha_i) + r\int_0^{\alpha_i - \theta_{i-1}} \frac{q_{\alpha_i}\alpha_i}{\pi/3}\sin(\theta_{i-1} + \alpha_i)d\alpha_i \tag{5.22}$$

$$V_{\alpha_i} = qr\sin\alpha_i + r\int_0^{\alpha_i - \theta_{i-1}} \frac{q_{\alpha_i}\alpha_i}{\pi/3}\cos\alpha_i d\alpha_i \tag{5.23}$$

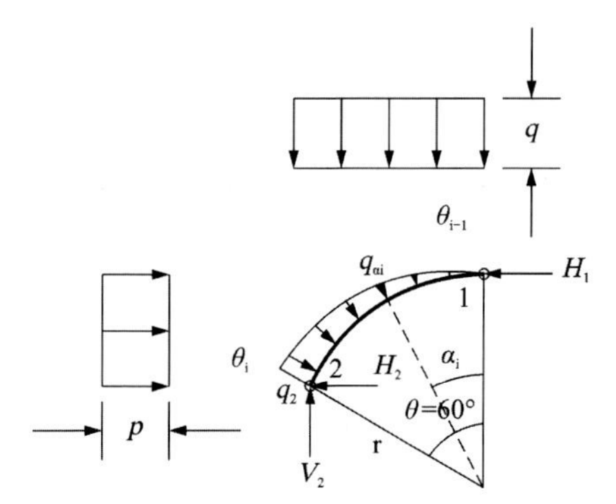

Figure 5.39 Schematic diagram of beam 1–2 calculation.

$$M_{\alpha_i} = \frac{(qr\sin\alpha_i)^2}{2} + \frac{p[r(1-\cos\alpha_i)]^2}{2} + \frac{3r^2}{\pi} q_{\alpha_i} \int_0^{\alpha_i-\theta_{i-1}} \alpha_i \sin(\theta_i - \theta_{i-1} - \alpha_i) d\alpha_i - H_1 r\sin\alpha_i \tag{5.24}$$

In the multihinged ring model, only one ring is used, the joint is treated as a hinge, and a stable structure is formed by reaction force of the surrounding ground. The calculation results are different due to different load conditions and setting methods of the subgrade reaction. Therefore the conditions of applicable subgrade should be fully studied. The model is applicable to stable ground while large deformation will occur in soft ground.

For the active earth pressure acting on the ring, the Winkler hypothesis is mostly used for the subgrade reaction resulting from ring deformation and displacement with the above conventional load cases. Using this calculation method, the section bending moment will be considerably reduced which could save the design cost. On the other hand, since the quality of the surrounding rock will make a decisive impact on the tunnel, it is necessary to conduct serious research on whether the tunnel surrounding ground and tunnel waterproofing will be damaged by nearby construction.

5.5.4 Elastic-hinge ring calculation method

The segment ring of the shield tunnel is spliced by multiple segments, and the joints between the segments are in various forms, so it is necessary to use connecting bolts. As a result, the segment lining with spliced joints cannot have the same stiffness as the integral cast-in-place reinforced concrete structure. In fact, there is an elastic hinge capable of bearing a part of the bending moment at each segment joint. Such hinge is neither rigid nor completely hinged. The amount of the bending moment borne by the elastic hinge is related to the joint stiffness (K_θ). In calculating the section internal force, the segment joint can be regarded as an elastic hinge structure. The joint stiffness is usually and comprehensively determined and confirmed by experiments and experience. The section internal force can be determined by the mechanical calculation model shown in Fig. 5.40. Since the structure load is symmetrical to the vertical axis, half of the structure is analyzed by a structural mechanics method, ignoring the influence of the axial force and the shear force on structure deformation.

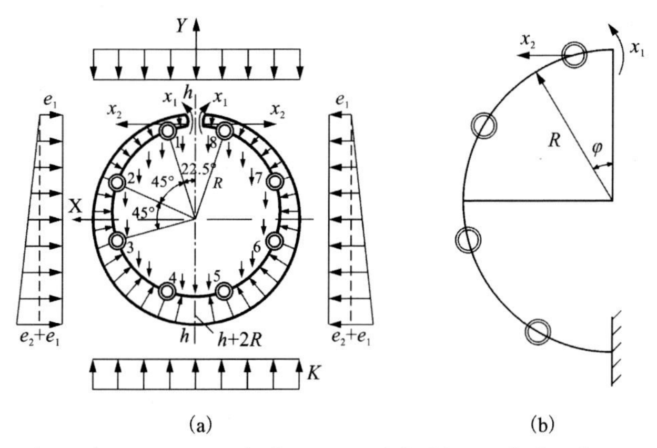

Figure 5.40 Elastic-hinge ring calculation model. (a) Load distribution model. (b) Semistructural calculation model.

The established equation of force method is

$$\begin{aligned} \delta_{11}x_1 + \delta_{12}x_2 + \Delta_{1p} = 0 \\ \delta_{21}x_1 + \delta_{22}x_2 + \Delta_{2p} = 0 \end{aligned} \tag{5.25}$$

where:$\delta_{11}, \delta_{12}, \delta_{21}, \delta_{22}, \Delta_{1p}, \Delta_{2p}$ are, respectively,

$$\delta_{11} = \frac{1}{EI}\int_0^{\pi} \overline{M_1^2} Rd\varphi + \sum_{i=1}^{4} \overline{M_1^i M_1^i} \cdot \frac{1}{K_\theta^i}$$

$$\delta_{12} = \delta_{21} = \frac{1}{EI}\int_0^{\pi} \overline{M_1 M_2} Rd\varphi + \sum_{i=1}^{4} M_2^i \overline{M_2^i} \cdot \frac{1}{K_\theta^i}$$

$$\delta_{22} = \frac{1}{EI}\int_0^{\pi} \overline{M_2^2} Rd\varphi + \sum_{i=1}^{4} \overline{M_2^i M_2^i} \cdot \frac{1}{K_\theta^i}$$

$$\Delta_{1p} = = \sum_{i=1}^{h}\int_0^{\pi} \frac{\overline{M_1} M_{p(j)}}{EI} Rd\varphi + \sum_{j=1}^{h}\sum_{i=1}^{4} \overline{M_1^i} M_{p(j)}^i \frac{1}{K_\theta^i}$$

$$\Delta_{2p} = = \sum_{i=1}^{h}\int_0^{\pi} \frac{\overline{M_2} M_{p(j)}}{EI} Rd\varphi + \sum_{j=1}^{h}\sum_{i=1}^{4} \overline{M_2^i} M_{p(j)}^i \frac{1}{K_\theta^i}$$

where i represents the number of segment joints; and j represents the number of load types, as obtained by solving with determinant:

$$\begin{aligned}
&\Delta = \delta_{11}\delta_{22} - \delta_{12}\delta_{21} \\
&\Delta x = \delta_{12}\Delta_{2p} - \Delta_{22}\Delta_{1p} \\
&\Delta y = \delta_{22}\Delta_{1p} - \Delta_{11}\Delta_{2p} \\
&x_1 = \frac{\Delta x}{\Delta}; x_2 = \frac{\Delta y}{\Delta}
\end{aligned} \tag{5.26}$$

where $\overline{M_1}, \overline{M_2}$ are the bending moments of the basic structure under unit load; M_p is the bending moment of the basic structure under load action; K_θ is joint stiffness of each join; and EI is structural stiffness. Then the internal force of an arbitrary section is

$$\begin{aligned}
M &= \overline{M_1}x_1 + \overline{M_2}x_2 + \sum_{j=1}^{k} M_{p(j)} \\
N &= \overline{N_1}x_1 + \overline{N_2}x_2 + \sum_{j=1}^{k} N_{p(j)}
\end{aligned} \tag{5.27}$$

5.5.5 Beam-spring model calculation method

The segment ring is assembled by connecting prefabricated segments with bolts. Such structural characteristics lead to large difference between the stress and deformation characteristics of the segment connection point (i.e., joint) and those of the segment body. Therefore two different calculation units must be adopted to describe the stress behaviour. The beam-spring model is a combination construction model (Fig. 5.41a) that

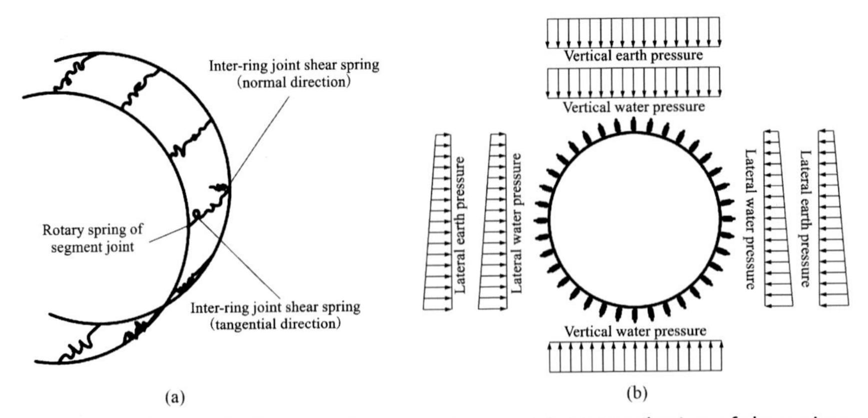

Figure 5.41 Schematic diagram of beam-spring model. (a) Distribution of the springs. (b) Calculation model.

simplifies the main section of the segment into a circular beam or linear beam structure, uses springs to simulate the joints, regards the segment joints as rotary springs, and regards the ring joints as shear springs (in the tangential and normal directions). The interaction of the segment ring and the surrounding soil layer is simulated by Winkler subgrade springs. There is no need to assume the distribution of the subgrade reaction, and the structural analysis is carried out on the elastic property of the segment ring by the finite element theory to calculate the section internal force, as shown in Fig. 5.41b.

5.5.5.1 Determination of spring coefficient

K_N, K_S, and K_θ represent the compression spring coefficient, the shear spring coefficient, and the rotary spring coefficient of the joint, respectively. At present, most of calculations based on joint test data are conducted by assuming that these spring coefficients are constants. However, in fact, the mechanical property of the structure is very complicated, and various spring coefficients are nonlinear, such that these spring coefficients must be determined on the basis of abundant joint loading test data.

For underground structures such as circular linings, the section shear force is small, owing to the support of the surrounding ground, and the section strength is mainly controlled by the bending moment and the axial force. Currently, there are few test data about joints, and the test methods still have shortcomings. Generally speaking, the spring coefficients should be taken with a larger value to make the design safer.

5.5.5.2 Establishment of element stiffness matrix

Beam element

For arc segments with small dimensions, no big error will be caused when a curved beam element is replaced with a straight beam element. To simplify the calculation, the straight beam element is used. As shown in Fig. 5.42, displacements $(\overline{u}_i, \overline{v}_i, \overline{\varphi}_i)$ and $(\overline{u}_j, \overline{v}_j, \overline{\varphi}_j)$ are respectively given to two ends of a micro beam element, and the generated beam end forces are $(\overline{X}_{ij}, \overline{Y}_{ij}, \overline{M}_{ij})$ and $(\overline{X}_{ji}, \overline{Y}_{ji}, \overline{M}_{ji})$, respectively.

The relation expression between the beam displacements and the beam end forces is established below. First, the end j is fixed, the end i respectively and separately generates displacements $\overline{u}_i, \overline{v}_i, \overline{\varphi}_i$, at the same time. The beam end forces respectively generated at the end i and the end j are shown in formula (5.28a); and then the end i is fixed, the end

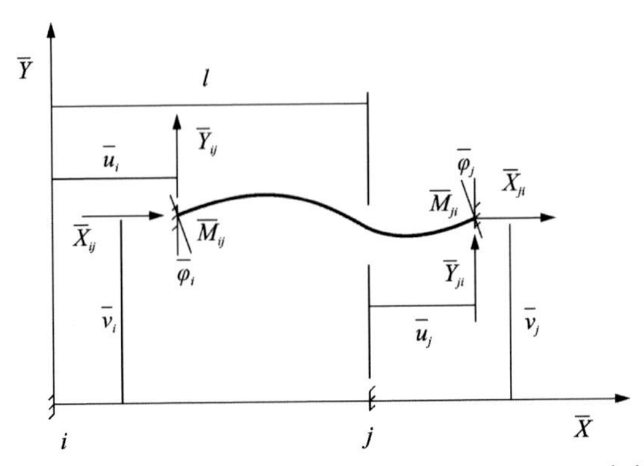

Figure 5.42 Schematic diagram of beam end forces and beam end displacements.

j respectively and separately generates displacements $\overline{u}_j, \overline{v}_j, \overline{\varphi}_j$, and a similar group of beam end forces can also be obtained, as seen formula (5.28b):

$$\begin{cases} \overline{X}_{ij} = \dfrac{EA}{l}\overline{u}_i \\ \overline{Y}_{ij} = \dfrac{12EI}{l^3}\overline{v}_i - \dfrac{6EI}{l^2}\overline{\varphi}_i \\ \overline{M}_{ij} = -\dfrac{6EI}{l^2}\overline{v}_i + \dfrac{4EI}{l}\overline{\varphi}_i \end{cases} ; \begin{cases} \overline{X}_{ji} = -\dfrac{EA}{l}\overline{u}_i \\ \overline{Y}_{ji} = -\dfrac{12EI}{l^3}\overline{v}_i + \dfrac{6EI}{l^2}\overline{\varphi}_i \\ \overline{M}_{ji} = \dfrac{6EI}{l^2}\overline{v}_i + \dfrac{2EI}{l}\overline{\varphi}_i \end{cases} \tag{5.28a}$$

$$\begin{cases} \overline{X}_{ij} = -\dfrac{EA}{l}\overline{u}_j \\ \overline{Y}_{ij} = \dfrac{12EI}{l^3}\overline{v}_j + \dfrac{6EI}{l^2}\overline{\varphi}_j \\ \overline{M}_{ij} = -\dfrac{6EI}{l^2}\overline{v}_j + \dfrac{2EI}{l}\overline{\varphi} \end{cases} ; \begin{cases} \overline{X}_{ji} = \dfrac{EA}{l}\overline{u}_j \\ \overline{Y}_{ji} = \dfrac{12EI}{l^3}\overline{v}_j + \dfrac{6EI}{l^2}\overline{\varphi}_j \\ \overline{M}_{ji} = \dfrac{6EI}{l^2}\overline{v}_j + \dfrac{4EI}{l}\overline{\varphi}_j \end{cases} \tag{5.28b}$$

On the basis of the superposition principle, the relation between the beam end forces and the beam end displacements can be obtained by superimposition of the corresponding beam end forces in the above two cases, which can be expressed as a matrix:

$$|\overline{N}| = |\overline{K}_L||\overline{\delta}| \tag{5.29}$$

$$
\text{where } \left|\overline{K}_L\right| = \begin{bmatrix} \frac{EA}{l} & 0 & 0 & -\frac{EA}{l} & 0 & 0 \\ 0 & \frac{12EI}{l^3} & -\frac{6EI}{l^2} & 0 & -\frac{12EI}{l^3} & -\frac{6EI}{l^2} \\ 0 & -\frac{6EI}{l^2} & \frac{4EI}{l} & 0 & -\frac{6EI}{l^2} & \frac{2EI}{l} \\ -\frac{EA}{l} & 0 & 0 & \frac{EA}{l} & 0 & 0 \\ 0 & -\frac{12EI}{l^3} & \frac{6EI}{l^3} & 0 & \frac{12EI}{l^3} & \frac{6EI}{l^2} \\ 0 & -\frac{6EI}{l^2} & \frac{2EI}{l} & 0 & \frac{6EI}{l^2} & \frac{4EI}{l} \end{bmatrix};
$$

$$
\left|\overline{N}\right| = \left|\overline{X}_{ij}, \overline{Y}_{ij}, \overline{M}_{ij}, \overline{X}_{ji}, \overline{Y}_{ji}, \overline{M}_{ji}\right|^T;
$$

$$
\left|\overline{\delta}\right| = \left|\overline{u}_i, \overline{v}_i, \overline{\varphi}_i, \overline{u}_j, \overline{v}_j, \overline{\varphi}_j\right|,
$$

E, A, I, l are the elastic modulus of the segment, sectional area, bending section modulus, and unit length of the segment, respectively.

Joint element

As shown in Fig. 5.41, the joint is simulated by three springs (the compression spring and the shear spring can resist only tension, and the rotary spring can resist only bending). Similar to the beam element deduction process, the relation expression between the joint forces and displacements is obtained as follows:

$$
\begin{cases} \overline{X}_{ij} = K_N\overline{u}_i - K_N\overline{u}_j \\ \overline{Y}_{ij} = K_S\overline{v}_i - K_S\overline{v}_j \\ \overline{M}_{ij} = K_\theta\overline{\varphi}_i - K_\theta\overline{\varphi}_j \end{cases} \begin{cases} \overline{X}_{ji} = -K_N\overline{u}_i + K_N\overline{u}_j \\ \overline{Y}_{ji} = -K_S\overline{v}_i + K_S\overline{v}_j \\ \overline{M}_{ji} = -K_\theta\overline{\varphi}_i + K_\theta\overline{\varphi}_j \end{cases} \tag{5.30}
$$

where K_N, K_S, K_θ are the compression spring coefficient, the shear spring coefficient, and the rotary spring coefficient, respectively. The equation coefficients of formula (5.30) are expressed as a joint element stiffness matrix:

$$\left|\overline{K_J}\right| = \begin{bmatrix} +K_N & 0 & 0 & -K_N & 0 & 0 \\ 0 & +K_S & 0 & 0 & -K_S & 0 \\ 0 & 0 & +K_\theta & 0 & 0 & -K_\theta \\ -K_N & 0 & 0 & +K_N & 0 & 0 \\ 0 & -K_S & 0 & 0 & +K_S & 0 \\ 0 & 0 & -K_\theta & 0 & 0 & +K_\theta \end{bmatrix} \tag{5.31}$$

If the shear spring coefficient and the rotary spring coefficient are simultaneously set to zero, this method is basically the same as the multi-hinged ring calculation method. If the shear spring coefficient is set to zero and the rotary spring constant is set to infinity, this method is the same as the calculation method of uniform stiffness ring. At the same time, the shear stiffness of the segment ring joint also can be used for representing the splicing effect of the staggered-joint. Therefore it is an effective method to explain the bearing mechanism of the segment ring from the perspective of the mechanical mechanism.

The loads used in this calculation method are basically all from conventional load systems, but there are also methods to convert all or part of subgrade reaction to subgrade springs for calculation. The beam-spring model can be configured to analyze the assembly method of any segment ring and the position of the joint, and the shear force generated on the ring joint can also be determined. Although the rotary spring coefficient and the shear spring coefficient can be obtained through experiment, they can also be obtained through calculation for general segment joints. If the value of the shear spring coefficient is too small, the calculated bending moment of the main section will also be too small, so a method of setting the value to infinity is often used for being in the safe side.

At present, this method is used in the design not only of large-section tunnels such as railway tunnels and roadway tunnels, but also of medium- and small-diameter tunnels with more complicated design conditions, using the routine calculation methods and modified routine calculation methods for comparing and checking.

5.5.6 Numerical simulation method

Owing to the diverse geometric shapes of the tunnel structure, the surrounding rock-soil body has various nonlinearities, and the interaction mechanism between the surrounding strata and the tunnel supporting structure is complicated, causing the mechanical calculation of the tunnel structure to be a problem of solving a statically indeterminate structure. Such a problem can be solved by the numerical simulation calculation method. The finite element method is a typical numerical method, and its idea is to divide the analysis object continuum into a finite number of elements that are connected at nodes. As a result, the whole continuum is replaced with an element aggregate. The forces acting on the elements are equally transferred to the nodes. Each element selects a displacement function to represent the distribution rule of a displacement component, and the relationship between the nodal force and the nodal displacement is established according to a variation principle. Based on the nodal equilibrium conditions, all element relations are set to form a set of algebraic equations with nodal displacement as an unknown quantity so as to figure out each nodal displacement. The basic steps of the solution are generally as discussed in the following subsections.

5.5.6.1 Problem and solution domain definition

The physical property and the geometric regions of a solution domain are figured out according to the actual problem.

5.5.6.2 Solution domain discretization

The solution domain is approximated as a discrete domain composed of a finite number of connected elements with different finite sizes and shapes, which is called finite element mesh division. Obviously, the smaller the element is, the better the approximation of the discrete domain is, and the more accurate the calculation the result is, but the calculation workload is increased.

5.5.6.3 Determination of state variables and control methods

Specific physical problems can be expressed by a set of differential equations containing boundary conditions of the problem state variables. To be applicable to finite element solution, the differential equation is usually transferred into an equivalent functional representation.

5.5.6.4 Element deduction

A suitable approximate solution is constructed for a element, and then the determinant formula of the element is deducted, including selecting a reasonable element coordinate system, establishing an element trial function, and giving a discrete relation of each state variable of the element in some way, thereby forming an element matrix. To ensure the convergence of the solution, the element deduction should follow the related principles.

5.5.6.5 Integral solution

The elements are finally assembled to form the global matrix equation of the discrete domain, reflecting the requirements for the discrete domain in the approximate solution domain, that is, the element function must meet a certain continuous condition. Final assembly is conducted at the node of adjacent elements, and the continuity of the state variables and their derivatives is established at the nodes and the boundary conditions.

5.5.6.6 Solution of simultaneous equations and result representation

The equations can be solved by the Gaussian elimination method, the direct method, the iterative method, and the random method. The solution result is the approximate value of the state variable at the element node. For the calculation results, whether the calculation needs to be repeated depends on the accuracy requirements.

In summary, the finite element analysis can be divided into three stages: preprocessing, processing, and postprocessing. In the preprocessing stage, a finite element model is established so as to complete element mesh division; in the processing stage, the finite element mode is calculated and analyzed; and in the postprocessing stage, the result of the analysis is acquired and processed. With the development of computers and advances in computing technologies, the finite element method has also been developed rapidly, which enables the finite element analysis become more than a calculation verification tool. It has developed into an effective tool for mechanical analysis of a complicated system. The visual display of the calculation result has evolved from simple static and dynamic display and color toning display of stress, displacement, and temperature fields to display of positions, shapes, and sizes (cracks) that may occur in a load object and other possible influences.

5.5.6.7 Load-structure model numerical simulation

The numerical simulation in the load-structure mode considers the supporting structure and the surrounding strata separately. In this model, the interaction between the tunnel supporting structure and the surrounding strata is reflected by restrictions imposed on the structure from elastic support, while the soil bearing capacity is indirectly considered when the soil pressure and the restraint ability of the elastic support are determined. The

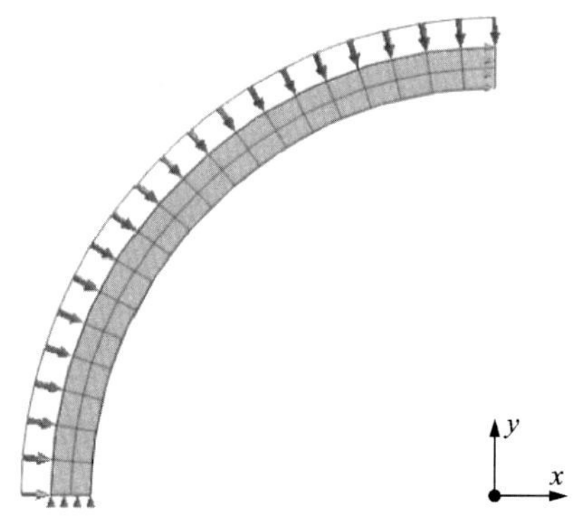

Figure 5.43 Numerical model of load-structure calculation.

supporting structure is a bearing body, and the soil acts as the source of the load and the elastic support of the supporting structure. The earth pressure is simplified equivalent to radial and tangential loads acting on nodes of the supporting structure. In most cases the tangential load is smaller than the radial load, the effect is ignored for the sake of simplicity, and only the discrete element of the supporting structure is analyzed.

For the structural symmetry of a circular shield tunnel structure, a quarter of the segment ring can be used as a simplified model (see Fig. 5.43). Plane elements are used to simulate the arc of the segment ring. The top of the model is constrained by horizontal displacement, the bottom is constrained by vertical displacement, and a normal load is applied to outside the arc of the elements. The load acting on the supporting structure can be calculated in accordance with the content of Section 5.4 and then converted into normal pressure acting on the elements.

5.5.6.8 Stratum-structure model numerical simulation

Since the tunnel structure is built underground, its engineering characteristics, design principles and methods are different from those of a ground surface structure. The deformation of the tunnel structure is restricted by the surrounding strata. In a sense, the strata is the load source of the underground structure and is also a part of the bearing structure itself. Therefore the stratum-structurecalculation method considering the interaction between the tunnel supporting structure and the surrounding strata and bearing jointly is much more in line with the actual stress state of the tunnel. However, the mechanical properties of the stratum media are very complicated. There are many factors affecting the stress and deformation of the stratum media, such as structures, pores, density, stress history, load characteristics, pore water, and time effect. Such complexity determines that the specific problems are often required to be simplified and hypothesized to build an analysis model for calculating tunnel and underground engineering problems.

Numerical simulation based on the stratum-structure calculation theory generally posits that the lining structure and the stratum are integrally stressed and deformed, and the internal force and deformation of the lining and the surrounding stratum can be calculated according to the principle of continuum mechanics. The common approach is to jointly model the strata and the shield tunnel lining, and the stress and the deformation of the tunnel lining and the surrounding strata are simulated by means of modern finite element or finite difference calculation software for shield tunnel design.

Segment ring simulation

The segment ring are mainly simulated by three dimensional solid element and shell element. However, with the adoption of the shell element, the interaction of the interface between the segment blocks cannot be simulated very well. Especially if the thickness of the segment used in the shield tunnel is large, it is not appropriate to simulate segment lining with the shell element, and the segment lining needs to be simulated with the solid element.

Segment joint simulation

In the segment ring system, the connecting bolt between the segments mainly shows the stress characteristics of tension and shear under the action of an external force, and compression and bending are mainly borne by the concrete segment at the joint. In the segment ring model, the homogeneous ring model uses only an equivalent reduction method to simulate the joint. The stress characteristics of the joints in the beam-spring model and the shell-spring model are concentrated, which is far from the actual situation. Therefore more and more intuitive bolt simulation methods have emerged in recent years, mainly including the beam element and the solid element. Compared with the solid element, the simulation using the beam element can more directly and accurately reflect the internal force characteristics of bolts such as tension, compression, bending, and shear.

The establishing of a discontinuous segment model can enable accurate simulation of the shield tunnel lining structure and improve the reliability of the analysis result. However, the segments, the bolts, and other components established in the model are relatively small in comparison with an integral model, resulting in a large amount of meshes divided. Meanwhile, many contacts as required in the modeling process which result in a complicated calculation process and a high calculation cost. Therefore a hybrid model is generally used for modeling. A discontinuous contact segment model capable of eliminating the boundary effect is established for the segment ring, while the equivalent homogeneous ring model is used for the other sections of lining segments.

To ensure the waterproofing effect of the shield tunnel lining structure, rubber waterstop strings are arranged on the side walls of the segments against the outer ring faces, and more elastic rubber gaskets are pasted on the other areas of the side walls. Therefore the rubber gaskets are mainly in contact with each other between the segments, and the area where the concrete interface is attached is small. In the numerical model, the corresponding contact relationship is established between adjacent segments, and the role of the rubber gaskets is reflected by setting a tangential Coulomb contact friction coefficient between the two segments.

Equivalent stiffness of segment joint

The modified routine method assumes that the decrease in the bending stiffness of the joint part is evaluated as the decrease in the bending stiffness of the integral ring. The segment ring is an equivalent homogeneous ring with the bending stiffness being ηEI, and the effective rate η of the bending stiffness is a ratio of the bending stiffness of the equivalent homogeneous ring to that of the segment body section. In this method, the bending moment distribution of the staggered-joint is also considered. The section bending moment is calculated from the equivalent homogeneous ring with the bending stiffness ηEI added or subtracted by a percentage of ζ. Then $(1+\zeta)M$ is set as the designed bending moment of the main section while $(1-\zeta)M$ is set as the designed bending moment of the joint. The parameter ζ is the ratio of the bending moment M_2 transferred to the segment adjacent to the joint to the bending moment M generated on the equivalent homogeneous ring. In normal cases, for the whole tunnel, the stiffness reduction coefficient is usually divided into a transverse stiffness reduction coefficient η_h and a longitudinal stiffness reduction coefficient η_z in consideration of the structure characteristics of the shield tunnel lining.

On the cross section of the tunnel, the modified routine method posits that the equivalent continuous model has the bending stiffness of η_hEI ($\eta_h \leq 1$, and EI is the bending stiffness of the cross section of the homogeneous tunnel lining). The size of the equivalent test model is the same as that of the assembled test model, so the reduction in the bending stiffness is reflected in the reduction in elastic modulus. In this way, the transverse elastic modulus can be reduced by the stiffness reduction coefficient of the cross section:

$$E_r = E_\theta = \eta_h E \tag{5.32}$$

where E is the elastic modulus of the assembled concrete segment.

In the longitudinal direction of the tunnel, referring to the concept of the transverse stiffness reduction coefficient of the segment ring, the tunnel with many inter-ring joints is equivalent to a homogeneous tunnel lining ring. The bending stiffness needs to be properly reduced to consider the influence of the joint on the integral stiffness of the homogeneous tunnel lining. The equivalent stiffness is $\eta_z EI$ where EI is the longitudinal bending stiffness of the homogeneous tunnel lining without considering the joint. When the maximum displacement of the assembled model is equal to that of the equivalent model, the stiffness of the two is regarded to be the same, and at the moment, the longitudinal stiffness reduction coefficient η_z of the tunnel can be obtained. From the above concept,

$$\eta_z = \frac{\Delta_{\text{uniform } ring}}{\Delta_{\text{real}}} \tag{5.33}$$

As in the transverse stiffness reduction method, the reduction of the longitudinal bending stiffness is reflected in the reduction of the longitudinal elastic modulus. The longitudinal elastic modulus can be reduced by the longitudinal bending stiffness reduction coefficient:

$$E_z = \eta_z E \tag{5.34}$$

where E is the elastic modulus of the assembled concrete segment.

The main Poisson's ratio is preferable according to the specification:

$$\upsilon_{r\theta} = \upsilon_{rz} = \upsilon_{\theta z} \tag{5.35}$$

The shear modulus $G_{r\theta}$ of the cross section is related to the elastic modulus E_r and E_θ of the cross section, while the longitudinal shear modulus $G_{\theta z}$ and G_{zr} are independent of the elastic modulus and can be reduced by the longitudinal stiffness reduction coefficient:

$$G_{\theta z} = G_{zr} = \eta_z G \tag{5.36}$$

where G is the shear modulus of the assembled concrete segment.

To obtain the stiffness reduction coefficient of the equivalent homogenous ring, two groups of transverse and longitudinal numerical values are set, and with loading simulation experiments, the transverse stiffness reduction coefficient η_h and the longitudinal stiffness reduction coefficient η_z can be theoretically determined. Another way is that one integral stiffness reduction coefficient η is determined from two groups of experimental results by a circulation experiment method on the basis of ensuring the calculation accuracy and saving calculation cost.

Material parameter selection

1. Stratum

   The thickness of each stratum layer of the shield tunnel is selected according to the actual condition of the calculated section. The analysis type is determined on the basis of the drainage condition of the stratum, and the corresponding physical and mechanical parameters of the stratum can generally be determined in accordance with the tunnel geological survey report and laboratory tests. Shields, linings, and grouting layers can be simulated with isotropic elastic materials, and the stratum layers is simulated with elastoplastic materials. For example, the commonly used constitutive models include the Duncan-Chang (DC) model, the Mohr-Coulomb (MC) model, the Drucker-Prager (DP) model, and the modified Cambridge Clay (MCC) model.
2. Shield machine

   The shield machine can generally be assumed to be an isotropic body. It is simulated with elastic elements and made of steel. In the model, the shield machine is simulated as a homogeneous ring with a certain thickness and a certain length along the tunnelling direction. The effect of a cutterhead and a soil chamber can be equivalent to the pressure of the soil chamber and the torque of the cutterhead measured on site.
3. Segment

   The segment is generally also assumed to be an isotropic body and is simulated with elastic elements. The structure parameters of the discontinuous lining are determined according to the segment concrete, and the strength of the equivalent lining structure is reduced on the basis of the aforementioned stiffness reduction coefficient. The material parameters of the circumferential and longitudinal bolts are determined in accordance with bolt models.
4. Grouting layer element

   The thickness of the grouting layer is the same as the shell of the shield machine. In view of its condensation effect, the physical and mechanical parameters of the grouting layer before hardening are

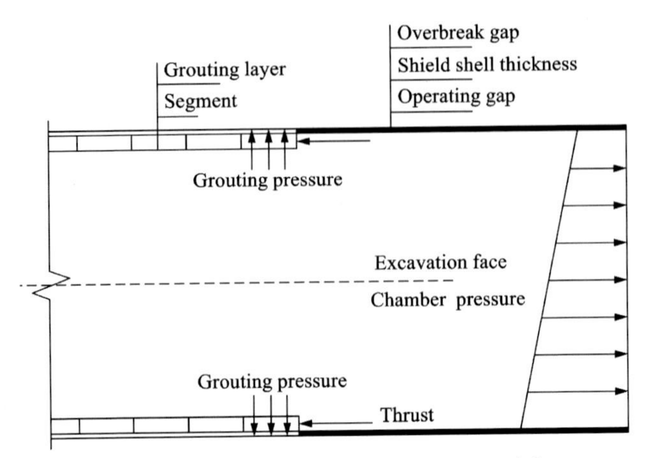

Figure 5.44 Spatial distribution of shield tunnel structure model.

determined according to mortar materials and after hardening determined according to cement-soil materials.

The spatial distribution of each structure in the numerical model is shown in Fig. 5.44.

Numerical value model establishment

The shield tunnel structure is analyzed by numerical simulation. To reduce the influence of boundary conditions on the calculation accuracy, the calculation range is generally three to five times the tunnel diameter (D), and the transverse, longitudinal and advancing directions of the shield tunnel are all about 5D. Then the numerical model is built on the basis of the buried depth, the inner diameter, the external diameter, and the segment lining structure. Generally, the boundary conditions are constraints from the surrounding and bottom normal direction, and the top surface is a free face. Fig. 5.45 shows the finite element analysis model and mesh diagram of a three-dimensional double-shield tunnel.

5.5.7 Case study of internal force calculation

5.5.7.1 Engineering overview

The section from Xiaocaiyuan Station to North Railway Station of Kunming Metro Line 4 starts from Xiaocaiyuan Station in the west. After leaving the station, the metro tunnel is laid along the meter-gage railway line (currently out of service) of Kunming-Shizui Railway, shifts to the southeast after passing through the Xiaocaiyuan overpass, crosses the

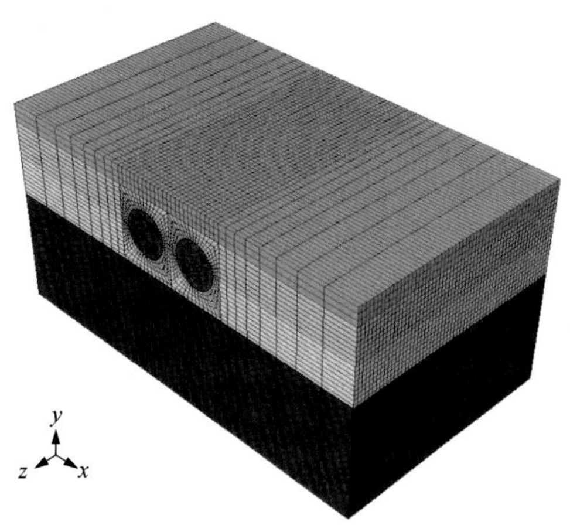

Figure 5.45 Three-dimensional meshed numerical model of double-shield tunnel.

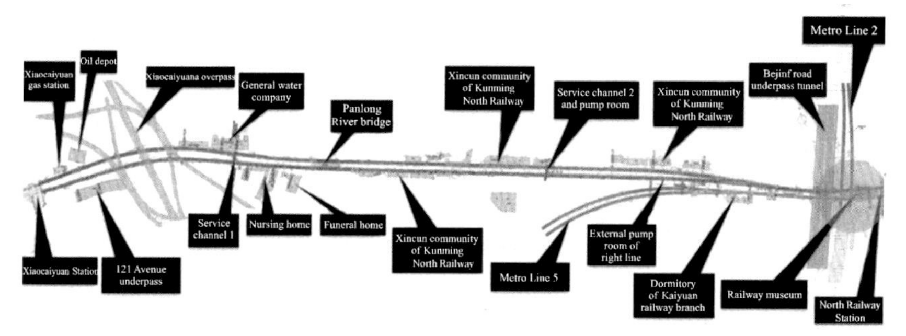

Figure 5.46 General layout of section between Xiaocaiyuan Station and North Railway Station.

Panlong River, and underneath passes about 300 m through Wanhua Road. In this section, the vertical spacing between the left line and the right line gradually increases, and the plane spacing gradually decreases. Finally, the section underneath passes through the tunnel culvert of the North Railway Station in an overlapping way (the left bound is above the right bound) and Kunming Metro Line 2 (in operation) and is connected to the North Railway Station (Fig. 5.46).

5.5.7.2 Geological conditions

The gravel ground is the main soil layer that must be traversed by the shield tunnel, while the clay and gravelly-sandy layer are traversed only in some local areas. There are small amounts of lenticular silt, sandy silt, and gravel sand in the gravel ground layer. The groundwater is abundant in this area and is closely related to the surface water. The mixed groundwater level of the area where the site is located is generally 1.3–9.1 m below the surface. The maximum buried depth of the shield tunnel section is selected as the calculated section, with the buried depth of the tunnel being 33.388 m. The geological distribution and the geotechnical parameters are respectively shown in Fig. 5.47 and Table 5.8.

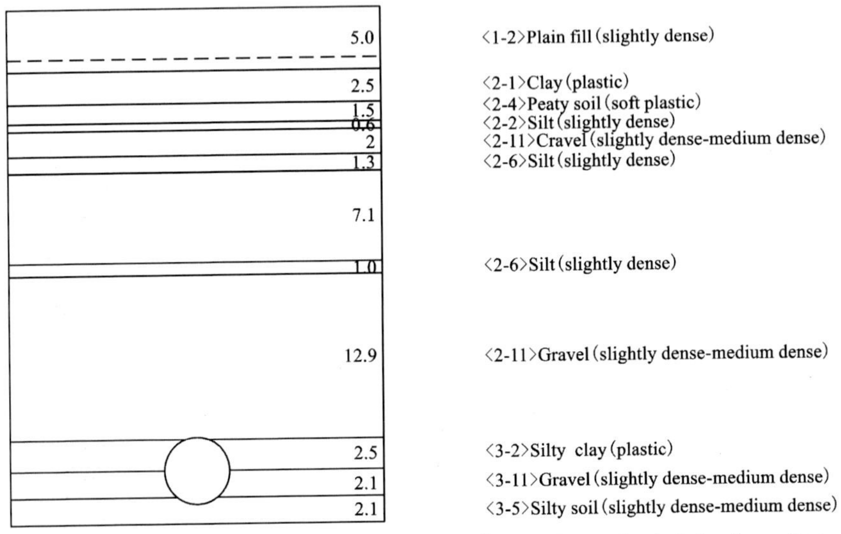

Figure 5.47 Schematic diagram of the strata at the maximum buried depth section.

Table 5.8 Geotechnical parameter tables of maximum buried depth.

Stratum name	Natural unit weight Υ (kN/m^3)	Cohesion (c/kPa)	Internal friction angle ϕ (degrees)	Compression modulus *Es* (MPa)	Poisson ratio *v*	Static lateral pressure coefficient K_0
Plain fill	18.9	/	/	4.0	0.38	0.6
Clay	18.6	22.0	3.0	5.0	0.35	0.53
Silty clay	19.2	20.0	3.0	5.0	0.31	0.45
Peaty soil	15.7	10.0	1.8	3.0	0.41	0.7
Silt	19.3	/	/	6.0	0.30	0.43
Gravel	21.0	/	32	/	0.26	0.33
Silty clay	18.9	22.0	3.0	5.0	0.31	0.45
Silt	19.3	15.0	16.0	6.0	0.3	0.43
Gravel	22.0	/	/	/	0.25	0.33

5.5.7.3 Segment parameters

The tunnel has an external diameter of 6.4 m and an inner diameter of 5.6 m. The segment is made of C60 concrete with a thickness of 0.4 m and a width of 1.5 m.

5.5.7.4 Internal force calculation of segment

Height calculation of the loosening ground arch

The thickness of the overlying formation of the tunnel is 33.388 m, which is more than twice the external diameter of the tunnel. In accordance with the code requirement, it is necessary to consider the influence of the unloading arch effect on the calculation results based on the Terzaghi theory when the overburden thickness of the tunnel is larger than twice the external diameter of the excavated tunnel. The calculation formula refers to formulas (5.2)–(5.4), and the calculation results are as follows.

The radius of the unloading arch is

$$B_1 = R_0 \cot\left(\frac{\pi/4 + \varphi/2}{2}\right) = 3.2 \times \cot\left(\frac{\pi}{8} + \frac{32\pi}{4 \times 180}\right) = 5.43\ m \quad (5.37)$$

The height of the loosening ground arch is

$$\begin{aligned} h_0 &= \frac{B_1\left(1 - c/B_1\gamma\right)}{K_0 \tan\varphi}\left(1 - e^{-K_0 \tan\varphi \cdot H/B_1}\right) + \frac{P_0}{\gamma} e^{-K_0 \tan\varphi \cdot H/B_1} \\ &= \frac{5.43 \times \left(1 - 2/5.43 \times 21\right)}{1.0 \times \tan 32^\circ}\left(1 - e^{-1 \times \tan 32^\circ \times 33.388/5.43}\right) + 0 \\ &= 8.35m < 2D = 12.8m \end{aligned} \quad (5.38)$$

Thus $h_0 = 2D = 12.8m$.

External load calculation of segment

In calculating the load, the structure with a segment width of 1 m is used for calculation. According to the design specifications, the lateral water pressure and the earth pressure of the sandy strata are calculated separately, and the water pressure and the earth pressure of the clayey strata are calculated in a combined way at the construction stage and calculated separately at the usage stage. In this case, only the load at the usage stage is considered, so the calculation method of the water pressure and the earth pressure in a combined way is used for all strata. In calculating the earth pressure of the tunnel, the design code stipulates that the earth pressure of the tunnel constructed by the shield method is calculated according to the static earth pressure. Therefore the load is calculated as follows.

1. Self-weight:

According to the previous design, the lining thickness is 0.4 m, and the concrete unit weight is 25 N/m^3, so its gravity is

$$g = 25000 \times t = 25000 \times 0.4 = 10\ kPa \tag{5.39}$$

2. Self-weight reaction p_g:

$$p_g = \pi \times g = 31.4kPa \tag{5.40}$$

3. Vertical load of upper part of tunnel:

The vertical earth pressure of the upper part of the tunnel is

$$p_{e1} = \sum \gamma_i h_0 = 140.8kPa \tag{5.41}$$

where γ' is the effective unit weight of the stratum where the height of the unloading arch is located, wherein the natural unit weight is used in the stratum above the groundwater level, and the floating unit weight kN/m^3 is used in the stratum below the groundwater level. In this example, the unloaded arch is all located in the gravel layer, and h_i is the loose height of the ground in meters.

The vertical water pressure acting on the upper part of the tunnel is

$$p_{w1} = \gamma_w H_w = 9.8 \times (33.388 - 3.8) = 289.96kPa \tag{5.42}$$

where γ_w is unit weight 9.8 kN/m^3 and H_w is the distance from groundwater level to the vault of the tunnel.

Thus the vertical load of the upper part of the tunnel is

$$p_1 = p_{e1} + p_{w1} = 140.8 + 289.96 = 430.76kPa \tag{5.43}$$

4. The vertical load at the bottom of the tunnel

$$p_2 = p_{w2} = \gamma_w (H_w + D) = 350.72kPa \tag{5.44}$$

5. The horizontal load at the top of the tunnel

The horizontal earth pressure of the top of the tunnel is

$$q_{e1} = \sum K_{0i} \gamma_i h_i = 140.8 \times 0.33 = 46.46kPa \tag{5.45}$$

where K_{0i} is the lateral pressure coefficient of the soil layer where the unloading ground arch above the lining is located

The horizontal water pressure at the top of the tunnel is

$$q_{w1} = p_{w1} = 289.96kPa \tag{5.46}$$

Therefore the horizontal load at the top of the tunnel is

$$q_1 = q_{e1} + q_{w1} = 46.46 + 289.96 = 336.42kPa \tag{5.47}$$

6. The horizontal load at the bottom of the tunnel

The tunnel traverses a total of 2.5 m of a silty clay layer, 2.612 m of a gravel layer and 1.288 m of a silt layer, so the horizontal earth pressure at the bottom of the tunnel is

$$q_{e2} = q_{e1} + \sum K_0 \gamma' D = 71.61kPa \tag{5.48}$$

The horizontal water pressure at the bottom of the tunnel is

$$q_{w2} = p_{w2} = 350.72kPa \tag{5.49}$$

so the horizontal load at the bottom of the tunnel is

$$q_2 = q_{e2} + q_{w2} = 71.61 + 350.72 = 422.33kPa \tag{5.50}$$

7. Subgrade reaction

In the modified routine calculation model, the subgrade reaction acting on the tunnel in the horizontal direction can be approximated as a triangular load distributed at two sides of the ring. The subgrade reaction is proportional to the horizontal displacement of the lining structure in the ground, where the horizontal displacement can be calculated according to the following formula:

$$\delta = \frac{(2p_1 - q_1 - q_2)R_c^4}{24\left(\eta EI + 0.045kR_c^4\right)} \tag{5.51}$$

where k is elastic resistance coefficient of lateral subgrade of the lining ring and here is 2×10^7 N/m^3; η is the bending stiffness reduction coefficient of the lining ring and here is 0.7; R_c is centroidal radius of the lining and here is $(6.4 + 5.6) \div 4 = 3$ m; E is elastic modulus of C60 concrete and here is 3.6×10^{10} N/m^2; and I is inertia moment of section and here is

$$I = \frac{1 \times 0.3^3}{12} = 0.00225m^4 \tag{5.52}$$

So the bending stiffness is

$$\eta EI = 0.7 \times 3.6 \times 10^{10} \times 0.00225 = 5.67 \times 10^7 N \tag{5.53}$$

Table 5.9 Calculated external load values of segments.

Load classification	Load name	Load value (kPa)
Vertical load	Self-weight g of lining	10
	Self-weight reaction p_g of lining	31.4
	Vertical earth pressure p_{e1} at the top of tunnel	140.8
	Vertical water pressure p_{w1} at the top of tunnel	289.96
	Vertical water pressure p_{w2} at the bottom of tunnel	350.72
Horizontal load	Horizontal earth pressure q_{e1} at the top of tunnel	46.46
	Horizontal water pressure q_{w1} at the top of tunnel	289.96
	Horizontal earth pressure q_{e2} at the bottom of tunnel	71.76
	Horizontal water pressure q_{w2} at the bottom of tunnel	350.72
Subgrade reaction	Subgrade reaction $k\delta$	52

Displacement is

$$\delta = \frac{(2 \times 430.76 - 336.42 - 422.33) \times 10^3 \times 2.95^4}{24\left(5.67 \times 10^7 + 0.0454 \times 2 \times 10^7 \times 2.95^4\right)} = 2.6 \times 10^{-3} m \tag{5.54}$$

So the subgrade reaction is

$$k\delta = 2 \times 10^7 \times 2.6 \times 10^{-3} = 52kPa \tag{5.55}$$

In summary, in calculating by the modified routine calculation method, the external load of the segment is shown in Table 5.9.

Internal force calculation of segment

According to the modified routine calculation formulae shown in Table 5.7, the internal forces of the segment caused by each load can be calculated, and the calculation results are shown in Tables 5.10–5.15.

By adding the internal forces caused by all the load components, the internal force of the segment structure can be obtained, as shown in the Table 5.15.

Table 5.10 Section internal force of segment caused by vertical load pressure.

Section location	*M* (kN · m)	*N* (kN)	*Q* (kN)
0	969.21	0.00	0.00
π/8	685.61	189.07	−456.71
π/4	0.00	645.63	−646.14
3π/8	−684.52	1102.48	−457.44
π/2	−969.21	1292.28	0.00
5π/8	−684.52	1102.48	457.44
3π/4	0.00	645.63	646.14
7π/8	685.61	189.07	456.71
π	969.21	0.00	0.00

Table 5.11 Section internal force of segment caused by uniform distributed lateral load pressure.

Section location	*M* (kN · m)	*N* (kN)	*Q* (kN)
0	−756.95	1009.26	0.00
π/8	−535.45	861.60	356.69
π/4	0.00	505.03	504.63
3π/8	534.60	148.23	357.25
π/2	756.94	0.00	0.00
5π/8	534.60	148.23	−357.25
3π/4	0.00	505.03	−504.63
7π/8	−535.45	861.60	−356.69
π	−756.95	1009.26	0.00

Table 5.12 Section internal force of segment caused by triangular distributed load pressure.

Section location	*M* (kN · m)	*N* (kN)	*Q* (kN)
0	−80.55	80.55	0.00
π/8	−62.20	74.08	30.67
π/4	−11.45	53.08	53.03
3π/8	53.37	21.48	51.76
π/2	96.62	0.01	16.21
5π/8	83.35	16.22	−39.27
3π/4	11.66	75.61	−75.79
7π/8	−74.26	145.73	−60.60
π	−112.77	177.21	0.00

Table 5.13 Section internal force of segment caused by lateral subgrade reaction.

Section location	*M* (kN · m)	*N* (kN)	*Q* (kN)
0	−52.85	53.21	0.00
π/8	−40.88	49.15	20.37
π/4	−6.87	37.61	37.61
3π/8	40.55	16.09	38.82
π/2	67.17	0.00	0.00
5π/8	40.55	16.09	−38.82
3π/4	−6.87	37.61	−37.61
7π/8	−40.88	49.15	−20.37
π	−52.85	53.21	0.00

Table 5.14 Section internal force of segment caused by self-weight.

Section location	*M* (kN · m)	*N* (kN)	*Q* (kN)
0	22.50	−3.69	0.00
π/8	16.83	−0.08	−4.74
π/4	2.18	9.68	−14.89
3π/8	−14.96	22.67	−27.48
π/2	−25.63	34.75	−38.43
5π/8	−80.02	36.66	46.26
3π/4	−21.29	25.07	24.10
7π/8	−2.19	10.26	8.65
π	27.57	3.69	0.00

Table 5.15 Section internal force of segment under all load combination.

Section position	*M* (kN · m)	*N* (kN)	*Q* (kN)
0	101.36	1139.33	0
π/8	63.91	1173.82	−53.72
π/4	−16.14	1251.03	−65.76
3π/8	−70.96	1310.95	−37.09
π/2	−74.11	1327.04	−22.22
5π/8	−106.04	1319.68	68.36
3π/4	−16.5	1288.95	52.21
7π/8	32.83	1255.81	27.7
π	74.21	1243.37	0

5.6 Reinforcement and structure design of segments

5.6.1 Design principle

The basic principle of segment structure calculation is as follows: The structure calculation of the lining must be carried out according to load cases in various conditions during construction and after completion. In calculating the nonstatic force and the elastic deformation of the concrete segment, the steel bars are generally not considered, but the whole section is regarded as an effective section of the concrete for calculation. The design load of the lining cross section must be determined on the basis of the most unfavorable conditions in the designed tunnel section and shall be conducted according to the following controlling cross sections [13]:

1. The cross section with the maximum/minimum thickness of overlying strata
2. The cross section with the highest/lowest groundwater level
3. The cross section with the largest surface overload
4. The cross section with asymmetric load
5. The cross section with an abrupt change on the surface
6. The cross section with adjacent existing tunnel or a new tunnel planned to be built.

5.6.2 Reinforcement calculation

After the internal force calculation of the lining structure in each working stage has been completed, the reinforcement must be configured to resist the internal force generated by the external load so as to ensure that the ring structure has enough strength, with checking whether the crack width exceeds the specified range. The reinforcement of the shield tunnel lining generally takes different overburden thicknesses as the calculation section. For example, there are four section types, i.e. ultrashallow buried ($h \leq 1.0D$), shallow buried ($1.0D < h \leq 1.5D$), moderately deep buried ($1.5D < h \leq 2.5D$), and deep buried ($h > 2.5D$) for internal force calculation of the segment structure according to the relationship between the effective overburden thickness and the diameter of the tunnel. Combined with control values at the construction stage and the usage stage, the reinforcement calculations in the ultimate limit state and the serviceability limit state are conducted, respectively, and reinforcement design is conducted according to a reinforcement envelope value.

Reinforcement design is mainly divided into circumferential main reinforcement, longitudinal reinforcement, hoop or tie bar reinforcement, and detailling reinforcement. In China, there are generally a beam method and a plate method for the configuration of the circumferential main reinforcement. The former is to set up a plurality of circumferential hidden curved beams along the longitudinal direction in the segment structure, and the hidden curved beams are connected through longitudinal steel bars; the latter is to design the segment structure in the form of shell plates, no stirrups are provided, and only ribs are arranged.

5.6.2.1 Circumferential main reinforcement

The segment structure is an eccentrically loaded member. Strength reinforcement and crack width check can be conducted for a single element as a short column (the calculation length can be the ring chord length). Meanwhile, the additional eccentric bending moment influence caused by poor roundness of the segment arch should be included. (If the internal force of the assembly structure has been included in structural mechanics, it can be ignored).

Considering the assembly modes of the segment rings and the requirements of reinforcement protection cover thickness, reinforcement is conducted for each block in the segment ring. For a certain deep-buried tunnel section, if the position of each block in the ring is relatively fixed in the longitudinal direction, reinforcement design can be done on the basis of the most unfavorable internal force combination where the maximum bending moment and the minimum axial force simultaneously act in an envelope diagram, according to the position of each block and its envelope diagram. If common wedge segments are assembled in a staggered manner, the position of each block in the ring cannot be fixed in the circumferential direction, and reinforcement design can be done on the basis of the most unfavorable internal force combination where the maximum bending moment and the minimum axial force simultaneously act in an envelope diagram, according to the internal force envelope diagram of each block.

5.6.2.2 Longitudinal reinforcement

Longitudinal reinforcement is calculated by treating the segment as a pure bending member according to the longitudinal internal force (mainly considering the bending moment, because except in seismic conditions, the longitudinal axial and shear force is generally very small), or reinforcement of the segment is conducted according to the minimum reinforcement ratio of the structure.

5.6.2.3 Hoop or rib reinforcement

In configuring the circumferential main reinforcement according to the beam method, the hoop reinforcement configuration amount should be calculated according to the shear force distributed on each hidden curved beam. In configuring according to the plate method, the steel ribs should be arranged according to the structural requirements.

For inter-ring end faces, radial ribs are generally configured from the end face adjacent to the jack from dense to sparse, according to the checking results of local compression capacity of the end face and the radial tensile capacity within the range of a distance equal to the segment thickness from the end face. The checking is conducted under the jack load at the construction stage. The ribs should be arranged in combination of rib reinforcement configuration in the plate reinforcement method.

For concrete around assembly holes or hoisting holes, the punching bearing capacity is checked according to the assembly force (the self-weight of the segment structure is multiplied by the dynamic coefficient) so as to configure stressed steel bars.

5.6.2.4 Detailling reinforcement

To prevent concrete around bolt holes, positioning holes (when assembling vacuum sucker), handholes, and corners of segments from being cracked or even falling off, splitting off, or chipping at the construction or usage stage, structural steel bars, such as spiral bars, hanging bars, wire meshes, and local reinforcing bars are configured at such parts.

5.6.3 Design of connection joints

5.6.3.1 Longitudinal joint calculation

In the calculation of the lining structure, the joint deformation and the joint strength need to be calculated separately at the basic service load stage. The joint strength needs to be calculated at the basic service load stage and the special load combination stage. The calculation method can be chosen according to the actual composition of the joint. The strength of the joint without elastic gasket on the joint face is generally calculated approximately according to the section of the reinforced concrete member. For the joint with a gasket, the joint strength can be calculated in consideration of the gasket effect.

5.6.3.2 Circumferential joint calculation

Owing to the complexity of the construction process, the extent of shield tunnelling influence and disturbance to the ground is also different in the longitudinal length of the tunnel. The deformation of the tunnel structure can be caused when the fabricated tunnel lining is built in such ground. Because of poor sealing of the joint of the fabricated segmental lining, causing water leakage and mud exposure at the bottom of the tunnel, differential settlement of the tunnel and the mutual dislocation of the ring faces would be generated. In addition, the effects of tunnels crossing buildings, three-dimensional intersections of tunnels, large eccentric loads caused by jacking forces of jacks during shield advancing, and instantaneous local dynamic loads will cause longitudinal deformation of the tunnel. Therefore, the circumferential joint structure of the lining must meet the requirements of various factors mentioned above, and the selection of longitudinal bolts is very important in the design of the circumferential joint structure.

5.6.4 Design of segment details

5.6.4.1 Segment end face joint design

Segment end face joints are divided into longitudinal joints and circumferential joints. The basic structures used are bolt joints, tongue-and-groove joints, plug/positioning pin joints, hinge joints, and insert-plug joints.

The bolt joints are commonly used joint structures, are applicable to segment joints and inter-ring joints, and can use arc-shaped bent bolts, straight bolts, oblique bolts, and other bolts to provide fastening forces. The plug/positioning pin joints are mainly used for joint structures of the inter-ring joints, there is no problem of weakening the local bending strength of the segment by the handhole location structure as a result of no handhole, and meanwhile, the shear force can be enabled to be transferred between the segment rings during staggered assembly. The hinge joints are generally used in good ground conditions but cannot be used in poor geological conditions and high groundwater levels. The tongue-and-groove joints are applicable to segment joints and inter-ring joints and can transfer forces by means of engagement because of bumps at the joints.

Improper use of the joint structure makes it difficult to assemble a fully reliable segment ring, which reduces the working efficiency, increases the construction difficulty, or causes a shortcoming in the lining structure. Therefore in determining details of the joint structure, it is necessary to study it from various aspects to fully understand the functions of the joint, and pay special attention to the assembly accuracy and operability. In normal cases the above several basic structures can be chosen in combination. For example, combinations of bolt plus positioning pin joint, or of bolts plus tongue-and-groove joint can be used for the joint structures between segments, and bolt plus positioning pins/plugs or bolts plus tongue-and-groove joints can be combined for the inter-ring joints.

The end face of the segment should be designed in consideration of factors such as whether to arrange buffer materials, the positions and sizes of waterproof grooves, and the shapes of caulking grooves.

5.6.4.2 Handhole structure

The size of the handhole should meet the process requirements of bolt installation and tightening; the edge shape and structure of the handhole should meet the demolding requirements of the segments; the spacing of handholes should not be too large to avoid poor layout of steel bars and not too small to avoid easy falling during demolding; and the distance from the handhole to the end of the structure should not be less than 150 mm. Otherwise the inner arc face between the handhole and the end face of the segment is easy to crack and break.

5.6.4.3 Bolt arrangement

Three types of bolts are used in the segment joint: arc-shaped bent bolts, straight bolts, and oblique bolts (Fig. 5.48). Currently, in China the segment rings of the medium-diameter shield tunnels mostly use arc-shaped bent bolts, and the segment rings of the large-diameter shield tunnels mostly use oblique bolts. Except for two deep and straight handholes in the connection part of adjacent segments, the straight bolt connection has an extra hole for bolt withdrawal in the outer side of one of the handholes, which greatly affects the strength and the stiffness of the segments and is rarely used nowadays.

The diameters of bolt holes should not be too large, generally in the range 6–9 mm, otherwise they will produce a larger dislocation.

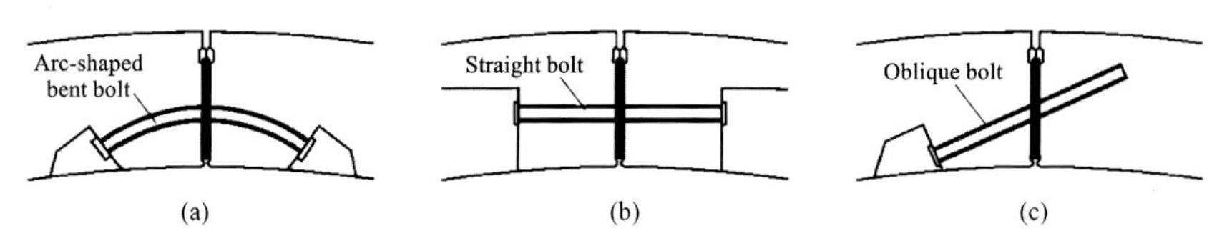

Figure 5.48 Bolt forms of segments. (a) Arc-shaped bent bolt. (b) Straight bolt. (c) Oblique bolt.

Circumferential bolt design calculation method

Under normal service conditions the static effects of water pressure, earth pressure, and surface ground overload around the shield tunnel structure, when acting on the cross section of the segment, produce section bending moment, shear force, and axial force. The main section can resist the bending moment, the shear force, and the axial force through steel bars and concrete, but the segment joint only resists the bending moment, the shear force, and the axial force at the section jointly through connecting bolts and section concrete. According to the static analysis of the segment, it can be seen that the sign of the bending moment are varying between positive and negative for the sections at the top, bottom, and waist of the segment; that is, the outside of the segment at the top and bottom is compressed, while the inside at the waist of the segment is compressed. The transverse bolt design calculation method is established in this context, and its calculation diagrams are shown in Figs. 5.49 and 5.50. The axial forces N and bending moments M of the joints at the top and bottom of the segment and joints at the waist of the segment are calculated as follows.

As shown in Fig. 5.49, the axial forces N and the bending moments M of the joints at the top and bottom of the segment can be obtained according to the section balance condition and the bolt centroidal moment:

$$N = \alpha f_c B x_b - A_b f_{by} \tag{5.56}$$

$$M = \alpha f_c B x_b (h - h_b - x_b/2) \tag{5.57}$$

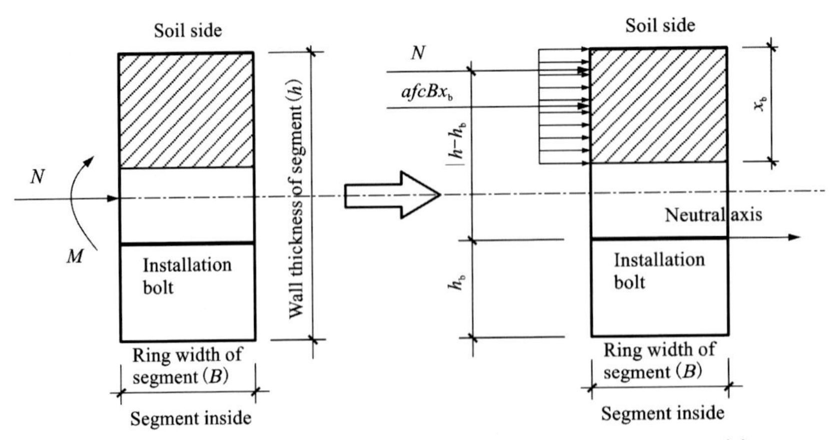

Figure 5.49 Static calculation diagrams of bolts at joints at the top and bottom of a segment.

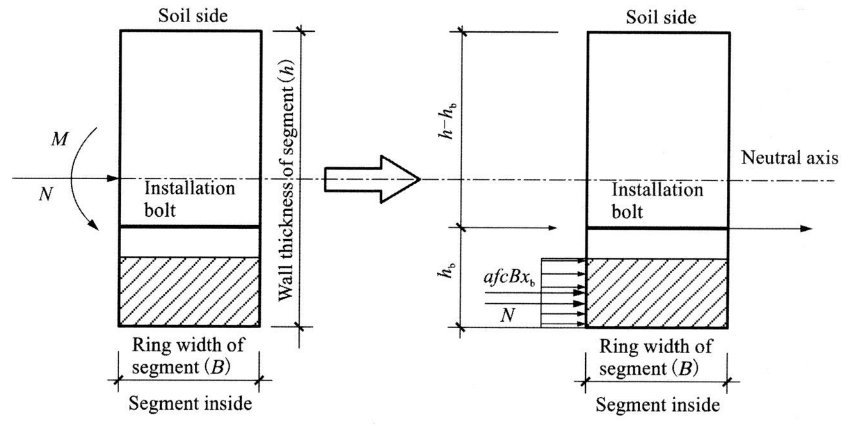

Figure 5.50 Static calculation diagrams of bolts at joints at the waist of a segment.

As shown in Fig. 5.50, the axial force N and the bending moment M of the joints at the waist of the segment can be obtained according to the section balance condition and the segment centroidal moment:

$$N = \alpha f_c B x_b - A_b f_{by} \tag{5.58}$$

$$M = \alpha f_c B x_b (h/2 - x_b/2) \tag{5.59}$$

In formulas (5.56)–(5.59), h and h_b are the thickness of the segment (i.e., the section height) and the distance between the bolt center and the segment inner the tunnel side (i.e., the installation height); B is the width of the segment ring; x_b is the converted height of the compressed area of segment concrete; α is the influence coefficient of concrete strength grade; f_c is the design value of axial compression strength of segment concrete; and A_b is the cross-sectional area of the bolt.

Longitudinal bolt design calculation method

Longitudinal bolts mainly resist differential settlement and the actions of longitudinal and horizontal earthquakes. Under the action of the longitudinal and horizontal earthquake action, the longitudinal bolts bear the tensile forces and also bear the bending moment. Under the action of the maximum bending moment $M_{max,}$ the centroidal axis of bolt groups is at the center of the innermost row of bolts. On the basis of the plane hypothesis and the torque balance relationship the calculation diagram of the longitudinal bolt design calculation method is established, as shown in Fig. 5.51,

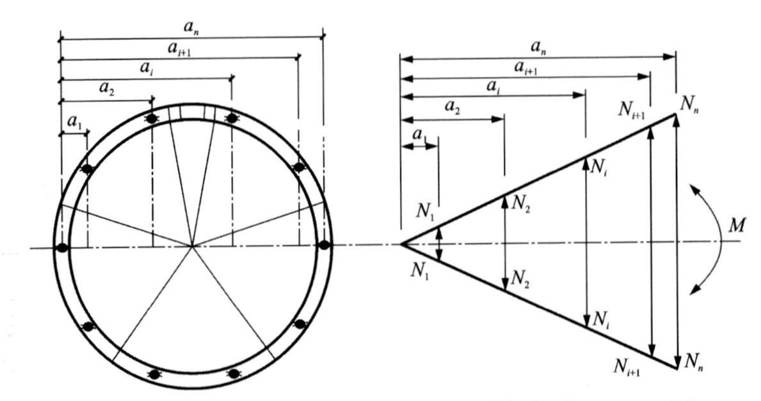

Figure 5.51 Static calculation diagram of longitudinal bolt of segment joint.

and the maximum tensile stress σ_S^M of the outermost row of longitudinal bolts is calculated as follows:

$$\sigma_S^M = \frac{N_n^M}{A_b} \leq \left[\sigma_{by}\right] \tag{5.60}$$

$$N_n^M = \frac{M_{\max}}{\sum a_i^2} \times a_n \tag{5.61}$$

where $[\sigma_{by}]$ is the allowable seismic stress value of the longitudinal bolt, which for the sake of safety can be 50%–70% of the design value of bolt strength; a_i is the distance between the ith bolt and the center of the bolt at the innermost row of the segment ring; and a_n is the distance between the centers of the innermost and outermost rows of bolts.

5.6.4.4 Grouting holes or assembled positioning holes

In small- and medium-diameter shield tunnel segments, hoisting holes or secondary grouting holes are often combined together. However, the hoisting hole should be formed in the center-of-gravity position to avoid instability due to additional bending moment during assembly. For small- and medium-diameter shield tunnel segments, polymer materials are preferred for hoisting holes or grouting holes to ensure the durability of the structure. For large-diameter shield tunnel segments, vacuum chucks are often used for hoisting, positioning holes and hoisting holes are generally designed separately, and the plane size, arc, and depth of the positioning holes and distance between two holes should all conform to the requirements of the vacuum sucker equipment.

5.6.4.5 Other detail designs

Marks such as mold number, block number, diameter or radius, bolt hole position, and staggered assembly (laser centering or arrow mark) should be provided on the segment as required.

5.7 Connection tunnel design

5.7.1 Design principle

The majority of traffic shield tunnels are located in lower semienclosed areas that are surrounded by ground mass. The shield tunnel structure has a better defense against external hazards but poor resistance against internal disasters. In the small underground space, people and equipment are densely positioned, and it will be very difficult to rescue and evacuate if a disasters occurs. From the historical lessons of tunnels in various parts of the world, fire is the form of internal disaster with the highest frequency and largest losses. For example, the Channel Tunnel runs between Foxton in the United Kingdom and Keles in France, with a diameter of 7.6 m and a total length of 50.45 km and consisting of two parallel operating tunnels in the north and south. A long service tunnel was built between the two operating tunnels, and 146 pedestrian passages were connected to the service tunnel. On the night of November 18, 1996, a piggyback shuttle train carrying freight trucks from France to England caught fire in the Channel Tunnel. The fire caused extensive damage to the southbound operating tunnel and burned five compartments at the rear of the train. Then the train stopped, and the fire spread to the front of the train. The tunnel control center immediately confirmed the fire alarm and took measures. A total of 34 truck drivers, passengers, and shuttle train attendants passed through the connection tunnels to the service tunnel in the middle in accordance with standby rescue measures. The ventilation system maintained a slightly higher air pressure in the service tunnel to prevent smoke from being discharged from the tunnel where the fire was. After 36 minutes, people in distress were evacuated by train in the northbound operating tunnel. The fire caused heavy damaged but no casualties. The transverse connection tunnel played an important role in the rescue.

Therefore the general double separated traffic shield tunnels are considered to design transverse connection tunnels as important passages for rescue and evacuation. Code for Metro Design (GB501572013) [14] in China stipulates that "When the continuous length of the tunnel is larger than 600 m between two single-track shield tunnels, a connection tunnel should be designed, and double-open A-level fire doors should be installed at both ends of the passage."

5.7.2 Key points for connection tunnel design

5.7.2.1 Structure design for special segments in tunnels

The main tunnel structure bears complex forces, the external loads at different stages are constantly changing leading to the changes in the bearing system accordingly. The tunnel structure should bear the external water and soil load during the shield driving, and the thrust of the shield also needs to be borne at the construction stage. The design of the main tunnel structure should not only meet the stress requirements under different working conditions, but also adapt to the process requirement that it is easy to be cut.

5.7.2.2 Structure design of connection tunnel lining

The connection tunnel is narrow in section with complicated surrounding environment, and it is easy to cause engineering accidents due to the limited construction area. During design, the impact of ground reinforcement on the surrounding ground pressure should be fully considered. Before construction, it must be ensured that the ground reinforcement conforms to the requirements, especially before opening the hole on the lining. In addition, the groundwater conditions and the reinforcement effect of the sand layer should be determined, and the necessary auxiliary measures for the existed shield tunnel structure should be taken to ensure construction safety.

5.7.2.3 Structure design of a T-shaped joint portal

A spatial curved face is arranged at the intersection of the connection tunnel and the main tunnel lining structure. With different structural stiffness, inconsistent deformation coordination, long-term bearing of train vibration load, and complicated stress characteristic, the spatial curved face is the weak link of the whole system. It is very important to take an appropriate portal structure form to ensure the durability and waterproof performance requirements of the joints.

5.7.3 Connection tunnel structure forms

5.7.3.1 Structure forms

1. According to the engineering geological and hydrogeological conditions, buried depth, and structural characteristics of the cross passage structure, the appropriate parameters are selected for calculation, then the tunnel lining and supporting parameters, or the thickness, and the inner/external diameter of the segment structure are determined after calculation of the most unfavorable load combination.

2. The lining structure form of a connection tunnel is shown in Fig. 5.52. When the conventional excavation method is adopted for construction, the tunnel lining and supporting parameters can be comprehensively determined through calculation, analysis, and optimization. The information from similar projects according to factors such as structure section, surrounding rock types, hydrogeological conditions, and structure stress characteristics should be considered as well. It is especially good to design waterproofing and advanced reinforcement measures before opening the hole on the segmental lining. When the cross passage is constructed mechanically, segment universal rings can be used. The rings are assembled in a staggered manner to determine parameters such as the standard ring width, block size, block numbers, and wedge size and other parameters. Width adjustment rings should be additionally designed to ensure that the relative position of the receiving ring and the main tunnel falls in the allowable range and facilitate the arrangement of portal joints. The segment blocks and the rings are all connected through bent bolts [15].

5.7.3.2 Structure forms of special segments in main tunnel

A part of the main tunnel structure must be destroyed for starting and receiving of the connection tunnel. If manual excavation is used, a large area of external strata needs to be reinforced to ensure independence and low permeability of the ground. If the segment is directly cut during construction, only a small amount of grouting is required in the external strata to stop water.

The special segment structure should be designed according to the structural size of the connection tunnel and the standard ring width of the shield tunnel. Special steek segments can be used; or glass fiber—reinforced concrete composite segment is used in cutterhead area and steel segments are used outside the cutterhead. High-strength steel bolts are still used, which are broken before

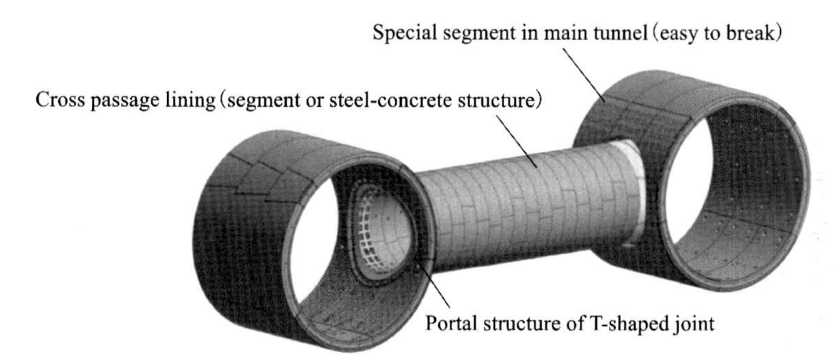

Figure 5.52 Schematic diagram of a connection tunnel between shield tunnels.

construction of the connection tunnel. Such special structure design is conducive to minimizing overall stress on the structure and convenient construction.

5.7.3.3 Structure forms of T-shaped joint portal

The joint part of the cross passage and the main tunnel lining structure is the weak link of the whole connection tunnel structure. The joint section not only bears complicated external loads, including train vibration load during operation, but also must meet the requirements of consistent deformation coordination of the main tunnel and the connection tunnel so as to ensure the waterstop effect. To effectively connect the main tunnel and the connection tunnel structure, steel segments are usually used for the design of the entrance and exit rings of the connection tunnel. The gap between the main tunnel and the connection tunnel segment is filled with grouting and welded with connecting steel plates, and reinforced concrete beams are poured on the outside, as shown in Fig. 5.53 [15].

5.8 Seismic design of shield tunnel

For the large-size underground tunnels, theoretical analysis can comprehensively and macroscopically grasp the vibration characteristics of the structure and structure response under different seismic inputs. Theoretical analysis is relatively low in cost and an indispensable research approach.

Wave theory and the finite element method are the two main methods of analyzing the earthquake resistance of underground structures. There are two methods to study the influence of stratum movement on the underground structures. One is a wave method, based on solving the wave equation, the underground structure is regarded as a reinforcement region of a hole in the infinite linear elastic (or elastoplastic) medium, and the whole system (including the

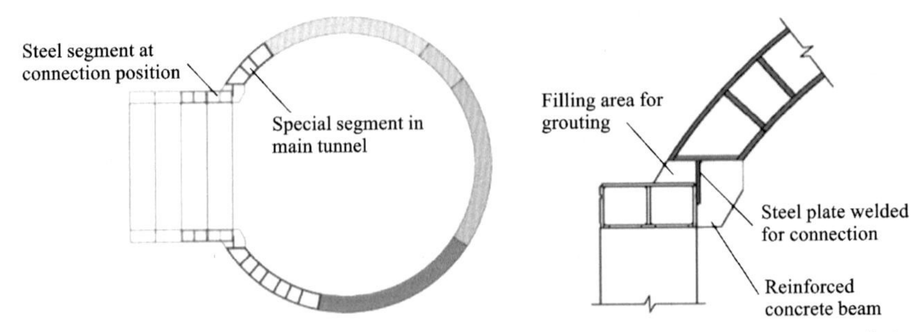

Figure 5.53 Schematic diagram of T-shaped joint between a cross passage and the main tunnel.

medium and structure) serves as an object for analysis so as to solve its wave field and stress field without separately researching the load. The other method is an interaction method, based on solving the structural motion equation, the effect of the medium is equivalent to springs and damping, and then they are applied to the structure to form the dynamic motion equation of the underground structure. Its solution method is the same as that of the ground structure.

On the basis of these two methods, some practical seismic analysis methods have been developed, such as the displacement response method, the surrounding ground strain transfer method, the subgrade reaction coefficient method, and the finite element method. They are all pseudo-static analysis methods. Their advantages are that the methods are relatively simple, but their disadvantages are that the influence of viscous damping of soil in the interaction cannot be considered. To conduct a comprehensive and in-depth study of the seismic characteristics of underground structures, the dynamic finite element analysis method of underground structures is required. The dynamic finite element analysis method is more reliable in calculation accuracy and can deal with homogeneity, anisotropy, nonlinearity, and complicated boundary conditions in the medium, with the disadvantage being that calculation is more complicated [16].

5.8.1 Seismic coefficient method

This method simplifies the dynamic earthquake effect to a static force for analysis. The force is applied to the center of gravity of the structure, and its size is the weight of the structure multiplied by the designed seismic coefficient, that is,

$$F = K_c Q \tag{5.62}$$

where F is the earthquake inertia force at the center of gravity of the structure; K_c is the seismic coefficient; and Q is the weight of the structure.

The seismic coefficient K_c can be determined by the seismic standard coefficient, the important engineering coefficient, the stratum property and type coefficient, and the buried depth coefficient. In addition, active earth pressure is applied onto one side of the tunnel side wall, and horizontal resistance is applied onto the other side. This method is widely used in engineering design owing to its simplicity in calculation.

5.8.2 Displacement response method

In the 1970s, Japanese scholars started with seismic observation and proposed a seismic design method for underground linear structures, namely, the displacement response method. The basic principle of this method is to simulate the underground linear structures by elastic foundation beams, take the foundation displacement during earthquake as a known condition to act on the elastic foundation so as to solve stress and deformation generated on the beams, thereby calculating the seismic response of the underground structures (tunnels, pipelines, shafts, etc.). The formula can be simplified to a pseudo-static calculation formula:

$$[K]\{U\} = [K_s]\{u_g\} \tag{5.63}$$

where the matrix $[K]$ comprises stiffness $[K_t]$ of underground structure and subgrade reaction $[K_s]$.

The key goal of this method is to determine the seismic displacement $\{u_g\}$ and the resistance coefficient $[K_s]$. Usually, $[K_s]$ is regarded as a diagonal matrix, and then $[K_s]$ is equivalent to the Winkler spring constant or a spring constant of foundation soil medium. The theoretical basis of this method is based on research results of response analysis on underground structures during earthquake; that is, the foundation deformation rather than the inertia force of the structure dominates the seismic response of the underground structures, and the displacement response method is first used in the seismic design of buried tunnels.

5.8.3 Surrounding stratum strain transfer method

The basic idea of seismic wave field analysis, the seismic observation results of pipelines, submarine tunnels, underground oil depots, and the like show that the strain waveform of the underground structure during earthquake is almost completely similar to the seismic strain waveform of the surrounding stratum medium, so the following relation can be established:

$$\varepsilon_s = a\varepsilon_g \tag{5.64}$$

where ε_s is seismic strain of underground structure; ε_g is seismic strain of surrounding stratum medium without influence of underground cave structure; and a is the train transfer coefficient, which can be regarded as a static coefficient. The value of a has nothing to do with the frequency and wavelength of ground vibration, only changes with the shape and stiffness of the underground structure and stiffness of the surrounding stratum, and can be determined by the static finite element analysis method.

The key issue is to determine surrounding stratum strain that is consistent with the designed seismic intensity. The following idea to invert the seismic wave and to input it from 300 m below the ground surface can be used. The shear strain is deduced according to the horizontal component of the seismic wave, and the vertical positive strain is deduced according to the vertical component of the seismic wave. The positive strain in the horizontal direction can simulate the processing method of the surface wave and can be obtained through dividing the velocity response value, which is generated by the horizontal component of the seismic wave at the center of the cave, by the P wave velocity of the surrounding rock.

5.8.4 Subgrade reaction coefficient method

The subgrade reaction coefficient method is a method of applying the interaction calculation method to the seismic response analysis of the cross section of the underground structure and can be applicable to buried or semiburied underground structures. The effect of the surrounding stratum medium is simulated by multipoint compression springs and shear springs, and the structure can be simulated by beam elements. The method includes three basic steps: (1) calculating the spring constant of the surrounding stratum medium, (2) calculating the seismic displacement of the surrounding stratum, and (3) calculating the seismic response of the underground structure. The static finite element method is used for approximate calculation of the resistance spring constant of the surrounding stratum, and a piecewise one-dimensional model or a plane finite element model is used for approximate calculation of the seismic displacement of the surrounding stratum.

5.8.5 Dynamic finite element method

The above methods are all pseudo-static methods. Although they are simple to calculate, the influence of various factors such as nonlinearity, homogeneity, and complicated boundary changes cannot be accurately considered, owing to the many assumptions. Therefore it is difficult to consider the dynamic interaction between the stratum and the structure. With the development of computers and calculation theories, numerical methods represented by the dynamic finite element method have emerged. By using these methods, the above theoretical defects are avoided, and powerful tools are provided for comprehensive and in-depth research on the seismic characteristics of the underground structures in various complicated situations.

References

[1] Zhang F, Zhu H, Fu D. Shield tunnel. Beijing: China Communications Press; 2004.
[2] Zhang Y. Study on mechanical behavior of double linings of railway shield tunnel. Chengdu: Southwest JiaoTong University;; 2010.
[3] Zhang Z. Research on reasonable application timing of secondary lining of underwater shield tunnel based on reliability theory. Railw Constr Technol 2016;00(08):36–41.

[4] Zhu M, Mu H. Comments on ECL (extruded concrete lining). Tunn Constr 2007;27(4):30−2 51.
[5] Pu A. Design research and engineering application of fiber reinforced concrete segment. Chengdu: Southwest JiaoTong University; 2007.
[6] Zhang F, Zhu H, Fu D. Shield tunnel construction manual. Beijing: China Communications Press; 2005.
[7] Huang H, Yan J, Xu L. Design suggestions on joints between longitude Liu Jian-Hang shield tunnel. Beijing: China Railway Publishing House; 1991.
[8] Jin S.T.E.C. International vision: big PK of segment quick connector in Japan[EB/OL]; 2019, p. 3-23. <https://mp.weixin.qq.com/s/5m41wGQfKbdGumzh1Kf6eQ>.
[9] (Japan). The Japanese Geotechnical Society. translated by Niu Q, Chen F, Xu H Shield method investigation, design and construction. Beijing: China Construction Industry Press; 2008.
[10] Liu J. Shield tunnel geology and exploration. Beijing: China Railway Press; 1991.
[11] TB 1003-2016. Code for design of railway tunnel. Beijing, China: China railway press; 2016.
[12] Xiao M, Feng K, Li C, Sun W. A method for calculating the surrounding rock pressure of shield tunnels in compound strata. Chin J Rock Mech Eng 2019;38(9):1836−47.
[13] Chen R. Discussion about the rules of tunnel cross passages in the code for metro. Tunn Constr 2005;25(2):7−9 51.
[14] GB50157-2013. Code for design of metro. Beijing, China: China Architecture & Building Press; 2013.
[15] Shen Z. Study on structural design of mechanical cross passage. Mod Urban Transit 2019;00(11):58−63.
[16] Shen H. Seismic response analysis of shield tunnel. Dalian University of Technology; 2006.

Exercises

1. What are the advantages and disadvantages of fabricated segment lining structures.
2. Please analyze the stress characteristics of segment joints.
3. What are calculation theories for vertical earth pressure of shield tunnels? What are their applicable conditions?
4. How is the subgrade reaction produced? How does the subgrade reaction affect the segment stress?
5. What are the applicable conditions of separate and combined calculation of water pressure and earth pressure in the shield tunnel load calculation.
6. What are the similarities and differences of various internal force calculation methods for shield segmental tunnel lining.
7. Suppose a shield tunnel construction crossing a composite strata composed of silty soil, coarse gravel sand, and medium sand; its underlying stratum is a medium sand layer; the buried depth of the tunnel is about 12 m; the external diameter of the shield tunnel is 10.8 m; and the segment has a thickness of 0.35 m and a width of 1.5 m. Please calculate the surrounding ground pressure acting on the shield tunnel and the internal force of the segment lining structure by the modified routine method (Related parameters are presented in Table 5.16).

Table 5.16 Physical and mechanical parameter table of strata.

Stratum	Depth (m)	Natural unit weight (kN/m^3)	Deformation modulus (MPa)	Cohesive force (kPa)	Internal friction angle (degrees)	Static lateral pressure coefficient	Subgrade coefficient (MPa/m)
Artificial filling layer	1.0	18	5	3.0	5.2	0.46	6.2
Silty soil layer	14.6	16.9	2	11.4	7.6	0.67	3.5
Coarse gravel layer	17.8	19.2	30	—	35	0.30	13
Medium sand layer	28.5	19.2	25	—	34	0.30	11
Cohesive soil layer	36.0	18.8	5	23.5	12.5	0.50	13

CHAPTER 6

Launching and receiving of shield machines

Shield tunnel construction can be divided into three stages: trial tunnelling, normal tunnelling, and shield machine receiving. In the trial tunnelling, the shield machine on a launching bench penetrates the ground from the portal in the launching shaft wall by using negative (temporary) segments, reaction frames, and other equipments, and then tunnels are driven along the designed route. The advancing distance of the shield machine during the trial tunnelling is consistent with the advancing distance required to remove the negative segments in the launching shaft. This ensures that the segments have enough rings to provide sufficient frictional resistance from the excavated ground to prevent the segments from moving backward after the negative segment rings of segments are removed. The shield machine drivers and technical personnels analyze the field data to determine the appropriate tunnelling parameters. Usually, the advancing distance of the trial tunnelling ranges from 50 to 100 m. At the trial tunnelling stage, shield machine launching (Fig. 6.1) is a significant starting process for shield tunnelling. To achieve successful shield machine launching, some works should be carried out as follows: ground improvement at the tunnel portal section, installation of the launching pedestal, shield machine assembly and debugging, installation of the reaction frames, cutting off the retaining wall in the scope of the tunnel portal, installation of the tunnel portal seal, checking the position of the shield machine, assembling the negative segments, and so on [1].

In the shield machine receiving (Fig. 6.2), the shield machine approaches the destination, and the retaining wall in the scope of the portal is cut from the inner side of the receiving shaft. Then the shield machine reaches the receiving pedestal in the shaft.

The stage between the trial tunnelling and the shield machine receiving is the normal tunnelling stage. In the shield machine launching and receiving, some accidents often happened in actual projects. These two stages are very significant for successful shield tunnelling.

Shield Tunnel Engineering
DOI: https://doi.org/10.1016/B978-0-12-823992-6.00006-0

Figure 6.1 Shield machine launching.

Figure 6.2 Shield machine receiving.

6.1 Working shafts and ground improvement for shield machine tunnelling

6.1.1 Working shafts

The working shafts are used for shield machine assembly, disassembly, U-turning, segment vertical transport, and muck bucket conveying. They include launching shaft, receiving shaft, and inspection shaft. The shafts used for shield machine assembly and disassembly are called as the launching shaft and receiving shaft, respectively. The shaft used for inspecting the working status of shield machine in long tunnels is called as the inspection shaft.

Usually, these working shafts are designed as vertical columnar structures with rectangular shape built from the ground surface. They have two main functions. First, the working shafts are used as construction operation stations for lifting up or down of shield machine, its assembly or disassembly, its launching or receiving, transportation way of materials such as segments, construction equipments, and others. Second, after tunnelling is finished, the working shafts can be used as permanent underground structures such as ventilation shafts, drainage shafts, or facility storage rooms. For metro or municipal road tunnels, the ends of the open-cut section are normally used as the working shafts. For long tunnels, considering the construction period, sometimes working shafts are built in the middle of the tunnel. For tunnel route design, if there is no open-cut deep excavation, it is necessary to build special shafts for launching, inspection, and receiving of shield machine.

To ensure the smooth shield tunnelling, the following requirements should be considered for the shaft dimensions and other aspects [2]: (1) The length of the launching shaft is recommended to be 3 m longer than that of the main shield machine, and its width is recommended to be 3 m larger than the diameter of the shield machine. (2) The net dimensions in the plane of the receiving shaft must meet the requirements for the receiving, disassembly or U-turning of the shield machine. (3) Portal cavities are normally reserved in the (secondary) support structures of working shafts for avoiding the need of being cut for launching and receiving the shield machines, and the portal scopes in the retaining walls can be built with glass faber reinforce concrete for easily being cut. (4) The elevations of the bottom plates of the launching and receiving shafts should be lower (generally by 700 mm) than the elevations of the bottom of the launching and receiving portals and should meet the minimum operating space requirement.

The main retaining walls to stabilize ground during the deep excavation of the shafts include retaining piles, diaphragm walls, caisson enclosures, and so on. The retaining piles mainly include steel sheet piles and column piles. The construction method of caisson enclosures mainly include drainage sinking, undrained sinking, and pneumatic caisson.

6.1.2 Ground improvement closed to the shafts

Because the chamber pressure of shield machine gradually increases during launching and decreases during receiving the shield machine, it is likely that the pressure balance on the excavation face cannot be achieved and the ground settlement is relatively large in the area closed to the shafts. In such a condition the excavation face is prone to collapse. To avoid the excavation face instability caused by the inflow of groundwater and soil closed to the shafts, it is significant to strengthen the ground closed to the launching and receiving shaft considering factors such as geological conditions, buried depth of the tunnel, and the surrounding environment. It is necessary to ensure that the method and area of the ground improvement are appropriate and that the improvement effect is good so that the ground deformation surrounding the shield machine is controllable when the pressure balance does not exist on the excavation face during shield machine launching and receiving.

6.1.2.1 Spray grouting

In the spray grouting method, a substance such as cement slurry is injected into the ground under high pressure, and the ground soil is solidified with the filling, seepage, compaction, and splitting off of grout to reduce ground permeability and increase ground strength [3,4]. Consequently, the groundwater carrying soil is prevented from invading the shafts from the gap between the surrounding ground and the shield shell. The spray grouting method is classified into perforated steel pipe grouting and sleeve valve pipe grouting according to the grouting pipe, and it can also be classified into horizontal deep-hole grouting and vertical deep-hole grouting according to the grouting direction.

Differently with perforated steel pipe grouting, the grout for the sleeve valve pipe grouting is injected into the weak ground under relatively high pressure. The high pressure can open the flexible positioning rings (Fig. 6.3), and the stop rings above and below the grouting core pipe openings (holes) can realize segmented and layered grouting, and continuous or skip grouting can be selected according to construction needs.

If there is significant groundwater seepage in ground, cement-sodium - silicate grout are normally used because of its rapid solidification. The cement-sodium silicate grout is injected into the soil by using two-liquid grouting pumps and channels to drive away the water and gas in the soil particles by filling, infiltration, and compacting. The minerals contained in

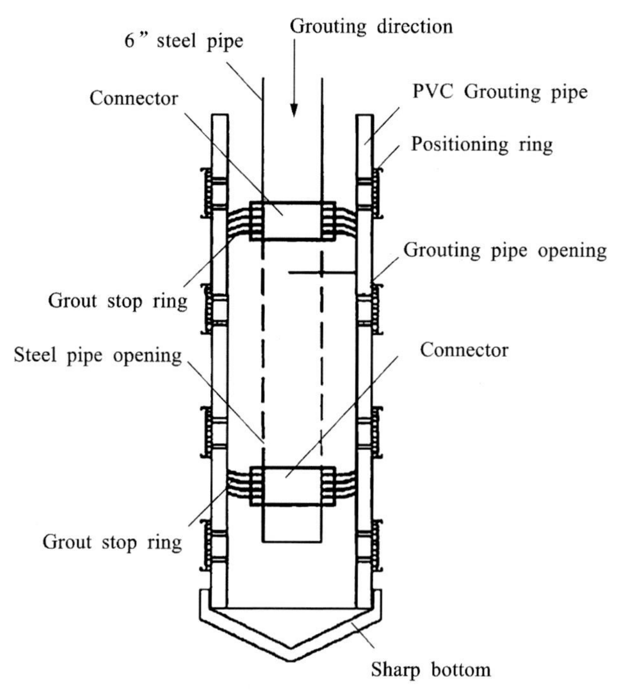

Figure 6.3 Schematic diagram of one-way sleeve valve tube grouting. PVC: Polyvinyl chloride.

the cement, water, and soil are hydrolyzed, hydrated, and aggregated to form suspended colloids and aggregates. After hardening, a cemented soil with high strength, low compressibility, low permeability, and good self-stability is formed. The double-liquid grout has a short gel time, which can more quickly impede water flow in time for the strongly permeable soil. The grouting effect in the soil layer is reliable, and the construction needs just a small operating space. It has high flexibility in emergency reinforcement construction. However, the ground uplift is not easy to control, and sometimes cracks may be generated in the ground. The construction period is normally long using the sleeve valve pipe grouting.

6.1.2.2 High-pressure jet grouting pile

As shown in Fig. 6.4, the high-pressure jet grouting pile method was developed on the base of chemical grouting using high-pressure water jet cutting technology. It changes the grout type and technological measures of the chemical grouting. It uses cement as the main raw material to strengthen the ground. It has powerful functions such as increasing the ground strength and preventing water seepage [5]. Since the 1970s, high-pressure jet

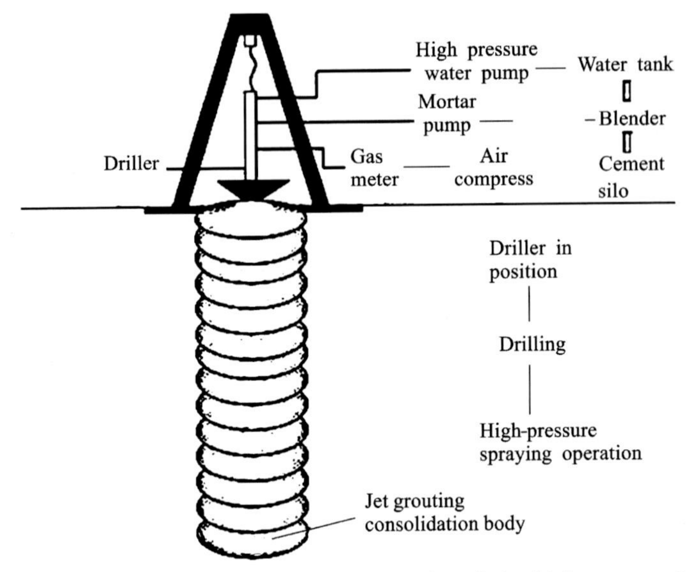

Figure 6.4 Schematic diagram of working principles of the high-pressure jet grouting pile method.

grouting pile technology has been widely used in China. It forms a reinforced ground area on the outside of deep excavation of the shaft for shield machine launching and receiving. It can effectively reduce the horizontal earth pressure and water pressure on the retaining wall when the retaining wall in the tunnel portal scope is cut off.

Grout amount

There are two methods of determining the required volume of grout: the filling volume method and the injecting volume method. The larger one is used as the required grout volume. According to the calculation of the required grout volume and the designed water-to-cement ratio, the amount of cement used can be determined.

The calculation formula of the filling volume method is

$$Q = \frac{\pi De^2}{4} K_1 h_1 (1 + \beta) + \frac{\pi {D_0}^2}{4} K_2 h_2 \tag{6.1}$$

The calculation formula of the injecting volume method is

$$Q = \frac{H}{v} q (1 + \beta) \tag{6.2}$$

where Q is the required grout volume, m^3; De is the diameter of the jetting grouting pile body, m; D_0 is the diameter of the grouting pipe, m; K_1 is the filling rate in the stirred zone, generally from 0.75 to 0.9, mainly depending soil void ratio; h_1 is the length of rotary jet, m; K_2 is the filling rate of the soil in the unstirred zone, generally from 0.5 to 0.75; h_2 is the length of the unstirred pile, m; β is the loss coefficient, generally from 0.1 to 0.2; v is the lifting speed, m/min; H is the injection length, m; and q is the unit volume of grout, m^3/m.

The high pressure jet grouting maybe induce grout bleeding on the ground surface, and the key points for ensuring the good performance of high-pressure jet grouting pile are as follows:

1. Little grout bleeding. A common reason for grout bleeding is that in the reinforced soil layer the particle sizes are too large and there exist many pores. The measures to avoid little grout bleeding can be taken as follows: (a) The grout concentration is increased, for example, from 1.1 to 1.3, to continue spraying; (b) Clay slurry or fine sand or medium sand are injected to first fill the pores in the coarse soil; (c) Aggregate is added to the grout; (d) After the grouting holes are sealed with clay balls, jet grouting is continued; and (e) After injecting cement mortar as the grout, the cement mortar is replaced with clay slurry for jet grouting.
2. Serious bleeding. Usually, the effective injection range is related to the inadaptability of the grout volume. The following measures can be taken: (a) The spraying pressure is increased appropriately; (b) The nozzle diameter is reduced appropriately; and (c) The lifting speed is increased appropriately. Since the bleeding materials contains a mixture of soil particles, groundwater, and grout, it is difficult to carry out the separation and recycling of the grout in the bleeding materials.

6.1.2.3 Metro Jet System

The Metro Jet System (MJS) is also called as an omni-directional ultra-high pressure jet grouting method. On the base of traditional high-pressure jet grouting, the MJS method uses a unique perforated pipe and front-end monitoring device to achieve forced mortar discharge in the hole and monitoring of ground pressure. The MJS method controls the ground pressure by adjusting the injection amount of pressurized mortar, greatly reducing the impact on the environment and ensuring the pile diameter.

Construction characteristics

1. A dedicated slurry pipe is equipped with the jet system, and so the slurry trough can be omitted.
2. The amount of slurry discharged can be adjusted.
3. The pressure of grout in the ground can be measured and adjusted.

Construction procedures

1. Work base (cap platform)

   If the special drilling rig is equipped with a special operating platform, the construction site should be cleared out before construction. The bearing platform of the special drilling rig can be placed, and then the special drilling rig is installed.
2. Installation of a special drilling rig

   Use a level to measure and confirm the slope of the steel rod. The theodolite installed at the back of the special drilling rig measures the construction measurement points and compares them with the designed baseline, and the special drilling rig is set up on the basis of the comparison results.
3. Drilling

   Double-pipe drilling rods (section length: 1.5 m) with outer diameters of ϕ165 mm (outer rod) and ϕ118 mm (inner rod) are used to drill holes. A target with a signal light is inserted at the drill tip, and the position of the target with a theodolite installed at the back of the drilled hole is measured to confirm the accuracy of the drilling.
4. Pulling out of the inner rod

   After drilling to a predetermined depth, the inner rod is pulled out in sections on a special machine.
5. Insertion of the porous rod

   The ϕ130 cm porous rods are connected section by section (1.5 m length per section) to make the nozzle reach the predetermined position.
6. Reinforcement construction

   After the porous rod has been inserted, the outer rod and the inner rod are simultaneously pulled at predetermined pulling and rotation speeds. In the meantime the grout is injected under the ultra-high pressure.
7. Inspection after completion

   After the injection has been completed, the performance of injection should be checked by suitable method.

6.1.2.4 Three-rod stirring pile

The three-rod stirring pile method uses a long screw pile machine, and there are three drilling rods at the same time during construction. It is an effective method in soft ground improvement. A mixing pile machine is used to inject cement into the soil and fully stir it to cause a series of physical and chemical reactions between the cement and the soil, to harden the soft soil, and to increase the ground strength. As shown in Fig. 6.5, the three-rod stirring pile method mainly adopts the hopping reinforcement method and the continuous reinforcement method for pile construction, and the hopping reinforcement method is usually used for the ground improvement closed to the launching and receiving shafts in shield tunnelling.

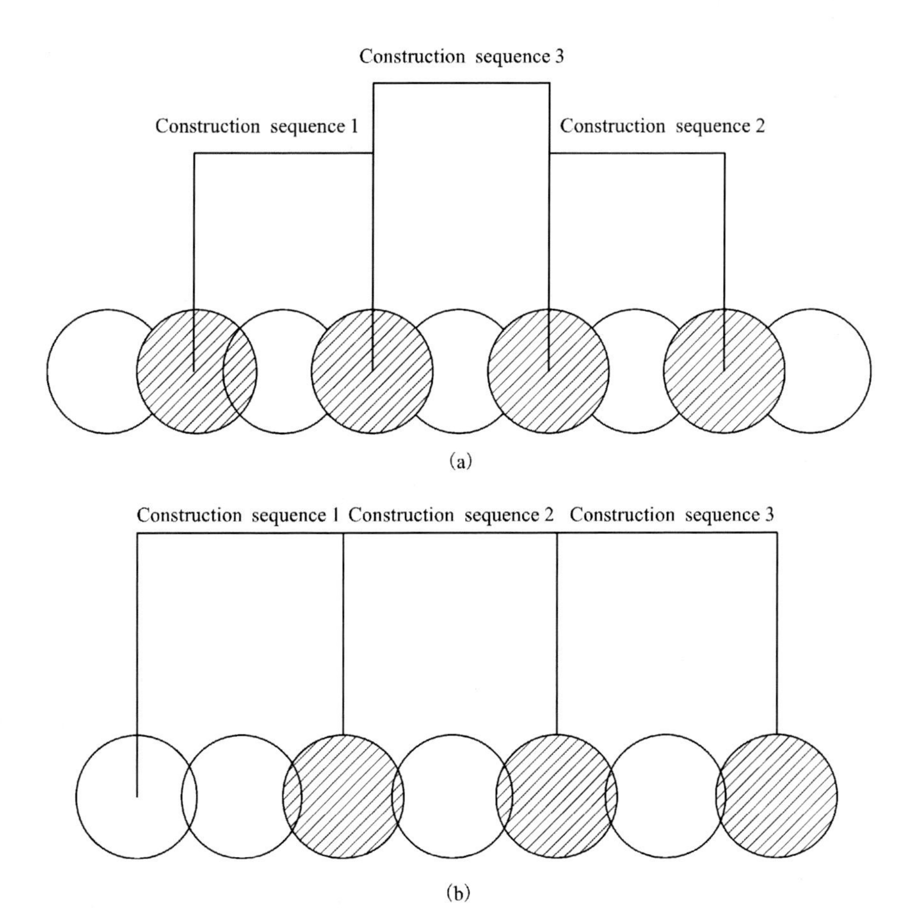

Figure 6.5 Schematic diagram of construction of the three-rod stirring pile. (a) Hopping reinforcement. (b) Continuous reinforcement.

Before construction of the three-rod stirring pile, no fewer than two pile tests should be carried out to determine the injected grout volume, drilling speed, lifting speed, mixing times, and so on. The overall construction process is shown in Fig. 6.6. To ensure the strength and uniformity of the

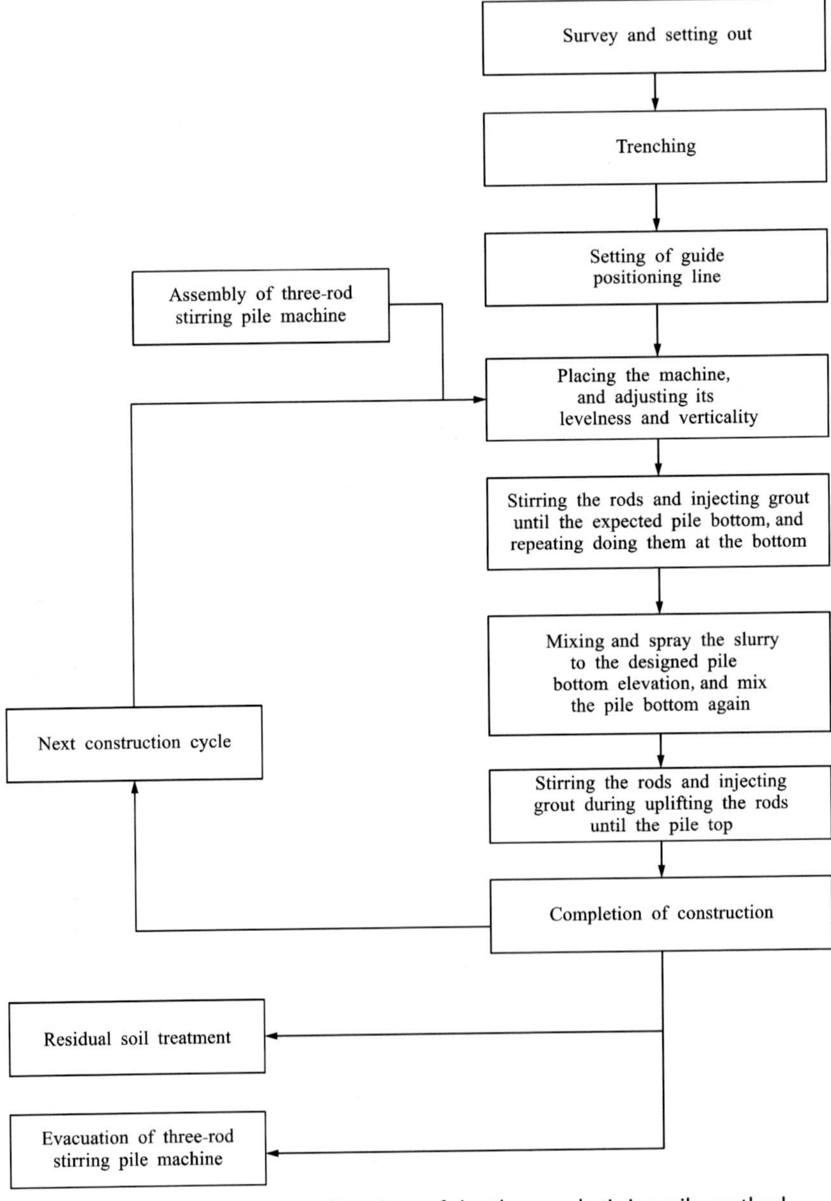

Figure 6.6 Construction process flowchart of the three-rod stirring pile method.

reinforcement, the following control measures can be adopted: (1) The grout should continuously be provided during the grout injection process. (2) The flow rate of grout and the lifting speed of mixing rods should be controlled appropriately, and the error should not be greater than ± 10 cm/min for the later. (3) The sinking and lifting speed during the repeated stirring should be controlled to ensure that each depth within the reinforcement range is fully stirred. (4) During the prestirring, the soft soil should be completely chopped to achieve uniform stirring the soil and grout. (5) The grout should be configured in accordance with the designed ratio, and the agglomeration in the cement should be screened in advance. To prevent the cement slurry as grout from segregating, it can be continuously stirred in the mortar mixer and poured into the collecting hopper before the grout is pressed.

6.1.2.5 Freezing method

The freezing method is an artificial method to temporarily make water freeze into ice in the ground in the prereinforced area, and the soil is cemented with ice to form a closed frozen zone. The freezing method is usually used to improve weak ground or to build high-strength water-retaining walls or load-bearing walls [6–8]. The freezing method has strong adaptability. It is suitable for soil and weak rock with a water content greater than 10%, a salt content of groundwater less than 3%, and a groundwater flow velocity less than 40 m/day. When the flow velocity is too high, other water isolation schemes can be considered to reduce the flow velocity in advance. At present, the most common artificial freezing technology uses liquid nitrogen, which is directly vaporized in the freezing tube to absorb the heat in the ground to achieve freezing. Fig. 6.7 shows the liquid nitrogen freezing tube inserted into the soil. In the launching and receiving of the shield machine, especially with a large diameter, large buried depth, and high water pressure, it is appropriate to adopt the freezing method to improve the tunnel ground closed to the shafts. The freezing method is a special auxiliary construction method. More details can be referred to the monograph "Ground Freezing Method" [9].

6.1.2.6 Temporary wall cutting method

In recent years a new technology has been developed to directly break the shaft wall by the shield machine. That means the shield machine can directly passes through the shaft wall [10]. The representative construction methods are the novel material shield machine-cuttable tunnel–wall

Figure 6.7 Schematic diagram of the freezing method.

system (NOMST) construction method and the super packing safety system (SPSS) construction method.

NOMST method

The NOMST method uses a new type of high-strength concrete to build the retaining wall. This new type of concrete replaces steel bars with fiber-reinforced resins such as carbon fiber and aramid and uses limestone as the coarse aggregates. The retaining wall built with such materials can be easily cut by the cutters of the shield machine.

SPSS method

The SPSS method uses superseals in the tunnel portals to prevent groundwater and slurry from flowing into the shaft. The superseal is generally an annular rubber tube reinforced with nylon fiber. Once the shield machine passes, the tube is grouted or inflated to make it bulge and play a sealing role. The shield machine can be advanced while cutting to achieve the shield machine launching and receiving.

This construction method has the following characteristics: (1) The amount of grout required is small. (2) There is no need to disassemble the retailing wall in advance, so the stability of the excavation face is maintained. (3) The construction period is short. However, it should be noted that because the pressure on the excavation face increases during the shield machine launching, the leakage of the entrance seals should be avoided. Also, measures must be taken to prevent wearing abrasion of cutters. Furthermore, when there are protrusions such as grouting equipment behind the synchronous wall outside of the shield machine, the leakage of the entrance seals must be avoided and the protrusion must be protected.

6.1.3 Safety calculation of the reinforced ground

On one hand, the shield machine must be able to smoothly cut off the reinforced soil, so the strength of the reinforced soil cannot be too high. On the other hand, to control the ground surface settlement and avoid water leakage in the portals, the reinforced soil should have high strength and low permeability [11]. There are currently the following methods to determine the required strength of the reinforced soil.

6.1.3.1 Strength checking method

As shown in Fig. 6.8, the reinforced soil is assumed to be a freely supported elastic circular plate with a thickness of t under water and earth pressures on the circular plate. According to elastic theory, the tensile

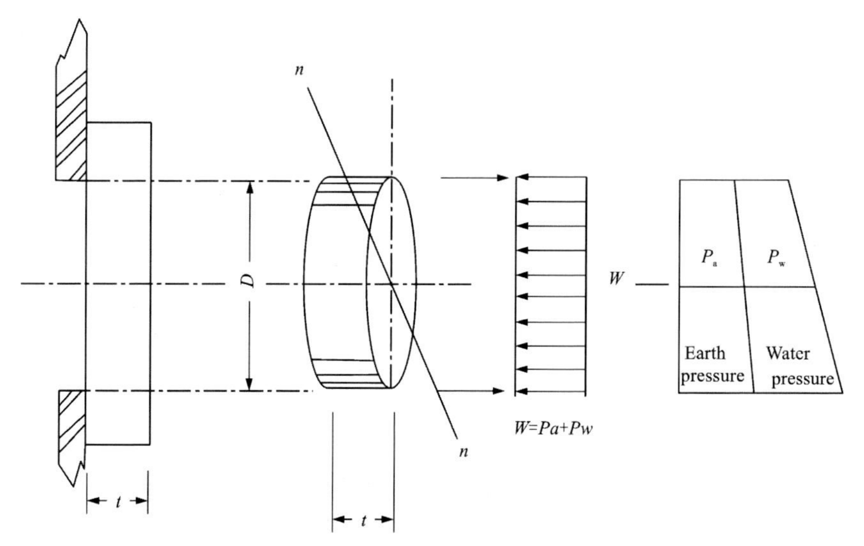

Figure 6.8 Schematic diagram of strength check.

strength and shear resistance of the soil can be obtained. The strength checking formula is as follows [12]:

$$\sigma_{\max} = \frac{P \cdot D^2}{4t^2} \cdot \frac{3}{8}(3+\mu) \leq \frac{\delta_t}{k_1} \tag{6.3}$$

$$\tau_{\max} = \frac{P \cdot D}{4t} \leq \frac{\tau_c}{k_2} \tag{6.4}$$

where $\sigma_{\max}$ is the maximum tensile strength; $\tau_{\max}$ is the maximum shear strength; D is the radius of the tunnel portal; t is the reinforced soil thickness; δ_t is the ultimate tensile strength of the reinforced soil, generally taken to be 10% of the unconfined compressive strength (*UCS*); k_1 and k_2 are the safety factors, generally k_1, $k_2 = 1.5$; τ_c is the shear strength of the reinforced soil, generally $\tau_c = 1/6$ of the *UCS*; P is the sum of water pressure and later earth pressure acting on the portal center; and μ is Poisson's ratio of the reinforced soil, generally $\mu = 0.2$.

6.1.3.2 Modified strength checking method

The strength checking method simplifies the lateral earth pressure (trapezoidal load) as a uniform pressure. Although the relationship between the strength of the reinforced ground and the longitudinal reinforcement range can be approximated with the strength checking method, it cannot reflect the actual situation of lateral earth pressure at the tunnel face.

To reflect the real stress status of the reinforced soil at the tunnel portal, the strength characteristics and failure pattern of the reinforced soil, Luo et al. [13] and Jiang et al. [10] analyzed and summarized the advantages and disadvantages of the traditional simplified model and established an equivalent model of reinforced soil within the allowable range of elasticity. The lateral trapezoidal pressure is equivalent to the superposition of the uniform pressure and the triangular antisymmetric pressure; that is, the asymmetric problem is equivalent to the superposition of a symmetry pressure and an antisymmetric pressure.

As shown in Fig. 6.9, the internal forces of the reinforced soil under the uniform pressure and the triangular antisymmetric pressure are respectively solved, and then the deflection and internal force of the reinforced soil under the trapezoidal pressure are obtained through the superposition of internal forces, which is

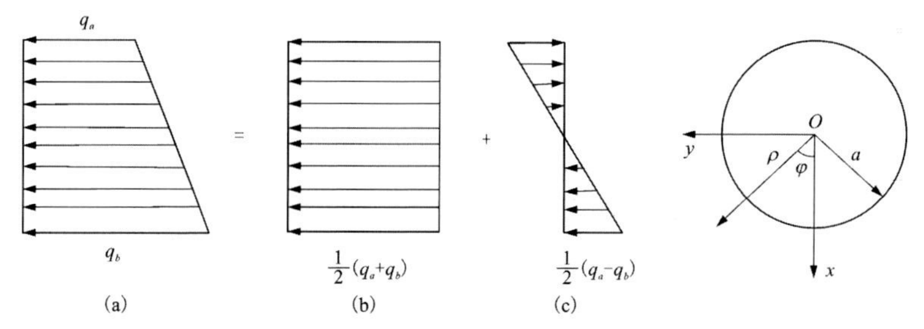

Figure 6.9 Schematic diagram of equivalent trapezoidal pressure.

$$\begin{cases} \omega_{\text{Trapezoid}} = \omega + \omega' \\ M_{\rho_{\text{Trapezoid}}} = M_\rho + M_\rho^{'} \\ M_{\varphi_{\text{Trapezoid}}} = M_\varphi + M_\varphi^{'} \end{cases} \tag{6.5}$$

where $\omega_{Trapezoid}$, ω, and ω' are the deflections of the reinforced soil plate, respectively by the lateral trapezoidal pressure, the uniform pressure, and the triangular antisymmetric pressure; where $M_{\rho_{Trapezoid}}$, M_ρ, and $M_\rho{}'$ are the radial moments of the reinforced soil plate, respectively by the lateral trapezoidal pressure, the uniform pressure, and the triangular antisymmetric pressure; where $M_{\varphi_{Trapezoid}}$, M_φ, and $M_\varphi{}'$ are the circular moments of the reinforced soil plate, respectively by the lateral trapezoidal pressure, the uniform pressure, and the triangular antisymmetric pressure.

Substituting Eq. (6.5), we get

$$\begin{cases} \omega_{\text{Trapezoid}} = \dfrac{q_0 D^3}{1024}\left(1 - \dfrac{4\rho^2}{D^2}\right)\left(\dfrac{5+\mu}{1+\mu} - \dfrac{4\rho^2}{D^2}\right) \\ \qquad + \dfrac{q_1 D^2}{1536}\left(1 - \dfrac{4\rho^2}{D^2}\right)\left(\dfrac{7+\mu}{3+\mu} - \dfrac{4\rho^2}{D^2}\right)\rho\ \cos\varphi \\ M_{\rho_{\text{Trapezoid}}} = \dfrac{(3+\mu)q_0 D^2}{64}\left(1 - \dfrac{4\rho^2}{D^2}\right) + \dfrac{q_1 D}{96}(5+\mu)\left(1 - \dfrac{4\rho^2}{D^2}\right)\rho\ \cos\varphi \\ M_{\varphi_{\text{Trapezoid}}} = \dfrac{q_0 D^2}{64}\left[(3+\mu) - (1+3\mu)\dfrac{4\rho^2}{D^2}\right] \\ \qquad + \dfrac{q_1 D}{96}\left(\dfrac{(5+\mu)(1+3\mu)}{3+\mu} - (1+5\mu)\dfrac{4\rho^2}{D^2}\right)\rho\ \cos\varphi \end{cases} \tag{6.6}$$

where q_0, q_1 are respectively equal to $(q_b + q_a)/2$, $(q_b - q_a)/2$.

According to the Eq. 6.6 of the reinforced soil, the maximum tensile stress and maximum shear stress can be obtained as follows.

Maximum shear stress

According to the geometric conditions of the circular thin plate and the characteristics of the shear stress, the location where the reinforced soil bears the largest shear stress can be known. According to the elastic theory, the maximum shear force is calculated as follow.

$$\begin{aligned}\left(Q_{\rho \text{Trapezoid}}\right)_{\max} &= -\left(\frac{q_0 D}{4} + \frac{3q_1 D^2}{32} - \frac{q_1 D}{24}\frac{5+\mu}{3+\mu}\right) \\ &= -\left[\frac{(q_a+q_b)D}{8} + \frac{3(q_b-q_a)D^2}{64} - \frac{(q_b-q_a)D}{48}\frac{5+\mu}{3+\mu}\right]\end{aligned} \tag{6.7}$$

According to Eq. 6.7, the corresponding maximum shear stress can be obtained as follow.

$$\begin{aligned}\left(\tau_{\rho \text{Trapezoid}}\right)_{\max} &= -\left(\frac{q_0 D}{4t} + \frac{3q_1 D^2}{32t} - \frac{q_1 D}{24t}\frac{5+\mu}{3+\mu}\right) \\ &= -\left[\frac{(q_a+q_b)D}{8t} + \frac{3(q_b-q_a)D^2}{64t} - \frac{(q_b-q_a)D}{48t}\frac{5+\mu}{3+\mu}\right]\end{aligned} \tag{6.8}$$

For safety in view of shear stress, it is required to satisfy the following condition.

$$\left(\tau_{\rho \text{Trapezoid}}\right)_{\max} \le \frac{\tau_c}{k_1} \tag{6.9}$$

Maximum tensile stress

According to the characteristics of tensile stress, let $\begin{cases} \dfrac{\partial M_{\rho \text{Trapezoid}}}{\partial \rho} = 0 \\ \dfrac{\partial M_{\rho \text{Trapezoid}}}{\partial \varphi} = 0 \end{cases}$, substituting the second expression in Eq. (6.6) into the solution, and obtain the following:

At $\varphi = 0, \rho_1 = \left(-B_1 + \sqrt{B_1{}^2 - 4A_1C_1}/2A_1\right)$ the maximum radial bending moment of the reinforced soil is

$$\left(M_{\rho \text{Trapezoid}}\right)_{\max} = \frac{D}{96}\left(1 - \frac{4\rho_1{}^2}{D^2}\right)\left[\frac{3(3+\mu)(q_a+q_b)D}{4} + \frac{(q_a-q_b)(5+\mu)\rho_1}{2}\right] \tag{6.10}$$

The maximum radial bending stress on the reinforced soil at the portal is calculated as

$$\left(\sigma_{\rho\text{Trapezoid}}\right)_{\max} = \frac{D}{32t^2}\left(1-\frac{4{\rho_1}^2}{D^2}\right)\left[\frac{3(3+\mu)(q_a+q_b)D}{2}+\rho_1(q_b-q_a)(5+\mu)\right] \tag{6.11}$$

For safety in view radial bending stress, it is required to satisfy the following condition.

$$\left(\sigma_{\rho\text{Trapezoid}}\right)_{\max} \le \frac{\delta_t}{k_2} \tag{6.12}$$

where $A_1 = 3(5+\mu)(q_b-q_a)/D$, $B_1 = 3(3+\mu)(q_a+q_b)$, $C_1 = -D(5+\mu)(q_b-q_a)/4$.

In the same way, let $\begin{cases} \dfrac{\partial M_{\varphi\text{Trapezoid}}}{\partial \rho} = 0 \\ \dfrac{\partial M_{\varphi\text{Trapezoid}}}{\partial \varphi} = 0 \end{cases}$

At $\varphi = 0$, $\rho_2 = \left(-B_2+\sqrt{{B_2}^2-4A_2C_2}\right)/2A_2$ the maximum circumferential bending moment of the reinforced soil at the portal is

$$\begin{aligned}\left(M_{\varphi\text{Trapezoid}}\right)_{\max} &= \frac{q_0D^2}{64}\left[(3+\mu)-(1+3\mu)\frac{4{\rho_2}^2}{D^2}\right] \\ &+ \frac{q_1\rho_2D}{96}\left[\frac{(5+\mu)(1+3\mu)}{3+\mu}-(1+5\mu)\frac{4{\rho_2}^2}{D^2}\right]\end{aligned} \tag{6.13}$$

The maximum circumferential bending stress on the reinforced soil at the end is

$$\begin{aligned}\left(\sigma_{\varphi\text{Trapezoid}}\right)_{\max} &= \frac{3(q_a+q_b)D^2}{64t^2}\left[(3+\mu)-(1+3\mu)\frac{4{\rho_2}^2}{D^2}\right] \\ &+ \frac{(q_b-q_a)\rho_2D}{32t^2}\left[\frac{(5+\mu)(1+3\mu)}{3+\mu}-(1+5\mu)\frac{4{\rho_2}^2}{D^2}\right]\end{aligned} \tag{6.14}$$

For safety in view of bending stress, it is required to satisfy the following condition.

$$\left(\sigma_{\varphi\text{Trapezoid}}\right)_{\max} \le \frac{\delta_t}{k_3} \tag{6.15}$$

where $A_2 = (1+5\mu)(q_b-q_a)/(16D)$, $B_2 = (1+3\mu)(q_a+q_b)/16$, $C_2 = -(5+\mu)(1+3\mu)(q_b-q_a)D/[192(3+\mu)]$; k_1, k_2, and k_3 are the safety factors, generally k_1, k_2, $k_3 = 1.5$; q_a and q_b are the sum of lateral water and earth pressures acting on the top and bottom of the reinforced soil at tunnel portal.

Jiang et al. [10] compared and analyzed the two calculation models of strength and found that when the buried depth of the tunnel remains unchanged, the maximum shear stress and maximum tensile stress of the reinforced soil calculated from the two models increase with the increasing diameter of the shield machine. When the diameter of the shield machine is less than 10 m, the calculated results from the two models are close.

The improved model of equivalent trapezoidal load can better reflect the real stress state of the reinforced soil at the end, but the calculation process is more complicated. Hence it has not been widely employed in practice.

6.1.3.3 Overall stability checking for clayey soil

As shown in Fig. 6.10, the reinforced clayey soil is assumed to slide along a certain sliding surface under the self-weight of the upper soil and the ground surface load (p). There are different theories to check overall stability of the ground at the tunnel portal. Here, the ideal sliding model is

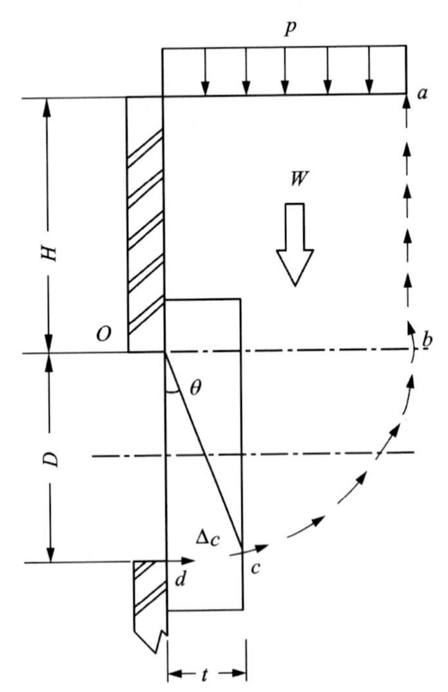

Figure 6.10 Schematic diagram of overall stability check.

introduced [10]. It is assumed that the sliding surface is an arc surface with the top point O of the tunnel portal as the center and the tunnel portal diameter D as the radius.

The moment that causes the sliding is

$$M = M_1 + M_2 + M_3 \tag{6.16}$$

where M, M_1, M_2, M_3 are the moments, induced respectively by all loads, the ground surface load, the self-weight of the upper soil, and the self-weight of the soil sliding along the arch surface. $M_1 = pD^2/2$, $M_2 = WD/2$, $M_3 = \gamma_1 D^3/3$.

The friction resistance is ignored for the clayey ground, and so the moment for resisting soil sliding is calculated as follow

$$\overline{M} = \overline{M_1} + \overline{M_2} + \overline{M_3} = c_u hD + c_u D^2\left(\frac{\pi}{2} - \theta\right) + c_{ut}\theta D^2 \tag{6.17}$$

where $\overline{M}$, $\overline{M_1}$, $\overline{M_2}$, and $\overline{M_3}$ are the anti-sliding moments, respectively in the whole sliding surface, in the sliding parts of *a-b*, *b-c*, and *c-d*.

The safety factor against the overall instability of the soil is calculated as

$$k_3 = \frac{\overline{M}}{M} \tag{6.18}$$

where c_u is the cohesion of the overlying soil; c_{ut} is the cohesion of the reinforced soil; H is the height of the overlying soil; and K_3 is safety factor, which should be larger than 1.5.

6.1.3.4 Overall stability checking for sandy soil

The model of sandy soil is different from that of cohesive soil. For a long time, many scholars all over the world have conducted model tests for studying the sliding failure modes of sandy soil slopes [14,15]. The research results show that the destruction process of sandy soil slopes is sudden, and an approximately linear sliding surface is formed from the top of the slope to the foot of the slope [16,17]. Terzaghi's loose earth pressure principle and centrifugal tests show that the failure surface of the sandy soil above the shield tunnel is a vertical sliding surface. The soil below the tunnel top is no longer a circular arc surface, but an inclined straight surface passing through the toe. Based on this, the failure model of the sandy soil in the tunnel portal section is established, as shown in Fig. 6.11.

It is assumed that the sandy soil has no cohesion and the vertical soil mass will slide and fail under the gravity of the overlying soil.

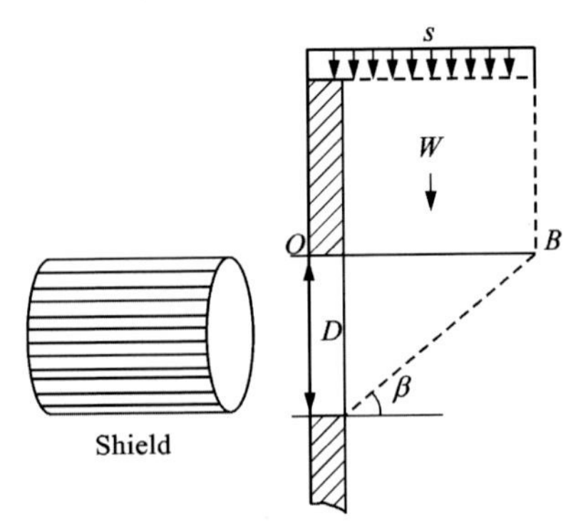

Figure 6.11 Schematic diagram of overall stability checking calculation for sandy soil.

Sliding force:

$$T_1 = W \cdot \sin\beta \tag{6.19}$$

where W is the weight of the overlying soil on the slip line; β *is sliding failure angle.*

Antisliding force:

$$T_2 = W \cdot \cos\beta \cdot \tan\phi \tag{6.20}$$

where T_2 is the sliding force; ϕ is the internal friction angle of sandy soil

Since it is assumed that the sliding surface of sandy soil is an oblique straight surface, when the vertical soil slope is in the limit equilibrium state, the sliding force and antisliding force on the potential failure sliding surface are in a static equilibrium state:

$$T_1 = T_2 \tag{6.21}$$

Substitute Eq. (6.19) and Eq. (6.20) for the two sides of Eq. (6.21) to obtain the static balance formula:

$$W \cdot \sin\beta = W \cdot \cos\beta \cdot \tan\phi \tag{6.22}$$

According to the static balance formula, the stability safety factor F_s of sandy soil is defined as

$$F_s = \frac{\tan\phi}{\tan\beta} \tag{6.23}$$

When $F_s = 1$ and $\beta = \phi$, the longitudinal slip range of the sandy soil at end is calculated as

$$OB = \frac{D}{\tan\phi} \tag{6.24}$$

where OB is the length of the longitudinal reinforcement range; D is the diameter of the tunnel.

It can be seen from Eq. 6.24 that the longitudinal reinforcement range of the sand layer is influenced only by the diameter of the tunnel and the internal friction angle of sandy soil. Hence, the influence of the buried depth on the longitudinal reinforcement range is ignored.

6.1.4 Reinforcement range at the end

The longitudinal range of soil reinforcement at the launching end of the shield machine should be determined according to the ground conditions, the length of the main shield machine, the soil strength at the tunnel portal, and the overall stability checking results.

6.1.4.1 Determination of the longitudinal range of soil reinforcement

Strength theory

According to the elastic solution for the small deflection bending problem of a thin plate [12], under uniform pressure, the bending stress at the center of the plate surface of the reinforced soil is the largest, and the shear stress at the peripheral support of the thin plate is the largest. As shown in Fig. 6.8, the longitudinal thickness (t_1) of the reinforcement range at the tunnel portal that meets the tensile requirements is calculated as

$$t_1 \geq \sqrt{\frac{3(3+\mu)k_1 pD^2}{32\sigma_t}} \tag{6.25}$$

The longitudinal thickness (t_2) of the reinforcement range at the tunnel portal that meets the requirements of shear resistance is

$$t_2 = \frac{k_2 PD}{4\tau_c} \tag{6.26}$$

The soil mass at the tunnel portal should meet both requirements of tensile and shear strengths, and hence the longitudinal thickness of the reinforcement range is calculated as

$$t = \max\left\{\sqrt{\frac{3(3+\mu)k_1 PD^2}{32\sigma_t}}, \frac{k_2 PD}{4\tau_c}\right\} \tag{6.27}$$

where t is the thickness of the reinforced soil and the other parameters are the same as in Section 6.1.3 Safety calculation of the reinforced ground.

Modified strength theory

In Fig. 6.12 the strength of the reinforced soil at the tunnel portal can be calculated for determining the longitudinal reinforcement range [10,13].

1. Maximum shear stress

According to the theory of maximum shear stress, for $\begin{cases} \varphi = 0 \\ \rho = D/2 \end{cases}$ the maximum shear stress is calculated as

$$\begin{cases} \tau_{\max} = \dfrac{\beta_1}{t_1} \le \dfrac{\tau_c}{k_1} \\ \beta_1 = -\left[\dfrac{(q_a + q_b)D}{8} + \dfrac{3(q_b - q_a)D^2}{64} - \dfrac{(q_b - q_a)D}{48}\dfrac{5+\mu}{3+\mu}\right] \end{cases} \tag{6.28}$$

The longitudinal thickness of the reinforcement range that limits the maximum shear stress is calculated as

$$t_1 \ge \frac{\beta_1 k_1}{\tau_c} \tag{6.29}$$

2. Maximum tensile stress

According to the theory of maximum tensile stress, for $\begin{cases} \varphi = 0 \\ \rho = \left(-B_1 + \sqrt{B_1{}^2 - 4A_1C_1}\right)/2A_1 \end{cases}$ the maximum radial tensile stress is calculated as

$$\begin{cases} \sigma_{\rho\max} = \dfrac{\beta_2}{16t_2{}^2} \le \dfrac{\sigma_t}{k_2} \\ \beta_2 = 2\left(1 - \dfrac{4\rho^2}{D^2}\right)\left[\dfrac{3(3+\mu)(q_a + q_b)D^2}{4} + \dfrac{(q_a - q_b)(5+\mu)D\rho}{2}\right] \end{cases} \tag{6.30}$$

The longitudinal thickness of the reinforcement range that limits the maximum tensile stress is calculated as

$$t_2 \ge \sqrt{\frac{\beta_2 k_2}{16\sigma_t}} \tag{6.31}$$

where $A_1 = 3(5+\mu)(q_b - q_a)/D$, $B_1 = 3(3+\mu)(q_a + q_b)$, $C_1 = -D(5+\mu)(q_b - q_a)/4$.

Similarly, for $\begin{cases} \varphi = 0 \\ \rho = \left(-B_2 + \sqrt{B_2{}^2 - 4A_2C_2}\right)/2A_2 \end{cases}$ the maximum circumferential tensile stress is

$$\begin{cases} \sigma_{\varphi\max} = \dfrac{\beta_3}{16t_3{}^2} \le \dfrac{\sigma_t}{k_3} \\ \beta_3 = \dfrac{3(q_a + q_b)D^2}{4}\left[(3+\mu) - (1+3\mu)\dfrac{4\rho^2}{D^2}\right] \\ \qquad + \dfrac{(q_b - q_a)\rho D}{2}\left[\dfrac{(5+\mu)(1+3\mu)}{3+\mu} - (1+5\mu)\dfrac{4\rho^2}{D^2}\right] \end{cases} \tag{6.32}$$

$$t_3 \ge \sqrt{\frac{\beta_3 k_3}{16\sigma_t}} \tag{6.33}$$

where $A_2 = (1+5\mu)(q_b - q_a)/(16D)$, $B_2 = (1+3\mu)(q_a + q_b)/16$, $C_2 = -(5+\mu)(1+3\mu)(q_b - q_a)D/[192(3+\mu)]$; β_1, β_2, β_3, k_1, k_2, and k_3 are all calculated coefficients; and the remaining parameters are the same as in Section 6.1.3 Safety calculation of the reinforced ground.

In summary, in the launching and receiving of the shield machine, to avoid the failure of the reinforced soil caused by the lateral earth pressure and water pressure, the reinforced soil must meet both requirements of tensile strength and shear strength. Thus, the longitudinal thickness of the reinforcement range is calculated as

$$t = \max\left\{\frac{\beta_1 k_1}{\tau_c}, \sqrt{\frac{\beta_2 k_2}{16\sigma_t}}, \sqrt{\frac{\beta_3 k_3}{16\sigma_t}}\right\} \tag{6.34}$$

3. Modified strength theory

As shown in Fig. 6.10, the overall stability of a cohesive soil at the tunnel portal can be calculated as follows for determining the longitudinal reinforcement range[18]:

$$\theta = \frac{K_3(M_1 + M_2 + M_3) - M_d}{C_{ut} \cdot D^2} \tag{6.35}$$

$$t = D \cdot \sin\theta \tag{6.36}$$

where t is the thickness of the reinforcement soil, and the other parameters are the same as in Section 6.1.3 Safety calculation of the reinforced ground.

4. Modified strength theory

According to the overall stability checking method in Section 6.1.3.4, the longitudinal thickness of the reinforced sandy ground can be determined as follows

$$t \geq OB = \frac{D}{\tan\beta} = \frac{D}{\tan\phi} \tag{6.37}$$

where t is the thickness of the reinforced soil and the other parameters are the same as in Section 6.1.3 Safety calculation of the reinforced ground.

6.1.4.2 Determination of the lateral range of soil reinforcement at the tunnel portal

Reinforcement range of the upper and lower sides of the tunnel portal section

Shield tunnelling breaks the stress balance in the soil. It disturbs the surrounding soil, causing stress concentration around the tunnel portal. When the maximum shear stress exceeds the shear strength of the soil, the soil around the tunnel will be unstable. The disturbed zone gradually spreads from the tunnel portal to the further area, forming a plastic circle as shown in Fig. 6.12a. The appearance of the plastic circle significantly reduces the stress within a certain range of the circle due to the stress

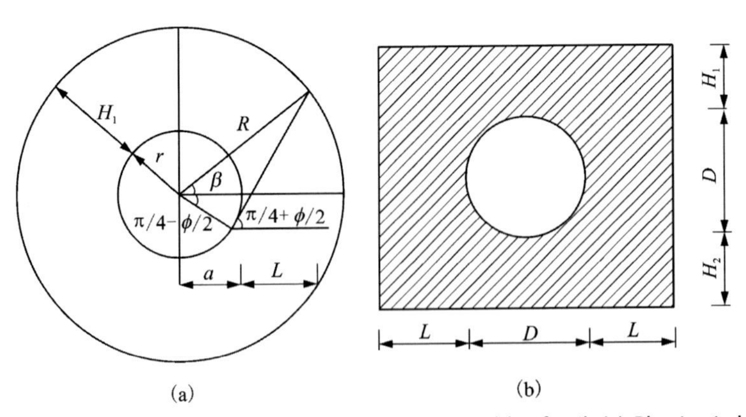

Figure 6.12 Plastic zone and lateral reinforcement width of soil. (a) Plastic circle calculation diagram. (b) Lateral soil reinforcement diagram.

release, and the maximum stress concentration location moves from the original tunnel portal to the junction of the plastic circle and the elastic circle [19–21]. Therefore, to ensure the stability of the lateral ground at the tunnel portals during the launching and receiving of the shield machine, the lateral ground at the tunnel portals must be reinforced in advance. The schematic diagram of the lateral ground reinforcement is shown in Fig. 6.12b.

The equilibrium equation in differential form can be expressed in polar coordinates as follows.

$$\begin{cases} \dfrac{\partial \sigma_r}{\partial y} + \dfrac{1}{r}\dfrac{\partial \tau_{r\theta}}{\partial \theta} + \dfrac{\sigma_r - \sigma_\theta}{r} = 0 \\ \dfrac{1}{r}\dfrac{\partial \sigma_\theta}{\partial \theta} + \dfrac{\partial \tau_{r\theta}}{\partial r} + \dfrac{2\tau_{r\theta}}{r} = 0 \end{cases} \quad \text{(Ignore self-weight)} \tag{6.38}$$

As shown in Fig. 6.12, it is assumed that the reinforcement height of the upper soil is H_1, and a circular cavity with a diameter of D is bored in a homogeneous and isotropic soil. The radius of the plastic circle formed after excavation is r, and the original stress in the soil is σ_m. The soil obeys the Mohr-Coulomb failure criterion. According to the stress balance of the plastic ring and the soil failure conditions, the equilibrium function is built as follow.

$$\sin\phi = \frac{\sigma_\theta - \sigma_r}{\sigma_\theta + \sigma_r + 2c\cdot\cot\phi} \tag{6.39}$$

where c and ϕ are the cohesion and internal friction angle of the soil, respectively.

The lateral disturbance of the tunnel can be simplified to an axisymmetric problem. A equilibrium function that ignores the self-weight can be given in a differential form:

$$\frac{\partial \sigma_r}{\partial r} + \frac{\sigma_r - \sigma_\theta}{r} = 0 \tag{6.40}$$

Formula Eq. (6.39) can be transformed into a clearer form as follows:

$$\sigma_\theta - \sigma_r = (\sigma_\theta + \sigma_r)\cdot\sin\phi + 2c\cdot\cot\phi\cdot\sin\phi \tag{6.41}$$

Eq. 6.41 can be transformed to

$$\frac{\sigma_\theta + c\cdot\cot\phi}{\sigma_r + c\cdot\cot\phi} = \frac{1+\sin\phi}{1-\sin\phi} \tag{6.42}$$

Transforming formula Eq. (6.41) into the following formula

$$\sigma_\theta = \frac{r\partial\sigma_r}{\partial_r} + \sigma_r \tag{6.43}$$

Substituting formula (Eq. 6.43) in (Eq. 6.42), we get

$$\frac{\frac{r\partial\sigma_r}{\partial r} + \sigma_r + c\cdot\cot\phi}{\sigma_r + c\cdot\cot\phi} = \frac{1+\sin\phi}{1-\sin\phi} \tag{6.44}$$

Simplifying Eq. 6.44 as follow

$$\ln(\sigma_r + c\cdot\cot\phi) = \frac{2\sin\phi}{1-\sin\phi}\ln r + A \tag{6.45}$$

Substituting the boundary conditions $\begin{cases} r = D/2 \\ \sigma_r = 0 \end{cases}$ into Eq. 6.45, so

$$A = \ln(c\cdot\cot\phi) - \frac{2\sin\phi}{1-\sin\phi}(\ln D - \ln 2) \tag{6.46}$$

Substituting Eq. 6.46 into Eq. 6.45, so

$$\ln(\sigma_r + c\cdot\cot\phi) = \frac{2\sin\phi}{1-\sin\phi}\ln r + \ln(c\cdot\cot\phi) - \frac{2\sin\phi}{1-\sin\phi}(\ln D - \ln 2) \tag{6.47}$$

Substituting the boundary stress of the loosening zone σ_m into Eq. 6.47, when $\sigma_r = \sigma_m$, the radius of the loosening zone can be obtained:

$$R = \left(\frac{D}{2}\right)\cdot \sqrt[\frac{2\sin\phi}{\sin}]{\frac{\sigma_m}{c\cdot\cot\phi} + 1} \tag{6.48}$$

The height of the reinforced soil at the upper part of the tunnel portal is

$$H_1 = H = k\left(R - \frac{D}{2}\right) \tag{6.49}$$

where k is the safety factor. Usually, the reinforcement thickness (H_2) of the soil below the tunnel is $H_2 = H_1$.

Reinforcement range on the left and right sides of the tunnel portal section

As shown in Fig. 6.12, the soil width on the left and right sides of the tunnel portal section that needs to be reinforced is L. According to the Rankine earth pressure theory, the angle between the shear failure surface and the direction of the maximum principal stress is $45° - \phi/2$. The angle with the maximum principal stress is $45° + \phi/2$. According to the distribution characteristics of the principal stress in the plastic circle, as shown in Fig. 6.12a, the triangle *Omn* is a right triangle, so the angle β can be calculated as

$$\beta = \arccos\left(\frac{D}{D + 2H_1}\right) - \left(\frac{\pi}{4} - \frac{\phi}{2}\right) \tag{6.50}$$

From the geometric conditions in Fig. 6.12b, the reinforcement range on both sides of the tunnel is

$$L = \left(\frac{D}{2} + H_1\right) \cdot \cos\beta - \frac{D}{2} \tag{6.51}$$

6.1.4.3 Considering groundwater

For soils that are not or little influenced by the groundwater (e.g., clay and silty clay), the parameters of the reinforcement range can be obtained directly on the basis of the calculation results of the strength checking, the overall stability checking, and engineering experience. For soils that are greatly influenced by the groundwater (e.g., sand and sand pebble), in addition to the results of the strength checking and the overall stability checking, the water and soil flowing into the launching shaft along the gap between the shield machine shell and the excavated ground should be considered. When the launching end of the shield machine has poor stability and is greatly influenced by the groundwater (especially when confined water exists), the end reinforcement length should be at least the length of the main shield machine plus two to three rings of segments. When the reinforcement length of the end is greater than the length of the main shield machine (as shown in Fig. 6.13a), after the shield tail enters the tunnel portal and starts synchronous grouting, the shield cutterhead has not yet left the reinforcement zone. Due to the sealing effect of the synchronous grouting, no water and soil will flow into the launching shaft along the gap between the shield machine shell and the ground. If the reinforcement length is less than the length of the main shield machine (as shown in Fig. 6.13b), when the shield tail has not yet entered the seal ring of the tunnel portal, the shield cutterhead has already left the reinforcement zone, and tail void grouting cannot be used. The water and soil (especially the sand layer or silt layer) in front of the reinforcement area may enter the launching shaft along the gap between the shield machine shell and the ground, causing soil erosion and large surface settlement.

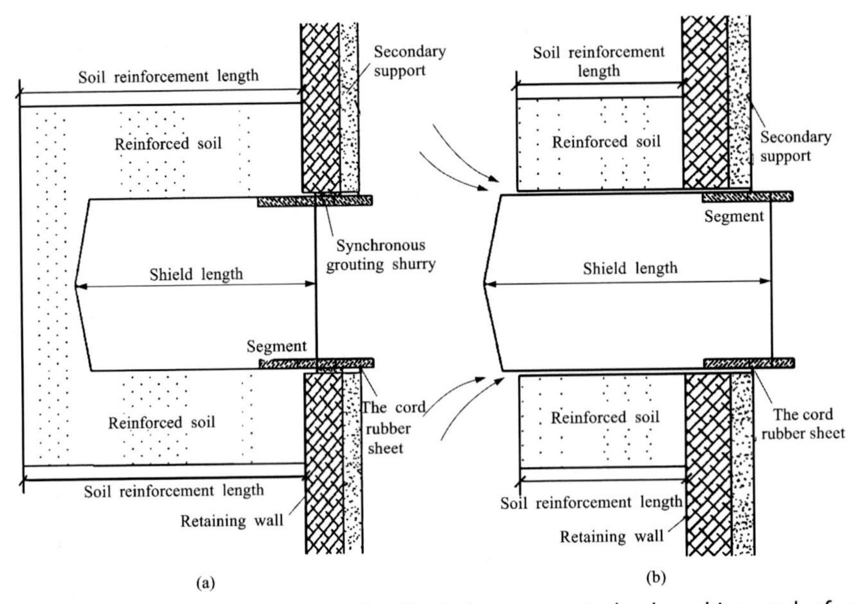

Figure 6.13 Schematic diagram of soil reinforcement at the launching end of a shield machine. (a) The reinforced length is greater than that of the main shield machine. (b) The reinforced length is less than that of the main shield machine. *Modified from Jiang Y, Yang Z, Jiang H, et al. Discussion on the reasonable range of reinforcement at the starting and reaching end of the earth pressure balance shield. Tunnel Constr 2009;29(3):263–266.*

The reinforcement range of the soil at the tunnel portal closed to the receiving shaft of the shield machine should also be determined according to the ground conditions and the length of the shield machine. For soil layers that are not affected or only little affected by groundwater (e.g., clay, silty clay layer), the reinforcement parameters can be simply selected according to engineering experience and construction plans. For strata with low stability and abundant groundwater (e.g., sand and pebble layers), in addition to engineering experience, consideration should be given to the flowing of water and soil into the receiving shaft along the gap between the shield shell and the ground. For soil with low stability and abundant groundwater (especially when there is confined water), as shown in Fig. 6.14a, the length of the reinforcement at the receiving end should be not less than the length of the main shield machine plus the width of two to three rings. In this way, when the cutterhead of the shield machine contacts the retaining wall (pile), synchronous grouting for two to three rings has been carried out. The solidified grout can block the inflow of water and sand and ensuring the construction safety. If the

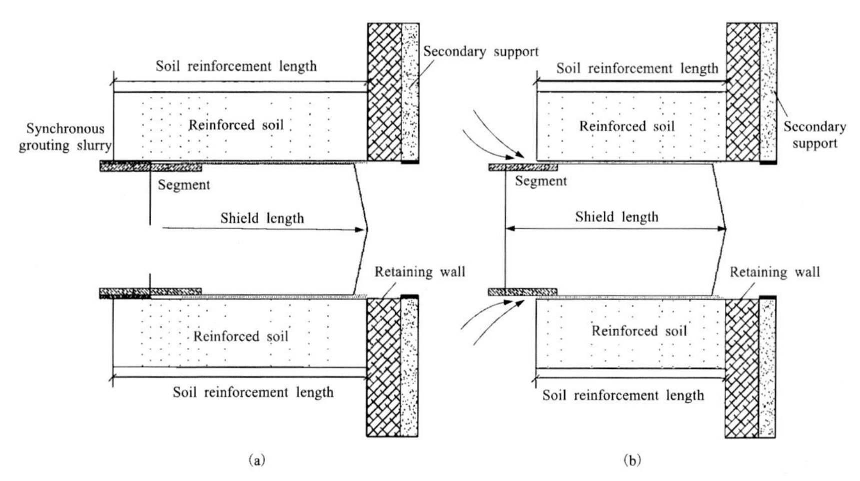

Figure 6.14 Schematic diagram of soil reinforcement at the receiving end of a shield machine. (a) The reinforced length is greater than that of the main shield machine. (b) The reinforced length is less than that of the main shield machine. *Modified from Jiang Y., Yang Z., Jiang H., Discussion on the reasonable range of reinforcement at the starting and reaching end of the earth pressure balance shield. Tunn Constr 2009;263-266.*

length of the end reinforcement is less than the that of the main shield machine, as shown in Fig. 6.14b, groundwater and soil are likely to flow into the receiving shaft along the gap between the shield shell and the reinforced soil, causing ground surface subsidence, segment failure, and even structural instability of the whole tunnel.

6.1.5 Soil reinforcement inspection

Generally, after reinforcement for about 2 weeks, the reinforced soil should be excavated. The strength, water permeability, continuity, and uniformity of the samples taken from the reinforced ground need to be inspected. The drilling can be conducted vertically on the ground surface and horizontally along the tunnel excavation line as follows [22,23]:

1. Vertical drilling inspection: The soil samples are drilled out in the reinforced ground to confirm their continuities and uniformities. The *UCS*s and permeabilities of the samples are determined according to laboratory tests and check whether they satisfy the design requirements.
2. Horizontal drilling inspection: No fewer than 6 holes are horizontally drilled along the tunnel excavation line and one hole is drilled in the center to take samples for checking soil seepage. In general, the amount of water seepage depends on the consolidation and water seepage. The flow rate of groundwater from each hole should not exceed 30 L/h.

6.1.6 An example of safety calculation

A shield tunnel with a diameter (D) of 6 m is designed with a buried depth (H) of 9.3 m in weak ground at the the tunnel portal (Fig. 6.15). From top to bottom, the soil layers are <1−2> plain fill, <2-2> silty fine sand, <2−4> silty sand, and so on. There is a overload of 70kPa on the ground suface. The strata parameters are shown in Table 6.1. The soil layer <2-4> silty sand is reinforced with a longitudinal thickness of 6 m at the tunnel portal.

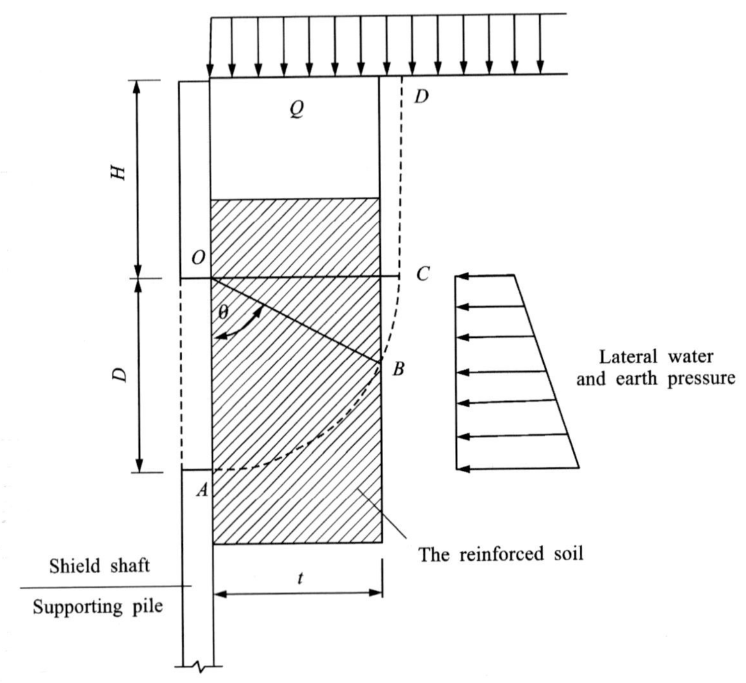

Figure 6.15 Schematic diagram of the soil reinforcement range at the tunnel end.

Table 6.1 Formation parameter table.

Formation code	Unit weight (kN/m^3)	Thickness (m)	Coefficient of lateral earth pressure
1−2	18.8	2	0.51
2−2	19.6	5.5	0.57
2−4	19.7	12	0.44

For calculation convenience here, the cohesions of all original soils are 12 kPa. For the reinforced layer <2-4> silty sand, the cohesion and unit weight are taken to be 150 kPa and 20 KN/m^3, respectively. The unconfined compressive strength q_u of the reinforced soil is 1.0 MPa, and the Poisson's ratio is taken to be 0.2.

6.1.6.1 Strength checking

There are the following conditions for the strength checking of the ground closed to the shield receiving shaft.

$$D = 6m, \quad t = 6m, \quad \mu = 0.2, \quad K_1 = K_2 = 1.5,$$

$$\sigma_t = \frac{q_u}{10} = 0.1MPa = 100kPa, \quad \tau_c = \frac{q_u}{6} = 0.1667MPa = 166.7kPa$$

Lateral earth and water pressures at the center of the excavation face

$$P = \sum_{i=1}^{3} K_0(q + \gamma_i h_i) = 0.44 \times (70 + 18.8 \times 2 + 19.4 \times 5.3 + 19.0 \times 5)$$

$$= 134.38KPa$$

Checking the maximum tensile stress σ_{max}

$$\sigma_{max} = \frac{P \cdot D^2}{4t^2} \cdot \frac{3}{8}(3 + \mu) = \frac{134.38 \times 6^2}{4 \times 6^2} \times \frac{3}{8}(3 + 0.2)$$

$$= 41.57KPa \leq \frac{\delta_t}{k_1} = \frac{100}{1.5} = 66.7KPa$$

Checking the maximum shear stress τ_{max}

$$\tau_{max} = \frac{P \cdot D}{4t} = \frac{134.38 \times 6}{4 \times 6} = 33.60KPa \leq \frac{\tau_c}{k_2} = \frac{166.7}{1.5} = 111.13KPa$$

Therefore, the strength of the reinforced soil at the tunnel portal section meets the safety requirements.

6.1.6.2 Overall stability check

There are the following conditions for the overall stability checking of the tunnel portal closed to the shield receiving shaft.

$q = 70KN/m$, $\gamma = 20KN/m^3$, $H = 9.3m$, $\theta = \arcsin\frac{t}{D} = \arcsin\frac{6}{6} = \frac{\pi}{2}$, $C_u = 12KPa$, $C_{ut} = 150KPa$

$$W = \sum_{i}^{n} \gamma_i h_i \cdot D = (18.8 \times 2 + 19.4 \times 5.3 + 19.0 \times 2) \times 6 = 1070.52KN$$

Sliding moment

$$M = M_1 + M_2 + M_3 = \frac{pD^2}{2} + \frac{WD}{2} + \frac{\gamma D^3}{3}$$
$$= \frac{70 \times 6^2}{2} + \frac{1070.52 \times 6}{2} + \frac{20 \times 6^3}{3} = 5911.56KN/m$$

Antisliding moment

$$\overline{M} = \overline{M_1} + \overline{M_2} + \overline{M_3} = C_u hD + C_u D^2(\frac{\pi}{2} - \theta) + C_{ut}\theta D^2$$
$$= 12 \times 9.3 \times 6 + 150 \times 6^2 \times 0 + 150 \times \frac{\pi}{2} \times 6^2 = 9147.6KN/m$$

Safety factor

$$k_3 = \frac{\overline{M}}{M} = \frac{9147.6}{5911.56} = 1.547 > 1.5$$

Therefore, the overall stability of the reinforced soil at the tunnel portal meets the safety requirements for the shield machine launching.

6.2 Configurations and technical controls for shield machine launching

6.2.1 Shield machine launching configurations

The configurations for launching shield machine include launching pedestal, reaction frame, temporary negative segment ring, sealing ring at tunnel entrance, and so on (Fig. 6.16). Their functions and compositions are briefly discussed below.

6.2.1.1 Launching pedestal

The launching pedestal is to use for shield machine assembling and for keeping the shield machine in an ideal launching position (elevation and

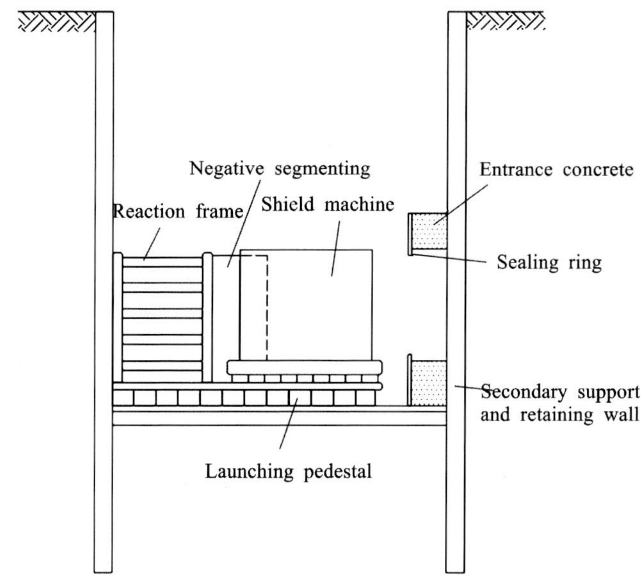

Figure 6.16 Launching configurations for shield machine.

direction). Therefore it has the following requirements: (1) The pedestal structure must be appropriate to ensure the the shield machine assembly process. (2) The pedestal components must have high rigidity and high strength, since shield machines normally are hundreds of tons in weight. (3) It must be fixed with the shaft bottom plate, avoiding any shaking and movement to ensure the expected position of shield machine.

Once the shield machine has been placed on the pedestal, it is necessary not only to ensure that the axis of the shield machine coincides with the design axis, but also to consider the change need of the axis after the shield machine is launched and the attitude control of the shield machine. The pedestal should be positioned following the surveying baseline in the shaft bottom.

The launching pedestal for the shield machine can be classified into reinforced concrete pedestal, steel pedestal, and composite pedestal.

Reinforced concrete pedestal

The reinforced concrete pedestal is usually a combination of multiple reinforced concrete structures and has two types, including cast-in-place pedestal and prefabricated pedestal. It has a strong stability and high compression strength.

Steel pedestal

The steel pedestal (as shown in Fig. 6.17) has two types of installation, including field assembly and integral installation. The field assembly is more preferred due to its flexibility, but the integral installation is quicker if there is a ready-made pedestal. Generally, the advantage of the steel pedestal is that the installation processing is short and the adaptability is strong, and it is the commonly used launching pedestal at present.

Composite pedestal

The composite pedestal combines reinforced concrete components and steel components, and it is used more popularly than other two types. Usually, the composite pedestal is assembled with I-shape steel beams and rails (Fig. 6.18). It is composed of a reaction base and a temporary assembly pipe ring. The shape is determined by conditions such as the transport of the segment and the dump space, and the position is determined by the starting position of the formal lining segments. The reaction frame should have sufficient strength to bear thrust and rigidity to resist deformation during advancing.

The reaction frame base is installed through the I-beam, and the temporary assembly segment ring is spliced with easy-to-handle steel or high-strength cast-iron pipe segments. The accuracy of the temporary segment affects the roundness of the formal segments, so more attention should be paid.

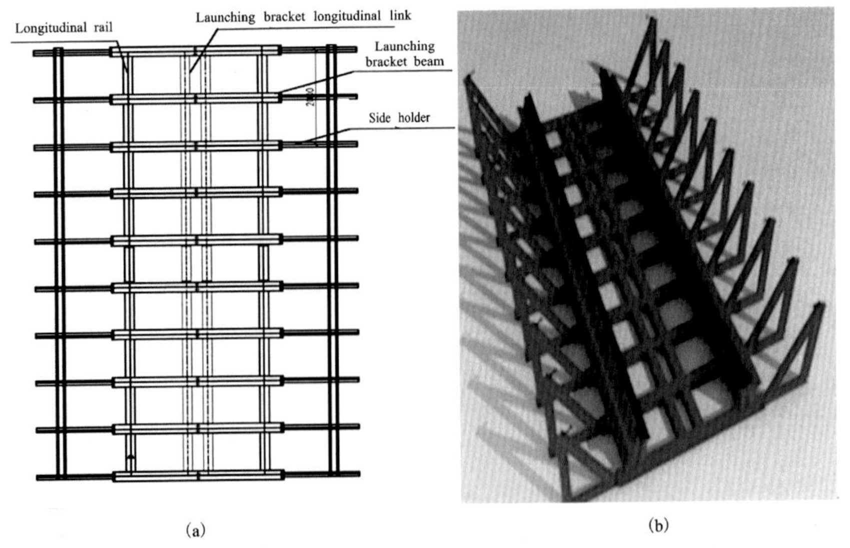

Figure 6.17 Integral steel structure base. (a) Diagrammatic sketch. (b) Photo.

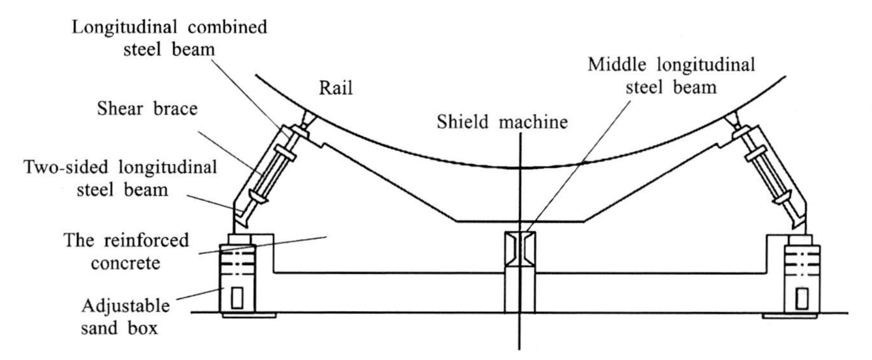

Figure 6.18 Complex launching pedestal for the shield machine.

6.2.1.2 Shield machine assembly

Before shield machine is assembled, it is necessary to make detailed plans in line. Since shield machine is large in both volume and weight, the cutterhead, front shield, middle shield, tail shield and other components are usually designed and manufactured in multiple stages and transported to the construction site in batches for the shield machine assembly. The sequence of lifting and assembling the shield machine components is as follows: (1) rear supporting trailer; (2) backup system; (3) connecting bridge; (4) screw conveyor, (5) middle shield, (6) front shield machine, (7) cutterhead, main engine, (8) segment ejector, (9) shield shell. It should be noted that the screw conveyor should be lifted down to the shaft at the step (4) but should be mounted after the installation of the shield shell.

6.2.1.3 Backup system assembly

Since the frames of the backup system are light and need large space to be assembled, they are preferred to be assembled on the ground, if their dimensions allow to be lifted in the shaft.

6.2.1.4 Mounting of reaction frame and negative segment rings

The reaction frame at the launching shaft is needed to provide the reaction force to push shield machine forward. The segment rings at the outside of shield tunnel are called as temporary negative segment rings for limiting the launching attitude of shield machine and transmitting the thrusts of jack cylinders at the shield tail. Generally, when enough segment rings (normally 100 rings for easy engineering management in field) have been asssembled in tunnel and the friction resistance is large enough between the segment rings and the ground to prevent them from moving back, the reaction frame and temporary negative segment rings can be removed.

The number of negative rings directly depends on the width of lining segment, the position of the segment starting ring (normally called as No. 0 ring), and the reference ring of the reaction frame. Assuming that the base ring of the reaction frame has a distance of L_1 from the tunnel portal, the 0# ring has a extending distance of L_2 from the tunnel portal, and the segment width is W, the required number of negative rings is $(L_1 - L_2)/W$.

Owing to the unavoidable axial deviation in tunnelling, the position determination of the starting ring mainly considers the designed size range of the tunnel portal at the launching shaft and the position of the reference ring in the reaction frame. In addition to the length of the shield machine, two other conditions must be considered to determine the position of the reference ring in the reaction frame: (1) The central axis of the reference ring is the same as that of the shield machine, and (2) the reference ring surface must be vertical and must not be tilted back.

The retaining wall of launching shaft in the scope of tunnel portal can be cut off manually. There should be sufficient space (usually about 1 m long in the longitudinal direction) between the cutterhead and the portal to set up a scaffold (usually about 1 m); and there should be correct space between the shield tail brush and the reference steel ring for coat with shield tail grease (about 0.2–0.3 m in thickness).

The position of the reaction frame is back-calculated according to the number of negative rings and the position of the No. 0 segment ring, and then, the steel plate is embedded in the bottom plate of launching shaft according to the reaction frame width, the inclining angle of the inclined struts, and the surveying baseline. The embedded steel plates and the steel bars fixing the pedestal are connected by anchor welding.

6.2.2 Reaction frame for launching

Normally, there are three working conditions for the launching of shield machine: launching in a shaft, launching in a metro station (deep excavation), and launching in a underground tunnel. When the shield machine is launched, it is necessary to install a reaction frame, which is like the "back" of the shield machine. It provides a sufficient reaction force for the shield machine to move forward [24,25].

Commonly, the reaction frame is a steel structure. The reaction frame can be fixed with different methods according to the construction environment. The reaction frame should be installed very stably and firmly. It is necessary to ensure that the reaction frame does not deform or move backward under the high thrust of the shield machine. The reaction frame is usually designed as a left-right symmetrical structure that consists of a portal frame and several support rods. A good reaction frame directly determines the success and safety of the shield machine launching.

6.2.2.1 Design principles

The reaction frame design for the shield machine launching should obey the following principles [18]:

1. The reaction frame should meet the requirements for structural strength and rigidity under the reaction force to minimize the overall deformation of the reaction frame.
2. There should be no gap between the reaction frame and the shield machine. The perpendicular relationship between the reaction frame and the shield machine axis should be kept during the shield machine launching. The reaction frame should be in line with the launching pedestal.
3. Based on the specific structural characteristics of the launching shaft, the shaft wall and the bottom plate are used as the supporting bases of the reaction frame.
4. Interference between the reaction frame and the shield machine and the tunnel portal should be avoided. The reaction frame should be easy to produce and transport.

6.2.2.2 Reaction frame type

There are generally three types of reaction frames, including straight bracing acting on the shaft wall, diagonal bracing acting on the shaft bottom plate, and their combinations. Regardless of any type, the reaction frame has components as follows.

Main beam parts

A reaction frame that satisfies the repeated uses for shield machine launching of different tunnels depends on the structure design of its main beam. If the reaction frame is used for different cross sections of shield machine, the main beams of the reaction frame should be divided into vertical components, horizontal components, and splayed-shape components. The vertical main beam is usually welded by double fashioned iron structures, and the horizontal components and the splayed-shape components of the main beams are usually welded by thick steel plates.

Supporting part

One end of each supporting structures for the reaction frame is welded to the main beam, and the other end is welded to the embedded steel plate in the bottom plate or supporting base on the launching shaft wall. Inclined supports can be used to prop up the main beams. If there is not enough space to place the inclined supports, the main beams can be directly fixed on the shaft wall with some steel fashioned iron structures. Sometime, both ways can coexist depending on the site conditions. The specific supporting method needs to be determined according to the site conditions, and each method can have a large scope of application.

Embedded parts

The embedded parts are used to fix the reaction frame. They sets the embedded ribs and embedded steel plates according to the forces required by the reaction frame.

Steel ring

The steel ring is a closed structure which is welded into a box shape. To meet the requirements of convenient installation and transportation, the steel ring needs to be reinforced with steel plates. The steel ring is the working surface of reaction frame to support the shield machine, and its planeness should be less than 5 mm.

6.2.3 Technical controls for shield machine launching

6.2.3.1 Preparation for shield machine launching

The preparation work for shield machine launching includes assembly of launching pedestal and shield machine, mounting of sealing system in tunnel portal, mounting of the reaction frame, the trial operation of shield machine, and so on.

Sealing system at tunnel portal for shield machine launching

A sealing system at tunnel portal must be installed in advance to avoid the leaking of groundwater, grout for synchronous grouting, and even soil into the launching shaft through the gap between shield machine and ground. The sealing system consists of rubber ring and steel gasket. Especially after the slurry shield machine is launched, the slurry pressure must be maintained in the excavation chamber. Thus, to prevent failure of the sealing system at the tunnel portal, the material, shape, and size of the rubber ring and steel gasket for the sealing system must be carefully designed and mounted.

A steel plate ring with bolts needs to be embedded in the secondary support of the shaft to mount the rubber ring and steel gasket. For the ring plate to be firmly embedded in the secondary support of the shaft, the steel plate ring is firmly welded to the steel bars in the secondary supports of the launching shaft.

When the steel bars in the secondary support are bounded to the tunnel portal position, the steel plate rings that have been divided into pieces are accurately placed and welded to the steel bars, and then the templates of the end walls and tunnel portal are mounted for casting concrete. During the construction process, it should be ensured that the longitudinal deviation of the steel plate position cannot be greater than 5 mm. The overall structure of the sealing system at the tunnel end is a circular ring, and the steel plate ring is turned outwards in an L-shape near the tunnel end.

6.2.3.2 Cutting off the retaining wall part at tunnel portal

The shield machine is often launched in a deep excavation or shaft with a retaining wall. The retaining wall parts in the tunnel portal scope can be cut off before launching shield machine or excavated through by shield cutterhead. If it is directly excavated by shield cutterhead, the retaining wall parts should be casted with glass fiber reinforced concrete in the tunnel portal scope. If it is cut off ahead of shield machine launching, the steel bars should be cleared out to prevent from tangling the shield cutterhead and screw conveyor.

Since the initial excavation of shield tunnel easily cause excavation face collapse and groundwater ingress, it is necessary to divide the retaining wall in the tunnel scope into multiple small parts, and these parts should be removed from bottom to top to ensure the ground stability. The removing process should be fast and continuous.For example, as shown in Fig. 6.19, the retaining wall in the tunnel portal scope is divided into nine zones to be cut off.

6.2.3.3 Initial excavation

After the retaining wall part in tunnel scope has been removed, the shield machine should be pushed in immediately. If a slurry shield is used, the concrete residues should be cleaned out to avoid blocking the slurry

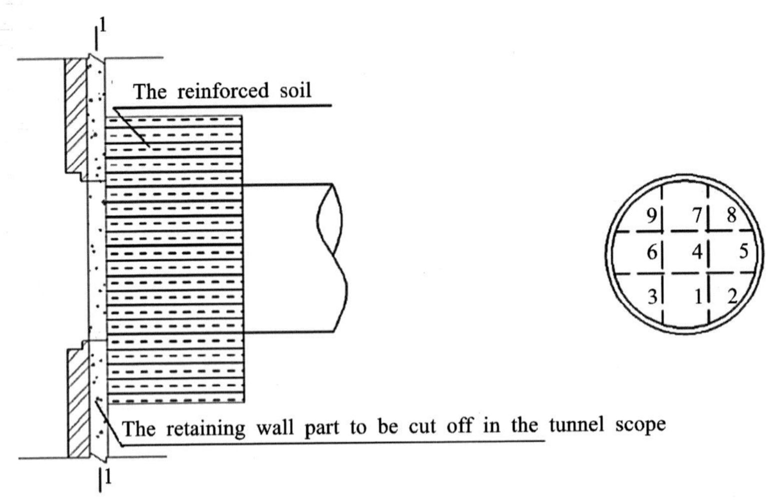

Figure 6.19 Schematic diagram of cutting off the retaining wall in the scope of the tunnel portal. (a) Lateral view. (b) Cross-section 1-1 of the retaining wall.

circulation pipelines for slurry shield machine. After the shield machine penetrates the ground, the condition of the sealing gasket should be monitored carefully. The chamber pressure should be increased slowly until the desired pressure value is reached. When the shield tail passes through the sealing gasket, the sealing gasket is easy to turn into a reverse state. After the shield tail has passes through the tunnel portal, synchronous grouting is performed to stabilize the portal ground. The shield machine launching process is shown in Fig. 6.20.

6.3 Configurations and technical controls for shield machine receiving

6.3.1 Shield machine receiving configurations

The configurations for shield machine receiving includes a receiving pedestal, a sealing system at tunnel portal, and so on. They are briefly discussed as follows.

6.3.1.1 Receiving pedestal

The centerline of the receiving pedestal should be consistent in plane with the designed centerline of the tunnel. The shield machine positioning should also be taken into account. The elevation of the rail surface on the receiving

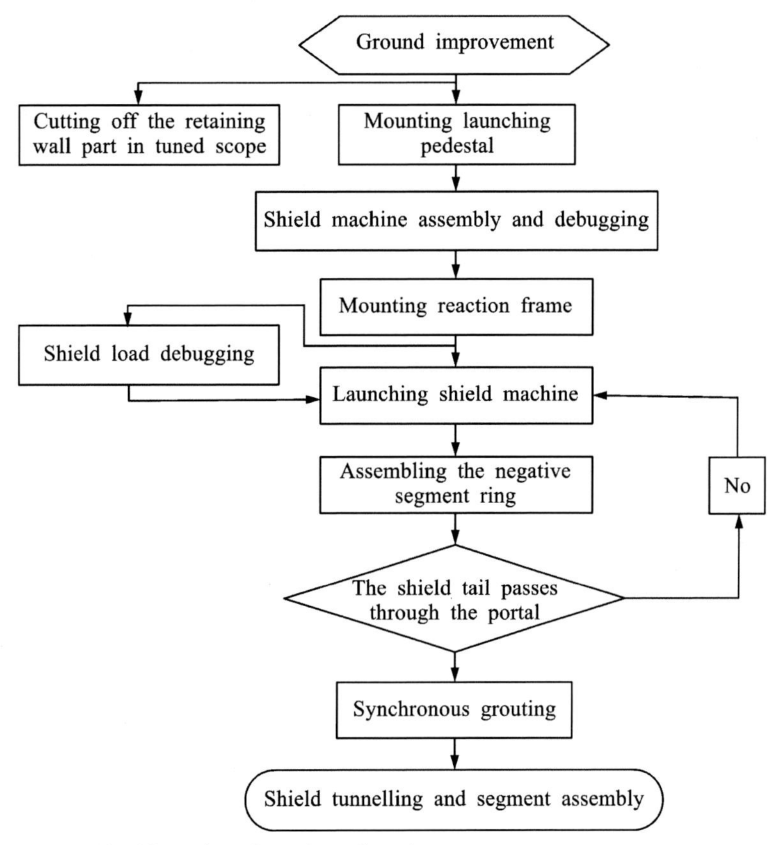

Figure 6.20 Shield machine launching flowchart.

pedestal can be adjusted appropriately in addition to adapting to the tunnel line conditions so that the shield machine can be smoothly pushed onto the base. To get enough reaction force for the assembled segments after the cutterhead of shield machine enters the shaft, the receiving pedestal is installed with a slope of +3‰ in the advancing direction. To fix the receiving pedestal, heavy steel plates are laid in the receiving shaft and are welded for connecting the receiving pedestal. Moreover, if necessary, the receiving pedestal and heavy steel plates are supported on the receiving shaft wall and bottom plate with expansive bolts, I-beams, and other materials. Longitudinal reinforcement should be strengthened to ensure that the shield machine can reach the receiving pedestal smoothly. Fig. 6.21 shows the shield machine receiving pedestal [26].

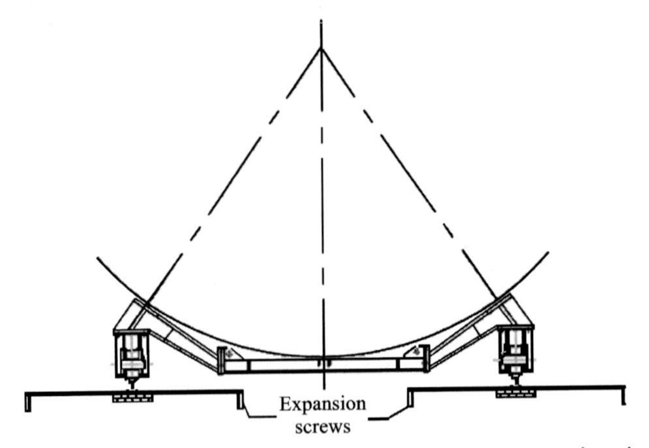

Figure 6.21 Schematic diagram of the shield machine receiving pedestal.

The longitudinal spacing between the pedestal end and the tunnel portal wall should not be more than 100 mm to ensure the stability of shield machine when its front shield passes the tunnel portal. After the pedestal has been placed, its position should be checked again. If it meets the requirements, the pedestal should be welded and fixed with the steel plates to avoid any movement.

6.3.1.2 Sealing at tunnel portal for shield machine receiving

Before the shield machine arrives at the receiving shaft, the sealing system consisting of rubber ring and steel gasket is mounted on the portal wall of receiving shaft. The rubber ring must touch the shield shell and the segment ring, as shown in Fig. 6.22.

6.3.2 Technical controls for shield machine receiving

Depending on the order difference between cutting off the retaining wall part in tunnel scope and receiving shield machine, there are two methods in receiving the shield machine:

1. Cutting off the retaining wall in the tunnel scope after the shield machine touches the wall.

   As shown in Fig. 6.23, the shield machine is pushed through the tunnel portal closed to the receiving shaft after the ground is reinforced and the retaining wall is cut off. If this method is used to remove the portal wall, the gap between the cutterhead and the receiving shaft is small when the cutterhead touches the wall. This method is widely used because of its strong self-stability and simple procedures. It is suitable for a tunnel excavated by a shield machine with a small or medium diameter, especially in a good stratum with low water pressure.

2. Cutting off the retaining wall part in the tunnel scope and building the partition wall early before shield machine touches the wall.

As shown in Fig. 6.24, this method requires removal of the retaining wall part in the tunnel scope in advance. Hence, the ground improvement has to be done before the removal, and a steel partition wall that can be easily removed is constructed in the receiving shaft. Then the retaining wall is removed from bottom to top, and cemented soil or poor proportion mortar is used to fill the stratum and the gap between the reinforcement and the partition wall in sequence. Finally, the shield machine is advanced to the front of the partition wall, the partition wall is removed, and the shield

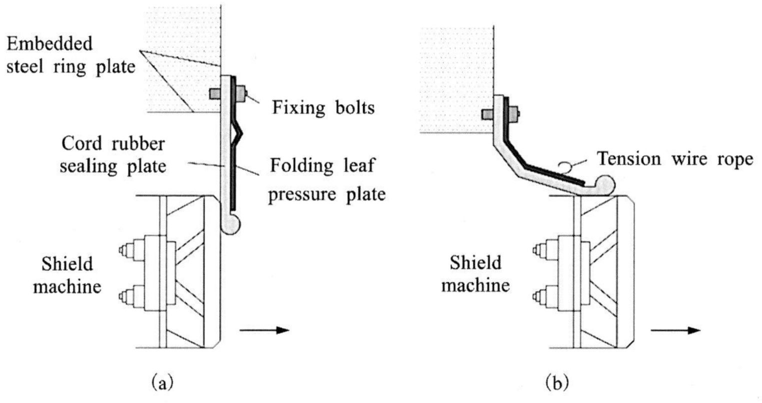

Figure 6.22 Sealing system for shield machine receiving. (a) When touching the sealing ring. (b) During passing through the sealing ring.

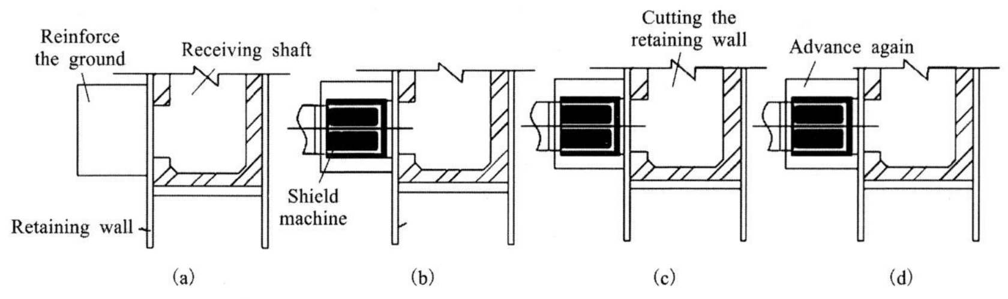

Figure 6.23 Construction procedures for the shield machine receiving method by cutting off the retaining wall in the tunnel socpe after the shield machine touches the wall. (a) The ground improvement is conducted in the receiving section. (b) The shield machine touches the retaining wall. (c) The retaining wall part in the tunnel scope is cut off. (d) The shield machine advances and is received.

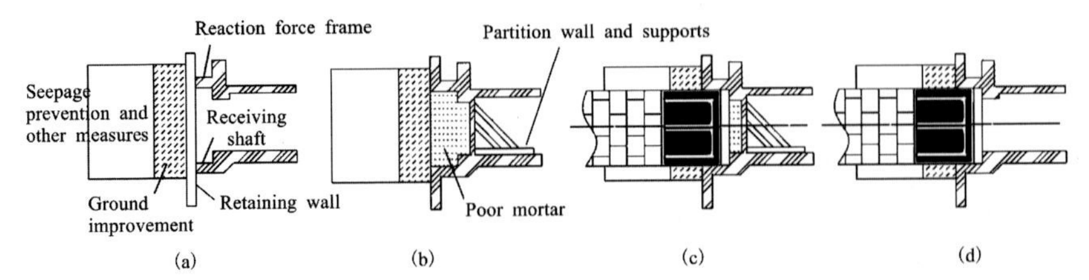

Figure 6.24 Construction procedures for the shield machine receiving method by cutting off the retaining wall part in the tunnel scope and building the partition wall early before shield machine touches the wall. (a) The ground improvement is conducted in the receiving section. (b) The partition wall and supports are mounted, the retaining wall part is cut off, and the poor mortar is injected. (c) The shield machine is received. (d) The partition wall and supports are removed.

machine receiving is completed. Because the shield machine is not allowed to advance again, this method can prevent the stratum from collapsing, and the water spewing does not easily happen at the tunnel entrance. However, as the scale of stratum reinforcement is increased and partition walls need to be installed, the receiving space should be increased in the shaft. This method is used mostly for a large-diameter and deeply buried tunnel, especially in ground with high water pressure.

6.4 Shield machine launching and receiving under special conditions

6.4.1 Steel sleeve—aided technology for shield machine launching and receiving

The shield machine launching and receiving is influenced by the geological conditions, especially in a weak ground with high water pressure [27]. For example, when there are buildings (structures) adjacent to the launching and receiving shaft, steel sleeve-aided are widely used for shield machine launching and receiving.

For steel sleeve—aided shield machine launching, a sealed steel sleeve is mounted in the launching shaft. It is connected with the seal ring of the portal and is filled with sand and other materials. Then grouting with poor grout such as bentonite slurry is performed to fill the pores in the loose soil in the steel sleeve. This method enables the shield machine to be launched in a closed environment before entering the tunnel and to maintain the pressure balance on the excavation face, thereby significantly reducing the risk of large deformation and collapse of ground.

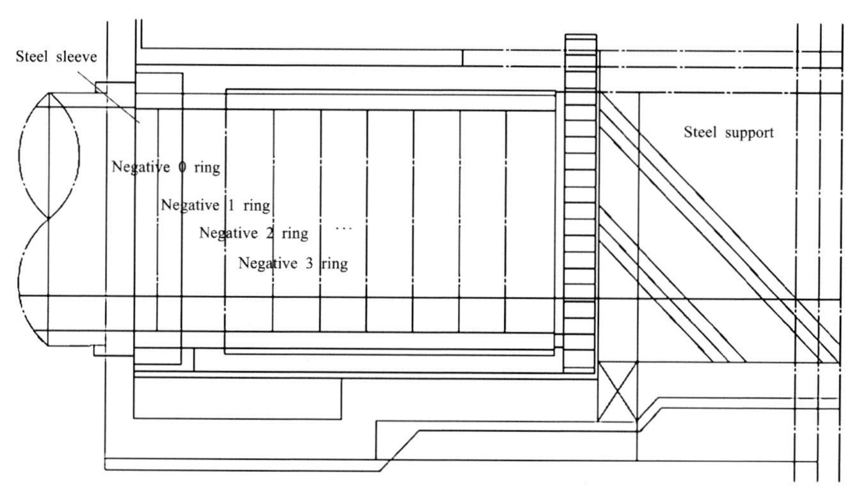

Figure 6.25 Schematic diagram of the structure of a steel sleeve-aided technology for shield machine launching.

As shown in Fig. 6.25, the steel sleeve—aided launching method is complicated. The key technical procedures of such a method include as follows: (1) removal of the concrete protection layer of steel bars on the tunnel portal wall; (2) positioning and mounting of the bottom semicircular plates of the transition ring, the bottom semicircle of the steel sleeve, and the reaction frame; (3) mounting of the steel rails in the steel sleeve; (4) sand filling in the bottom of the steel sleeve for the first time; (5) assembling and debugging of the main shield machine in the steel sleeve; (6) connection of the top and bottom semicircles of the transition ring with the steel ring at the portal wall; (7) mounting and connection of the top and bottom semicircles of the steel sleeve; (8) inspection of the reaction frame and the steel sleeve; (9) the negative ring assembly and the shield machine advancing to the excavation face in the tunnel portal; (10) filling sand and poor grout in the steel sleeve; (11) inspection of soil pressure in the steel sleeve; (12) mounting of more negative segment rings and synchronous grouting; and (13) launching of the shield machine in the steel sleeve [28,29].

Similarly, the main principle of the steel sleeve-aided shield machine receiving method is to mount a cylindrical container (Fig. 6.26) that can supply a closed environment to remain the earth pressure balance at the excavation face for safe shield machine receiving. The steel sleeve is filled with sand, and then grout is injected into the sand, resulting in that the earth pressure conditions can be built in the steel sleeve. When the shield machine

Figure 6.26 Steel sleeve–aided shield machine receiving site.

approaches the tunnel exit, the tunnel portal is sealed with the steel sleeve to prevent water and soil from flowing into the receiving shaft.

The steel sleeve system is composed of cylinder (including brackets), rear cover plate, reaction frame, push roller, and supports. The shield machine receiving method with this steel sleeve system has many advantage, such as wide applicability, recyclability, small operating space, and little impact on surrounding environments, and even it can avoid the improvement of the ground closed to the receiving shaft. It also has some shortcomings, such as a long mounting period, high cost, difficult transportation and storage, and high requirement of shield machine receiving control [30].

To ensure successful shield machine receiving, the steel sleeve–aided technology should be conducted with the following precautions:

1. Both the mounting accuracy of the steel sleeve and tunnelling parameters control through the retaining wall and into the steel sleeve are the key influence factors for the success of this technology. The pre-control of construction is important, and the careful management of key procedures such as shield machine control, grouting management for the sand in the steel sleeve, and shield machine attitude monitoring should be strengthened. It is significant to ensure that the centerline of the portal coincides with the centerline of the steel sleeve. After the steel sleeve has been assembled, it should be filled with pressurized air to check its sealing [31].

2. A transitional connecting ring plate should be mounted between the steel sleeve and the portal ring plate. The portal ring plate and the transition connecting ring plate are connected by welding, and the flange end of the steel sleeve and the transition connecting ring plate are connected by bolts.
3. The position of the shield machine should be corrected according to the measured portal elevation to ensure that the shield machine can enter the steel sleeve smoothly.
4. According to the measured pressure on the top of the steel sleeve, the thrust must be adjusted in time to avoid excessive thrust of jack cylinders at shield tail and any leakage out of the steel sleeve. When the pressure is too high, the outlet on the backplate cover of the steel sleeve is opened to reduce pressure.

In addition, when the shield machine uses a steel sleeve for launching and receiving, the backfilling materials in the steel sleeve must meet the following requirements: On one hand, it must be able to enhance the chamber pressure control for safe control of the shield machine launching and receiving. On the other hand, it must have a certain bearing capacity to prevent the front shield from subsiding and directly contacting the inner wall of the receiving steel sleeve. According to construction experience, the steel sleeve filling materials should be selected based on the soil layer conditions and the shield machine type. The sand with injected inert mud and the thick slurry are filled in the whole steel sleeve for the launching of both EPB and slurry shield machines. The sand is filled in the bottom with a height of 1/3 the steel sleeve for the receiving of both EPB and slurry shield machines, but the sand with injected inert mud and the thick slurry are filled in the top, respectively for the receiving of EBP and slurry shield machines [28].

6.4.2 Passing station technology

The passing station technology of shield machine is sometimes applied due to the construction need of the next metro tunnel section and the completion delay of its launching shaft construction in metro station. Thus, the shield machine used for the previous metro tunnel section passes the metro station directly and to be used again for the next tunnel section. In practice, one of the options such as an arched guide platform, hydraulic clamping rail, and backfilled station can be selected for the shield machine passing the station. Fig. 6.27 shows the technical process of the shield machine passing the station using the arched guide platform. Details are as follows.

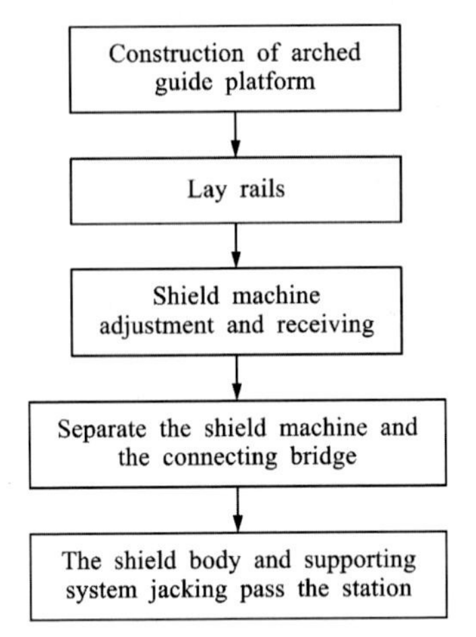

Figure 6.27 Flowchart of the receiving technology using an arched guide platform to pass the station.

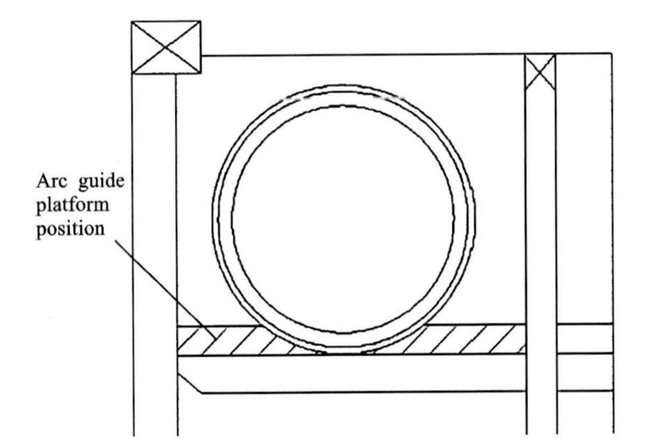

Figure 6.28 Schematic diagram of an arched guide platform.

6.4.2.1 Preparation for the arched guide platform

When the shield machine approaches the metro station, a concrete guide platform is constructed at the bottom structure based on the centerline of the design route and the rail surface. Two steel rails are placed on the concrete guide platform. A arched guide platform is shown in Fig. 6.28.

6.4.2.2 Shield machine receiving

Normally, the shield machine is received first, then passes the station, and then is launched for the next tunnel section. Therefore the construction accuracy of the arc guide platform at the receiving and launching positions is required to be high. If a long distance has been tunneled before being received, to eliminate the influence of measurement errors, it is recommended to place a receiving bracket with the same length as a guide platform for easy adjustment of the bracket and smooth receiving of the shield machine.

Before the shield machine enters the receiving pedestal, the main shield machine is moved to the receiving bracket or the rail on the arched guide platform by the guide rail. Through the last ring segment in the tunnel to provide reaction force, the main shield machine is fully pushed onto the receiving bracket or arched guide platform with jack cylinders at the shield tail.

6.4.2.3 Separation of the shield machine from the connecting bridge

After the main shield machine has been fully pushed onto the receiving bracket or arc guide platform, the connecting bridge need to be disconnected with the main shield machine before translation, and the disconnection is selected at the front and back ends of the connecting bridge. It is also necessary to separate various pipelines from the main shield machine, and the pipelines should be marked before separation.

6.4.2.4 Construction technology of shield machine passing the station

The shield machine passing the station

1. Shield machine translation

   After the preliminary preparations have been completed and the shield machine is received, the receiving bracket is separated from the embedded steel plate in the station bottom. The shield machine and the receiving bracket are welded as a whole, and four corbels are welded on the shield machine. The hydraulic pumps are started to extend the jack cylinders uniformly and steadily to push the shield machine. The station rails are placed below, jack cylinders are used to push the brackets, and so the horizontal translation of the shield machine is started.

2. Shield machine jacking

   After the main shield machine is moved to the experted direction, the rail position at the station is adjusted. A reaction support is welded at the end of the rail, and then jack cylinders are placed on the reaction support. A cyclic operation process is performed: advancing for 1 m, retracting the jack cylinders, placing a 1 m long jacking supports, and advancing again. There is 1 m long for each cycle operation process. After one cycle is completed, the main shield machine is jacked up and then the rails is pulled 2 m forward. The operation cycle is repeated until the main shield machine passes the station.

The backup system passing the station

3. Laying rails for the backup system

While the shield machine is advancing, it is also necessary to start laying the rails for the backup system. Since the shield machine is translated at the receiving end for passing station, the backup system may fail to connect with the main shield machine. Therefore, during laying the rails, it is necessary to measure and adjust the position of the rails in time, ensuring the smooth passing of the backup system. The front end of the connecting bridge is supported on the segment carriage, and a battery locomotive with a pulling capability of 45 tons is used to pull the entire backup system to move forward to complete its passing the station.

References

[1] Chen K, Hong K, Wu X. Shield Constr Technol. Beijing: China Communications Press; 2016.

[2] Technology and Industrialization Development Center of Ministry of Housing and Urban Rural Development, China Railway Tunnel Group Co., LTD. Code for construction and acceptance of shield tunnelling method (GB 50446-2017) 2017.

[3] Yu T. Application of advancing horizontal grouting: case study on Tianjin soft soil. Tunn Constr 2011;31(S2):157–61.

[4] Hu Z. Soil mechanics and environmental geotechnical. Shanghai: Tongji University Press; 1997.

[5] Gao C, Liang X. Application of high-pressure spiral jet grouting piles in shield tunnel portal soil reinforcement engineering. Prospecting Eng 2009;2:69–72.

[6] Dai Y. Freezing reinforcement technology and application in shield-driven project. Construction & Design for Project 2015;134–9.

[7] Wang W. Combined application of freezing method and horizontal grouting in Tianjin subway shield receiving. Mod Tunn Technol 2013;50(3):183–90.

[8] Atsushi K, Niu Q, Chen F, et al. Investigation, design and construction of shield method. Beijing: China Building Industry Press; 2008.

[9] Chen X. Ground freezing method. Beijing: China Communications Press; 2013.

[10] Jiang Y, Wang C, Jiang H. Theoretical research and engineering practice of reinforcement at the beginning and end of shield tunneling. Beijing: China Communications Press; 2011.

[11] Li D, Wang H, Wang T. Analysis of soil reinforcement for the staring and arriving of shield machine in metro construction. J Railw Eng Soc 2006;1:87–90.

[12] Shi Z. Underground railway design and construction. 2nd ed. Xi'an: Shanxi Science and Technology Press; 2006.

[13] Luo F, Jiang Y, Jiang H. Theoretical model and sensitivity analysis of end reinforcement based on strength and stability. J Eng Geol 2011;19(03):364–9.
[14] Romo MP, Diaz MC. Face stabilty and ground settlement in shield tunneling. In: Soil mechanics and foundation engineering 10th international conference. A.A. Balkema; 1981. p. 357-359.
[15] Chaffois S, Lareal P, Monnet J. Study of tunnel face in a gravel site. Numerical methods in geomechanics. Swoboda; 1988. p. 1493–8.
[16] Zhou X, Pu J, Bao C. Study on the stability mechanism and loose earth pressure of tunnel excavation in sand. J Yangtze River Sci Res Inst 1999;04:10–15.
[17] Zhou X, Pu J. Centrifuge model test on ground settlement induced by tunneling in sandy soil. Rock Soil Mech 2002;(05):559–63.
[18] Chen P. Structural design and analysis of the reaction frame of the large-diameter slurry shield in the launch process. In: "Industrial Architecture" 2018 National Academic Annual Conference; 2018. p. 419–423.
[19] Jiang Y, Yang Z, Jiang H, et al. Discussion on the reasonable range of reinforcement at the starting and reaching end of the earth pressure balance shield. Tunn Constr 2009;29(3):263–6.
[20] Zhang Q, Tang Y, Yang L. Research on design and construction technology of grouting reinforcement for shield tunnel entrance and exit Undergr. Eng Tunn 1993;93–101.
[21] Wu T, Wei L, Zhang Q. Research on stability of reinforced soil at large scale shield departure area. Chin J Undergr Space Eng 2008;477–82.
[22] Cao H. Reinforcement design and construction of subway shield tunnel's terminal. Soil Eng. and Foundation 2010;24(3):1–3.
[23] Chen D. Ground improvement and inspection analysis at the shield tunnel portal. Railway Survey and Design 2010;3:16–20.
[24] Zhu Q, Li X. Analysis of the safety performance of the starting reaction frame of the large-diameter shield machine at Chengdu south railway station. Constr Technol 2018;47(S1):791–4.
[25] Zhao B, Wang Y, Yue C, et al. Stress monitoring and safety evaluation of reaction frame during shield launching process. Eng Mech 2009;26(9):105–11.
[26] Gao H, Chen K, Wang Z, et al. Research on the key technology of large diameter shield machine dismantling. Constr Mechanization 2015;36(11):70–4.
[27] Wang L. Receiving technology of shield tunnel construction. Tianjin Constr Technol 2016;26(03):32–5.
[28] Wu W, Zhu H, Zou Y, et al. Research on the key technology of launching and receiving of shield steel sleeve. Tunn Constr 2017;(07):97–102.
[29] Ma T. Shield construction technology and risk control of balanced launch and arrival. China Eng. Consulting 2015;60–2.
[30] Chen S. Analysis of the use of shield arrival and reception aids. Tunn Constr 2010;30(4):492–4.
[31] Zhao L. Auxiliary construction receiving technology of soil pressure balance shield reaching steel sleeve. Railw Stand Des 2013;(8):89–93.

Exercises

1. Briefly explain the procedures for launching and receiving shield machine.
2. Briefly explain the purpose of ground reinforcement for launching and receiving shield machine. How is the range of ground reinforcement determined?

3. What configurations are needed for launching and receiving shield machine?
4. Please give an example to report the main details of the design and construction of steel sleeve–aided technology, respectively, for shield machine launching and receiving.

CHAPTER 7
Shield tunnelling and segment assembling

7.1 EPB shield tunnelling technology

EPB shields utilize the excavated soil (muck) to balance the earth pressure and water pressure on the excavation face. As the shield advances, the strata on the excavation face are cut off by the cutterhead. Subsequently, the excavated soils enter the excavation chamber through the cutterhead openings. Afterward, the excavated soils are discharged through a screw conveyor. EPB shields are initially suitable for low permeable cohesive strata. As soil-conditioning technology has developed, the applicable range of EPB shields has been significantly extended. Now EPB shields are also used for tunnelling in high permeable cohesionless strata with muck conditioning. The EPB shield tunnelling mainly includes shield advancing, soil conditioning, segment assembly, synchronous grouting, muck transportation, and so.

During tunnelling, appropriate determination of the tunnelling parameters is important to achieve a high advancement speed, low abrasion of cutters, good performance of the cutterhead, stable excavation face, and small ground displacement. The tunnelling parameters of an EPB shield include advancement speed, rotation speed and torque of the cutterhead, rotation speed and torque of the screw conveyor, chamber pressure, and shield thrust. The rotation speed of the cutterhead, rotation speed of the screw, and shield thrust can be actively adjusted by the operators, whereas the other tunnelling parameters are indirectly controlled under the influence of the soil conditioning and geological conditions.

7.1.1 Calculation and selection of tunnelling parameters

Tunnelling parameters such as chamber pressure, shield thrust, advancement speed, rotation speed and torque of the cutterhead, and rotation speed of the screw are introduced as follows.

7.1.1.1 Chamber pressure

Chamber pressure directly influences the stability of the excavation face. Normally, earth pressure gauges are installed on the bulkhead to monitor

Shield Tunnel Engineering
DOI: https://doi.org/10.1016/B978-0-12-823992-6.00007-2

the chamber pressure. Since the chamber pressure on the excavation face is greater than that on the bulkhead, the muck can flow from the excavation face into the excavation chamber. The measured pressure therefore will not exactly reflect the chamber pressure acting on the excavation face due to the influence of cutterhead and muck. The ratio (α) of chamber pressure on the excavation face to that on the bulkhead represents the chamber pressure transfer ratio through the excavated muck:

$$P = \alpha P_0 \tag{7.1}$$

where P is the chamber pressure at the bulkhead; P_0 is the earth pressure at the excavation face; and α relates to the opening layout and ratio of the cutterhead and the physical properties of the muck.

Wang [1] proposed a formula to determine α:

$$\alpha = 1 - \frac{L}{\xi^2 D}\left[\frac{4c}{k_1 K_0 \gamma H} + 2\left(1 + \frac{1}{K_0}\right)\frac{\tan\phi}{k_2}\right] \tag{7.2}$$

where L is the length of the excavation chamber; ξ is the opening rate of the cutterhead; D is the cutterhead diameter; c and ϕ are the cohesion and the internal friction angle of soil, respectively; k_1 and k_2 represent the reduction factors of the cohesion and internal friction angle, respectively; γ is the bulk density of the soil; H is the buried depth of tunnel; and K_0 is the coefficient of lateral earth pressure.

The chamber pressure should be appropriate to stabilize the excavation face and control the ground displacement. The excavation face collapse or water spewing due to insufficient chamber pressure, poor muck conditioning and the ground blow-out should be avoided. Moreover, the excessive pressure may lead to large shield thrust and cutterhead torque and serious muck clogging.

The chamber pressure is a crucial parameter for controlling the pressure balance on the excavation face. For EPB shield tunnelling, the chamber pressure is determined by the muck discharging rate and shield advancing speed. The desired chamber pressure is maintained by adjusting the muck discharging rate and shield advancing speed. Face collapse and large ground settlement indicate that the chamber pressure is insufficient, and the operator should reduce the discharging rate or increase the advancing speed to increase the chamber pressure. When the total face support pressure exceeds the in situ earth pressure, the ground blow-out may happen. The operator should reduce the chamber pressure by increasing the discharging rate or reducing the advancing speed.

Before tunnelling, the range of chamber pressure should be determined. The earth pressure is classified into active, passive, and static ones. The static earth pressure is induced by stress release due to excavation, and it is equal to the in situ earth pressure. Theoretically, there is no ground displacement if the chamber pressure at the excavation face is equal to the static earth pressure. The active earth pressure is generated when the excavation face deforms toward the excavation face, and it is equal to the required minimum value of the chamber pressure. As was mentioned previously, the chamber pressure should be as low as possible for the muck entering the chamber. The static earth pressure is thus selected as the upper limit value of the chamber pressure, and the active earth pressure is chosen as the lower limit value. The upper and lower limit values are calculated according to the geological conditions and tunnel parameters within a certain distance (e.g., 20 m) from the excavation face.

The range of chamber pressure should consider the construction conditions. A lower chamber pressure is adopted when the surrounding ground has good self-stability, and a higher value is adopted when the ground easily fails or some adjacent buildings or underground pipelines exist. The maximum value is calculated as

$$P_{max} = \text{groundwater pressure} + \text{static earth pressure} + \text{marginal pressure} \tag{7.3}$$

The minimum value is calculated as

$$P_{min} = \text{groundwater pressure} + \text{active earth pressure} + \text{marginal pressure} \tag{7.4}$$

The marginal pressure is to compensate for the pressure loss possibly induced by ground leakage. It usually ranges between 20 and 50 kPa.

All these formulas are applicable to cohesionless soils and separately consider earth pressure and water pressure, and effective stress principle is applied. For cohesive soils, the formulas should consider water pressure and earth pressure together and the total stress is applied.

7.1.1.2 Shield thrust

The jack cylinders in the shield tail supply thrusts to push the shield machine forward and overcome the friction between the shield and the surrounding strata and the earth and water pressures on the excavation face. There are several groups of jack cylinders arranged on the middle beam of the ring inside the shield. The operator can control each group

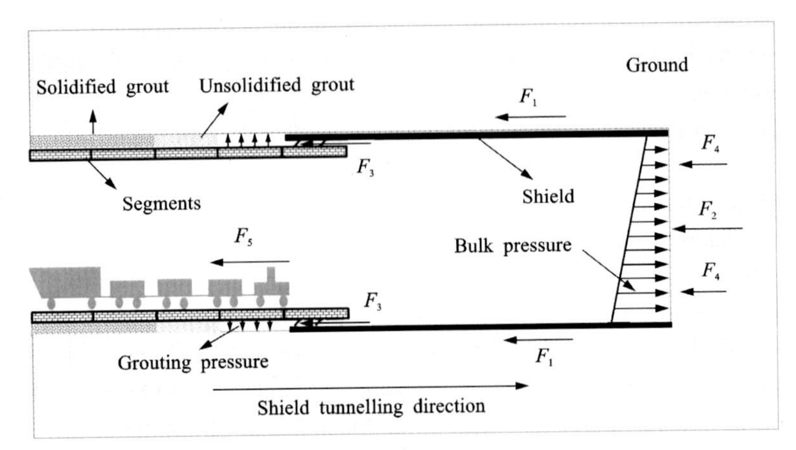

Figure 7.1 The composition of the thrust force.

of jack cylinders (usually four groups: up, down, left, and right) separately to adjust the shield attitude. The shoes of the jack cylinders are placed against the ends of lining segments. As shown in Fig. 7.1, there are five kinds of resistances that have to be overcome for shield advancing. The shield thrust (F_d) is calculated as

$$F_d = F_1 + F_2 + F_3 + F_4 + F_5 \tag{7.5}$$

where F_1 is the friction between the shield and the surrounding ground; F_2 is the ground resistance on the excavation face; F_3 is the friction between the shield tail and the lining; F_4 is the resistance to the penetration of the cutterhead; and F_5 is the drag force of the backup system.

The friction between the shield and the surrounding strata, F_1

For shield advancing, as shown in Fig. 7.1, the shield is subjected to various pressures. Soil suction is not considered for calculation convenience here. The average normal stress is taken for the determination of friction:

$$F_1 = \left(q'_{e1} + q'_{e2} + q'_1 + q'_2\right)\pi D_e L\mu/4 \tag{7.6}$$

where q'_{e1} is the vertical effective earth pressure at the top of the shield; q'_{e2} is the vertical effective earth pressure at the bottom of the shield; q'_1 is the horizontal effective earth pressure at the top of the shield; q'_2 is the horizontal effective earth pressure at the bottom of the shield (the weight of shield should be considered); and μ is the friction coefficient between the shield and the surrounding strata (which can be roughly estimated to be tan ϕ', if not tested).

For shield advancing in clay, the friction can be calculated as follows:

$$F_1 = \pi D_e L c' \tag{7.7}$$

where c' is the effective cohesion of clayey soil.

Zhang [2] pointed out that for shield tunnelling in a complex geological condition, $c' \neq 0$ and $\phi' \neq 0$. Therefore the results calculated by using the above methods are often greatly different from the actual case. Zhang deduced a calculation method considering the complexity of the ground and the shear strength of the soil according to the Mohr-Coulomb criterion.

The ground resistance on the excavation face, F_2

$$F_2 = \pi D_i^2 \left(q_e' + p_w + \alpha\right)/4 = \pi D_i^2 q_e'/4 + \pi D_i^2 (p_w + \alpha)/4 = F_{2.1} + F_{2.2} \tag{7.8}$$

where q_e' is the average effective earth pressure at the excavation face $= \left(q_{e1}' + q_{e2}'\right)/2$; p_w is the average water pressure at the excavation face; α is the pressure margin for avoiding face collapse; $F_{2.1}$ is the resistance determined by the shield cutting $= \pi D_i^2 q_e/4$; and $F_{2.2}$ is the resistance determined by the chamber pressure $= \pi D_i^2 (p_w + \alpha)/4$.

The friction between the shield tail and the lining, F_3

$$F_3 = n_1 W_s \mu_1 \tag{7.9}$$

where n_1 is the number of segment rings in the scope of the shield tail; W_s is the total weight of a segment ring; and μ_1 is the friction coefficient between the sealing brush and the segment ring, $\mu_1 = 0.3-0.5$.

The resistance to the cutterhead penetration into strata, F_4

For shield tunnelling in sand, the friction is calculated as,

$$F_4 = \pi\left(D_e^2 - D_i^2\right) q_4/4 + \pi D_e Z K_p P_v \tag{7.10}$$

where q_4 is the back pressure when the cutterhead penetrates into the strata; Z is the penetration depth; K_p is the positive earth pressure coefficient; and P_v is the average pressure acting on the outer edge of the cutterhead.

For shield tunnelling in clay, that is

$$F_4 = \pi\left(D_e^2 - D_i^2\right) q_4 + \pi D_e Z c' \tag{7.11}$$

The drag force of the backup system, F_5

$$F_5 = W_5 \cdot \mu_5 \tag{7.12}$$

where W_5 is the weight of backup system; and μ_5 is the friction coefficient between the backup system and the rails.

The resistance induced by shield turning also exists for the curved tunnelling. According to a great number of calculation results, $F_1 + F_2$ accounts for 95%–99% of total thrust force (F_d), while F_3, F_4, and F_5 contribute little to the total thrust. The sum of F_3, F_4, and F_5 is about 300–400 kPa times the area of the excavation face [2].

The designed shield thrust of jack cylinders should consider the safety coefficient, and it is normally equal to 3–4F_d [2].

Shield thrust directly affects the ground displacement through chamber pressure. On one hand, when the thrust is large, the strata in front of the excavation face may move forward and cause ground heave. A large thrust can also increase the friction between the cutterhead and the excavation face, leading to too high torque of the cutterhead. On the other hand, a low thrust may reduce the advancing speed and construction efficiency. Furthermore, the ground in front of the excavation face may move backward the shield, and the excavation face collapse may happen. Therefore the thrust of jack cylinders should be set appropriately.

7.1.1.3 Advancing speed

Advancing speed is an direct indication of tunnel construction efficiency. It is governed by the thrust and torque of the cutterhead and geological conditions. Generally, the advancing speed should be controlled under 20 mm/min during the entire shield trial tunnelling and receiving. During normal tunnelling, the advancing speed is less than 40 mm/min. For cases of small ground displacement and good work conditions of the shield machine, the advancement speed could reach as high as 80 mm/min.

The followings should be noted:

1. The shoes of the jack cylinders should be placed closely against the lining segments, and the advancing speed should increase gradually to avoid an impact load on the ground.
2. Large fluctuation of advancing speed should be avoided.
3. The advancing speed should be considered with ensuring the efficiency of synchronous grouting.
4. Attention should be paid to the geological conditions and adjacent buildings or pipelines to avoid unconventional abrasion of shield

cutters and excessive disturbance to the surrounding ground. The advancing speed is set as a low value during shield tunnelling underneath sensitive buildings. However, considering the delay of ground displacement, some engineers prefer to push the shield machine in a high speed as well as efficient synchronous grouting.

7.1.1.4 Rotation speed of the cutterhead

The rotation speed and torque of the cutterhead are relevant to the geological conditions surrounding the shield machine. The rotation speed is low and torque is high when tunnelling in hard strata, but the rotation speed is high and torque is low when tunnelling in soft strata. The rotation speed ranges from 0 to 3.5 rpm in accordance with the design requirements. The constant torque control is taken when the rotation speed is low, and the constant power control is taken when the rotation speed is high. For the pattern of constant power control, if the chamber pressure is too large, the torque will increase and the rotation speed will decrease.

7.1.1.5 Torque of the cutterhead

The torque of the cutterhead is a critical parameter for tunnelling. Its sharp fluctuation is caused when muck discharging is not smooth. If the cutterhead torque exceeds its capacity, the shield machine will operate abnormally. Therefore it is important to control the cutterhead torque.

The cutterhead torque includes the friction torque between the cutterhead and the strata, the cutting torque of the cutters, the stirring torque of the stirring rods, and so on. Here, only these three types of torques are considered:

$$T_d = T_1 + T_2 + T_3 \tag{7.13}$$

where T_1 is the friction torque between the cutterhead and the strata, which is generated by the friction between the front plate, back plate, and edge of the cutterhead and the surrounding ground; T_2 is the cutting torque of cutters; and T_3 is the stirring torque of stirring rods.

T_1 can be calculated by using the following formula:

$$T_1 = \frac{2\pi\mu_1 R^3 P_c(1-\xi)}{3} + \frac{2\pi\mu_1 R^3 P_w(1-\xi)}{3} + 2\pi\mu_1 RBP_z \tag{7.14}$$

where μ_1 is the friction coefficient between the cutterhead and the strata; ξ is the opening ratio of the cutterhead; R is the cutterhead radius; P_c is the earth pressure on the front plate of cutterhead; P_w is the chamber pressure; P_z is the average earth pressure around the outer edge of the cutterhead; and B is the thickness of the cutterhead edge.

A variety of cutters are equipped on the cutterhead, such as scrapers, peripheral cutters, central cutters, advance cutters and so on. During excavation, the cutting torque of the cutters is calculated as

$$T_2 = \sum_{i=1}^{n} T_{2i} = \sum_{i=1}^{n} F_{2i} L_i \quad (7.15)$$

where T_{2i} is the resistance torque of No. i cutter; F_{2i} is the resistance force of No. i cutter; and L_i is the distance between No. i cutter and the cutterhead center.

T_3 can be calculated as follows:

$$T_3 = \gamma H_0 D_b L_b R_b f \quad (7.16)$$

where γ is the bulk density of muck in the excavation chamber; H_0 is the buried depth of the stirring rods; L_b is the length of the stirring rod; R_b is the distance between the stirring blade and the cutterhead center; and f is the friction coefficient between the muck and the stirring rod.

According to these formulas, the main factors influencing the torque of the cutterhead are analyzed as follows.

The radius, opening ratio, and type of cutterhead

The contact area between the cutterhead and the surrounding ground affects the friction torque. If the opening ratio is constant, the contact area and friction torque increase with increasing radius. The greater the opening ratio, the smaller the contact area and torque. For a cutterhead with the same radius, the torque of the spoke cutterhead is lower than that of the panel cutterhead because the opening ratio of the former is normally lower.

The lateral earth pressure

According to the Coulomb's friction criterion, the cutterhead torque is positively related to the lateral earth pressure, which is mainly determined by the tunnel buried depth and geological conditions. The torque on the back of the cutterhead is affected by the chamber pressure and conditioned muck.

The mechanical property of strata

The friction coefficient, cutting pressure of the cutters, and torque of the stirring rods are determined by the mechanical properties of the strata and cutterhead. Muck conditioning can promote the fluidity of muck in the excavation chamber, significantly affecting the cutterhead torque.

Cutterhead penetration

Cutterhead penetration is defined as the advancing distance per rotation circle of the cutterhead as follow:

$$\text{Cutterhead penetration} = \text{Advancing speed}/\text{Rotation speed} \quad (7.17)$$

The cutterhead penetration is controlled by the shield thrust and the cutterhead rotation speed. The cutting torques of the cutters are determined by the penetration and layout of the cutters on the cutterhead. The cutting torque increases with an increase in cutterhead penetration for the same geological conditions.

Layout of mixing blades

The stirring rods cause additional torque on the cutterhead when it rotates. Therefore the muck conditioning and shield tunnelling modes have great impacts on the stirring torque.

Screw rotation speed

The screw conveyor is used to discharge the muck out of the excavation chamber. The screw rotation speed is important for maintaining a stable chamber pressure and controlling the ground settlement. If the chamber pressure increases, the operator should increase the rotation speed to increase the muck discharge rate for maintaining constant chamber pressure. Otherwise, the operator should decrease the screw rotation speed to reduce the discharge rate.

The tunnelling parameters also depend on other factors such as the shield type and geological conditions. The calculated values are just a first estimation, and timely adjustment should be made according to the actual geological conditions.

7.1.2 Pressure balance control of the excavation face

The pressure balance state of the excavation face can be evaluated and controlled by ground surface settlement, chamber pressure, discharging ratio, and earth pressure balance ratio. The control methods are shown in Table 7.1.

Table 7.1 Pressure balance control methods of the excavation face for EPB shield tunnelling.

Methods	Standards	Discrimination of balance		Standard values	Real values
		State	Criterion		
Ground settlement	No settlement	Overpressure	Ground heave $u > 0$	Measurable	Measurable
		Balance	No ground displacement $u = 0$		
		Underpressure	Ground Settlement $u < 0$		
Chamber pressure	Chamber pressure being P_0 for balancing excavation face	Overpressure	$P > P_0$	Calculable but theoretical	Measurable
		Balance	$P = P_0$		
		Underpressure	$P < P_0$		
Discharging ratio	1	Overpressure	$R_d < 1$	Measurable	Hardly measurable
		Balance	$R_d = 1$		
		Underpressure	$R_d > 1$		
Earth pressure balance ratio	$(N/v)_b$	Overpressure	$<$	Calculable	Measurable
		Balance	$\left(\frac{N}{v}\right) = \left(\frac{N}{v}\right)_b$		
		Underpressure	$>$		

7.1.2.1 Controlled by ground settlement

The balance between the chamber pressure and the static earth pressure can be judged by the ground settlement. When the ground settlement is observed, the chamber pressure on the excavation face could be lower than the static earth pressure, and the excavation chamber is under underpressure. If a ground heave is observed, it is indicated that the chamber pressure is larger than the static earth pressure and the soil chamber is in the state of overpressure. In the ideal situation, the chamber pressure on the excavation face is equal to the static earth pressure, and the ground displacement is 0. Since a time is required for the transfer of pressure distribution and ground deformation from the excavation face to the ground surface, resulting in a delay of ground surface settlement, sometimes, it is late to control the chamber pressure through the ground settlement. However, since the ground settlement is visual and measurable, it is usually to adjust the muck discharging rate on the basis of the time-dependent ground settlement in practice.

7.1.2.2 Controlled by chamber pressure

As was mentioned in Section 7.1.1, the chamber pressure is designed to the sum of effective lateral earth pressure, water pressure, and marginal pressure for balancing the earth pressure and water pressure on the excavation face. However, the theory for calculating effective lateral earth pressure has its own limitations, and the chamber pressure measured by sensors on the bulkhead is also affected by the cutterhead type, the cutterhead opening ratio, the muck properties, and so on. As a result, it is not a easy task to determine the ideal pressure on the bulkhead. The theoretically calculated value of the chamber pressure can only be used for a first estimation during the shield machine launching. The chamber pressure should be adjusted in a timely fashion according to the on site conditions.

7.1.2.3 Controlled by discharge rate

The discharging ratio (R_d) is the ratio of muck entering rate to the muck discharging rate, and the amount of injected materials for muck conditioning should be accounted for the muck entering. The discharging ratio is one of the important factors for tunnelling control. To maintain a stable chamber pressure, the discharging ratio needs to remain 1. However, it is hard to achieve this goal, especially for a large-diameter shield, because it is affected by many factors, such as the rotation speed of cutterhead, rotation speed of the screw, and advancing speed.

In general, the discharged muck amount can be evaluated with mass or volume. For the evaluation with volume, the loosening coefficient of muck should be considered, since the strata will swell owing to stress releasing induced by excavation. The loosening coefficient is defined as the ratio of the volume of loose muck in carriage to that of the excavated strata, and it is generally measured at the job site. Thus, for muck discharging management, the volume of loose muck should be the volume of excavated strata times the coefficient, additionally plus the volume of conditioning agents. For the evaluation with mass, there are some errors induced by groundwater flowing into the muck. Therefore it is more reasonable to determine the discharged muck amount by both mass and volume. The theoretical discharged volume per a rotation cycle of screw (Q) is calculated as

$$Q = \frac{\pi}{4}\eta\left(D_2^2 - D_1^2\right)P \tag{7.18}$$

where η is the discharge efficiency, relating to the advancement speed, rotate speed, bulk pressure, and so on; D_2 is the inner diameter of the screw conveyor cylinder; D_1 is the diameter of the screw cylinder shaft; and P is the spacing between two adjacent blades.

The muck loosening parameters include the natural loosening coefficient K_1 and the compacting loosening coefficient K_2. They can be calculated as follows:

$$K_1 = \frac{V_2}{V_1} \tag{7.19}$$

$$K_2 = \frac{V_3}{V_1} \tag{7.20}$$

where V_1 is the natural volume of soil in the strata; V_2 is the muck volume in the loosest state; and V_3 is the muck volume in the densest state.

Theoretically, the maximum volume (Q_{max}) and the minimum volume (Q_{min}) of the discharged muck can be calculated as follows:

$$Q_w = \frac{1}{4}\pi D^2 vt \tag{7.21}$$

$$Q_{max} = K_1 Q_w \tag{7.22}$$

$$Q_{min} = K_2 Q_w \tag{7.23}$$

where Q_w is the theoretical ground volume per segment ring; v is the advancing speed; D is the cutting diameter of the shield machine; t is the advancing time for one segment ring.

The discharged muck volume should be controlled between Q_{min} and Q_{max}. Otherwise, instability of the excavation face may be caused. The muck volume is also influenced by the compression stress and compaction method. In practice, it is hard to keep the discharging rate to be 1 by controlling the screw rotation speed and shield advancing speed.

7.1.2.4 Controlled by earth pressure balance ratio

The earth pressure balance ratio (N/v) is the ratio of the rotation speed of the screw (N) to the advancing speed (v). When the muck volume entering chamber is equal to the muck discharging volume for one segment ring, the ratio is called as ideal earth pressure balance ratio, $(N/v)_b$. The ratio can be used as a critical index to evaluate whether there is the deviation of the earth pressure balance state or not. The N and V are recorded by the shield machine computer. When the ratio is less than the ideal value, the excavation chamber will on the way to the state of overpressure, and the N should be increased for remaining the earth pressure balance at the excavation face. Otherwise, when the ratio is greater than the ideal value, the state is on the way to underpressure, and the N should be reduced.

The control methods mentioned above are hard to work out independently. The setting of tunnelling parameters should refer to two or more of those controlling methods. The goal is to control the ground displacement for EPB shield tunnelling. The chamber pressure and earth pressure balance ratio are controlled to ensure that the tunnelling parameters are in a good status. Ideal tunnelling parameters are obtained after continuous trials and adjustments.

7.1.3 Selection of tunnelling modes

One advantage of the EPB shield is its flexibility of tunnelling mode. There are three modes for EPB shield tunnelling: open mode, semiopen mode, and closed mode. The selection of EPB tunnelling modes depends on the geological conditions and surrounding environments.

7.1.3.1 Open mode

The open mode means that the shield tunnelling is carried out without any pressurized muck and air in chamber, and it is suitable for tunnelling in self-stable ground with little groundwater. The muck is removed

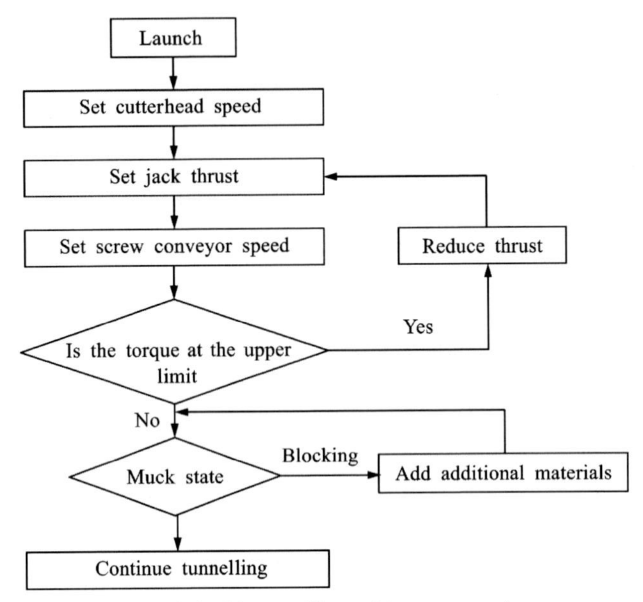

Figure 7.2 The flowchart of shield tunnelling with open mode.

directly by the screw conveyor when it flows into the excavation chamber. For open mode, there is no excavation pressure against the excavation face, and the chamber pressure is just the atmospheric pressure. The control flowchart is shown in Fig. 7.2.

7.1.3.2 Semiopen mode

For the semiopen mode, the upper part of the excavation chamber is filled with compressed air against earth and water pressures on the excavation face, and the lower part is filled with muck. The cutterhead torque in the semiopen mode is smaller than that in the full EPB mode (i.e., closed mode), and the risk of clogging in semiopen mode is lower than that in EPB mode, and the tunnelling efficiency and the capacity to break rock are higher. The semiopen mode is suitable for stable ground under low water pressure. The pressurized air is used to prevent the groundwater from flowing into the shield machine. When shield tunnel is excavated under the groundwater level and encounters a locally unstable ground or a fault crushed zone, the operator should increase the advancement speed so that shield can go through this area as fast as possible and temporarily closes the muck discharging gate to fill the lower part of the excavation chamber with muck. At the same time, a conditioning agent (e.g., bentonite slurry) and compressed air are injected onto the excavation face and chamber. The conditioning agent penetrates the ground in front of the shield and forms a low permeable mud film under pressure. This is helpful for avoiding face collapse and water flowing into the excavation chamber.

7.1.3.3 Closed mode

The closed mode means that the shield tunnelling is carried out with excavation chamber full of pressurized muck. It is suitable for tunnelling in weak ground or high permeable soft ground with high water pressure. For this mode, the muck should completely fill the excavation chamber, and the excavation pressure should be transferred through the muck onto the excavation face to balance the earth and water pressures on the excavation face.

Chamber pressure can be controlled by changing the shield advancing speed and the screw rotation speed. The excavation rate should stay in balance with the discharging rate so that the chamber pressure can efficiently transfer onto the excavation face. The chamber pressure is measured by the pressure sensors that are equipped on the bulkhead, and the discharging ratio can be controlled according to these pressure sensors. For this closed mode, the soil is cut off by the cutters and the shield machine drives at a low rotation speed with large torque. The chamber pressure should be a little greater than the sum of the earth pressure and water pressure to control the ground settlement for the safety of adjacent buildings or pipelines. If tunnelling is occurring in a high permeable weak ground such as coarse-grained soil, the muck needs to be mixed with foam or slurry to improve the flowability of muck. In addition, the risk of water spewing through the screw conveyor discharging gate and loss of chamber pressure can be avoided by mounting a pressure-keeping system in the screw conveyor outlet.

7.2 Slurry shield tunnelling technology

The slurry shield uses a pressurized slurry to support the tunnel face. Water and bentonite are mixed in a high-speed mixer to produce slurry, and then the slurry is pumped into the slurry chamber through the feeding pipelines. Jack cylinders in the shield tail push the shield machine forward when the excavation chamber is fully filled with slurry, and the slurry pressure is transferred to the excavation face. Some slurry shields such as those made by Germany companies adjust the chamber pressure by controlling the air pressure in the air chamber, which is equipped behind the excavation chamber. The concentration and pressure of slurry increase gradually and the slurry forms a filter cake to balance the earth and water pressures at excavation face. During tunnelling, the excavated soil and groundwater flow into the excavation chamber through the openings in the cutterhead. After being mixed in the excavation chamber, the slurry containing excavated soil flows into the separation system through the discharging pipelines. The excavation face is stabilized by controlling the balance between the volumes of pumped-in and pumped-out slurry. The tunnelling parameters, such as the shield thrust and the cutterhead torque, are similar to those of the EPB shield. Therefore this section focuses only on the control of slurry pressure and the volume of the excavated soil.

7.2.1 Control of slurry pressure

The slurry pressure is equal to earth pressure plus water pressure and marginal pressure. It should be between its upper and lower limits and be adjusted appropriately considering the geological conditions and the adjacent buildings. Differently from EPB shield, the flow rate of excavated soil and the opening ratio have little impact on the pressure on the excavation face and chamber pressure. The chamber pressure monitored on the bulkhead is therefore nearly equal to the chamber pressure on the excavation face. However, similarly to the EPB shield tunnelling, the upper limit of slurry pressure should consider static earth pressure, and the lower limit should consider passive earth pressure. The marginal pressure is to ensure that the slurry can penetrate the ground at the excavation face and form a filter cake. The value of marginal pressure depends on the buried depth of the tunnel, the groundwater pressure, and the geological conditions. Normally, it is set to in the range of 20–50 kPa.

There are two methods of controlling slurry pressure, including direct control and indirect control. Most slurry shields with direct control are initially developed in Japan. For this kind of shield machine, an impermeable filter cake is formed on the excavation face, and the slurry pressure is transferred onto the soil skeleton. The slurry pressure is maintained by controlling the slurry circulation. Slurry shields with indirect control are initially made in Germany. The slurry system is composed of the slurry loop and compressed air loop. The slurry pressure is controlled by air pressure in the air chamber behind the slurry chamber.

The loss of slurry and the change in advancing speed may influence the balance between the slurry feeding and the slurry discharging. In the meantime the slurry pressure in the excavation chamber may fluctuate due to the influence. Slurry shields with indirect control solve this problem more easily compared to those with direct control, since the air cushion can act as a buffer and the pressure of compressed air is easy to adjust. Therefore the indirect control system is more beneficial to control the excavation face stability and ground settlement.

In addition, if the slurry discharging is not smooth, the slurry discharging system should change to the state of the bypass discharging. When the suction ports of the discharging pipes are blocked, reverse washing could help to remove blocks in the suction ports. However, the slurry pressure in the excavation chamber should be remained in this process. For the exchange of normal discharging, reverse washing and bypass discharging, the deviation of slurry pressure should be controlled within ± 40 kPa.

7.2.2 Management of excavation volume

Since the excavation face stability for the slurry shield is invisible, the volume of the excavated soil and the excavation face stability are normally evaluated and controlled by the flowmeter and densitometer on the discharge pipeline. It can be achieved according to the volume of excavated wet soil (discharging volume - feeding volume) and volume of excavated dry soil (discharging volume - feeding volume) [3], as shown in Fig. 7.3.

7.2.2.1 Control of the volume of excavated soil

For the tunnelling without overexcavation, the theoretical volume of the excavated soil Q (m^3) can be calculated as follow:

$$Q = \frac{\pi}{4} \times D^2 \times S_t \tag{7.24}$$

where D is the cutting diameter of shield, m; and S_t is the advancing distance, m.

Within a advancing distance of S_t, the measured volume of the excavated soil Q_3 (m^3) is

$$Q_3 = Q_2 - Q_1 \tag{7.25}$$

where Q_1 is the feeding volume, m^3; and Q_2 is the discharging volume, m^3.

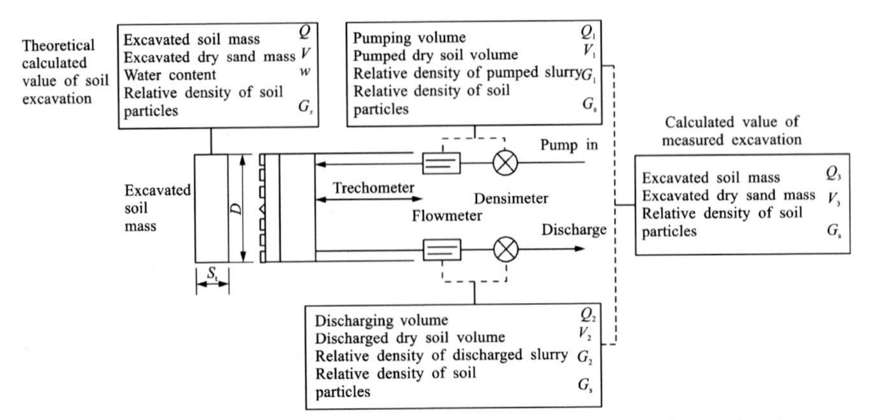

Figure 7.3 Measurement of excavation volume [3]. *Modified from The Japanese Geotechnical Society. Shield method investigation, design and construction. Translated by Niu Q, Chen F, Xu H. Beijing: China Construction Industry Press; 2008.*

The earth pressure balance conditions on the excavation face can be analyzed by comparing Q with Q_3. If $Q > Q_3$, the slurry or the water would escape from the excavation chamber and permeate the ground. Instead, when $Q < Q_3$, the chamber pressure is lower than the groundwater pressure, and the groundwater will flow into the excavation chamber. However, the collapsing soil is mixed with slurry when the excavation face collapses, meaning that there are no apparent changes of Q and Q_3. In practice, more cases of slurry escaping without the face collapse were found.

7.2.2.2 Control of the dry soil volume

The dry soil volume is the volume of soil particles in the feeded and discharged slurry. The dry soil volume can be calculated by assuming that the natural densities of soil particles in ground, pumped slurry, and discharged slurry are identical. It can be calculated as follow:

$$V = Q \times \frac{100}{100 + G_s w} \tag{7.26}$$

where V is the dry soil volume; G_s is the specific gravity of the soil particles; and ω is the water content of the soil.

When the water content of slurry in the feeding and discharging pipelines are unknown, the water content can be estimated by relative density and specific gravity of the soil:

$$w_1 = \frac{[100(G_s - G_1)]}{[G_s(G_1 - 1)]} \tag{7.27}$$

where w_1 is the water content of slurry in feeding pipelines; and G_1 is relative density of slurry.

The calculation of the dry soil volume in the feeding pipelines is similar to Eq. 7.26, and the water content should be calculated according to Eq. 7.27. That is,

$$V_1 = \frac{Q_1(G_1 - 1)}{(G_s - 1)} \tag{7.28}$$

where V_1 is the dry soil volume in the feeding pipelines.

Similarly, the dry soil volume in the discharging pipelines can be calculated as follow:

$$V_2 = \frac{Q_2(G_2 - 1)}{(G_s - 1)} \tag{7.29}$$

where V_2 is the dry soil volume in the discharging pipelines (m^3); and G_2 is the relative density of slurry in the drainage pipelines.

The dry soil volume per unit tunnelling distance can be calculated as follow:

$$V_3 = V_2 - V_1 = \frac{1}{G_s - 1}[(G_2 - 1) \times Q_2 - (G_1 - 1) \times Q_1] \tag{7.30}$$

where V_1 is the feeded dry soil volume; V_2 is the discharged dry soil volume; G_1 is the relative density of the feeded slurry; and G_2 is the relative density of the discharged slurry.

The result based on Eq. 7.30 is calculated for unit advancing distance. The real volume for an advancing distance of S_t should be calculated by integral computation. If $V > V_3$, the slurry would escape from the excavation chamber and penetrate into the ground. If $V < V_3$, the discharged slurry would exceed the feeded one in volume, resulting in possible failure of the excavation face.

7.3 Segment transport and assembling

7.3.1 Segment storage and transport

During the storage of segments, some measures should be taken to avoid damage and erosion of segments. The segments are heavy and so should be transported carefully. Segments are generally stacked layer by layer, as shown in Fig. 7.4. The gap between segments should be padded by using wood sleepers. In addition, to avoid erosion from rain and sunlight, the

Figure 7.4 The storage of segments.

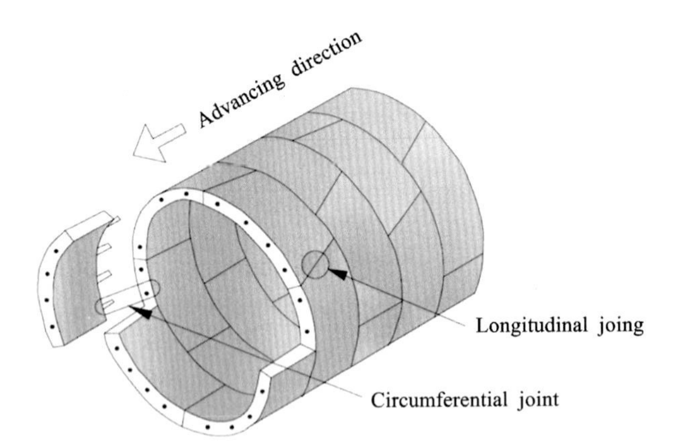

Figure 7.5 Segment assembly.

segments should be covered by plastic sheets. The segments of each ring have to be marked and placed together to make their subsequent delivery and assembly convenience.

7.3.2 Segment assembly

7.3.2.1 Segment assembly method

As shown in Fig. 7.5, segments are installed immediately by an erector at the shield tail once the shield machine has finished the excavation distance of one segment ring by cutterhead. There are two methods for segment assembly: straight-joint assembly and staggered-joint assembly. The straight-joint assembly method is sometimes applied for the small-radius tunnelling route or the case of position correction, and the staggered-joint assembly is more preferred normally due to its beneficence to the overall stability of tunnel supports. The segment assembly begins from the bottom to the top until the capping block is installed. The sealing strip on the keystone should be lubricated. There are two methods for inserting the capping block, including radial inserting and longitudinal inserting. The radial inserting is to initially place the block with about two-thirds width and to adjust the block attitude, and then the block is pushed longitudinally. After the block has been assembled, the shoes of the jack cylinders should be placed on the segment blocks within a short time, and the thrust should be sufficient to stabilize the entire segment ring. Afterward, segments are bolted manually, and radial bolting should be done ahead of longitudinal bolting. The bolts are screwed again by pneumatic screwdriver after one ring is placed. Then, the shield machine recovers its advancing. Since the thrust and the earth and water pressure may make the segment ring loose after the shield tail moves away, the bolts should be screwed one more time.

7.3.2.2 Roundness retaining

Due to the earth pressure and segment assembly quality, it is possible that the segment ring deforms and has a shape changing from a circle to an ellipse. The length difference between the major axis and the minor axis of the deformed segment ring is called as roundness. A circle is a ring with a roundness of 0. Roundness retaining for the segment ring is important for shield advancing direction, segment waterproofing, and ground displacement. When the shield tail escapes from the segments, the earth pressure, water pressure, and segment self-weight that act on the segment ring may lead to its deformation. If the deformation is large, some dislocations will occur between the segment blocks completed and those being assembled. In this case, longitudinal bolting will be difficult to be conducted. Therefore a roundness-retaining system can be used to decrease the deformation during the shield tail escaping from the segment ring. A roundness-retaining device is shown in Fig. 7.6. It is equipped with jacks and two arched supports on the top and bottom. The supports can move along steel beams. Once the assembly of a segment ring has been completed, the roundness-retaining system is moved into this ring immediately, and the shield machine continues to advance with the shoes of the jack cylinders are placed on the segment ring. However, this system is not very popular because it occupies a large construction space in the tunnel.

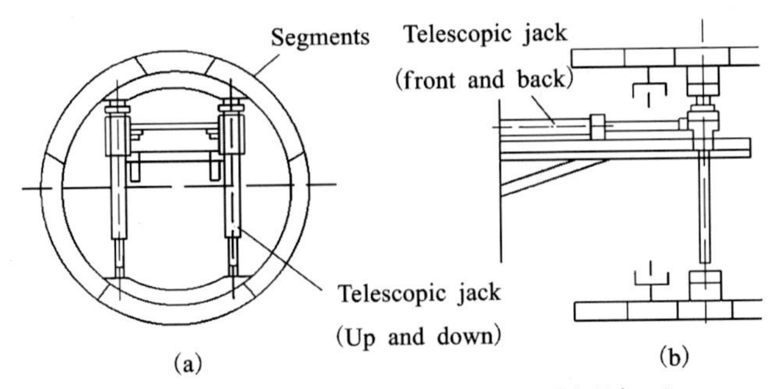

Figure 7.6 Roundness-retaining system. (a) Front view. (b) Side view.

7.3.3 Selection of segments

The shape and type of segments should match the designed tunnelling route. The segment selection should consider segment types and assembly locations of capping blocks of segment. The segment selection influences the segment dislocation, joint leakage, and segment breakage. It is important to take into account overall arrangement of segments, locations of the capping blocks and shield tail gap, and so on.

7.3.3.1 Overall arrangement of segments

The curved tunnel route is achieved with a combination of standard rings and turning rings or an application of wedge-shaped rings. The turning rings and wedge-shaped rings can also be used for correcting the shield attitude and tunnelling route, since both of them have wedges. An overall arrangement of segments will be designed ahead of tunnel construction, and the engineers should figure out the numbers of turning rings and standard rings in the turning section and the planned location of capping blocks of segment in each ring (Fig. 7.7). The overall arrangement should consider the exact location of connection tunnels between uplink and downlink. Otherwise, the location of the connection tunnel has to be changed during the shield tunnelling.

The segment layouts in circular curve and transition curve of the tunnel route are introduced in the following sections.

Layout of segments in a circular curve

First, the deflection angle is calculated as follow:

$$\theta = 2\gamma = 2 \times arctg(\delta/D) \tag{7.31}$$

where θ is the deflection angle of turning ring; δ is the half of maximum wedge; and D is the diameter of segment ring.

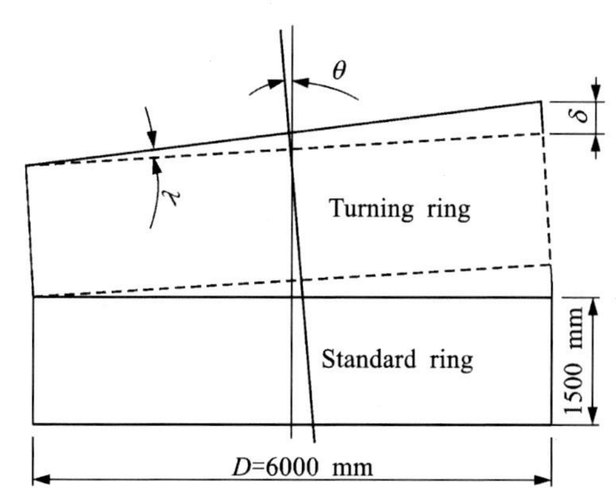

Figure 7.7 The arrangement relationship between standard rings and turning rings of segments.

The curve length for assembling each turning ring can be calculated with $\theta = \alpha$ as follow:

$$L = \frac{\theta \pi R}{180} \tag{7.32}$$

where L is the curve length for assembling one turning ring; and R is the radius of circular curve.

This formula (Eq. 7.32) shows that one turning ring should be used for each tunnelling distance of L in the circular curve with a radius of R.

The number of turning rings in a length of a circular curve can be calculated as follow:

$$N = \frac{L'}{L} \tag{7.33}$$

where N is the number of segments for a circular curve; and L' is the length of the circular curve.

Layout of segments in the transition curve

1. Number of turning rings

   The basic parameters of the transition curve include the curve radius (R), the curve length (L_s), and the tangent angle of curve (β_0) (Fig. 7.8).

   First, the tangent angle of any point in the transition curve (Fig. 7.8) is calculated as follow:

$$\beta = \frac{L^2}{2RL_s} \tag{7.34}$$

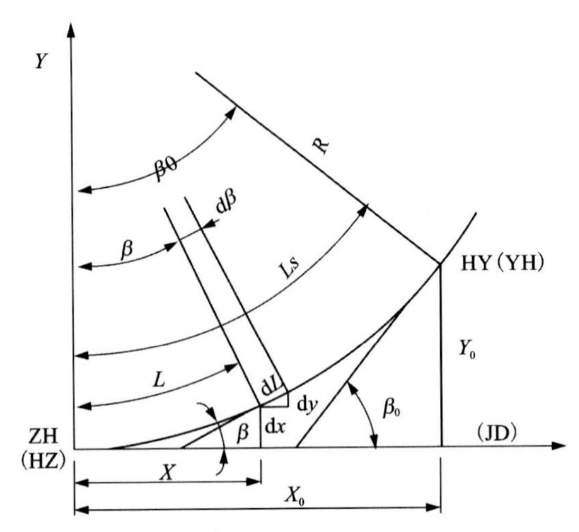

Figure 7.8 The basic parameters of transition curve.

where β is the tangent angle of any point on the transition curve; L is the length of the curve part from the starting point ZH to any point on the transition curve; L_s is the total length of the transition curve; and R is the curve radius at the end point HY.

When $L = L_s$, the tangent angle at the end point can be calculated as follow:

$$\beta_0 = \frac{L_s}{2R} \tag{7.35}$$

where β_0 is the tangent angle at the end point HY.

In the transition curve, the deflection angle ratio of the curve is the number required for arranging turning rings. That is,

$$N = \frac{\beta_0}{\theta} \tag{7.36}$$

where N is the number of turning rings for the transition curve; and θ is the deflection angle calculated in Eq. 7.31.

2. The positions of turning rings

Generally, a turning ring is arranged when the accumulative value of β at any point exceeds the value of θ. The turning ring is started to be arranged at the point with an accumulative value of β equal to 0.5θ. Therefore the lengths of the curves at $\beta = 0.5\theta$, 1.5θ, 2.5θ, and 3.5θ can be calculated to determine the positions of the turning rings.

Table 7.2 The basic parameters of a transition curve of metro tunnel.

Parameters	Value	Sign	Mileage
R (m)	450	ZH	K6 + 273.459
Ls (m)	60	HY	K6 + 333.459

Examples:

The basic parameters of a transition curve of metro tunnel are shown in Table 7.2. Please determine the numbers and positions of the turning rings. The key parameters of turning rings are as follows: The outer diameter D is 6000 mm, and the wedge is 38 mm.

1. The number of turning rings:

   The tangent angle of any point on the transition curve is

$$\beta_0 = \frac{L_S}{2R} = \frac{60}{2 \times 450} = 0.067$$

   The deflection angle of each turning ring is

$$\theta = 2 \times arctg(\delta / D) = 2 \times arctg(19/6000) = 0.3629^\circ$$

   The number of turning rings is

$$N = \frac{\beta_0}{\theta} = \frac{0.067 \times 180}{0.3629\pi} = 10.58$$

   The result shows that the number of turning rings is 10.58. With taking the whole number, there should be 11 turning rings for the transition curve.

2. The position of turning rings

   First the curve lengths at the points with $\beta = 0.5\theta$, 1.5θ, 2.5θ, 3.5θ, ... are calculated. Then the specific positions of the turning rings can be deduced according to the chainage of the transition curve. Table 7.3 shows the chainage of the center of each turning ring.

7.3.3.2 Assembly positions of capping blocks

The tunnel route is achieved by assembling the capping blacks at different positions and using turning rings with a wedge. The capping block positions and the segment wedge can control the relative deflections of the turning rings of segments to ensure that the axis of the constructed shield tunnel is consistent with the designed one.

Table 7.3 Mileage of the turning rings.

Ring number	Tangent angle β	The distance between the turning ring and the point of ZH (m)	Chainage of turning ring
1	0.5θ	13.077	K6 + 286.536
2	1.5θ	22.650	K6 + 296.109
3	2.5θ	29.241	K6 + 302.700
4	3.5θ	34.599	K6 + 308.058
5	4.5θ	39.232	K6 + 312.691
6	5.5θ	43.372	K6 + 316.831
7	6.5θ	47.150	K6 + 320.609
8	7.5θ	50.648	K6 + 324.107
9	8.5θ	53.919	K6 + 327.378
10	9.5θ	57.002	K6 + 330.461
11	10.5θ	59.927	K6 + 333.386

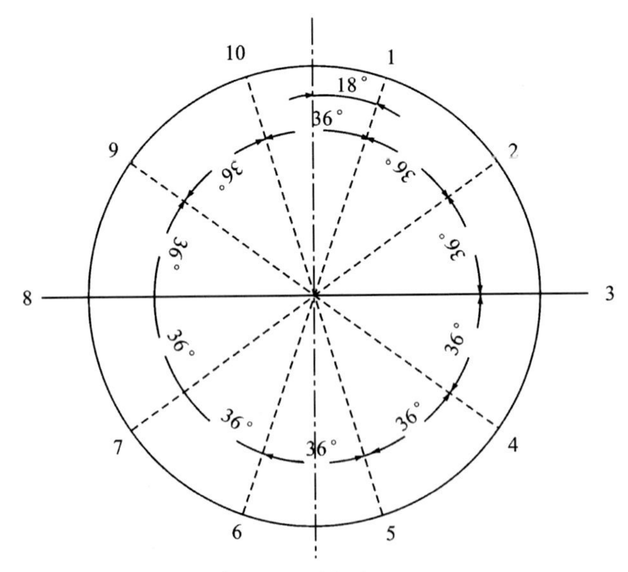

Figure 7.9 Assembly positions of capping block.

A segment ring normally has 10 assembly positions for the capping blocks, as shown in Fig. 7.9. For segments with a width of 1.5 m, Table 7.4 shows the calculated wedges of the right-turning segments.

7.3.3.3 Shield tail gap

There is a shield tail gap between the segment ring and the surrounding strata. If the gap is too small, the shield tail may contact the segment ring,

Table 7.4 The right-turning wedge (mm).

Installation position	Length of left side	Length of right side	Wedge	Length of upper side	Length of lower side	Wedge
10	1518.24	1481.75	Right 36.49	1491.91	1508.09	Up 16.18
1	1518.24	1481.75	Right 36.49	1508.09	1491.91	Down 16.18
2	1512.56	1487.43	Right 25.13	1516.04	1483.96	Down 32.08
9	1512.56	1487.43	Right 25.13	1483.96	1516.04	Up 32.08
8	1500.00	1500.00	0	1481.00	1519.00	Up 38.00
3	1500.00	1500.00	0	1519.00	1481.00	Down 38.00

generating friction between the shield tail and the segment ring and inducing cracking or damaging of segments. In this situation the resistance to shield advancing may increase, and the advancing speed decreases and even dislocation happens at the joints between segment blocks. The gap at one side of shield tail decreases while that of the opposite side increases. The grout may flow into the tunnel when the tail brush is damaged by grouting. Therefore the operators should measure the gap thickness around the shield tail before selecting the positions of the capping blocks. If the gap thickness at one position is less than 50 mm, the operators should select the opposite position as the assembly position of the capping block. If the gap thickness is decreasing, the gap will be adjusted by controlling the travel difference between the jack cylinder groups.

7.3.3.4 Travel differences of jack cylinders at the shield tail and at shield articulation

The shield advancing is achieved by jack cylinders pushing against the segments at the shield tail. The jack cylinders are generally separated into upper, lower, left, and right groups. The travel differences between different groups of jack cylinders indicate the spatial relationship between the shield shell and the segment ring and the change trend of the gap thickness in the next segment ring. If the segment ring centerline is not in line with the tail shield axis, the travel distances of jack cylinders will be different between different groups. In this case, large radial forces will be generated due to the travel differences, affecting the attitudes of the assembled segments and shield machine. As shown in Fig. 7.10, if the standard segment is assembled in the next ring, the gap thickness will continue to decrease at the bottom of shield tail.

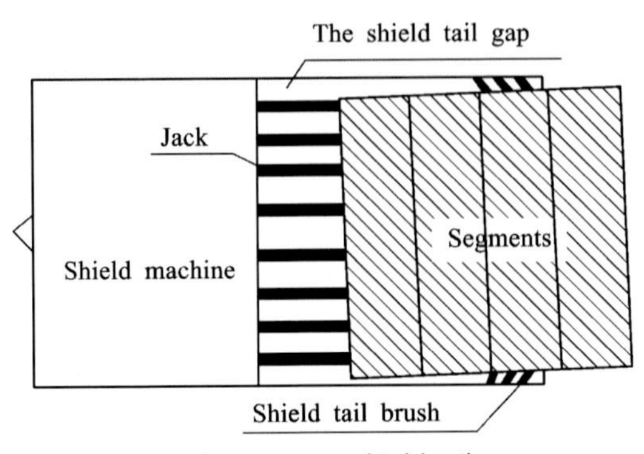

Figure 7.10 The spatial relationships among shield tail, segments, and jack cylinders.

Generally, the wedge-shaped segments will be used, if the travel difference between two groups of jack cylinders in opposite positions exceeds 40 mm at the shield tail.

Commonly, the tail shield and middle shield are connected by an articulation system with jack cylinders. This system is helpful for correcting the shield attitude. When the travel difference between two groups of jack cylinders in the opposite positions of articulation system is too large, there is an angle between the tail shield and the middle shield. Such an angle will affect the judgment to the travel difference between the jack cylinder groups at the shield tail. Therefore the result is calculated as the selection reference of segments by subtracting the travel difference of jack cylinders at the articulation system from that at the shield tail.

7.3.4 Tunnelling errors and technical controls

7.3.4.1 Shield attitude control

The principles of shield attitude control

The shield attitude should be controlled according to the designed tunnelling route and the deviation trend of shield attitude. The correction process cannot damage the segments.

The control of shield advancing direction

Advancing direction control is to correct any deviation of advancing direction so as to keep the direction consistent with the designed route. The shield can turn left, turn right, head downward, head upward, and go straight by controlling the thrust and travel distance of each group of jack cylinders. For example, when the shield heads upward excessively, the operator can increase the travel distance of the upper jack cylinders to make the shield head downward and correct the shield attitude. The key points of shield advancing direction control are as follows:

1. Reference point: The position of the shield tail is set as the reference point for the shield attitude correction.
2. Correction speed: To keep the tunnel axis in a good shape, the correction value should be controlled within 5 mm per ring.
3. Adjustment to advancing trend: Adjustment to the advancing trend should be gradual, and excessive correction should be avoided.

The control of shell rotation

1. Change the cutterhead rotation to be in the opposite direction of the shell rotation.

2. Change the assembly sequence of segments at the left and right sides of tunnel.
3. Adjust the travel distance of the left and right jack cylinders so that the centerline of the jack cylinder ring is not parallel to the axis of the shield for avoiding the continuous shell rotation.
4. When the rotation angle is too large, the rotation angle can be adjusted by increasing the weight of the lateral side of the cutting ring or supporting ring of the shield machine.

7.3.4.2 Allowable errors in segment assembly

The shield attitude is the state of its position deviation from the designed tunnelling route to the actual tunnelling route. It is measured by an automatic monitoring system or a manual monitoring system. The main parameters of shield position include horizontal deviation, vertical deviation (angle of pitch), shield rotation degree, and so on. The horizontal deviation and vertical deviation (angle of pitch) reflect the horizontal difference and vertical difference, respectively, between the advancing path and the designed axis. The rotation degree represents the rotation degree of the shield machine around its central axis. The values of deviations between the constructed tunnel and the designed tunnel directly reflects the construction quality. The inside surface of segments should be smooth, and the bolts should be screwed tightly. Any penetrating cracks, surface cracks with a width larger than 0.2 mm, and concrete spalling are unacceptable.

The Chinese *Code for Construction and Acceptance of Shield Tunnel* (GB50446-2017) [9] stipulates the allowable deviation and inspection method of the shield tunnel axis (Table 7.5). Additionally, the norms in the process of segment assembly are shown in Table 7.6.

7.3.4.3 Shield attitude correction

The Chinese *Code for Construction and Acceptance of Shield Tunnel* (GB50446-2017) [9] recommends that the shield attitude should be corrected when the shell rotation angle exceeds 3° or the deviation between the actual shell axis and the designed tunnelling route is greater than 50 mm. Shield attitude correction involves controlling the tunnelling parameters to keep the shield tunnelling along or make it return to the designed route. The shield always departs from the designed route due to geological and construction factors. The shield position should be corrected to make the shield return to the designed route gradually. The

Table 7.5 Allowable deviation and inspection methods of shield tunnel axis (mm).

Inspection items	Allowable deviation						Inspection method	Frequency	
	Metro tunnels	Road tunnels	Railway tunnels	Hydraulic tunnels	Municipal tunnels	Oil and gas tunnels		Ring	Position
The horizontal deviation of tunnel axis	± 50	± 75	± 70	± 100	± 100	± 100	Total station	All	One location per ring
The vertical deviation of tunnel axis	± 50	± 75	± 70	± 100	± 100 (The bottom of tunnel)	± 100	Total station	All	

Note: The municipal tunnels in the table include water supply and drainage tunnels and power tunnels.
Translated from Technology and Industrialization Development Center of Ministry of Housing and Urban Rural Development, China Railway Tunnel Group Co., LTD. Code for construction and acceptance of shield tunnelling method (GB 50446-2017). 2017.

Table 7.6 Allowable deviations and inspection methods of segment assembly.

Inspection items	Allowable deviation						Inspection method	Frequency	
	Metro tunnels	Road tunnels	Railway tunnels	Water tunnels	Municipal tunnels	Oil and gas tunnels		Ring	Position
Ovalization (%)	± 5	± 6	± 6	± 8	± 5	± 6	Profiler and total station	Every ten rings	—
Longitudinal joint dislocation height (mm)	5	6	6	8	5	8	Ruler	All	Four location per ring
Circumferential joint dislocation height (mm)	6	7	7	9	6	9	Ruler	All	

Translated from Technology and Industrialization Development Center of Ministry of Housing and Urban Rural Development, China Railway Tunnel Group Co., LTD. Code for construction and acceptance of shield tunnelling method (GB 50446-2017). 2017.

attitude correction must be carried out in a planned way and be done step by step. The key points of attitude correction are as follows:
1. The changes of shield attitude should be less than 5 mm per ring of segment.
2. The tunnelling parameters should be suitable for the geological conditions.
3. Reasonable types of segments should be selected, and significant impacts of operation on the position of the shield should be avoided. The assembly quality of segments is also a crucial factor for correcting the shield attitude.
4. Attention should be paid to the shell rotation at all times during the attitude correction.
5. The shield advancing speed should be slowed down for attitude correction.
6. The axis of the actual tunnelling route should be less than 20 mm away from the designed one to avoid large gap and leakage at the shield tail.

7.4 Technologies for opening excavation chamber and replacing cutters

7.4.1 Reasons for opening excavation chamber and changing cutters

When shield tunnelling is being done in the ground with boulders, gravel, or hard rock, serious wearing abrasion of cutters may occur. If the abraded cutters are not replaced in time, the cutterhead might be damaged and have to be scrapped. The durability of the shield is important in tunnelling in a complex geological conditions. The most important factor is the durability of the cutters. Serious wearing abrasion and damage of cutters occur frequently during shield tunnelling in ground where a soft soil layer lies above a hard rock layer [4,5]. In this situation the advancing speed will decrease, and the load of the cutterhead will increase. Sometimes the shield has to be stopped, and hence the excavation chamber has to be opened and checked. This takes a lot of time and is expensive [6,7].

Thus, it is necessary to timely open the soil chamber and replace cutters to ensure smooth shield tunnelling. If the muck temperature is too high, the cutterhead torque and shield thrust are too large, the advancing speed is too low, or other abnormal situations occur, the cutters should be checked if the stability of the excavation face can be maintained with some measures. The constructors should organize experts and experienced engineers to determine the reasonable location for opening excavation chamber and replacing cutters.

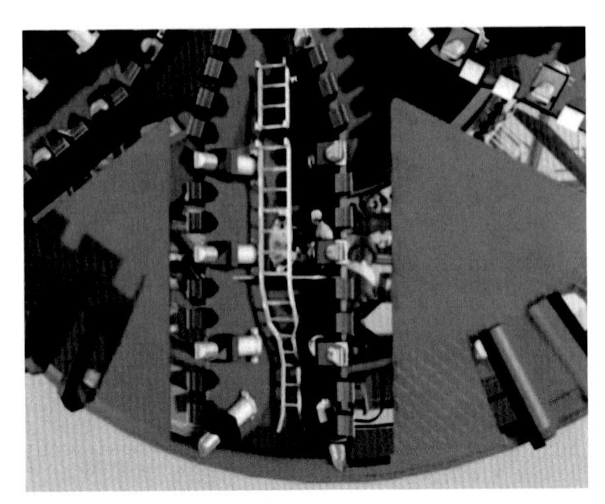

Figure 7.11 The working space for changing cutters at the atmospheric pressure.

7.4.2 Methods of replacing cutters

7.4.2.1 At the atmospheric pressure

Workers enter the excavation chamber at the atmospheric pressure and check the cutters from the cutterhead openings. Then the worn cutters are changed. If the natural ground is stable and has low permeability, the replacing cutters at the atmospheric pressure will have a lower cost, a safer work environment, and higher efficiency compared to replacing cutters under an excessive pressure. In addition, this method can maintain the stability of the excavation face and avoid injuring the workers. Therefore replacing cutters under the atmospheric pressure is the preferred option. Fig. 7.11 shows the operation space for replacing cutters under the atmospheric pressure.

The method of replacing cutters at the atmospheric pressure is suitable for the ground with strong stability. If the ground stability is low, it should be improved from the ground surface. In general, it is not applicable to the shield tunnel projects without the conditions for the ground improvement, for example, when there are numerous buildings on the ground surface or the groundwater pressure is high.

7.4.2.2 Under an excessive pressure

For avoiding the ground improvement in the weak soil, replacing cutters is carried out under an excessive pressure in excavation chamber. A low permeable filter cake layer is required to be formed on the excavation face ahead of worker entering excavation for replacing cutters under the excess pressure. The filter cake helps to avoid the water seepage and pressurized air leakage at

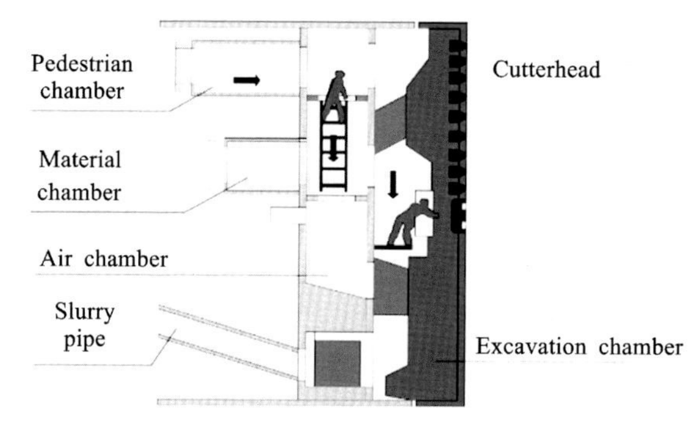

Figure 7.12 The working space for replacing scrapers under the excessive pressure.

the excavation face and to stabilize the excavation face. Taking a slurry shield with an air chamber as an example, in the excavation chamber the pressurized air is used to balance the water and earth pressures on the excavation face. The workers enter the air chamber first and then enter the excavation chamber with pressurized air through the pedestrian chamber to check, maintain, and replace cutters when the air pressure in the air chamber is increased and equal to the pressure in the excavation chamber. Fig. 7.12 shows the operation space for replacing cutters under the excessive pressure.

The technology for replacing cutters under the excessive pressure can be used for shield tunnelling crossing rivers, lakes, straits, and urban areas and in the ground with poor stability and high groundwater pressure. It should be ensured that there is no great air leakage in the ground surrounding the cutterhead, and there is no groundwater entering the shield machine.

7.4.3 Techniques for replacing cutters

7.4.3.1 At the atmospheric pressure

The location for replacing cutters at the atmospheric pressure should be in good geological conditions with strong self-stability and low permeability. If the geological conditions cannot meet the requirements and the cutters replacement is urgent, the ground improvement surrounding shield cutterhead should be carried out in advance. The technical process of replacing cutters is shown in Fig. 7.13.

Common problems and their solutions:

1. The emergency stop button of cutterhead should be pressed for the work in the excavation chamber, and there should be at least one worker monitoring the work in the control room.
2. Rotation of the cutterhead should be forbidden when there are workers in the excavation chamber.

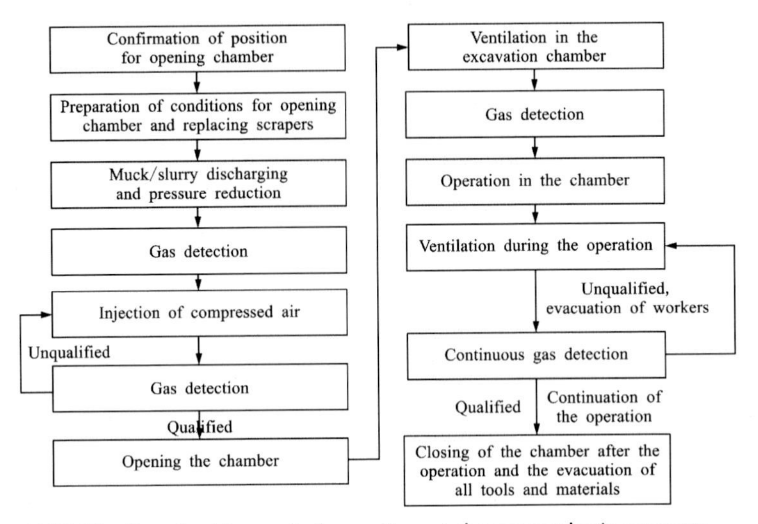

Figure 7.13 The flowchart for replacing cutters at the atmospheric pressure.

3. Toxic gas detection should be done once every hour, and the relevant data should be recorded.
4. The workers must not stand under the cutters when they are being transported in the excavation chamber.
5. The lifting and positioning of cutters should be done by grasping forceps, which should be checked carefully ahead of use.

7.4.3.2 Under an excessive pressure

In the process of replacing cutters under an excessive pressure, to build a safe space for operating, the ground ahead of the excavation face is reinforced for the weak ground in advance. Then compressed air is applied to the excavation chamber to balance the earth and water pressures on the excavation face. A flowchart for replacing cutters under an excessive pressure is shown in Fig. 7.14.

Common problems and their solutions:

1. The airtightness of the ground before the excavation face has to be checked. If air leakage is detected, the excavation face should be reinforced.
2. The ground around the shield should be reinforced as well, since groundwater may flow into the excavation chamber from the shield back.
3. The tools used under pressure should be checked carefully before opening chamber, and the shield should be equipped with a spare air compressor and a spare electric generator.

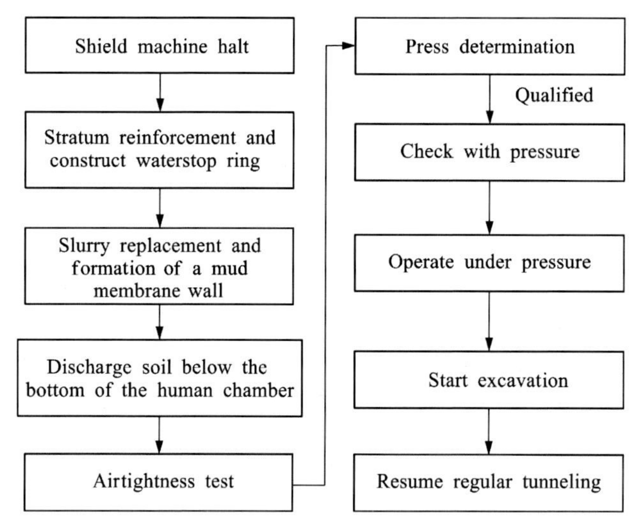

Figure 7.14 Flowchart for replacing cutters under excess pressure.

4. The gate between the air chamber and the excavation chamber always leaks air in the process of reducing pressure, so the pressure in the air chamber cannot perfectly be decreased to 0. In practice, if the air pressure in the air chamber can be decreased to 0.3 bar, the work conditions are safe. Otherwise, the gate sealing should be handled under pressure.

7.5 Connection tunnels between two main tunnels

7.5.1 Functions of connection tunnels

Connection tunnels should be built in accordance with any emergency during the operation of main tunnels. Generally, one connection tunnel with pump stations are built in the middle of the tunnel chainage, but more than one are needed for long tunnels. Connection tunnels have the functions of connecting main tunnels, water drainage, fire safety, and so on. There are two patterns of connection tunnels, including with and without a pump station. Fig. 7.15 shows a typical connection tunnel with a pump station for metro tunnel.

7.5.2 Construction method of connection tunnels

Connection tunnels are usually constructed after the main tunnels have been built. The NATM or shield tunnelling method is normally adopted to construct the main tunnels. The influence on the surrounding

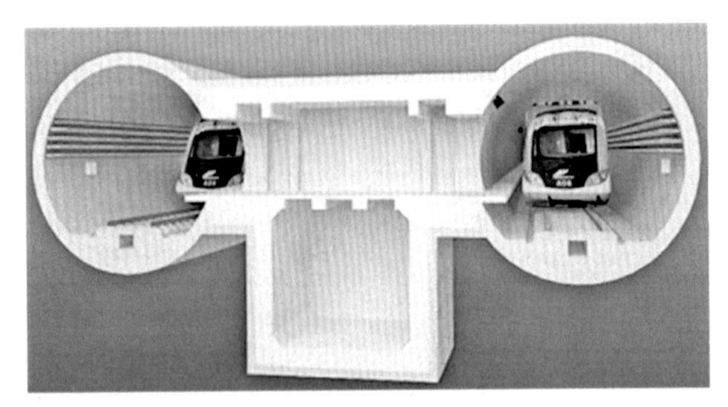

Figure 7.15 The connection tunnel of a subway.

environment and main tunnels is inevitable during the construction of the connection tunnels. Therefore the ground surrounding the connection tunnels should be reinforced, especially for ground with high permeability and low bearing capacity.

Various tunnelling methods for the connection tunnels have been developed for more than 100 years since the first metro was built. At present, the main tunnelling methods include the New Austrian tunnelling method (NATM), pipe jacking, and so on. These methods were adopted in Great Britain and Germany in the 19th century and in Germany, Japan, the United States, and France in the 20th century. In the 1990s China began to introduce and use shield machines and advanced shield tunnelling technologies. Metros were rapidly constructed in the large cities, and so the construction technologies for the connection tunnels rapidly developed from then on.

7.5.2.1 New Austrian tunnelling method

The NATM is suitable for the ground with strong stability and low groundwater pressure (maybe with dewatering measure). To ensure the safety and stability of the main tunnels, inner supports are used in the main tunnel sections closed to the connection tunnels and emergency protection gates are installed at the starting ends of the connection tunnels. When the preparation for construction is finished, one temporary steel segment of the main tunnel sections at the connection tunnels can be removed. Then the excavation face stability is evaluated before the other temporary steel segments are removed. After that, the tunnelling should obey the NATM rules. In addition, the pump station should be built at the same time. A flowchart for constructing the connection tunnel adopting the NATM is shown in Fig. 7.16.

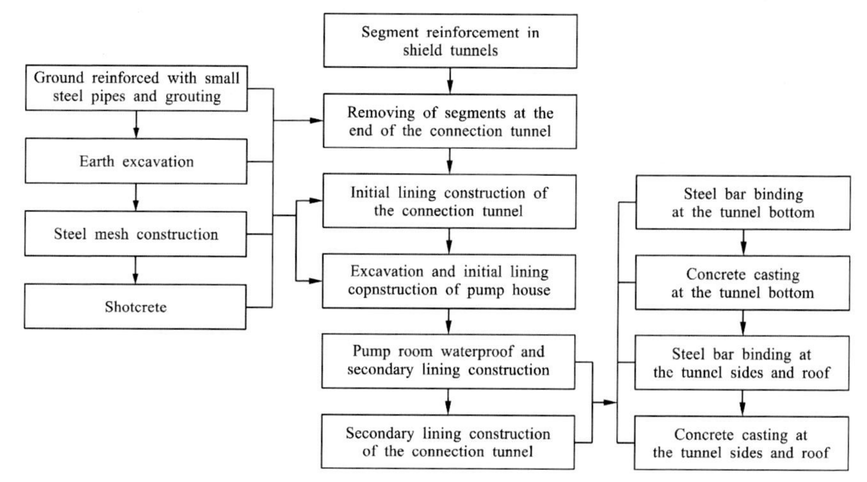

Figure 7.16 The flowchart for constructing the connection tunnel adopting the NATM.

7.5.2.2 Pipe jacking method

Pipe jacking method is widely used in underground passage and utility tunnels in soft ground. This method has also been used in some connection tunnels in Shanghai, Nanjing, and Wuxi. First, the ground is improved by grouting through the reserved holes in temporary steel segments. The pipe jacking begins when the strength of the improved ground reaches the desired value. A flowchart for constructing connection tunnel adopting the pipe jacking method is shown in Fig. 7.17.

7.5.2.3 Freezing reinforcement

Freezing reinforcement is sometimes used for the ground with abundant groundwater. Some condensate pipes are embedded in the ground first. Then cold salt brine circulates in the pipes and the cooling system. In the process, the heat in the soil is transferred into the salt brine, and then the heat is transported to outside. The ground with abundant water changes into a frozen soil when the temperature of the soil is decreased to $-28°C \sim -30°C$. After that, the tunnelling process is similar to that of the NATM. The freezing reinforcement is suitable for tunnelling in the fault zone, a flowing sand stratum, a silt layer, and other geological conditions that are prone to collapse and have abundant groundwater. A flowchart for freezing reinforcement is shown in Fig. 7.18.

7.6 Monitoring

Monitoring for shield tunnelling should take into account the construction conditions and geological conditions. Table 7.7 shows the common

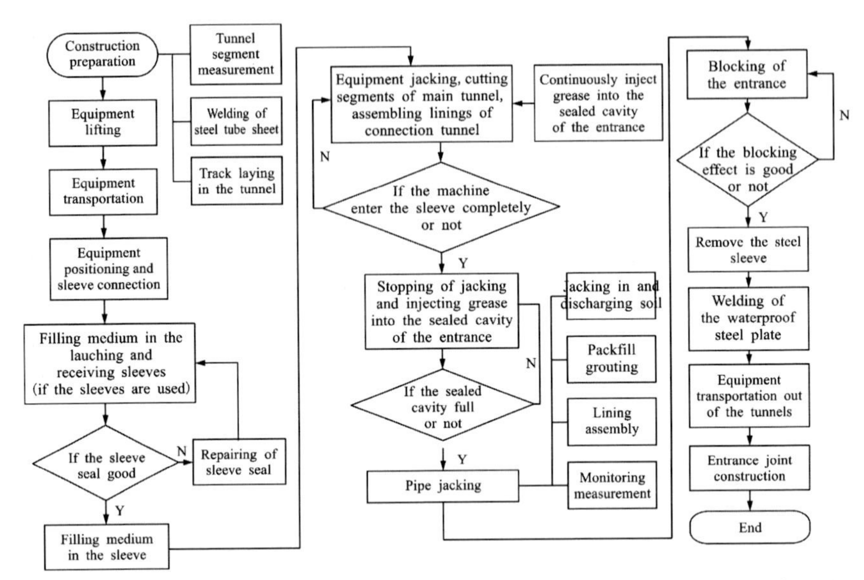

Figure 7.17 The flowchart for constructing connection tunnel adopting the pipe jacking method.

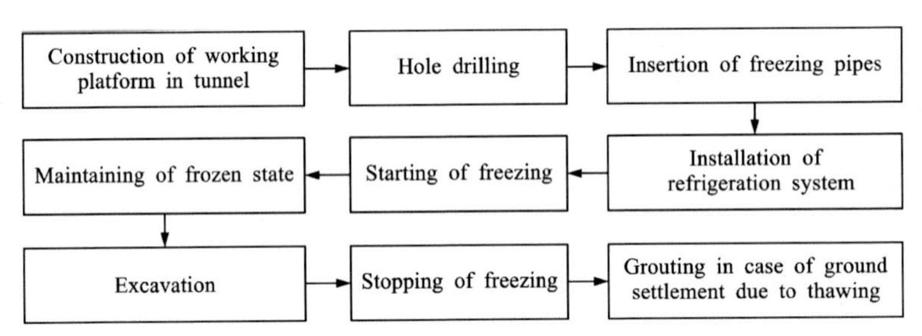

Figure 7.18 The flowchart for freezing reinforcement adopted for constructing connection tunnels.

monitoring items, classified into the required ones and the optional ones. The required ones include ground surface settlements, deformations of adjacent buildings and pipelines, and structural deformations of tunnel linings, and the optional ones include horizontal displacements and vertical displacements at different ground depths, internal forces of segments, and contact stresses between ground and segments. According to the difference in monitoring targets, the monitoring can also be classified into monitoring of environments around tunnel and monitoring of tunnel structure. The limit values of monitoring items refer to the Chinese *Code for Monitoring Measurement of Urban Rail Transit Engineering* (GB 50911-2013) [10].

Table 7.7 Monitoring items.

Category	Monitoring items
Required items	Ground surface settlements, deformations of adjacent buildings and pipelines
	Structural deformations of tunnel linings
Optional items	Horizontal displacements at deep ground and vertical displacements at different depths
	Internal forces of segments
	Contact stresses between ground and segments

Translated from Beijing Urban Construction Exploration & Design Research Institute Co., LTD, Code for Monitoring Measurement of Urban Rail Transit Engineering (GB 50911-2013). Beijing: China Architecture Industry Press; 2013.

Table 7.8 Arrangement range of monitoring points.

Buried depth (m)	Longitudinal spacing (m)	Lateral spacing (m)
$H>2D$	20−50	7−10
$D<H<2D$	10−20	5−7
$H<D$	3−10	2−5

Translated from Beijing Urban Construction Exploration & Design Research Institute Co., LTD, Code for Monitoring Measurement of Urban Rail Transit Engineering (GB 50911-2013). Beijing: China Architecture Industry Press; 2013.

A warning message is sent when the monitoring values exceed two-thirds of the limit values.

7.6.1 Monitoring of environment around tunnel

Monitoring of environment around tunnel mainly includes ground surface settlements, deformations of adjacent buildings and infrastructure. The monitoring points of the ground surface settlements should be arranged with cross-sections with a desired longitudinal spacing along the centerline of the tunnel, and the monitoring points should reflect the total ground displacement in the sensitive area to tunnelling. The monitoring points can be reduced if there are few buildings surrounding the tunnel. Table 7.8, in which H is the tunnel buried depth, and D is the excavation diameter, shows a reference for the arrangement of monitoring points. The deformation of adjacent buildings and infrastructure should be arranged according to the structural status and importance. If possible, monitoring for underground pipelines should be performed too. In addition to the deformation of buildings, railway, bridges, and pipelines, the displacement of the surrounding ground should also be monitored. Continuous monitoring should be carried out in advance of tunnelling until the ground settlement is stable. In addition, the monitoring area should be larger than the sensitive area of tunnelling.

7.6.1.1 Monitoring method

Monitoring of ground surface settlement

Monitoring of ground surface settlement is used to judge the ground stability and safety of the adjacent buildings. First, the monitoring points are arranged along the designed route and the standard of monitoring points are implemented in accordance with the technical requirements of the national second-class survey. The monitoring cross section should be perpendicular to the designed tunnel route, and one is arranged every 50 m. The monitoring points of the cross section should be greater than 13, and the monitoring range should reach at least 30 m in horizontal distance from the tunnel centerline. Fig. 7.19 shows a typical arrangement of monitoring points. In addition, the longitudinal monitoring points are arranged every 10 m along the tunnel axis.

Observation of the ground surface

Observation of the ground surface includes the recordings of ground surface cracks, ground displacement, inclination of adjacent buildings, and so on. The monitoring frequency must be at least once a day.

Monitoring of adjacent buildings

The monitoring of adjacent buildings includes settlement, inclination, and displacement. The monitoring is carried out by using electronic level, total station, and joint meter. The monitoring points should be arranged

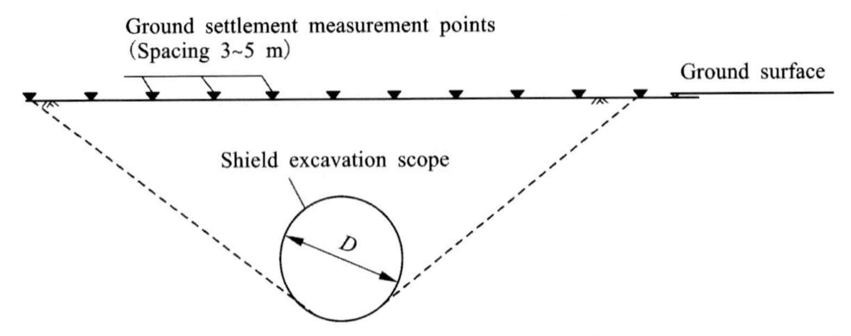

Figure 7.19 Arrangement of points at the cross section for monitoring ground surface settlement.

at the corner of the exterior walls of buildings and bearing structures, and the points can be arranged only on the ground floor for buildings with fewer than five stories. Otherwise, the bottom, middle, and upper parts of the buildings should be monitored. If the side length of the building is greater than 50 m, the monitoring points should be arranged every 10 m from the side center. The inclination value needs to be monitored only for buildings with more than eight stories. The buildings that need to be monitored are those within 30 m away from the tunnel axis.

Monitoring of deep ground displacement

To monitor the ground displacement and analyze its mechanism and influence on the surrounding environment and adjacent buildings, and the inclinometers are used to measure the displacement of the deep ground. They are arranged in accordance with the relative distance between the tunnel and adjacent buildings. In practice, one to four boreholes for inclinometers are arranged in each cross section. The boreholes on the side of the tunnel should have at least the same depth as the bottom of the tunnel, and the boreholes above the tunnel should be at least 1 m above the top of the tunnel. The monitoring points in the boreholes are arranged every 1 m in the vertical direction. Fig. 7.20 shows the arrangement of monitoring of deep ground displacement.

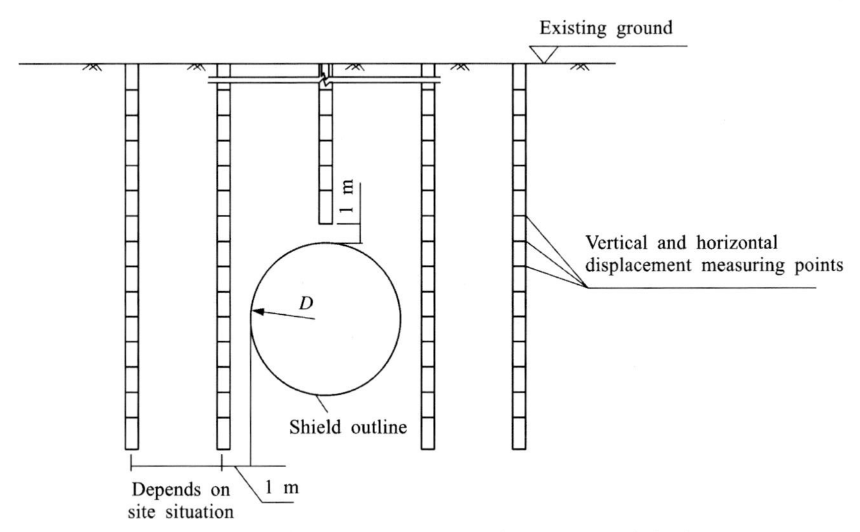

Figure 7.20 Arrangement of monitoring points of deep ground displacement.

The elevation of the borehole orifice is measured by electronic level in accordance with the requirement of the national second-class survey, and the horizontal displacement is measured by an inclinometer. In addition, settlement signs should be buried with inclinometer pipes, and the displacements at different depths are measured by a settlement meter.

Monitoring of groundwater level

The groundwater level is monitored for the whole tunnelling process, and the monitoring spacing can be increased appropriately. The distance between monitoring boreholes for groundwater and the tunnel scope should be larger than 3 m, and the borehole bottom should be at least 3 m below the tunnel bottom. A water level pipe is placed in each borehole, and the groundwater is measured by a water level indicator.

Monitoring of underground pipelines

The monitoring of underground pipelines includes horizontal displacement and vertical displacement. The constructor should survey the layout of underground pipelines in as much detail as possible before construction. The underground pipelines within 30 m in horizontal distance from the tunnel axis should be monitored, especially for water supply pipes and oil and gas pipes. Generally, the monitoring points are arranged every 10 m along the pipelines, and the pipe joints must be monitored. The monitoring measurements of underground pipelines are done by an electronic level and a total station. The monitoring measurements of underground pipelines are classified into direct and indirect measurements in accordance with the materials, types of joints, and inner pressure and requirements of codes. In principle, the monitoring points should attach to pipelines directly. If it is hard to monitor the displacement of pipelines directly, the ground surrounding the pipelines should be monitored, the distance between the monitoring points and the pipelines should be less than 0.5 m, and the monitoring points should be as deep as the bottom of the pipelines.

7.6.1.2 Monitoring frequency

The monitoring frequency should meet the requirement of the Chinese *Code for Construction and Acceptance of Shield Tunnelling Method* (GB 50911-2013) [9]. They are selected according to Table 7.9.

Table 7.9 Monitoring frequency.

Monitoring positions	Monitoring items	The distance between excavation face and monitoring point	Frequency
Ahead of excavation face	Surrounding ground and environment	$5D < L \leq 8D$	Once every 3–5 days
		$3D < L \leq 5D$	Once every 2 days
		$L \leq 3D$	Once per day
Behind excavation face	Tunnel structure and surrounding ground and environment	$L \leq 3D$	Once or twice per day
		$3D < L \leq 8D$	Once every 1–2 days
		$L > 8D$	Once every 3–7 days

Notes: 1. D (m) is the excavation diameter, L (m) is the horizontal distance between excavation face and monitoring cross section. 2. The deformation of the tunnel structure and clearance convergence are monitored when the inner space is intervisible after shield tail escapes from segments. 3. The monitoring frequency is once every 15–30 days when the monitoring value stabilizes.
Translated from Technology and Industrialization Development Center of Ministry of Housing and Urban Rural Development, China Railway Tunnel Group Co., LTD. Code for construction and acceptance of shield tunnelling method (GB 50446-2017). 2017.

7.6.1.3 Deformation control standard

Deformation control standard of surrounding buildings

The building cracks induced by tunnelling should be controlled within 1.5 mm, and the accumulative development of cracks every two times of monitoring and the maximum settlement of buildings should be less than 20 mm [10].

The inclination of masonry buildings should be less than 0.002; the difference between adjacent pillars should be less than 2‰ of the spacing between columns. When the height of buildings (H_g), which is calculated by taking the outside ground as a reference plane, is less than or equal to 24 m, the total inclination should be less than 0.004. When H_g is higher than 24 m and less than or equal to 60 m, the total inclination should be less than 0.003; when H_g is greater than 60 m, the total inclination should be less than 0.0025.

If the monitoring value is less than 50% of the requirements above, tunnelling is normal.

If the monitoring value is greater than 50% of the requirements above, the monitoring frequency should be increased, and the grouting works should be conducted with the tunnelling.

If the monitoring value is greater than 75% of the requirements above, additional protection measures should be carried out on the basis of the original design.

If the monitoring value is greater than or equal to the requirements above, the construction groups should activate the emergency plan and evacuate people if necessary.

Control standard of ground surface settlement

When the value of ground heave is less than or equal to 10 mm or that of settlement is less than or equal to 30 mm, the tunnelling is normal; when ground heave ranges from 10 to 15 mm or settlement ranges from 30 to 40 mm, the monitoring frequency should be increased, and the construction groups should pay close attention to the process of tunnelling; when ground heave is greater than or equal to 15 mm or settlement is greater than or equal to 40 mm, the construction should be postponed, the construction process should be inspected, and the emergency plan should be activated [10].

Control standard of deep ground deformation

Monitoring of the deformation of deep ground is a kind of auxiliary measurement, and the pile foundation of adjacent buildings can be estimated from the monitoring data. Generally, engineers take 10 mm as the control standard. If deformation exceeds the standard, the monitoring frequency should be increased, and grouting should be conducted with the tunnelling. If the deformation increases continuously, the tunnelling should be stopped, and the construction process should be inspected. The tunnelling can restart after the deformation becomes stable.

Control standard of groundwater level and that of deformation of underground pipelines

The accumulative change of groundwater level should be within 1 m, and the change rate should be within 0.5 m/d. Chinese *Technical Standard for Monitoring of Building Excavation Engineering* (GB50497-2019) [11] recommends that the control value of the accumulative deformation of pressure rigid pipelines should range from 10 to 30 mm, and the change rate should not exceed 3 mm/d; the accumulative deformation of non-pressure pipelines should not exceed 40 mm, and the change ratio should not exceed 5 mm/d. If the pipelines are flexible, the accumulative deformation should be controlled between 10 and 40 mm, and the change ratio should range from 3 to 5 mm per day. There is a wide range of underground pipelines, and the structures, types of joints, and nondeformability of different pipelines are different. Therefore the control standards of different types of pipelines vary significantly, and the correct control standard is not drawn up until the detailed conditions of pipelines have been surveyed and the allowable deformation is determined by calculating the allowable minimum curvature radius of the pipelines.

If the deformation of pipelines is large and even exceeds the control standard, the monitoring frequency should be increased, and the construction process should be optimized. In addition, the grout amount of synchronous grouting could be increased, and the ground around the pipelines could be reinforced, or pipelines could be hanged. If the pipelines are used to supply water or for oil or gas transportation, the pipelines should be closed and resume operations after the deformation has been stabilized with ground improvement.

7.6.2 Monitoring of tunnel structure

The monitoring items for the tunnel structure include tunnel settlement and segment ring ovalization. If necessary, the inner force of segments should be measured as well. The deformations of segments such as horizontal convergence, vault settlement, and floor heave are monitored for obtaining the tunnel ovalization. The internal forces are measured by strain gauges to understand the change mechanism of stress in the structure during construction. The monitoring starts after 12 hours of grout solidification for synchronous grouting.

Different monitorings for the tunnel structure should be performed in the same cross section, and the best longitudinal spacing of the monitoring cross section is 50 m. If possible, a tunnel laser profiler is a good tool for structural monitoring. The precision should be less than 1 mm.

7.6.2.1 Monitoring method

The vertical displacement of the lining is measured by using an electronic level. The precision is the same as in monitoring on the ground surface. The segment ring ovalization is measured by an aluminum ruler with a length of 4 or 5 m. In the monitoring of ovality, the ruler is held horizontally, and the coordinate of the center is measured by the total station. Then the horizontal deflection is measured by comparing the coordinate changes of the designed tunnel center, and the changes of the ovalization

Table 7.10 The control standard of tunnel structure deformation (Translated from the reference [9]).

Items		A limit of cumulative value (mm)	Changing rate (mm/d)
Segment settlement	Stiff - medium stiff soil	10–20	2
	Medium soft - soft soil	20–30	3
Settlement difference of segment		0.04%*Ls*	-
Segment ring convergence		0.2%*D*	3

of the lower part of the lining can be deduced. Additionally, the horizontal convergence is monitored by a convergence meter.

7.6.2.2 Monitoring frequency

If the distance between segments and the excavation face is shorter than 20 m, monitoring should be performed once per day; if the distance ranges from 20 to 50 m, the frequency of monitoring is once or twice every one to two weeks; if the distance is longer than 50 m, the frequency of monitoring is once per month, and the monitoring measurement cannot be finished until the deformation has stabilized.

7.6.2.3 Deformation control standard

As shown in Table 7.10, the segment settlement cannot exceed 10–20 mm and 20–30 mm, respectively for the tunnelling in stiff - medium stiff soil ground and in medium soft - soft soil ground. The settlement difference of segment cannot exceed 0.04%*Ls* (*Ls* - the longitudinal spacing between two monitoring points). The segment ring convergence cannot exceed the 0.02%D.

7.7 Construction techniques and case studies for shield tunnelling in special conditions

7.7.1 Special conditions for shield tunnelling

Shield tunnelling may encounter some special geological conditions, such as multiple layers of soil (e.g., a hard layer overlaid by a soft layer), sand-gravel, bedrock, boulder, karst caves, and so on. In addition, the design of the tunnelling route needs to consider some factors such as crossing rivers or existing tunnels and small-radius curved tunnels.

The construction progress, safety, and quality of the structure are affected by these special conditions. Therefore in special conditions, the control methods of tunnelling mode, tunnelling parameters, and ground stability are complicated. The constructors should strengthen organization and management and pay attention to risk management and prevention technology. In addition, they should take advantage of auxiliary construction measures and extend their application ranges.

Specifically, the following items should be considered:

1. Detailed site investigations should be made ahead of construction. Constructors should use relevant construction methods and auxiliary measures and should prepare specific construction machines and materials. The construction organization should be executed strictly so that the project progresses safely and efficiently.
2. A tunnel section is recommended to be used for tunnelling test. The ground displacement and tunnelling parameters should be analyzed frequently in this section. The analysis and experience of the trial zone can be used as a reference for the subsequent construction.
3. The chamber pressure, thrust force, rotation speed and other parameters should be set reasonably according to the geological and other conditions so that the muck discharging ratio and shield attitude can be ensured.
4. Muck-conditioning technology is often required for EPB shield tunnelling. The ideal mechanical properties of muck are low shearing strength, high compressibility, low permeability, low adhesion strength, and suitable flowability. If the mechanical properties of the muck are not good for tunnelling, then foam, bentonite slurry, and other additives should be used to condition the muck so that the stability of chamber pressure and smooth muck discharging can be ensured.
5. The grout materials, grouting pressure, and grout flow rate should be planned reasonably. During tunnelling, the grout flow rate is adjusted by analyzing the monitoring. Compensation grouting should be carried out if necessary.
6. The cutters should be selected in accordance with the special geological conditions, and the shield thrust and cutterhead torque should be monitored during the whole process of tunnelling. In addition, the constructors should make plans to open the excavation chamber and check the cutter abrasion. The cutters should be replaced scientifically. The work of checking and replacement of cutters should be done in stable ground, and the weak ground should be reinforced before shield tunnelling.

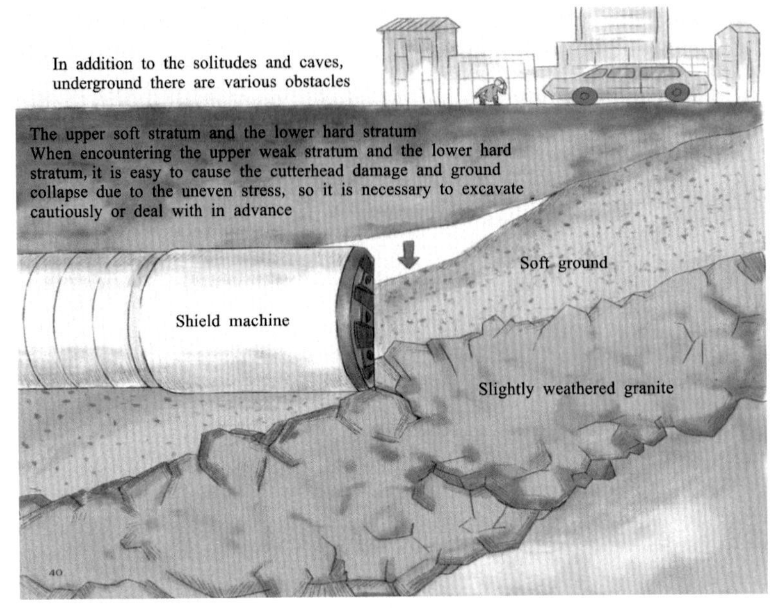

Figure 7.21 Shield tunnelling in a hard rock layer overlaid by a soft soil layer. *Modified from Guangzhou Metro. How was the subway built? Guangzhou: New Century Publishing House; 2014.*

7.7.2 Shield tunnelling in multilayer ground

The tunnelling route is limited by buried depth, geological conditions, and existing barriers. For a large-diameter shield, multilayer ground is always encountered [8] (see Fig. 7.21). The properties of soil and rock are different, such as mineral composition, strength, and permeability. Shield tunnelling in this type of ground (especially for a hard layer overlaid by a soft layer) will encounter a variety of problems.

7.7.2.1 Keypoints and difficulties

Control of shield position is difficult

Tunnelling in uniform ground easily causes the shield to be in the head-down position. However, when shield tunnelling is done in ground where a soft soil layer lies above a hard soil layer, the resistance to the lower part of the cutterhead is larger than that to the upper part, causing the shield to head upward. This shield attitude is difficult to correct. The nonuniform force of the jack cylinders, joint dislocation, and damage of segments may also occur.

Overlying strata and adjacent buildings are susceptible to disturbance

The strength of upper-soft ground at the excavation face is lower than that of lower-hard ground, and the overlying strata are susceptible to disturbance induced by shield tunnelling. Therefore the displacement of the overlying soil layer tends to be large, and the adjacent buildings are at risk of cracking.

Serve abrasion of cutters and the cutterhead

The strength and hardness of lower-hard rock on the excavation face are high, especially for ground with a high content of quartz (The Moh's hardness of quartz is 7, and it is harder than iron, stainless steel, and titanium). Therefore the abrasion of cutters is high, and it is easy for them to be broken when shield tunnelling is done in ground with a high content of quartz. In addition, the cutters rotate in the hard rock and soft soil strata alternately. In the process of the cutters moving from soft soil to hard rock, the cutters suffer from impact and collision when the cutters start to cut hard ground, and this leads to brittle failure of the cutters. The abrasion of the cutters is determined by the linear velocity of the cutters. The cutters that are arranged on the edge of the cutterhead are more likely to be abraded.

Clogging of the cutterhead

If the hard rocks at the lower part of the excavation face are mudstone or siltstone with a high content of clay mineral, the flowability of the muck is poor, and some muck will stick on the cutters and form mud cake, putting the cutters out of action. In addition, the friction between cutters and hard rock generates a lot of heat. The rising temperature will harden the mud cake, and it will form large blocks, making it difficult for the muck to flow into the chamber.

Tunnelling with low efficiency

The high temperature induced by the large amount of friction will lead to clay clogging, which in turn will exacerbate the temperature increase. The abrasion of cutters on the edge of the cutterhead will reduce the diameter of the excavation face and increase the resistance against the shield from the ground. As a result, the shield thrust and cutterhead torque will increase.

Difficulty of changing cutters

The permeability of the excavation face are high during shield tunnelling in ground where a sand-gravel layer is lying above a layer of hard soil with fractured rock. Therefore it is hard to maintain the stability of the excavation face, and it is impossible to replace cutters at the atmospheric pressure. Moreover, the air pressure in the air chamber is hard to maintain; hence it is challenging to change cutters under an excessive pressure.

7.7.2.2 Solutions and measures

Increasing monitoring frequency

The displacements of ground surface and deep ground should be monitored. According to the monitoring data analysis, the grout amount for synchronous grouting should also be ensured. The constructor should establish an efficient management system.

Paying attention to the work conditions of screw conveyor

The torque of the screw is an important parameter. Low torque highly possibly indicates muck spewing. If the torque is large, the screw tends to be blocked, and the discharging rate is low. The torque of the screw is adjusted by muck conditioning and the rotation speed of the screw. In addition, it is better to reduce the standstill time, since the soil particles with large sizes may fall down in the excavation chamber.

1. Monitoring of soil conditioning

   The flowability of muck can be estimated by a slump test, and the temperature should also be measured immediately when muck is discharged. The soil conditioning should be done in a timely fashion so that the muck has good flowability with a low temperature.

2. Reasonable selection and changing of cutters

   The opening ratio of the cutterhead and the types of cutters should be suitable for the geological conditions. It is necessary to stop the shield machine and check the abrasion of cutters by opening the chamber before the shield machine is driven to ground with an upper-soft soil layer and a lower-hard soil layer. The cutters need to be checked and replaced according to the shield position and variation of tunnelling parameters. A long standstill period should be avoided.

7.7.2.3 Case study

Project introduction

In a section of Nanchang Metro Line 1, China, the soil types in the upper part of the excavation face are gravelly sand and fine sand, whereas the soil types in the middle and lower parts are moderately weathered argillaceous siltstone. The shield is tunnelling under the water level.

Tunnel design and selection of shield types

The lining adopted precast reinforced concrete segments. The inner and outer diameters of the segment were 5.4 and 6.0 m, respectively. The assembling method was a staggered assembling of wedge-shaped rings. An EPB shield was used, and the opening ratio was 34%. Three foam injection holes and two bentonite slurry injection holes were equipped in the cutterhead. The shield was also equipped with 6 active stirring rods, 6 disc cutters, 40 front scrapers, and 12 edge scrapers.

Problems encountered and solutions

1. The ground at the upper part of the excavation face was water-rich sandy soil, and the lower part was moderately weathered argillaceous siltstone. Therefore the muck was a mixture of two types of excavated soil. The injection ratios of conditioning agents was difficult to determine, and the groundwater diluted the conditioning agents.

   Solution: The properties of the conditioning agents, the conditioning parameters, and the status of muck were investigated before tunnelling. Then the conditioning parameters were applied at the construction site to maintain the chamber pressure and the balance between muck entering and being discharged.
2. The clay mineral content of moderately weathered argillaceous siltstone is high. Unreasonable conditioning parameters and a long standstill period caused solidification of the muck, so the cutterhead was clogged.

   Solution: The state of the muck and the torque of the cutterhead were monitored so that the injection of water and foam could be adjusted in real-time. The foam injection system was continuously checked. In addition, the standstill period was made as short as possible, and the chamber pressure was decreased appropriately during standstill.
3. Spewing occurred when the shield was tunnelling in the fault fracture zone or the ground contained a small amount of clay and small particles so that the water pressure on the excavation face increased abruptly.

   Solution: Bentonite slurry was used to deal with the spewing because it could reduce the permeability of the muck greatly, since the bentonite particles blocked the flow channels in the muck. An appropriate chamber pressure was maintained to prevent the groundwater from flowing into the chamber. A chamber pressure that is too high may lead to clogging.

7.7.3 Shield tunnelling in a sandy gravel stratum

The sandy gravel stratum, which contains a high content of large particles, is encountered frequently in China. For example, this kind of stratum is widely distributed in Chengdu, Lanzhou, Beijing, and so on. The sandy gravel stratum is highly unstable. The muck with large particles and poor flowability presents great challenges to the soil conditioning for EPB shields and slurry treatment for slurry shields. Generally, the sandy gravel stratum is under the groundwater level. Therefore the problems of water spewing and water leakage need to be solved urgently. Fig. 7.22 shows the shield tunnelling in a sandy gravel stratum [13].

7.7.3.1 Keypoints and difficulties

Low stability stratum

The sorting characteristics and uniformity of a sandy gravel stratum are poor, and its permeability is high. The cohesion is almost zero. The chamber pressure should be controlled strictly.

Large ground settlement

The ground surrounding the excavation face is squeezed, and the friction between the shield and ground may cause loosening and large displacement

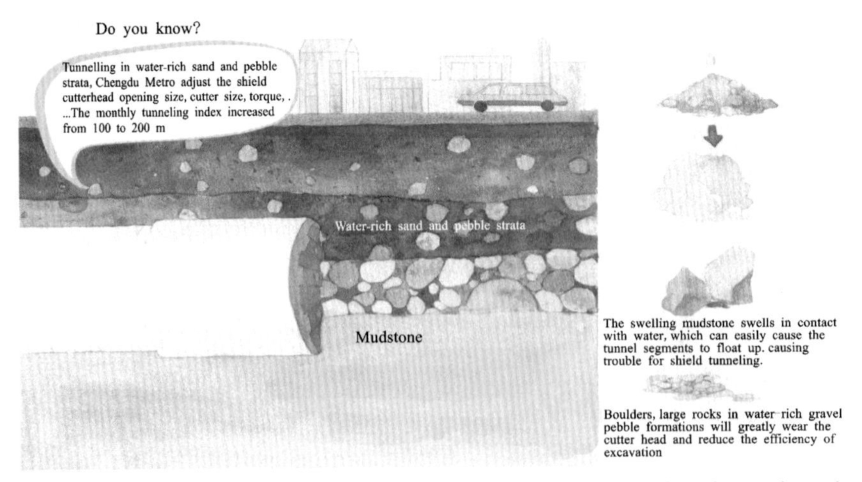

Figure 7.22 Shield tunnelling in a sandy gravel stratum. *Modified from Chengdu Metro Group Corporation. Chengdu metro was built in this way [EB/OL]. [2021-02-18]. [https://www.sohu.com/a/83823729_355523].*

of the ground, since the sandy gravel is uncompact and cohesiveless. The adjacent buildings are vulnerable to damage. Therefore segment assembly and synchronous grouting should be done in a timely fashion so that the radius of the plastic zone is decreased as much as possible.

Quicksand and piping

The interlayer of silty sand is distributed randomly within the sandy gravel. The distribution of water pressure and change of flow path induced by shield tunnelling may cause quicksand and piping so that the ground surface collapses and the linings become unstable.

Antifloating

When shield tunnelling is done in water-bearing sandy gravel, the floating of linings should be given with close attention in the process of construction. If the lining structure cannot meet the requirement of antifloating, some antifloating measures should be taken.

Muck discharging

The flowability of muck is poor because there are too many large grains. The chamber pressure is difficult to control, so the support pressure on the excavation face cannot reach the ideal value. At the same time, the muck is also difficult to discharge, since large grains may block the screw conveyor for EPB shields and slurry pipelines for slurry shields. Furthermore, the most difficult problem is the serious abrasion of the cutterhead and screw conveyor and frequent cutters replacing.

Water spewing

The permeability of the sandy gravel is high, and the muck plug is hard to form in the screw conveyor. Therefore water spewing may happen frequently.

Uncertain Tunnelling parameters

The cutting mechanism in sandy gravel is uncertain, and the ranges of tunnelling parameters are hard to determine.

7.7.3.2 Solutions and measures

Strengthen monitoring of construction and ground settlement

Synchronous grouting and compensatory grouting should track tunnelling behaviors to prevent the collapse of the ground surface and deformation of surrounding buildings.

Reasonable selection of cutters and cutterhead

The shield should have the ability to deal with large-radius pebbles and boulders. The cutterhead should have enough stiffness, driving torque, and opening ratio. The cutters should have the ability to cut hard rock, and the double disc cutters should be able to break boulders. Teeth cutters are used to scrape ground into the chamber and decrease the disturbance of the ground. The cutters should have abrasive resistance and antiwear measures. In addition, a full-featured human chamber should be equipped to allow breaking of big boulders manually under pressure.

Improve the waterproof sealing performance of the shield

The articulation system and shield tail sealing system should have the ability to prevent water under high water pressure. At the same time, the chamber pressure should be increased to 1−1.2 times to maintain the stability of the excavation face.

Reasonable design of the muck discharging system

The screw conveyor of EPB shields should be capable of discharging pebbles and broken boulders with a certain radius. To prevent water and muck spewing, the screw conveyor can be lengthened to decrease water pressure. A sealing gate or pressure-maintaining pump systems installed on the discharging gate of a screw conveyor are also reliable measures.

Reasonable soil conditioning

Bentonite slurry, foam, and high-molecular polymer are used to improve the flowability and impermeability of muck. The water pressure of groundwater and conditions of discharging should be monitored frequently to adjust the conditioning agents and conditioning parameters.

7.7.3.3 Case study

Project introduction

The bored tunnels of Section 1 of Line 10 of the Chengdu subway system pass through pebble, pebble with an interlayer of fine sand, and fine sand. The type of shield used was an EPB shield.

Allocation of cutters

The shield was equipped with four 17-inch double-edge disc cutters, twenty-two 17-inch single-edge disc cutters, ten 17.6-inch double-edge disc cutters, 28 front scrapers, eight edge scrapers, and 20 advance cutters.

Table 7.11 Main tunnelling parameters of shield tunnelling in the first section of Line 10 of the Chengdu subway system.

Items	Parameters
Chamber pressure	1–2 bar
Rotation speed of cutterhead	1.0–1.5 rpm
Thrust force	900–1200 kN
Advancing speed	50–70 mm/min

The shield did not need to replace cutters for an advancement distance of 500 m.

Main tunnelling parameters

Table 7.11 shows the main tunnelling parameters for the shield tunnelling.

Problems and solutions

1. The flowability of muck will be poor if the conditioning is improper. Consequently, the torque of the cutterhead is large, and the advancement speed will be low. The abrasion of the cutterhead, cutters, and blades of the screw conveyor is serious. At the same time, spewing and abnormal rotation of the screw may be caused. The stacking of soil particles with large sizes in the excavation chamber aggravated the abrasion of the cutterhead and cutters.

   Solutions: The torque of the cutterhead was controlled within 2500–3500 kN · m, and the torque of the screw was controlled within 10–20 kN · m. The torque of the cutterhead and screw were predicted and adjusted in time when the chamber pressure was stable. There were seven injection holes in the cutterhead and two injection holes in the bulkhead that were used to inject foam and one injection hole in the cutterhead that was used to inject bentonite slurry and polymer to improve the flowability and impermeability and to lower the abrasion of the cutterhead, cutters, and screw and to impede water invasion. It should be noted that improper injection of bentonite slurry may cause clogging.
2. Ground settlement and collapse

   Solutions: Tunnelling should be efficient, continuous, and uniform. The settlement after shield tail passing is often obvious. Generally, the double-liquid grout (cement - sodium silicate) is injected from holes

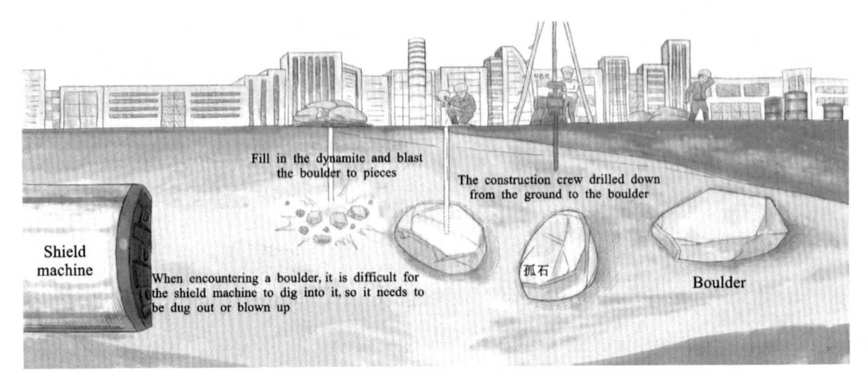

Figure 7.23 Shield tunnelling in ground with boulders. *Modified from Guangzhou Metro. How was the subway built? Guangzhou: New Century Publishing House; 2014.*

in the fourth segment ring after the shield tail to accelerate the initial coagulation of grout so that the grout will fully fill the tail gap. If a large settlement occurs, a compensatory grouting should be carried out until the settlement is stable. When the discharged volume of muck is larger than the theoretical value, the compensatory grouting is made after the shield tail moves away from the corresponding ring or the cavity due to over-excavation is filled with grout mortar from ground surface to prevent the additional ground subsidence.

7.7.4 Shield tunnelling in ground with bedrock or large boulders

Boulders are often generated as a result of nonuniform weathering of granite. Their mechanical properties are different from those of the surrounding soils. The unconfined compressive strength of a boulder is high (normally above 100 MPa and even up to 200 MPa). As shown in Fig. 7.23, the shield encounters boulders with large particles [8]. The cutting head has difficulties in breaking the hard rock, and it may even be stuck there. Therefore bedrock and boulders have significant negative impacts on safe, high-quality, and fast tunnelling.

7.7.4.1 Keypoints and difficulties

1. The boulders move with the rotation of the cutterhead and impede the advancement of the shield. Sometimes the cutterhead becomes stuck.
2. The tunnelling position and direction are hard to control.
3. Serious abrasion of the cutterhead leads to reduction in stiffness of the cutterhead.

4. The main bearing and sealing of the cutterhead are broken due to the nonuniform stress in the cutterhead.
5. The cutters experience serious abrasion.
6. Tunnelling behavior causes intense vibration of the ground that has a negative effect on surrounding buildings.

7.7.4.2 Solutions and measures

Reasonable selection of shields

1. A cutterhead reinforced with high abrasion resistance is adopted to deal with the high stiffness of the bedrock and boulders. The cutterhead can be equipped with rock-breaking and soil-cutting cutters together.
2. The shield should have enough thrust and driving torque to prevent the cutterhead from becoming stuck.
3. The shield is easily deviated from the correct tunnelling direction in bedrock or when it encounters boulders. Therefore the shield should be equipped with a precision guidance system to ensure the correct tunnelling direction and position.

Measurements on the ground surface

1. Underground blasting through the ground surface borehole

   When the underground distribution of spheroidal weathered granite has been accurately determined, blasting is used to break the stone from the surface boreholes, which are drilled into the spheroidal weathered granite. After blasting, the boreholes are filled with clay, and then grout is injected into the ground to make the mechanical conditions of ground meet the requirement for shield tunnelling.
2. Breaking rock by punching hammer

   When the tunnel is being built in deep ground, the spheroidal weathered granite can be broken by a punching hammer. The size of the hammerhead, hole spacing, and number of holes are determined by the size of the spheroidal weathered granite. The induced holes is filled with clay, and then grout is injected into the ground to make the mechanical conditions of ground meet the requirement for shield tunnelling.
3. Underground measures
   1. Manual rock-breaking at the excavation face

      Workers can enter the excavation chamber and break spheroidal weathered granite stones by blasting or machines when the stability of the excavation face can be guaranteed. The broken pieces are transported through the human chamber.

2. Breaking rock by shield cutters

   The cutterhead is stuck when the spheroidal weathered granite stones stack in front of the shield. In this situation the ground around the spheroidal weathered granite is reinforced by grouting through an advanced borehole to bind the surrounding ground with the boulders so that the spheroidal weathered granite stones cannot move with the rotation of cutterhead and the shield can cut the rock-soil body normally.

7.7.4.3 Case study

Project introduction

Taishan Nuclear Power Plant is a nuclear power station with a 6×1000 MW pressurized water reactor. The hydraulic tunnels are located in the sea area between the land area of Waigutsui and Dabao Island. The tunnels are twin tunnels with a length of 4330.6 m. The buried depth ranges from 11 to 29 m, and the spacing between the two tunnels is about 29 m. The tunnels in the soft ground were built by a mixed slurry (S-551) shield with a large radius, and the tunnels in the hard ground were built by NATM. The excavation radius of the shield is 9.030 m. The opening ratio of the cutterhead is 34%. The first 300 m tunnelling was the bottleneck of entire project, owing to the protuberance of bedrock and numerous boulders.

Key points of the construction scheme

1. The bedrock and boulders were broken by a percussion drill machine and undersea blasting

   The blasting holes were drilled by a combination of a down-the-hole (DTH) drilling machine and a geological drilling machine. The DTH drilling machine bored holes and cased drivepipes, and the geological drilling machine drilled into bedrock through drivepipes. Then, the geological drilling machine drilled to designed elevation.

   Undersea boulders were dealt with by a geological drilling machine installed on a ship, and the spacing of the holes was 1×1 m. The boreholes, which were not drilled in boulders, were sealed by cement slurry with a W:C ratio of 1:1.

2. The boulders were buried deeply, and the sizes of the boulders were large and uniform. Therefore it was difficult to break a boulder by a

single blast. To facilitate the construction and improve the blasting effect, blasting work was done in the front holes first, then the free surface formed by the squeezing function of front hole blasting was used to carry out the following blasting. The bedrock was blasted into pieces having a diameter of 30 cm. The blasting areas were reinforced by sleeve valve pipe grouting to maintain the stabilities of the excavation face and slurry pressure.

3. According to the analysis of the adaptability of the cutters to the stratum, the abrasion resistance of the cutters was reformed, and alloy wear resistance ring was welded and inlaid in the outer edge of the cutterhead to improve their wear resistance. Shell cutters were installed around the cutter carriers of the disc cutters and scrapers on the edge of the cutterhead to prevent the cutter carriers from being worn since the heights of the shell cutters was higher than those of the cutter carriers.
4. Abrasion-monitoring devices were installed to monitor the working conditions of the cutters.

7.7.5 Shield tunnelling in a karst stratum

The construction of a bored tunnel encounters great challenges in shield tunnelling through a karst stratum [8] (Fig. 7.24). Some karst caves are prone to collapse due to surrounding fractured rock. Some tunnels are built above karst caves, and the fillers in caves are loose and the buried depth of caves is deep, so the treatment of the tunnel basement is hard. Sometimes water-carrying soil from the fillers may flow into the shield

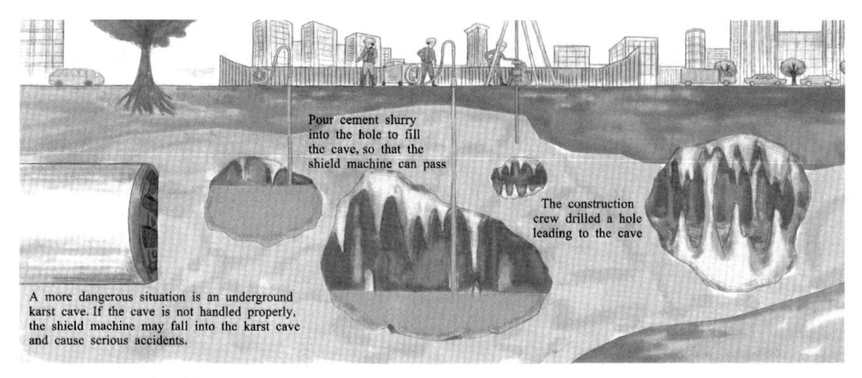

Figure 7.24 Shield tunnelling in a karst stratum. *Modified from Guangzhou Metro. How was the subway built? Guangzhou: New Century Publishing House; 2014.*

machine. It is hard to deal with this problem. Ground settlement and cracks in the ground may be observed. The shield may encounter big water sacs and submerged rivers. In this case the water-carrying soil will flood into the shield machine. The distributions of karst caves and submerged rivers are sometimes circuitous, branching, and complex.

7.7.5.1 Keypoints and difficulties

1. The existence of a karst cave weakens the bearing capacity of the tunnel foundation, increases the unstable factors of the surrounding rock, and reduces the safety and reliability of the tunnel structure. The collapse of the roof of the karst cave will cause the shield tunnelling to stop.
2. Caves above tunnels easily form a seepage funnel connected with the surface water system, which can result in large ground settlement and even ground surface subsidence.
3. In shield tunnelling through karst caves under high water pressure, a flood of soil into the tunnel may be triggered.
4. Caves beneath tunnels may cause floating of the tunnel. The loose fillers in caves and submerged rivers bring numerous troubles.
5. The loose fillers in caves subside easily and thus redistribute the soil stress. This is negative for the stress distribution of the tunnel structure.
6. The tunnelling position is hard to control, and staggered segments and serious loss of grouting materials are caused when shield tunnelling is taking place in a karst stratum.

7.7.5.2 Solutions and measures

1. The geological forecast of the tunnelling route is carried out by means of geophysical and prospecting techniques to determine the distribution of karst caves in detail.
2. The design and selection of the shield machine are improved.
3. The designed driving torque of the cutterhead should be increased to significantly promote the torque of the cutterhead at a relatively higher speed.
4. The shield should be equipped with heavy cutters and cutter carriers, and the spacing of cutters should be shortened to strengthen the rock-breaking ability.
5. Pretreatment of the ground should be carried out in places where there are the risks of water and mud bursting. The measures should be in accordance with local conditions and comprehensive treatments.

6. Suitable grouting materials should be selected to fill karst caves.
7. Synchronous grouting and compensatory grouting should be carried out to strengthen the stability of the surrounding ground and decrease ground settlement.

7.7.5.3 Case study

Project introduction

The lengths of the tunnels of one section in the third line of the Changsha Metro are 1400 m. The tunnel section crosses the Xiangjiang River. As shown in Fig. 7.25, the highly permeable weathered zone contains gravel, sand, and faults. Completely weathered rock and residual silty clay overlay the bedrock. The permeability coefficient of the covering layer is 2.4×10^{-4} m/s. Part of the tunnelling region is a karst distribution area, which develops in a beaded shape, and most of it contains fillings. The karst fissure water is confined to water and linked with the river water. The type of shield machines used were slurry shields with an excavation diameter of 6480 mm, and the outer diameters of the segment rings were 6200 mm.

Solutions

1. In view of the complex distribution of karst caves, the detection of caves adopted a combination of intensive boreholes and geophysical prospecting. The geophysical prospecting was done first, and then intensive geological boreholes were used to verify the distribution of caves. In addition, the three-dimensional space distribution of caves

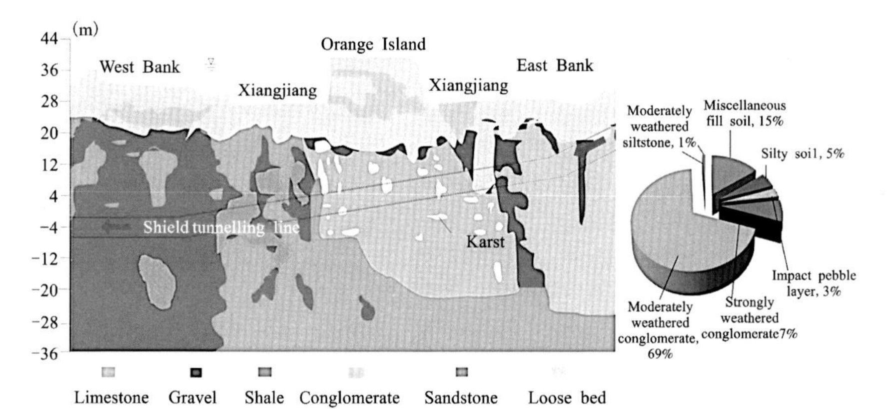

Figure 7.25 The geological profile of the Xiangjiang shield tunnel of the third line of the Changsha subway system.

and construction process were visualized by building information modeling technology.

2. The technology for precontrolling of the shield attitude was adopted to deal with the risks of subsidence and shield heading down. When the shield was close to caves, the thrusts of lower-part jack cylinders were larger than those of upper-part jack cylinders. The tunnelling position was monitored over time so that abnormal situations of positioning could be detected and solved in a timely manner.
3. The karst cavity wall was irregular, and the mechanical property was uniform, owing to water erosion. The shield tunnelling parameters adopted low shield thrust and rotation speed of the cutterhead to decrease the impact of the karst on the cutters.
4. The Xiangjiang River basin is a reserve of drinking water. Grout must be made of environmental friendly materials. The water quality was monitored and evaluated throughout the project.

7.7.6 Shield tunnelling in gassy ground

Gas in a ground is often poisonous, flammable, and explosive. In shield tunnelling in gassy ground, the gas may flow into the tunnel through gap between the segments and surrounding ground or the joints of segments. The increase in gas concentration during the shield tunnelling may cause accidents such as poisoning or asphyxiating of workers, fire, and explosion.

7.7.6.1 Keypoints and difficulties

1. The volume fraction of the gas exceeds the limit

   According to Chinese standards, the volume fraction of combustible gas in the tunnel construction should not exceed 5‰ of the total volume of air. If the volume fraction of combustible gas is more than 5‰, fire or even explosion accidents may be triggered.
2. The whole tunnel passes through the gassy stratum.

   The stratum with gas will be released completely around shield tunnel, and most gas will flow into the tunnel. It is hard to dilute the gas in the tunnel, owing to the confined space.
3. Ground displacement is induced by loss of gas

   Gas is distributed mainly in groundwater and ground below the water level in the form of clumps. Release of a large amount of gas caused by shield tunnelling may cause ground settlement and thus damage to the surrounding structures and underground pipelines.

7.7.6.2 Solutions and measures

Geological forecasting ahead of tunnelling

The construction workers should develop plans for geological forecasting and tunnel construction, and the plans should be checked by experts and supervisors. In addition, the geological forecast should be completed and analyzed in time to carry out the corresponding measures.

Soil conditioning

During tunnelling in gassy ground, muck discharge and sealing of the shield tail gap are important. The first flow path of gas is the discharge gate of the screw conveyor. Therefore workers should pay close attention to the discharge conditions of the screw and adjust the conditioning parameters so that the muck can form an impermeable plug in the chamber and screw conveyor.

Improving the assembly quality of segments

The second main flow paths are joints at the shield tail and cracks and joints of segments. Therefore the management of segment assembly should be strengthened. The transportation of each segment should be done carefully. Joint dislocations and segment cracks should be avoided.

Gas concentration monitoring

Infrared methane sensors should be equipped in the built tunnel every 100 m, and an automatic monitoring and alarm system should be established that can transmit the gas concentration to the control center so that the construction workers can take timely measures. In addition, the ventilation system of the tunnels should have the ability to dilute gas. The ventilation system should work all day, and the workers should be evacuated in time if the ventilation system is closed.

Construction principle

The construction workers should obey the principles that control shield tunnelling and muck discharging, paying close attention to ventilation and gas monitoring frequently.

Control of tunnelling position

As was mentioned earlier, the gas would flow into the tunnel through gap around the shield tail. Therefore the gap around the shield tail should be controlled uniformly to avoid large gap at one side of shield tail and failure of the shield tail sealing.

7.7.6.3 Case study

Project introduction

The length of the tunnel sections between Jiashan Station and Wuhan Road Station of Line 1 of the Chengdu Metro is about 1067 m. The shield machine bored mainly in moderately weathered mudstone and moderately weathered sandstone. The geological survey detected the presence of methane and carbon monoxide (gas) in the tunnelling area at a buried depth of between 50 and 70 m. The concentration of gas increased with depth, and groundwater was distributed in faults in the bedrock. The permeability coefficient of the soil around the tunnel was 2.89×10^{-3} cm/s.

Solutions

The main solutions to avoid accidents are to decrease the gas concentration, to avoid any fire, and to keep monitoring.

1. To decrease the gas concentration: Gas-releasing boreholes were drilled around both tunnels. The diameter of the boreholes was 0.108 m, and the spacing distance was 5 m. All boreholes were arranged 3 m in horizontal distance from the tunnel lining. The borehole bottom was in the same level with the bottom of the tunnel. Steel pipelines with small holes were installed in the boreholes to avoid hole collapse and groundwater invasion. The level of the steel pipe tops was 1 m higher than the ground surface. Each tunnel was equipped with two ventilation system. One system was used as standby equipment. Four fans were installed in each trolley train.
2. To avoid any fire in the tunnels: Therefore all cables used for lighting, communication, signal, and control were replaced by antistatic and flame-retardant cables. All lighting equipment was explosion-proof.
3. To keep monitoring: Three infrared methane sensors, one hydrogen sulfide sensor, and one carbon monoxide sensor were equipped on each trolley train. In addition, the gas concentration at the muck discharging outlet, recirculated air, and any gas gathering places was checked by a portable gas detector and light-sensitive gas detector every 60 minutes. Thus the gas concentration was monitored in different ways, and when alarms occurred, the muck discharging gate was closed and the shield tunnelling was paused.

7.7.7 Construction of parallel bored tunnels with small spacing

Tunnels for urban metros are often designed as two parallel tunnels. The spacing between two tunnels is often small, owing to the space limitation of the metro line in urban. The interaction between two tunnels affects the ground displacement, tunnel structures and adjacent buildings and infrastructures.

7.7.7.1 Keypoints and difficulties

Ground displacement superposition

Construction of double-line tunnels significantly disturbs the ground and causes ground displacement. If the ground improvement is not appropriate, the adjacent structures may be damaged.

High risk during tunnelling close to advance tunnel

Tunnelling in the disturbed ground may cause collapse of the excavation face and ground subsidence.

Deformation of advance tunnel induced by the construction of later tunnel

The squeezing effect and stress release of ground induced by shield tunnelling may cause offset of the centerline of the advance tunnel, deformation and cracks of segments, and even fracture of bolts.

7.7.7.2 Solutions and measures

Monitoring the displacement of ground and advance tunnel

When shield tunnelling is being done close to the advance tunnel, besides the ground displacement, the deformation, cracks, and movement of the advance tunnel should also be monitored.

Reasonable setting of tunnelling parameters

The shield thrust, chamber pressure, and synchronous grouting should refer to the monitoring data.

Grouting reinforcement of multiperturbed ground

To maintain the stability of the surrounding ground and advance tunnels, the ground could be reinforced by grouting through the grouting holes reserved in the advance tunnel.

Temporary supports in advance tunnel

To improve the bearing capacity of the tunnel structure and its longitudinal stiffness and overall stability, cross steel supports are installed in the advance tunnel, or a hydraulic wheeled trolley is used for tunnel reinforcement (Fig. 7.26).

7.7.7.3 Case study

Project introduction

The tunnels in the third section of the Guangzhou-Foshan interurban railway were about 3566 m long. EPB shield machines were used for the tunnels. The excavation diameter of the shield tunnels was 8.8 m. The minimum buried depths at the launching section and receiving section were 6.46 and 13.6 m, respectively. The maximum buried depth in the middle section was about 30 m. The spacing between two tunnels at the launching section was 2.1 m. The ground consisted of silt, mucky silt, and fine sand. The ground surface conditions were complex, such as the tunnel undercrossing highways, rivers, highway bridges, and high-speed railway. There were also some underground pipelines distributing in this area.

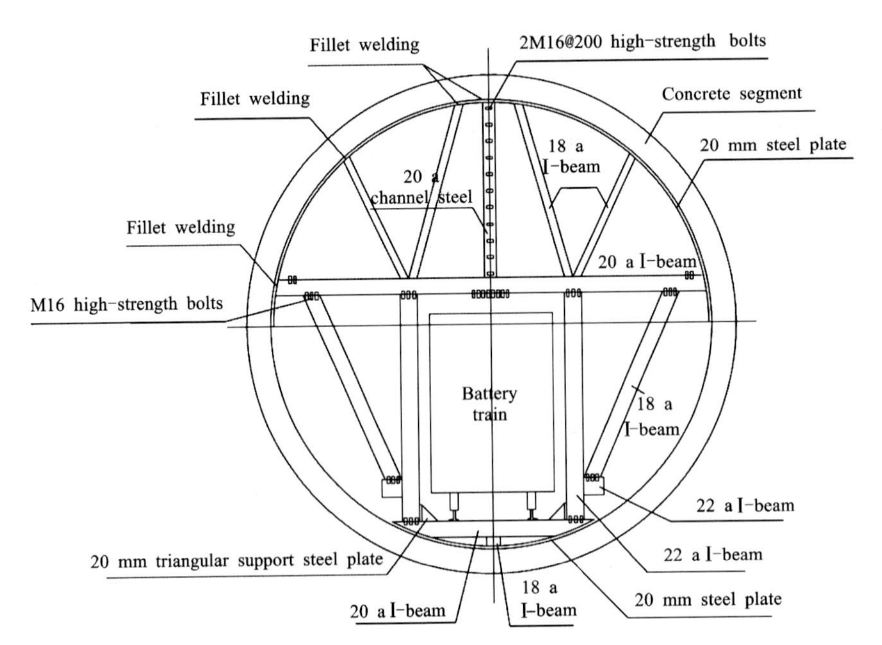

Figure 7.26 Temporary steel support.

Construction technology

The right line tunnel was constructed first. The deformation of adjacent structures and ground displacement were monitored over time to control synchronous grouting and complementary grouting. The second-time settlement, displacement, and stress of segments of the advance tunnel were monitored during construction of the left line tunnel. In addition, to ensure the safety of the advance tunnel and receiving the shield machine, rotary-jet grouting piles were used to reinforce the ground for the shield machine launching and receiving.

Problems encountered and solutions

1. The first tunnelling cannot disturb the ground around the second-built tunnel excessively, and the tunnelling route of the first-built tunnel cannot affect the tunnelling behavior of the second one.

   Solutions: The construction of the first tunnel adopted the EPB mode, and the soil conditioning agents were bentonite slurry and foam. The attitude correction of shield machine was controlled in a reasonable range to avoid affecting the tunnelling of the second tunnel. In addition, the synchronous grouting materials incorporated double-liquid (cement - sodium silicate) grout to improve the sealing capability of the tunnel and the strength of the ground between the tunnels. The ratio of grout is shown in Table 7.12.
2. The construction of a second-built tunnel cannot cause large displacement of the first-built tunnel, water leakage, and deformation of segments, large settlements, and displacement of the ground.

   Solutions: A low advancing speed of the second-built tunnel would cause serious disturbance of ground, and a high one would influence the stability of chamber pressure. Therefore the advancing speed should be controlled appropriately, normally within 10–20 mm/min. In addition, ultrasonic detection was used to detect the cavities behind the first-built tunnel to evaluate whether to carry out compensatory grouting. The deformation and stress of the first-built tunnel structures

Table 7.12 The ratio of grout.

Items	Sodium silicate	Water cement ratio	Stabilizing agent	Water-reducing agent	Ratio of A to B
C-S	35Be	0.8–1.0	2%–6%	0%–1.5%	1:1–1:0.3

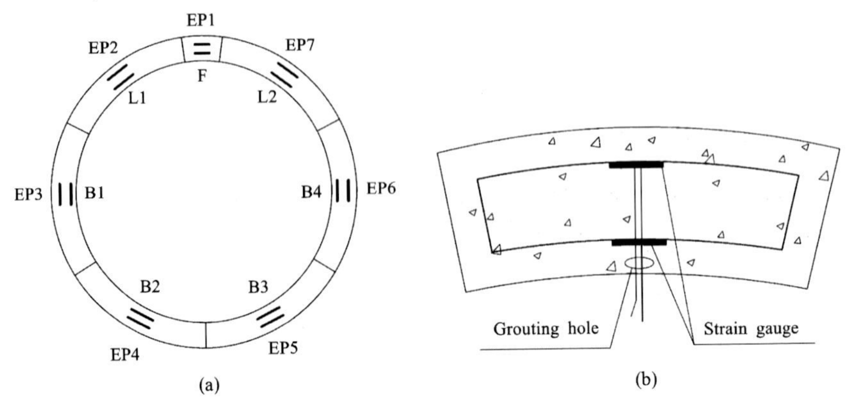

Figure 7.27 The installation positions of sensors in the segments of the first-built tunnel. (a) The installation positions of sensors in the segment ring. (b) Partial enlarged detail.

were monitored to direct the setting of the tunnelling parameters for the second-built tunnel. Fig. 7.27 shows the installation positions of sensors in the segments of the first-built tunnel.

7.7.8 Shield tunnelling with a small-radius curve

The Chinese *Code for Design of Metro (GB50157-2013)* [12] determines the design standard of the tunnelling route with a small curvature radius according to the types of trains and geological conditions (Table 7.13). It is hard for the shield to turn, owing to the large length and high stiffness of the shield. Control of the shield attitude greatly affects the segments and the surrounding ground. The requirements for shield tunnelling are therefore high in a small curve radius.

7.7.8.1 Keypoints and difficulties

Movements of tunnel segments toward the outside of the curve

An angle will be generated between the end face of the segments and the advancing direction of the shield machine when the shield tunnelling in a curve with a small radius. The thrust of the jack cylinders acting on the segments would generate a lateral force. The tunnel segments would move toward the outside of the curve, owing to such a lateral force. In addition, the synchronous grouting difficultly fill the tail gap with grout. If the early-stage strength of the grout is low, the segments will be pushed toward the outside of the curve by the lateral earth pressure.

Table 7.13 The minimum radius of circular curve (m).

Types of train	A		B	
Route	**General section**	**Difficult section**	**General section**	**Difficult section**
Main line	350	300	300	250
Entrance and exit of depot line, connect line	250	150	200	150
Depot line	150	—	150	—

Increase in ground loss

The tunnelling route in the curve with a small radius is composed of a series of broken lines. The excavation radial scope of tunnel in the outside of curve is larger than that in the inside of curve, leading to a larger ground loss in the outside of the curve. In addition, the profiling cutter, which is equippeded for turning the shield, is activated. Therefore the actual excavation face is an ellipse rather than a circle, and the actual amount of excavated soil exceeds the theoretical value normally calculated according to a circular excavation shape. Ground loss is inevitable even though the tunnelling parameters are well controlled, and the ground loss is related to the length of shield during the tunnelling in a small-radius curve. Compared to the tunnelling in a straight line, the ground loss increases with the decrease of the tunnelling curve radius.

Excessive disturbance of ground induced by frequent shield attitude correction

To bring the broken lines close to the smooth curve with a small radius, continuous shield attitude correction is required during tunnelling. The smaller the curve radius and the longer the shield machine, the larger the attitude correction and the lower its sensitivity. Frequent attitude correction aggregates the disturbance of the ground and even causes large ground settlement. If the strength of the tunnel and ground is low, the displacement of the tunnel would be large.

Damage of segments

In shield tunnelling with a small-radius curve, the tunnelling attitude changes greatly, and the shoes of the jack cylinders may slip laterally (or have the tendency to slip). Local cracks and breakup may occur due to excessive local stress. In addition, the segments would squeeze the ground to the outside, affecting the stress state of the stratum and tunnel and leading to the shield becoming stuck and the breakup of segments.

Risk of seepage and large ground settlement

The gap is not uniform around the shield tail in shield tunnelling with a small-radius curve. The brushes equipped in the shield tail may lose elasticity and thus the ability to seal the gap. At the same time, joint dislocations are easily caused between segment blocks, and thus the brushes cannot wrap the whole segment ring. Groundwater would flow into the tunnel through the shield tail gap. In addition, the groundwater leakage would cause large ground settlement and even stop the shield tunnelling.

7.7.8.2 Solutions and measures

Equipping an articulation system

To control the tunnelling route with a small radius and improve the sensitivity of shield attitude correction, the shield should be equipped with an articulation system. Such a system increases the sensitivity of shield machine and can decrease the overexcavation scope of the shield cutterhead.

Utilization of profiling cutters

The excavation face is not a circle when the articulation system is used. As a kind of auxiliary measure, it needs to combine the construction measures such as profiling cutters, wedge-shaped segment rings, and thrust difference in jack cylinders. The profiling cutters directly influence the performance of the articulation system. Excessive overexcavation would significantly disturb the surrounding ground, and small overexcavation would not make use of the profiling cutter completely, and hence the tunnelling route does not meet the design requirement.

Reasonable selection of segments

1. Reasonable selection of wedge-shaped segment rings

   Wedge-shaped segment rings are used to counteract the displacement difference in jack cylinders in the shield tail so that the differences can be decreased as much as possible to ensure the correct tunnelling direction of the shield.
2. Decreasing the width of segments

   Numerous cases show that small-width segments are good for decreasing the risk of segment breakup and improve the curve fitting. Segments that are 1.0 or 1.2 m wide are most beneficial for segment assembly.

Increasing the longitudinal stiffness of segments

The shield thrust is related to the segment stiffness. To ensure an appropriate thrust, it is necessary to strengthen the stiffness of the segments and adopt ground reinforcement measures.

Shield attitude adjustment

1. Reserving deviation of tunnel

   The segments move toward the outside of the tunnelling curve due to lateral force of segments. To constrain the deviation of the tunnel axis within a reasonable range, a certain deviation of the excavation axis is reserved in advance. Shield tunnelling along the secant direction of the curve and the axis of segments assembled is in the inside of the curve so that the deviation is generated when the shield tail escapes from the segments. The deviation can be determined by theoretical or empirical calculation considering the geological conditions.
2. Strict shield attitude correction

   The deviation of the shield axis should be controlled within 2–3 mm/m, and the end surface of wedge segments should be placed in the radial plane of the tunnelling curve.

Control of tunnelling parameters

1. Control of tunnelling parameters

   The advancing speed should be controlled within 10–20 mm/min, and excessive lateral force should be avoided by decreasing shield thrust.
2. Control of chamber pressure

   The chamber pressure should be controlled strictly so that the earth pressure and water pressure on the excavation face can be balanced. A ground surface heave in the range of 0.5–1.0 mm is recommended ahead of the excavation face to partially offset the ground settlement generating at the shield tail.
3. Control of synchronous grouting

   The synchronous grouting at the shield tail should be controlled strictly, since tunnelling with a small-radius curve would increase ground loss, and frequent shield attitude correction would significantly affect the surrounding ground.
4. Adjustion of tunnelling parameters based on monitoring

   To reduce the risk for shield tunnelling with a small-radius curve, the ground loss should be controlled. Tunnelling parameters are optimized by constantly analyzing the monitoring data.

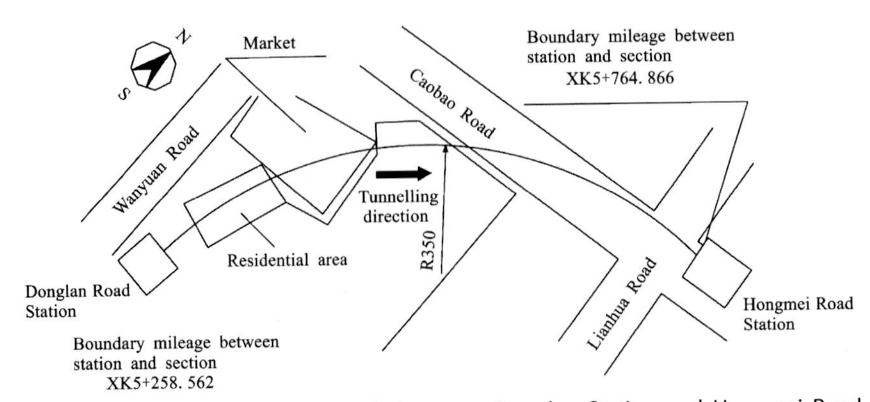

Figure 7.28 The plan of the tunnels between Donglan Station and Hongmei Road Station.

7.7.8.3 Case study

Project introduction

The Section 5 between Donglan Station and Hongmei Road Station on Line 12 of the Shanghai Metro is about 48 m long. Each tunnel has 425 segment rings. and there are a range from Ring #35 to Ring #230 under existing buildings. The centerline of the tunnel is a small-radius curve, and the radius is 350 m. The minimum and maximum buried depths of tunnels are 8.87 and 11.23 m, respectively (Fig. 7.28).

Key points of construction scheme

1. The pressure at the cutterhead was controlled strictly, and the pressure at the center of cutting wheel was set according to the buried depth and geological conditions. The fluctuation of chamber pressure was controlled between −0.02 and 0.02 MPa. The tunnelling parameters in the range undercrossing existing buildings were set on the basis of the monitoring data and previous experience.
2. The shield advancing speed was controlled within 20–40 mm/min.
3. The articulation system was activated to improve the flexibility of the shield. This measure lowered the impact on the adjacent buildings.
4. The pressure balance on the excavation face was controlled by controlling muck discharging to decrease the impact on the surrounding ground.

7.7.9 Shield tunnelling undercrossing existing buildings

Tunnelling undercrossing existing buildings and other infrastructures is inevitable in urban environments and causes a lot of problems. The undercrossed infrastructures are tunnels, railways, subway stations, bridges, pipelines, and embankments. Risks in tunnelling may be reduced by controlling tunnelling behaviors.

The key problems of EPB tunnelling are the ground squeezing surrounding the shield machine and insufficient support pressure on the excavation face. Therefore an appropriate chamber pressure is crucial for the safety of existing buildings and other infrastructures.

7.7.9.1 Keypoints and difficulties

Maintaining the performance of existing tunnels

High smoothness, low deviation of rails and small elevation difference among rails must be guaranteed.

Control of the excavation face stability

The excavation face stability will be reduced when the support pressure is lower than the sum of earth and water pressures in ground. When the support pressure is larger than the sum, the ground will uplift. In contrast, when the sum is larger than the support pressure, the ground will settle.

Ground disturbance

Ground disturbance is inevitable during shield tunnelling. When the tunnelling route is a curve or snakelike curve, the ground is vulnerable to disturbance, and it is hard to control ground displacement and avoid cracks in the lining segments.

Control of tunnelling parameters

The requirements for control of tunnelling parameters are strict when shield tunnelling is undercrossing buildings and infrastructure.

Monitoring

Monitoring is crucial for the tunnelling behaviors, especially when undercrossing existing buildings and other infrastructures. In urban areas the factors that affect the monitoring are complex.

7.7.9.2 Solutions and measures

Information construction

Manual and automatic monitoring run simultaneously. The monitored items are the vertical and horizontal ground displacements, the elevation difference of rails, deformation and the convergence of existing tunnels, and so on.

Survey for existing tunnels

The performance, stress state, internal conditions, and environment surrounding the existing tunnels should be investigated in detail. The safety of existing tunnels should be evaluated.

Ground reinforcement

The ground where the undercrossing tunnelling is located should be reinforced by grouting or other measures to reduce soil permeability and improve the bearing capacity of the ground.

Maintenance of construction equipment

The failure of the main configurations for the shield tunnelling including frame cranes, mortar mixer, compensatory grouting, trolley, and others should be avoided. The construction equipments and systems should be checked ahead of excavation.

Summary of previous tunnelling construction

Previous cases, especially in similar ground, should be summarized, and the successful experiences should be used to guide the current project.

7.7.9.3 Case study

Project introduction

The tunnels of the Metro Line 4 project in Kunming, China, was bored in a round gravel stratum with clay, gravelly sand, round gravel with an interlayer of silt, fine sand, and gravelly sand. The tunnels undercrossed existing tunnels of Metro 2 (Fig. 7.29). For the Line 4, the right tunnel was located below the left tunnel. In the undercrossing zone, the minimum vertical spacing was only 3.837 m between the existing tunnels (in Line 2) and the built tunnels (in Line 4). This zone was crucial for the successful tunnelling.

Risk analysis

1. The minimum vertical spacing was only 3.837 m between the existing tunnels and the built tunnels. The ground improvement and tunnelling parameters of the new tunnels would directly affect the ground displacement.
2. The tunnel was being driven mainly in a round gravel stratum in this zone. The soil was loose and significantly disturbed by tunnelling. In addition, the tunnels were bored in highly permeable soil under the

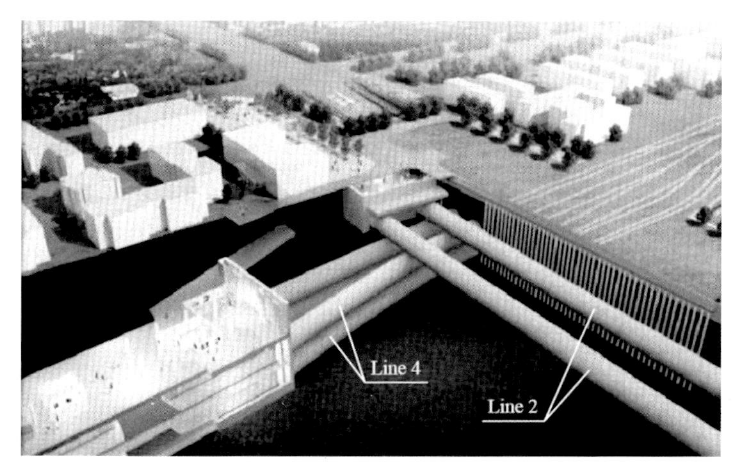

Figure 7.29 Model of shield tunnelling beneath the existing tunnels.

water level. Therefore insufficient muck conditioning would posed the risk of groundwater spewing in the outlet of screw conveyor.

3. The requirement for the prevention of damage to existing tunnels was strict. The challenges for the monitoring were huge, since the stress, deformation, and movement of existing tunnels also had to be measured.

Solutions

1. Construction control

   A tunnelling test section, which has geological conditions similar to those of the undercrossing zone, was selected to simulate the tunnelling and optimize tunnelling parameters, such as the rotation speed and torque of the cutterhead, shield thrust, the rotation speed of the screw, and conditioning agents. In the test section, the ground displacement, discharge rate of muck, and displacement of deep ground were monitored. The discharge rate was controlled by muck volume and weight simultaneously, and the ground loss was controlled within 0.5%. The synchronous grouting adopted double-liquid (cement - sodium silicate) grout materials, and it was done as long as the shield tail escaped from the ring segments. The compensatory grouting was early conducted in the zone than that in normal zones, and sometimes, the compensatory grouting was repeated, depending on the conditions.
2. Injection of clay slurry into the gap between the shield shell and the surrounding ground

A kind of special clay slurry was used to compensate for the ground loss in the gap between the shield shell and the ground. The slurry was injected through a hole in the top of the shield, and the workability of clay slurry was verified in the tunnelling test section.

3. Grouting reinforcement in the tunnel

   The lining segments in the undercrossing zone had special grouting holes, which were used to install grouting pipelines. The surrounding ground was improved by grouting through these holes of the segments in the right tunnel. The grouting pipelines of the right line tunnel adopted microsteel grouting pipelines, and the grouting material was pure cement-based grout. In addition, the stiffness and bearing capacity of the lining segments were improved to ensure structural safety.

4. Protection of the right line tunnel

   The right line tunnel was constructed first. A support system was used in the right tunnel during the undercrossing construction of the left line tunnel.

5. Monitoring for existing tunnels

   The deformation and stress for existing tunnels were monitored by an automatic monitoring system. The measurements for existing tunnels included ground displacements, tunnel stresses and convergence, and the vertical displacement of the rails in the existing tunnels. The alarm system would be activated when abnormal data were detected so that the constructor could deal with the problem in time.

References

[1] Wang H. Influence of aperture ratio of cutterhead of EPB shield on earth pressure in the chamber. Chin J Undergr Space Eng 2012;8(1):89−93.

[2] Zhang F, Zhu H, Fu D. Shield tunnels. Beijing: Chinese Communication Press Co., Ltd; 2004.

[3] The Japanese Geotechnical Society. Translated by Niu Q, Chen F, Xu H. Shield method investigation, design and construction. Beijing: China Architecture Industry Press; 2008.

[4] Huang Q. Research on interaction with soil of TBM cutting-wheel tools and their type selection design in gravel stratum. Beijing: Beijing Jiaotong University; 2010.

[5] Wu J, Yuan D, Li X, et al. Analysis on wear mechanism and prediction of shield cutter. Chin J Highw Transportation 2017;30(8):109−16.

[6] Song K, Pan A. Operation principle analysis of cutting tools on shield. Constr Machinery 2007;(2):74−6.

[7] Zhao J, Gong QM, Eisensten Z. Tunnelling through a frequently changing and mixed ground: a case history in singapore. Tunn Undergr Space Technol 2007;22(4):388−400.

[8] Guangzhou Metro. How was the subway built? Guangzhou: New Century Publishing House; 2014.

[9] Technology and Industrialization Development Center of Ministry of Housing and Urban Rural Development, China Railway Tunnel Group Co., LTD. Code for construction and acceptance of shield tunnelling method (GB 50446-2017). 2017.
[10] Beijing Urban Construction Exploration & Design Research Institute Co., LTD. Code for Monitoring Measurement of Urban Rail Transit Engineering (GB 50911-2013). Beijing: China Architecture Industry Press; 2013.
[11] Jinan University Ronghua Construction Group. Technical Standard for Monitoring of Building Excavation Engineering (GB50497-2019). Beijing: China Plan Press; 2019.
[12] Beijing Urban Construction Exploration & Design Research Institute Co., LTD. Code for Design OF Metro (GB 50157-2013). Beijing: China Architecture Industry Press, 2013.
[13] Chengdu Metro Group Corporation. Chengdu metro was built in this way [EB/OL]. [2021-02-18]. [https://www.sohu.com/a/83823729_355523].

Exercises

1. List some methods to achieve the pressure balance on the excavation face for EPB shield tunnelling. Analyze and compare their features.
2. List all tunnelling modes of EPB shield machines and compare their features.
3. Describe the mechanism of the filter cake formation for slurry shields.
4. State the specific differences between the muck discharging for EPB shield tunnelling and the slurry discharging for slurry shield tunnelling.
5. The main parameters of a curve section of metro section are listed in the accompanying table. Calculate the required number of turning rings and their assembly positions. The outside diameter of the segments for the turning rings is 6200 mm, and their wedge value is 42 mm.
6. Main parameters of a curve section

Symbols	Values	Signs	Chainage
R	420	ZH	K8 + 372.115
Ls	80	HY	K8 + 452.115

7. Compare the adaptability and safety measures of changing the cutters of a cutterhead under excessive pressure and atmospheric pressure in chamber.
8. Analyze the challenges and give your solutions for the EPB shield tunnelling, respectively in a clay stratum, a sandy gravel stratum, an soft soil layer overlaid by a hard rock layer, a soft ground with boulders, bedrock, and a karst stratum.
9. Collect a geological profile and other relevant information of a tunnel project driven by a slurry shield, and determine the ranges of chamber pressure, shield thrust, and cutterhead torque.

CHAPTER 8

Backfill grouting for shield tunnelling

8.1 Introduction

With the advance of the shield machine during tunnelling, an annular gap (i.e., a tail void) with a thickness of 160−370 mm usually exists between the segment and surrounding strata after the tail passing, ascribed to the diameter difference between the cutterhead and tunnel lining. To fill the annular gap effectively, backfill grouting is necessary, during which fresh grout mortar is pumped into the annular gap at the shield tail under certain pressure. In 1864 cement grout was initially employed in the construction of the metro tunnels in London and Paris, and the first patent was obtained for grouting technology in the field of shield tunnels [1]. Thereafter, backfill grouting for shield tunnelling has been widely applied in practice.

8.1.1 Purposes of backfill grouting

When the gap between the shield segment and the surrounding strata is unsupported, a local collapse may occur. The ground deformation caused by a local collapse directly affects the magnitude of ground settlement, and the stability of shield tunnel structure will be unfavorably affected by stratum load and groundwater, affecting the use of the tunnel. Backfill grouting can fill the gap and limit these adverse effects. The main purpose of backfill grouting for the shield is shown in Fig. 8.1.

8.1.1.1 Stratum gap filling

The gap volume generated after the shield tail escapes from the segment ring accounts for about 3%−16% of the space volume of the tunnel excavation scope. When the gap is unsupported, the surrounding soil will be squeezed into the gap, causing deformation or local collapse, which will further cause loosening and destruction of the stratum in a larger area and will eventually cause the ground surface above the shield tunnel to settle and destroy existing urban pipelines and underground structures in

Shield Tunnel Engineering
DOI: https://doi.org/10.1016/B978-0-12-823992-6.00008-4

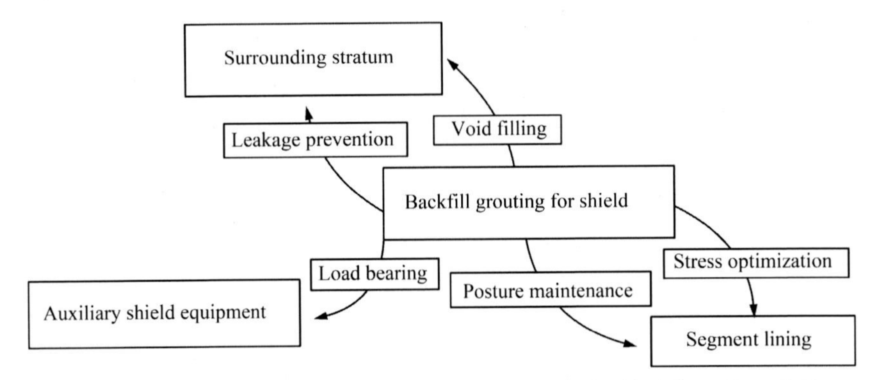

Figure 8.1 Purposes of backfill grouting for a segment. *Modified from Tang J, Huang J, Zhou Y, et al. Study progress of synchronous grouting material for shield tunneling method. Modern Urban Transit 2016;(04):89–92.*

adjacent strata, bringing potential hazards to complicated urban environments. Backfill grout can effectively fill the gap and consolidate in a short time. Therefore the backfill grouting method is an important means to compensate for ground loss and control stratum deformation.

8.1.1.2 Structure stress optimization

For shield tunnels, the uniform and integral interaction between the segment lining and the stratum ensure a reasonable stress state. When there is a gap between the segment and the stratum, an uneven load may occur to the segment structure, causing the stress state of the segment to deteriorate. However, the segment and the stratum can be strongly enveloped and supported by solidified grout, optimizing the stress of tunnel lining. Besides, the bearing load of shield tunnel auxiliary facilities can be reduced by a solidified grout layer.

8.1.1.3 Segment posture maintenance

During shield machine advancing, assembled segments suffer the thrusting force from the shield tail jacks. If the segment structure is suspended, the segment is prone to deflect and deviate at the spatial position under the additional thrust and gravity of the segment. Larger shear stress may be placed on the segment bolts that connects adjacent segments, reducing the structural stability. Notably, overall spatial deformation of tunnel can be limited by timely shield tail grouting, which will effectively avoids the stress on tunnel lining.

8.1.1.4 *Impermeability improvement*

The backfill grouting layer can serve as a protective layer to resist lining leakage. It effectively optimizes engineering problems, such as damage to the waterproofing layer and appearance of water seepage.

8.1.2 Classifications of backfill grouting for the shield

According to different grouting methods and grouting stages, the backfill grouting for the shield can be divided into the following classifications.

8.1.2.1 *Shield tail grouting*

Shield tail grouting is also known as synchronous grouting. In this method, when segment assembly is just completed behind the shield tail, several grouting pipes, placed in the shield tail, are used to fill the gaps and stabilize the segments and strata (Fig. 8.2).

The steps in shield tail grouting are as follows: First, the fresh grouts is prepared in a certain ratio, and transferred to a storage tank behind the shield machine. Then the grout is pressed by the grouting pump through pipelines (usually arranged symmetrically up and down) into the gap behind the segment wall. Finally, the grout solidifies and achieves gap filling. The whole grouting process is carried out simultaneously with the shield machine tunnelling. Generally, this method can better achieve the grouting filling effect, adjust the shield position in a timely manner, and control the deviation of the segment and the stratum deformation. It is applicable not only to a stable stratum and soft soil, but also to complicated geological environments with poor self-stabilization ability, such as sandy soil, cohesive soil, and a water-rich stratum.

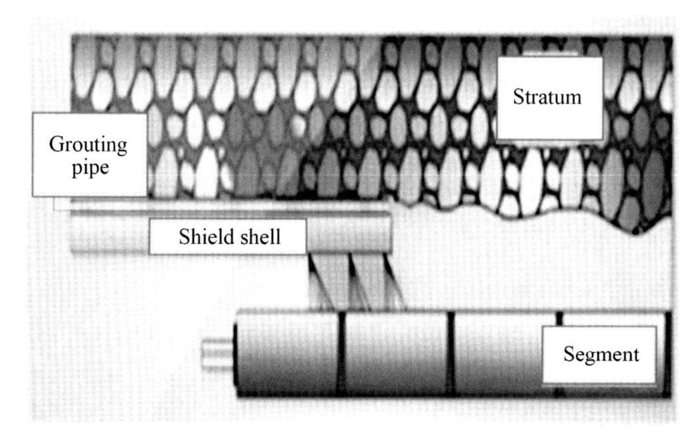

Figure 8.2 Schematic diagram of grouting at the shield tail.

8.1.3 Segment grouting

Segment grouting is another grouting method. In this method, the grout is injected into the gap through a prefabricated hole reserved in the segments, after the shield machine has advanced for a certain number of rings, to stabilize the segment and stratum (Fig. 8.3).

Segment grouting is not synchronized with the escaping of the shield tail from segments. Therefore when the tunnel is embedded in a complex geological location, such as a water-rich stratum and soft soil, there are often problems of segment floating, segment position deviation, and stratum subsidence. Therefore the segment grouting method is applicable only to stable strata.

8.1.3.1 Secondary grouting

Secondary grouting is used mainly to supplement parts with uneven filling and reduced first grouting volume. Secondary grouting also can be used to improve the impermeability of the grouting layer and reduce other construction effects.

8.2 Backfill grouting materials and performance demands

8.2.1 Backfill grouting materials and their applicability

The backfill grouting technology has a significant effect in coping with the surface settlement, segment leakage, and other problems. The grouting body that is filled behind the shield lining not only needs to fill ground loss caused by excavation, but also can serve as a protective wall to prevent corrosive substances such as groundwater from entering the

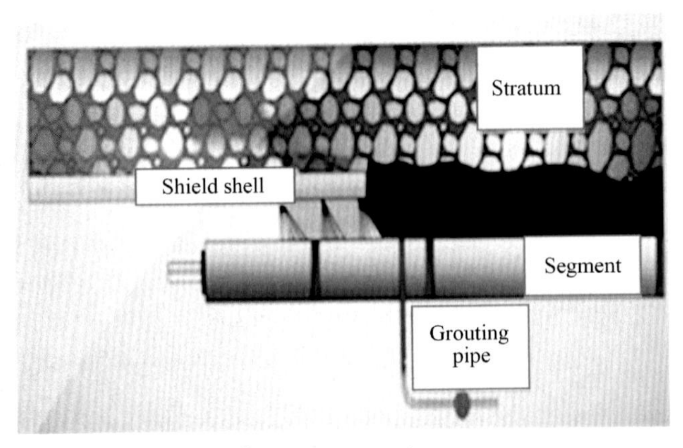

Figure 8.3 Schematic diagram of segment grouting.

tunnel [2]. However, different projects have different geological conditions, and the characteristic requirements for grout setting time, grouting body impermeability, unconfined compressive strength, and the like are also different. To apply shield tunnelling to a complex geological environment, it is necessary to thoroughly study the composition of backfill grout materials, optimal formulations, and its working performance.

8.2.1.1 Common backfill grouting materials and their classification

Backfill grouting materials for the shield can be classified according to their injection type: single-liquid types and double-liquid types (see Table 8.1). The single-liquid types of grout can also be divided into single-liquid inert grout and single-liquid active grout according to different materials.

Single-liquid inert grout is a mixture of fly ash, sand, lime putty, water, and additives without cement and other gel substances [3]. This type of material has the characteristics of a long setting time, low early strength, high bulk shrinkage and fluidity. Single-liquid grout applies to stable, dry and was widely used in shield tunnel projects in the middle and late 19th century. However, the only role of single-liquid inert grout is to fill the gaps between segments and the stratum. Single-liquid inert grout consolidated by water evaporation has a series of shortcomings, such as poor waterproofing, high bulk shrinkage, a long setting time, and low strength. Therefore it cannot assist in controlling stratum settlement in soft ground and so is not conducive to the early stability of the tunnel lining or minimizing seepage in the tunnel [4].

Table 8.1 Classification of injection types.

Classification criterion	Grout types	Grout materials
Injection types	Single-liquid type	Gypsum inert grout, cement-based active grout, etc.
	Double-liquid type	Cement-sodium silicate grout, etc.
Grout material properties	Inorganic type	Cement-based grout, etc.
	Organic type	Acrylamides, polyurethanes, lignin, epoxy resins, unsaturated esters, etc.

Single-liquid active grout has gradually gained recognition in the field of backfill grouting engineering, owing to its advantages of high early strength, controllable setting time, and pumping. Single-liquid active grout is a cement-based material without lime putty, formed by mixing cement, fly ash, sand, bentonite, water, and additives. Compared with inert grout, single-liquid active grout greatly shortens setting time, controls bulk shrinkage, and can be hardened in water, making it possible to conduct the shield technology in an unstable and water-rich strata. Besides, in consideration of the national economic level and construction requirements, active grout has been widely used because of its low requirement for construction management and economical grout materials [5].

Double-liquid grout is an active grout and also a cement-based material. It is made by mixing two parts of liquids A and B, where liquid A is an ordinary cement grout and liquid B is generally a sodium silicate grout. Highly active $Ca(OH)_2$ and $Na_2O \cdot nSiO_2$ are contained in the hydration grout and the sodium silicate grout, respectively. After being mixed, the two quickly react and produce calcium silicate gel ($CaO \cdot nSiO_2 \cdot mH_2O$) with a certain strength (see Eq. 8.1). The hardening time of the double-liquid grout can be controlled by regulating the different proportions of those two liquids. If necessary, a certain amount of accelerator or retardant can be added to adjust the setting time.

$$Ca(OH)_2 + Na_2O \cdot nSiO_2 + mH_2O \rightarrow CaO \cdot nSiO_2 \cdot mH_2O + NaOH \tag{8.1}$$

Cement-sodium silicate grout, which was applied to the south line project of the Yan'an Dong River tunnel in Shanghai in the early 1990s [6], has the advantages of a short setting time, high early strength, dispersion resistance, and difficulty in splitting under pressure—grouting conditions. However, the cement-sodium silicate grout also has obvious shortcomings. On one hand, the grout is poor in corrosion resistance and stability. On the other hand, the grout has high requirements for construction management when grouting is applied. The grout setting time is difficult to control, and grout delivery pipelines are easily plugged, which may further affect the project's progress.

According to the nature of the grout materials, backfill grouting materials can be divided into inorganic materials and organic materials. Inorganic grout materials include cement-based grout, and organic grout materials include chemical grouts of organic polymers (such as acrylamide, polyurethanes, lignin, epoxy resins, and unsaturated esters). The use of organic polymer chemical grout materials can solve engineering problems that ordinary cement grouts cannot solve (such as water-rich stratum). Secondary injection of segment grouting holes or cracks can play a role in strengthening the stratum, blocking water, and impermeableness. However, most chemical grouts are toxic and expensive, features that limit the development of organic grout materials, especially after chemical grouting pollution accidents in Japan. In 1974, the Japanese government issued relevant regulations on the use of chemical grout materials, explicitly prohibiting the use of any chemical grout materials except cement-sodium silicate grout.

There have been many studies on new types of grouts. The new ones can be obtained by using the grout mentioned above as a base liquid, with additives or replacement of some component materials. The new types of grout usually have engineering specificity, that is, the grout performance can maximally meet the site construction demands of the shield project. For example, some researchers use active slurry as the base liquid and add a certain amount of polyacrylamide polymer or xanthan gum to improve the pseudo-plasticity of the grout. Such new types of grout can effectively fill the shield tail gap, fix the lining, and resist the tunnel floating. To improve the grout performance, rapidly hardening sulfoaluminate cement can act as the main gel material, used with a water reducer and an accelerator. In addition, there are still many researchers using slag, steel slag, and waste products or shield tunnelling waste soil to replace traditional grout material components. In this way, these shield residues are handled from the perspective of green recycling and comprehensive utilization. It not only conforms to the current international and domestic development, but also handles the accumulated residue effectively.

In summary, with the rapid development of rail transit tunnel engineering, shield tunnel construction is facing increasingly complex geological conditions. It is still necessary to do more in-depth research on the durability of grout materials, environmental protection issues, and adaptability to different environments. New synchronous grouting materials with high comprehensive performance are needed. The future research directions for grout materials are as follows: green, environmental protection, impermeability, corrosion resistance, easy control of setting time, good economic benefits, and convenient construction [3].

8.2.1.2 Selection of backfill grouting materials

The selection of grouting materials is greatly related to factors such as ground conditions, construction environments, and construction cost. Therefore to select the appropriate grouting materials in actual engineering, getting familiar with their performance is essential. The basic performance and the applicable environments of single-liquid grout and double-liquid grout are analyzed in Table 8.2.

At present, grouting materials are often selected according to the ground conditions. If the stratum is relatively stable, single-liquid grout is usually used, and grouting and tunnelling do not need to be conducted at the same time. By contrast, if the stratum is silt with poor stability layers or sand layers that collapse easily, grouting and tunnelling need to be conducted at the same time. For water-rich gravel layers or sand layers, double-liquid grout is recommended, and single-liquid active grout with underwater dispersion resistance can also be selected. For silt layers or clay layers, a good effect can be achieved by using single-liquid active grout or double-liquid grout.

8.2.2 Performance indexes of backfill grouting materials

Through a large number of experimental studies and field-related data analysis, the ground injected behind the segments wall must quickly

Table 8.2 Performance and applicable environments of three main backfill grouting slurries.

Performance and the applicable environments	Single-liquid inert grout	Single-liquid active grout	Double-liquid grout
Setting time	$\geq$ 20 hours	5–15 hours	3–30 minutes
Filling property	Very good	Good	Good
Early unconfined compressive strength	Low	Relatively high	Very high
Ultimate unconfined compressive strength	Low	High	Relatively high
Pipe plug	Difficult to plug	Occasionally plugged	Easy to plug
Application cost	Low	Relatively low	Relatively high
Applicable environments	Stable stratum	Stable stratum, soft soil or water-rich stratum	Soft soil or water-rich stratum

and effectively fill the gap between the segment and the stratum. Generally, the grout materials need to meet the following requirements:

1. The grout should have good fluidity and low bleeding, and it should satisfy the requirements of pumping.
2. The grout should have good filling property and high stability, and it should fill up the shield tail gap quickly.
3. The grout should have a certain consistency and be difficult to dilute with muddy water.
4. The grout should solidify instantly after filling, with certain early strength and an adjustable setting time.
5. The grout should have a lowbulk shrinkage and permeability coefficient after hardening.
6. The grout should have a rich source of raw materials, be environmental friendly, have a low price, and be easy to clean when remaining in the grouting pipe.

Specific technical performance requirements are shown in Table 8.3.

Among the requirements, the filling property, fluidity, and early strength of the grout are the keys to achieve good performance of backfill grouting. However, it should be noted that there is a certain contradiction among these conditions. To coordinate the relationship among the properties of grout, appropriate proportion and additives are the keys to obtaining the best grouting performance.

Table 8.3 Performance indexes of synchronous grouting material.

Inspection items	Technical performance requirements
Initial setting time	3–12 hours
Consolidation strength	Not less than 0.2 MPa in 1 day, not less than 2.5 MPa in 28 days
Grout stability	No precipitation, no segregation, or less segregation when standing for the gel time, bleeding rate $< 5\%$
Consolidation rate	Consolidation shrinkage $< 10\%$
Mortar density	1.80–2.00 g/cm^3
Grout fluidity	The initial fluidity is 220–250 mm, the fluidity is not less than 200 mm within 4 hours, and it can still be pumped in 8 hours

8.2.3 Basic performance test of backfill grouting material

Currently, the existing code and industry standards test the performance of backfill grouting slurry from the following aspects: density, viscosity, bleeding rate, fluidity, consistency, setting time, reinforced stone shrinkage rate, and unconfined compressive strength.

8.2.3.1 Grout density

The grout density refers to a ratio of the mass of the grout to its volume. When the grout density is measured, the measurement should be completed after all components of the grout have been fully mixed and before the grout coagulates. If the grout coagulation time is too short, different components should be measured separately, and then the density is calculated according to the dosage of each component in the formula. The measurement can be conducted by the "Test Method for Performance on Building Mortar" (JGT/T70–2009) [7].

8.2.3.2 Grout viscosity

The viscosity is the key to the grouting performance. It determines the diffusion capability of the grout, grouting pressure, flow, and other parameters. For an ideal backfill grouting material, its viscosity should be low at the initial stage and then increase rapidly. The viscosity of the grout needs to be measured immediately after the components of the grout have been mixed. Commonly used measurement methods include rotary viscometers and funnel viscometers, which are used to express viscosity in centipoises (cp) in engineering practice. The rotary viscometer can measure the absolute viscosity of both Newtonian fluids and non-Newtonian fluids. However, for granular materials such as cement slurry, the funnel viscometer is more often used. The grout viscosity is expressed by measuring the time (usually expressed in seconds) of grout (500 mL) leaking out of a funnel.

8.2.3.3 Bleeding rate of grout

The bleeding rate of the grout is expressed by ratio of the volume of water, precipitated after grout coagulation, to the volume of the grout. During the test, the volumetric cylinder is first filled with grout. Then the height of water is read at every exact time interval. Consequently, the height reading shows the bleeding rate (expressed as a percentage). The test method can be conducted according to the bleeding rate test method of cement slurry under normal pressure in "Grouting Admixture for Prestressed Structure" (GB/T 25182-2010) [8].

8.2.3.4 Grout fluidity

The grout fluidity can reflect the fluidity of the static cement slurry. It can be measured with reference to Test Method for Fluidity of Cement Mortar (GB/T-2419-2005) [9].

8.2.3.5 Grout consistency

The grout consistency can reflect the solidity or fluidity of a material. It refers to the ability of the grout resisting deformation or damage caused by external forces, which is the most important physical state feature of a viscous fluid. It can be measured with reference to Test Method for Performance on Building Mortar (JGJ/T 70–2009) [7].

8.2.3.6 Grout setting time

The setting time comprises the initial setting time and the final setting time. The initial setting time is the time during which the cement slurry begins to lose its plasticity, and the final setting time is the time when the plasticity losing of the cement slurry is completed. The setting time is often measured by Vicat Apparatus, referring to the measurement method provided in "Test Methods for Water Requirement of Normal Consistency, Setting Time and Soundness of Portland Cement" (GBT1346–2001) [10].

8.2.3.7 Compressive strength

The strength of the reinforced stone, an important indicator, reflects the quality of backfill grouting and is related to many factors such as the cement grade and the water-cement ratio. For backfill grouting, it is generally hoped that the early strength is high and the later strength can match the surrounding strata. Measurement of compressive strength can be conducted with reference to "Test Method for Performance on Building Mortar" (JGT/T70–2009) [7].

8.2.3.8 Volume shrinkage

The volume shrinkage is the percentage of the ratio of the difference between the grout volume after solidifying and the grout volume to the grout volume after solidifying. For backfill grouting materials, a lower volume shrinkage will lead to denser filling and a better grouting effect.

8.2.4 Mixing proportion of grouting materials for typical shield engineering

The composition of the grouting material and the change of its ratio affect the performance of the grout directly. Compared with inert grout, single-liquid active grout has a shortened setting time, certain early consolidation strength, and reduced volume shrinkage. Compared with double-liquid grout, its application cost is greatly reduced, and it is less likely to cause grout pipeline blocking. Single-liquid active grout is still being used in most synchronous grouting construction of shield tunnels in China. Table 8.4 shows the mixing proportion of synchronous grouting materials using single-liquid active cement mortar in some projects in China.

8.3 Common equipments for backfill grouting

The backfill grouting equipments mainly consists of mixing equipment, injection equipment, and a control system. The main components of mixing equipment are grouting material storage equipment, metering equipment, mixer, and storage tank. The injection equipment contains a grouting pump and a grout delivery device. The control system is mainly composed of a jack speed measurer, a grout injection volume adjustment device, and an injection rate setting device. During construction a constructor first uses the mixing equipment to mix, stir, and prepare raw materials for the grout for storage according to the required grout proportion, then pumps the grout to the shield tail for injection by the injection equipment. Meanwhile, the control system monitors and controls the grouting parameters in real-time to avoid the problems of grout blockage and leakage and to ensure the safety of the construction [11].

8.3.1 Mixing equipment

8.3.1.1 Grouting material storage equipment

At present, silos are the main equipment used to store cement, fly ash, bentonite, and other grout raw materials at the construction site. There are two types of silos: vertical silos (Fig. 8.4) and horizontal silos. The choice of silos is based on-site conditions. The vertical silo has a smaller occupancy space than the horizontal silo and can have a storage capacity of 20T when the height of the silo is 10 m from the ground surface. However, owing to load concentration, the demand for reliability of the foundation is high. The transverse silo occupies a large area, can use the space under the silo, and is dispersed in load. Its simplicity in construction has made it popular in recent years.

Table 8.4 Mixing proportion using single-liquid active cement mortar in some projects in China.

Shield project name	Cement (kg)	Fly ash (kg)	Bentonite (kg)	Sand (kg)	Water (kg)	Additives (kg)
Chengdu Metro	120–260	241–381	70–80	779	460–470	\
Guangzhou Metro Line 3	122	223	248 (clay)	910	248	2.5 (water-reducing agent)
Guangzhou Metro Line 4	240	320	30	1100	470	2.5 (water-reducing agent)
Guangzhou Metro	120	381	54	779	465	—
Nanjing Metro Line 15 (portal section)	225	400	50	1000	245	0–2 (expansive agent)
Nanjing Metro Line 15 (interval section)	100	300	75	1350	225	0–2 (expansive agent)
Shanghai Metro Line 11	100	360	20	400	210	—
Shenzhen Metro Line 1	180	310	37	875	310	2.5 (water-reducing agent)
Chongqing sewage river-crossing tunnel (starting section)	120	381	54	779	344	—
Chongqing sewage river-crossing tunnel (interval section)	80–160	381	60	779	460	—
Chongqing sewage river-crossing tunnel (arriving section)	160	341	56	779	324	—
Yangtze River Cross Tunnel in Wuhan	80–260	241–381	50–60	779	460–470	—
Yellow River Tunnel Project in first-stage of middle route south-to-north water transfer project	187	313	37.5	770	375	4.25 (water-reducing agent)

Figure 8.4 Vertical silo.

Besides the silos that are widely used for storing powdered raw materials, tanks for storing liquid materials are also required for double-liquid grout. Iron silicate tanks are most commonly used and are of two types: cylindrical type and box type. The general liquid storage capacity is 6–10 m^3, and the diameter and the height are both 2–3 m.

8.3.1.2 Metering equipment

The function of metering equipment is to measure the mass and volume of the grouting material. Generally, the mass is measured for powder, and the volume is measured for liquid. When necessary, a special measurement system is used to measure the volume of powder.

8.3.1.3 Mixer

There are two types of mixer: stirring mixer and jet mixer (Fig. 8.5). The stirring mixer is more common. In actual construction the proper parameters for the mixer are expected to be 200–600 L in capacity and 250–500 r/minute in mixing speed.

8.3.1.4 Liquid storage tank

The grout mixed by the mixer needs to be stored in the liquid storage tank before use. The commonly used liquid storage tank with an agitator requires a capacity of 1.5–4 m^3 and a low mixing speed of 20–30 r/minute.

(a)

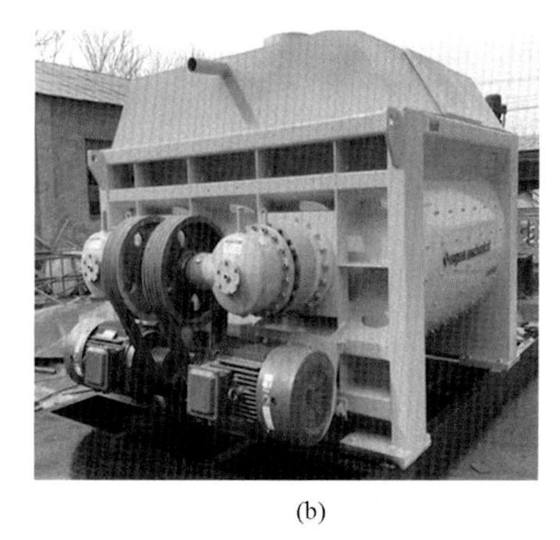
(b)

Figure 8.5 Stirring mixer and jet mixer. (a) Stirring mixer. (b) Jet mixer.

8.3.2 Injection equipment

8.3.2.1 Grouting pump

Grouting pumps are mainly divided into four types: hydraulic grouting pumps, piston grouting pumps, extruded grouting pumps, and screw grouting pumps. Pressure pumps are mostly piston types, and injection pumps are mainly hydraulic types. For different backfill grouting methods the pumps are arranged at different positions and have different operation characteristics. The advantages and disadvantages of different locations are shown in Table 8.5.

Generally, two grouting pumps are provided for a shield machine grouting system. Each pump contains two grouting (discharging) ports. The grouting flow can be adjusted through a solenoid flow valve. Each grouting pump has two adjacent cylinder pistons, a suction chamber, a discharging chamber, a hydraulically driven disk valve seat, a movable piston, a cylinder cleaning water chamber of the disk valve seat, an automatic stroke control device, and a hydraulic valve with a handle (which can be operated back and forth when no pressure is generated). The volume of grout being pumped is adjusted by the speed of the hydraulic cylinder. A pump assembly is installed on foundation support and connected to other foundations through the support.

8.3.2.2 Grout delivery device

The grout delivery device is composed of a grout delivery pipe and an injection rubber pipe. When the grouting material is transported from the mixer to the cave, an ordinary low-pressure steel pipe with 51 mm in diameter is used for a liquid mortar, and an iron pipe or a plastic pipe with a diameter of 19–38 mm is used for the B liquid (double-liquid grout). The design of the grout delivery pipe diameters is mainly based on construction experience or the actual press of the grout being delivered.

8.3.2.3 Shield tail grouting pipe

The shield tail is equipped with some built-in grouting pipelines, which are attached to the shell of the shield mostly in a built-in form and arranged symmetrically along the shield tail circle (Fig. 8.6). The pipelines are divided into four groups. Each pipeline group contains two adjacent pipelines: one for use and the other one for backup. (Only one pipelines is used in construction. The other serve as backup pipelines to prevent the blocking in pipeline.) After being assembled in the shield machine, the backup pipelines should be filled with yellow grease to prevent rusting and being blocked by debris. The distribution of the built-in grouting pipes is shown in Fig. 8.7. The grouting port is located in front of the

Table 8.5 Arrangement locations of pumps and its advantages and disadvantages.

Arrangement locations	Advantages	Disadvantages
Shaft	1. Omitting delivery operation of injection grout 2. Do not repeat operation near cutting face 3. Applicable to small-bore sections	1. Injection pumps and materials both limited 2. Difficult to manage injection 3. Many accidents such as blockage of grout delivery pipe
Trolley at the rear of cave	1. Easy to manage injection 2. Grout delivery pipe has few accidents and is easy to clean 3. Wide application range of injection material on pump	1. Obstacle to excavation and segment assembly operation 2. Increase delivery operation of injected materials
Special train behind cave wall	1. No trolley	1. Establishment of special line for backfill grouting 2. Difficult to connect cutting face and mixing

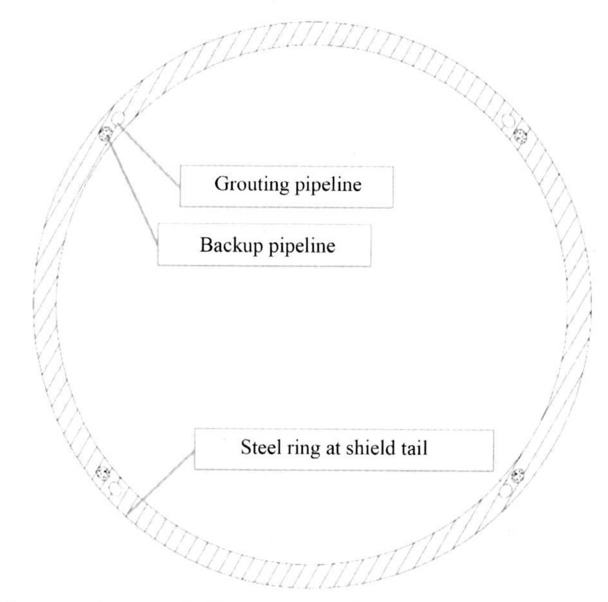

Figure 8.6 Cross section of a built-in grouting pipe.

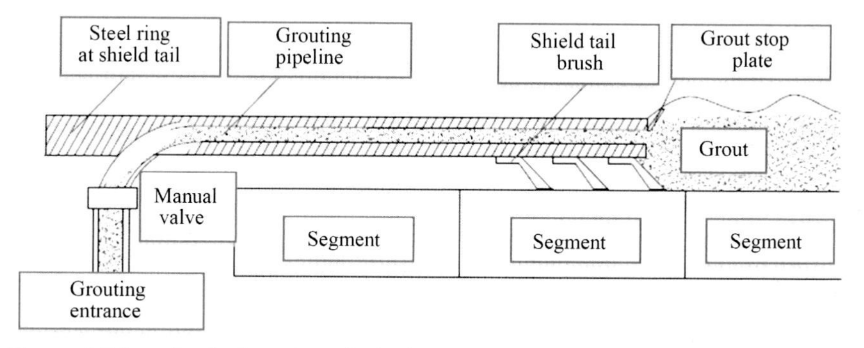

Figure 8.7 Longitudinal section of a built-in grouting pipe.

back shield, and the grout outlet port is located behind the back shield tail sealing brush. Each pipe is equipped with a manual valve at the grouting port [12].

8.3.3 Control system

The controller with a programmable logic controller (PLC) is located in the shield machine operation control room. The controller is the core part of the whole shield machine and is responsible for data collection and task control. The PLC communicates with a computer that controls an indoor data acquisition system through a series of communication interface to achieve data acquisition, storage, display, counting, and printing. Various digital data related to shield control are transmitted to the PLC through a digital input interface and an analog input interface. It is very important to control indicator lamps, contactors, relays, and solenoid valves to execute the corresponding control of parameters such as the dynamic grouting pressure, grout injection volume, and injection rate.

The backfill grouting control system includes the following devices: a jack speed measurement device, an injection volume adjustment device, an automatic injection rate setting device, a variable-speed motor, a pressure adjustment device, a recording device, an alarm display device, and an A liquid and B liquid proportion setting device. The related introduction of the main devices is as follows.

8.3.3.1 Jack speed measurement device

A jack speed measurement device aims to detect the advancing speed of a heading machine. It is installed on the shield jack and works with a speed converter that sends a speed signal to a digital calculator in a central monitoring panel, where the instant speed is displayed.

8.3.3.2 Injection rate and injection volume setting and calculation

The central monitoring panel is equipped with a calculation device specially designed for automatic backfill grouting, so that the optimal injection volume is calculated according to the following formula:

$$Q = \alpha \times S \times V \tag{8.2}$$

where Q is the optimal injection setting quantity (L/min); α is the injection rate (generally set at 100%–200%); S is the area (m^2) of a tail gap; and V is the speed (mm/min) of the jack that is fed back by the jack measurement device.

$$S = \frac{\pi}{4}(D_1^2 - D_2^2) \tag{8.3}$$

where D_1 the external diameter (m) of the shield tail; and D_2 is the external diameter (m) of the segment.

8.3.3.3 Pressure adjustment device

The pressure control device can automatically measure and control the injection flow, and it can also automatically control the injection pressure.

Each pumping cylinder is equipped with a counting indicator, a pump operator, and a shield driver who can read the grouting volume in each grouting pipe from the counter to the PLC. A pressure sensor is installed at the outlet of the grouting pipeline at the shield tail, and the pressure at the outlet of the pipeline during grouting can be seen in both the shield operating room and the grouting control box. The four pressure sensors monitor the grouting pressure of the four respective grouting pipelines, transmit the data to the PLC, and then pass the processed result to the screen of the control panel. The injection will be conducted at a preset injection rate. When the injection flow is beyond the optimal injection rate, the injection pressure rises.

8.3.3.4 Control panel

For general grouting systems the control panel will also display the injection ratio setting of the A liquid and B liquid if double-liquid grout is involved (Fig. 8.8).

8.3.4 Precautions

1. The grouting system involves the following steps: (1) transporting the grout to the mixing tank, and continuously stirring; (2) connecting the grouting pipeline to ensure pipeline smoothness; (3) starting the pumping station; (4) selecting a grouting control mode; (5) starting the grouting pipeline, wherein only a special pipeline is started for separately grouting if necessary; and (6) adjusting the grouting speed according to the advancing speed.
2. In the grouting process, attention should be paid to the changes in the number of strokes and the pressure value to locate the pipe and to determine whether the pipe is blocked or not. If the pressure value rises abruptly, the pipe at the shield tail may be blocked. If the pressure value and the alleviation number do not change, the pipes may encounter blockage between the shield tail and the pump or the mortar tank and the pump. The blocked pipe should be dredged in a short time to avoid difficulty in cleaning due to being blocked for too long. The pipeline should be washed and examined in a timely manner.
3. The grouting pressure and the shield advancing speed of each path are displayed on the grouting operating panel. The advancing speed should match the grouting speed to avoid the situation in which the grouting has not been completed at the end of advancing or the grout has been injected in advance.

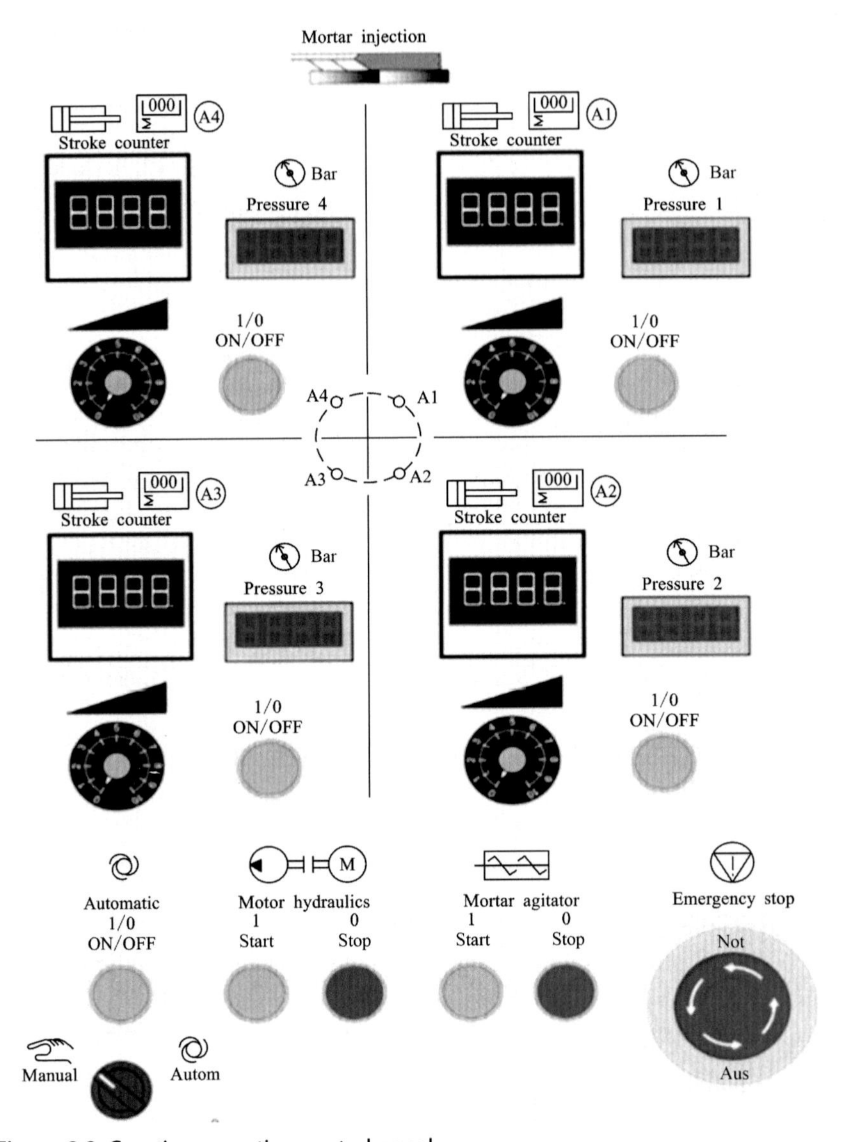

Figure 8.8 Grouting operation control panel.

4. In the grouting process, if the grouting pressure of a pipeline is high and the injection volume is small, the pipeline may be blocked, and it should be stopped immediately so that the pipeline can be cleaned.

8.4 Backfill grouting construction and control for shield

Backfill grouting for shield is an important step for ensuring the stability of the stratum and the segment structure. Therefore reasonable design, strict control, continuous monitoring, and timely adjustment of the injection flow are the keys to ensuring the quality and safety of backfill grouting.

8.4.1 Backfill grouting process parameter control

The main design parameters of the backfill grouting process include grouting volume, grouting pressure, grouting speed, and grouting ending standard.

8.4.1.1 Grouting volume

The determination of the synchronous grouting volume is closely related to the shield engineering. However, the injection rate should be appropriately increased to ensure that the clearance in the shield is filled tightly due to overexcavation and grout leakage caused by the strata, line, and tunnelling methods. Generally speaking, the theoretical grouting volume Q of each ring of the segment is conducted concerning the following formula:

$$Q = \lambda\pi\left(D^2 - d^2\right)L/4 \tag{8.4}$$

where λ is the filling coefficient, which should be determined comprehensively according to the geological conditions, construction conditions, and environmental requirements (the volume of the filling coefficient ranges is normally in the range of 1.3–1.8, but 1.5–2.5 in the stratum with fairly developed fractures or large amounts of groundwater); D is the contour diameter of the shield excavation; d is the external diameter of the shield segment; and L is the length of a single ring segment.

In the actual construction process, factors such as overexcavation and leakage must be checked if the injection volume increases continuously. When the injection volume is lower than the predetermined injection volume, this may be due to the proportion of the injected grout, the injection period, the injection location, or the improper and barrier releasing of the injection machine, and a careful check should be done, after which corresponding measures must be taken.

8.4.1.2 Grouting pressure

The grouting pressure is the pressure near the injection hole. However, in the actual construction process, it is not convenient to install a pressure gauge near the injection hole, and the injection pressure is often measured in the pumps instead. At present, the grouting pressure in the industry is mostly determined on the basis of a combination of empirical methods and theoretical calculations.

The empirical methods are mostly judged and selected on the basis of the water and earth pressure at the grouting position. Japanese researchers believe that the grouting pressure can be increased by 0.2 MPa on the basis of the water pressure at the grouting position, while some domestic engineering researchers think that the grouting pressure should be slightly larger than the sum of the earth pressure and water pressure of the stratum and generally can be 1.1–1.15 times the sum of the earth pressure and water pressure of the stratum or be approximately equal to the water pressure of the cutting face. In terms of theoretical analysis, it is generally believed that a split phenomenon will occur to cause grout loss along cracks if the grouting pressure is too large, and the overburden earth pressure is difficult to balance if the grouting pressure is too small, which easily causes stratum formation in the early stage of grouting. Therefore a reasonable range of grouting pressure can be determined on the basis of the upper limit of the grouting pressure determined by split grouting and the upper and lower limits of the grouting pressure determined by soil stability. Generally speaking, the grouting pressure should be between the active earth pressure and the passive earth pressure at the position of the grouting hole, and its upper and lower limits can be calculated as follows:

$$P_{\perp} = \gamma_f h_a \times \tan^2\left(\frac{\pi}{4} + \frac{\varphi}{2}\right) + 2c \times \tan\left(\frac{\pi}{4} + \frac{\varphi}{2}\right) \tag{8.5}$$

$$P_{\top} = \gamma_f h_a \times \tan^2\left(\frac{\pi}{4} - \frac{\varphi}{2}\right) - 2c \times \tan\left(\frac{\pi}{4} - \frac{\varphi}{2}\right) \tag{8.6}$$

where γ_f is the buried depth; h_a is the submerged unit weight of the soil layer above; φ is the internal friction angle of the stratum; and c is the cohesive strength of the stratum.

8.4.1.3 Grouting speed

The synchronized grouting speed should be matched with the tunnelling speed. The grouting volume of the current ring is completed within the time of shield tunnelling in a circle to determine the average grouting speed so as to achieve the purpose of uniform grouting. Generally, synchronous grouting begins when the shield machine advances to 0.1 m and stops at 0.1 m before the end of advancing. Besides, to prevent malpositioning, which causes misalignment and damage due to uneven force on the segment generating bias during grouting, it is very important to inject symmetrically and uniformly during synchronized grouting.

8.4.1.4 Grouting ending standard

Generally, the dual index control of grouting pressure and grouting volume is used as the grouting ending standard. In other words, grouting is stopped when the grouting pressure reaches the designed pressure and the grouting volume reaches more than 85% of the theoretical grouting volume. Grouting reinforcement is also needed at places with insufficient grouting or a bad effect of grouting to increase the compactness of the grouting layer.

8.4.2 Grouting construction organization and management

8.4.2.1 Preliminary preparation

Preliminary preparations are required before grouting construction. These mainly include technical preparation, material preparation, equipment preparation, construction personnel preparation, and preparation of the construction site conditions.

Technical preparation: Before construction, technicians and management staff are organized to survey and familiarize themselves with the construction site; check the construction drawings again; study related construction technical specifications; and compile construction operation instructions, technical data, and data record forms based on the actual situation at the site.

Material and equipment preparation: It is important to ensure that the materials for grouting are mixed according to the proportion designated for the construction. The relative density, consistency, and workability of the grout, the maximum particle size of debris, the setting time, the strength after coagulation, and the curing shrinkage rate of the slurry should all conform to the engineering requirements, and the grout should be easy to pressure-inject after mixing and cannot be segregated and precipitated during transportation. Relevant quality inspection should be conducted on the grouting equipment. Before grouting, the equipment for mixing, storing, and transporting should be prepared according to the requirements of grouting and should be tested. The apparatus for automatically recording parameters such as grouting volume, grouting pressure, and grouting time should be equipped to ensure real-time monitoring of the grouting process.

Construction personnel preparation: According to construction needs, detailed construction personnel demand plans are made for different construction stages before entering the site. All departments should be coordinated to ensure reasonable and efficient personnel scheduling.

Construction site condition preparation: The site conditions of transportation, water, electricity, and communications are prepared, with orderly management to ensure quick and safe construction.

8.4.2.2 Grouting construction

To fill the annular gap uniformly and prevent the lining from bearing uneven bias, pressure injection is conducted on four grouting holes reserved in the shield tail, and a pressure divider is arranged at the outlet of each grouting hole to facilitate detection and control on the grouting pressure and the grouting volume of each grouting hole so that symmetric and uniform pressure injection behind the segment is obtained. To ensure the grouting effect and the grouting quality, the *P-Q-t* (grouting pressure—grouting volume—grouting time) curve should be calculated during the construction process to analyze the grouting effect and feedback and guide the next grouting. Special attention should be paid to changes in the number of strokes and the pressure values fed back by the grouting systems so as to judge whether the pipe is blocked and, if so, to determine the position of the blocked pipe in time.

At the same time, information should be fed back in time based on monitoring results of the deformation of the segment lining in the tunnel and the deformation of the ground and surrounding buildings to correct the grouting parameters and the design and construction methods. During the grouting process, the grouting hole should be sealed to ensure no leakage, and maintenance of the grouting equipment and the supply of grouting materials should be done to ensure a smooth and continuous grouting operation. The synchronized grouting construction process is shown in Fig. 8.9.

8.4.2.3 Grouting ending

After grouting has been completed, the equipment needs to be disassembled according to the process and must be cleaned and maintained.

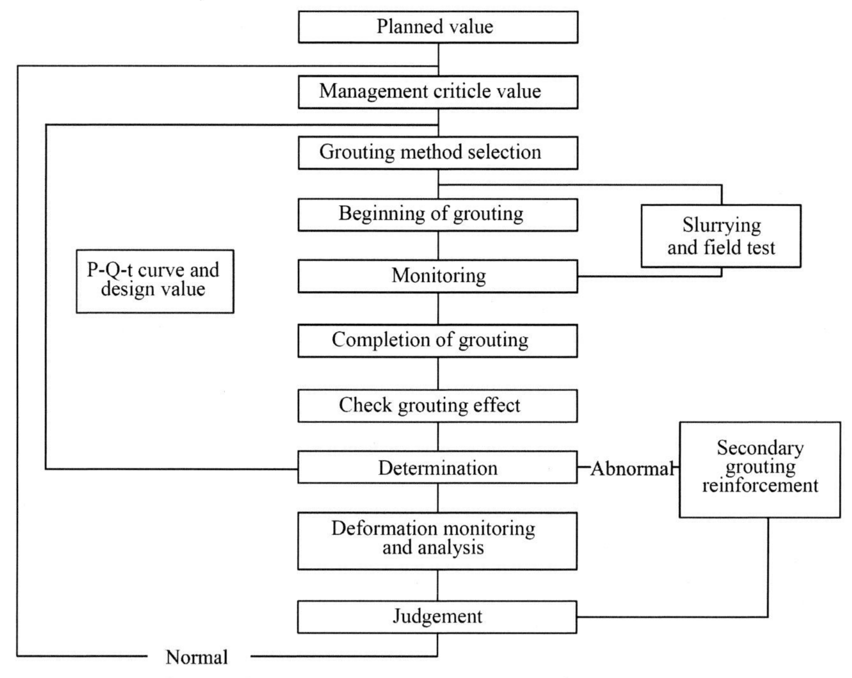

Figure 8.9 Synchronized grouting construction process flowchart.

After grouting with segment grouting ports, the grouting ports should be plugged. In addition, it is necessary to check the grouting effect, and grouting needs to be supplemented if the expected effect was not achieved.

8.4.3 Effect evaluation of backfill grouting for segment

The quality of the grouting effect of shield construction directly relates to the construction safety of the shield tunnel and the stability of the surrounding buildings. Therefore evaluation of the distribution of the grout has always been a key concern of engineers and technicians.

At present, the main methods for checking the effect of backfill grouting for a segment include an analytical method, a heat transfer method, an optical method, a radiation method, and an electromagnetic method. The analytical method is used for comprehensive analysis and judgment based on the *P-Q-t* curve and in combination with the deformation results of the lining, the ground surface, and the surrounding buildings. The heat transfer method, the optical method, and the electromagnetic method are nondestructive test methods and meet the accurate, rapid, and nondestructive development demands of the construction and detection technology, which are widely used in domestic and foreign engineering construction. Among many nondestructive detection technologies, the ground penetration radar (GPR) method is regarded as one of the best synchronized grouting detection methods for shields by a large number of engineers and technicians, owing to its advantages of high resolution, fast speed, nondestructive operation, and large area monitoring depth range [13].

GPR is a high-resolution electromagnetic scanning technology using reflection of high-frequency electromagnetic beams to detect invisible targets or underground interfaces so as to determine the shape or location of their internal structures. For backfill grouting for shield tunnelling, the detection range of GPR is the area within about 1 m behind the segment wall. Considering the electrical parameters of concrete segments, grouting materials, and soil behind the wall in targets detected by GPR, it is known that the electrical properties of these three media are quite different, and the electromagnetic wave will obviously be reflected between the segment and the grouting layer and between the grouting layer and the soil. Based on the reflection of the electromagnetic wave and the propagation speed of the electromagnetic wave in different media, good detection results can be obtained by using the GPR to emit and collect waveforms.

When using the GPR method to detect the effect of backfill grouting, on the one hand, it should pay attention to the distribution of grout in the circumference of the tunnel, since the grout is in a free-flowing liquid state before solidification; on the other hand, the grouting pressure and the grouting volume between segment rings are different, and grout is supplemented as a part of a segment during construction. Therefore survey lines should be laid out in the axial and circumferential directions of the tunnel to form a grid, and the distribution state of backfill grouting for a segment can be obtained by comprehensively analyzing the profile of each survey line. The grid-shaped survey line layout plan and detection results are shown in Fig. 8.10.

8.4.4 Common problems and solutions

8.4.4.1 The performance index of the grouting materials is not up to standard

When the performance index of the grouting material is not up to standard, it may cause substandard grouting and engineering accidents. The main reasons for poor performance are that the grouting material is not

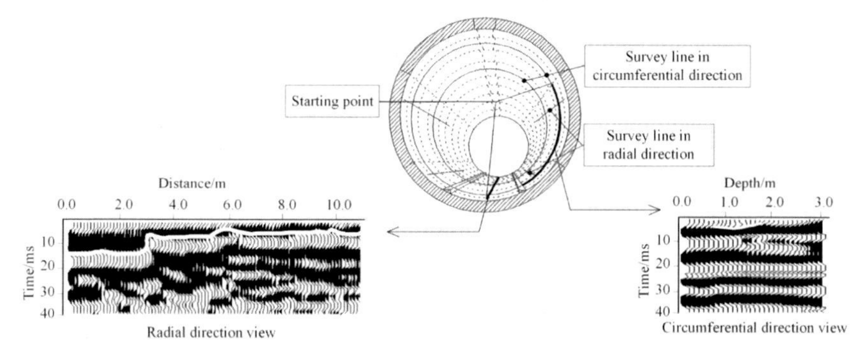

Figure 8.10 Grid-shaped survey line layout plane and detection results.

adaptable to the stratum, the proportioning of the grouting material is unreasonable, the grouting raw material is unqualified, or metering or transportation equipment is not up to standard. The main preventive measures include matching the grouting material with the stratum environment, performing quality inspection and effective quality control on the grouting material according to regulations, maintenance and repair of the metering equipment and the transportation equipment in a timely manner, and conducting performance evaluation and detection on the mixed grout.

8.4.4.2 The surface settlement is too large during construction

If the surface settlement is too large during the construction process, it can damage surface buildings or underground pipelines. The main reasons are that the backfill grouting is not timely or the grouting quality is not up to the standard, the grouting is not densely filled and the grouting volume is too small, the grouting control is unreasonable, or the grouting pressure and the grouting volume are substantially changed. The main preventive measures include ensuring that the grouting construction parameters are reasonably determined and the grouting is injected in a timely manner, the grouting ratio and the quality of the grouting material are controlled, and shield tail sealing grease is injected regularly to ensure the applicability of the TBM sealing wire brush.

8.4.4.3 The pipeline is blocked during grouting

If the grouting pressure increases, grouting cannot be conducted, or pipe enclosing occurs, the main reasons are that the grouting intermittent time is too long to make the grout remain in the pipeline and the pipeline becomes blocked, the content of sand or other particles in the grouting material is too high, the grout has accumulated in corners of the pipeline or three-way positions, the double-liquid grouting control is unreasonable, the pressure of the A pipeline and the B pipeline are quite different and cause a plug when the material at one side enters the pipeline at the other side, or the pipeline is not cleaned in place. The main preventive measures include use of a circulation loop to ensure liquefaction of the grout when grouting is temporarily stopped, cleaning the pipeline in a timely manner when grouting is stopped for a long time and bentonite grout is used for precirculation during cleaning, designing the proportioning of the grouting material and the grouting construction parameter in a reasonable way, and cleaning the corners of the pipeline or the three-way positions regularly.

8.5 Backfill grouting process optimization technology and cases

To ensure a better grouting effect in the process of backfill grouting for a shield, it is necessary to control the grouting material and technology to meet a series of control requirements that the grout has good stability, the grout smoothly flows in the pumping, and the grout can be solidified quickly after injection, which can reach a certain early strength. The relationship among these control measure is often complicated, and each affects the others. Therefore the backfill grouting process for a shield is often generalized into a mathematical model and solved by a multiobjective programming method.

8.5.1 Solution of optimal proportioning

The multiobjective programming method, an important branch of optimization theory and method, is a mathematical method that was developed to solve multiobjective decision-making problems on the basis of linear programming. The general ideas of optimization study are as follows: (1) The optimization objective is proposed and described in a certain form; (2) optimization variables and various constraint conditions are determined, that is, the solution space of the problem is given; and (3) the optimization model is formed through the above two steps, and an appropriate optimization algorithm is chosen to solve the model [14].

In multiobjective programming problems the standard form of the mathematical model is

$$\begin{cases} \text{V-min} \;\; F(x) \\ \text{s.t.} \qquad g_i(x) \geq 0 \;\; (i = 1, 2, \ldots, m) \\ \qquad\quad\; h_i(x) = 0 \;\; (i = 1, 2, \ldots, l) \end{cases} \tag{8.7}$$

where V-min $F(x)$ refers to minimizing objective functions in vector forms; and the objective function $F(x)$ and constraint functions $g_i(x)$ and $h_i(x)$ can be linear functions and can also be nonlinear functions.

An absolute optimal solution of multiobjective programming problems generally does not exist [14]. In multiobjective programming problems, the commonly used solution methods include a constraint method, an evaluation function method, and an efficiency coefficient method. The process of solving the optimal proportioning will be briefly discussed below by an ideal point method in the evaluation function method under the constraint conditions.

The general process of solving the multiobjective programming problems by the ideal point method is shown in Fig. 8.11.

The essence of the ideal point method is to construct an evaluation function $h[F(x)]$ by using each objective function $F(x)$ so that the multiobjective programming problem is converted into a single-objective programming problem for solution and the optimal solution x^* of the single-objective problem can serve as the optimal solution of the multiobjective programming problem. The concrete implementation method is as follows: The single-programming problem optimal solution x_i^* and its objective function value f_i^* of each objective function are solved in multiobjective programming, the vector F^0 consisting of these objective

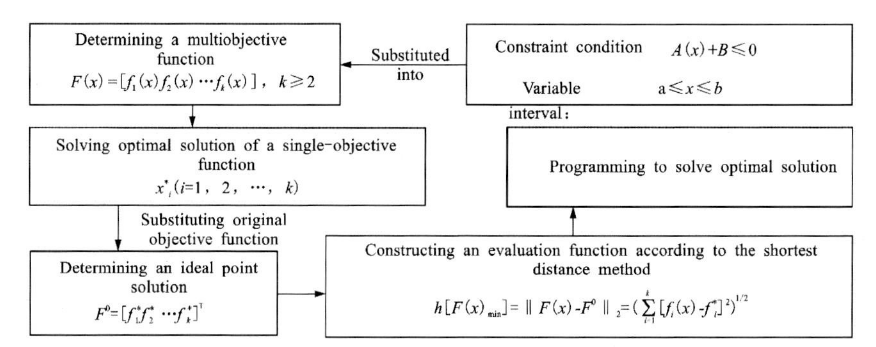

Figure 8.11 Solution process of the ideal point method.

functions is an ideal point of the multiobjective programming problems, and the ideal point is almost impossible to achieve. If a point x^* can be found in the feasible region R to make its corresponding $F(x^*)$ closest to the ideal point F^0, the optimal solution of the multiobjective programming problem will be achieved.

8.5.2 Cases

The river-crossing shield double-line tunnel on Wangjiang Road in Hangzhou, in China, has a total length of 1.837 km, and the buried depth is about 12—22 m. To alleviate the impact of the tunnelling waste soil on the urban environment in the process of shield tunnelling, a new backfill grouting material using shield waste soil was developed and applied in actual engineering. A multiobjective ideal point programming method was used to optimize and control the performance and proportioning of the grouting material. The overall idea is shown in Fig. 8.12.

At present, the mass ratios, including the water-binder ratio, cement-sand ratio, bentonite-water ratio, and ash-cement ratio, are the main influence factor of the single-liquid active grout. The water-binder ratio is the mass ratio of water to a gel material (cement and fly ash), the cement-sand ratio is the mass ratio of the gel material to sand, the cement-sand ratio is the mass ratio of the gel material to sand, and the ash-cement ratio is the mass ratio of fly ash to cement. To obtain the influence rule of each test factor of the grout, a multilevel test should be used, which is a significant amount of work. The cost will be lower if the uniform design test method is adopted, and this method is suitable for a multifactor multilevel test. For the uniform design method, please refer to the Uniform Design and Uniform Design Table written by Fang Kai-tai [15]. The site grout

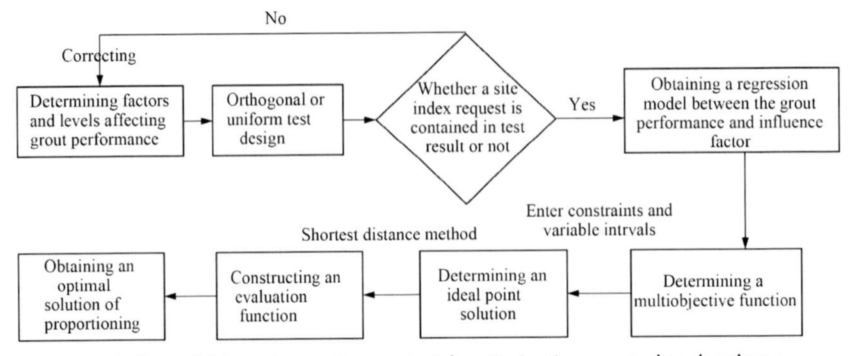

Figure 8.12 Overall idea of grouting material optimization control technology.

performance required value should be contained in each group of test results to ensure the correctness of the optimal proportioning results.

Step 1: In combination with the site proportioning of synchronous grouting slurry, it is determined that the test water-binder ratio ranges from 0.6 to 1.0, the cement-sand ratio ranges from 0.60 to 0.84, the bentonite-water ratio ranges from 0.08 to 0.24, the ash-cement ratio ranges from 1.8 to 2.4, and four-factor as well as five-level uniform tests are designed. From the uniform test results, the grout specific gravity of each test group is between 1.6 and 2.0, the shrinkage rate in 28 days is between 2.3% and 13.5%, the consistency is greater than 11.7 cm, the initial fluidity is between 11.5 and 37.0 cm, the bleeding rate in 3 hours is between 1.0% and 9.0%, the initial setting time is between 8 and 26 hours, and the unconfined compressive strength in 24 hours is between 0.17 and 0.92 MPa, in 7 days between 0.36 and 1.36 Mpa, and in 28 days between 0.54 and 2.70 MPa. By combining the performance results of the site synchronous grout, it is determined that almost all performance index required values are contained in the performance index test values of 15 test groups, which indicates that the optimal solution can be sought. The uniform design is reasonable, and the optimal mass ratio of the material can be obtained by further regression fitting as well as an ideal solution (Table 8.6).

Step 2: The relation equation needs to be established between the proportioning influence factor of the grouting material and each grout performance, which is the basis for subsequent optimization of the proportioning. Regression analysis, a fitting method, is widely used in fields of predictive modeling, relation seeking between variables, and so on. It

Table 8.6 Performance index requirements of backfill grouting slurry for tunnel on Wangjiang Road.

Inspection item	Technical performance requirements
Initial setting time	10–14 hours
Unconfined compressive strength	Greater than 0.2 MPa in 1 day, greater than 0.6 MPa in 7 days, and greater than 3.0 MPa in 28 days
Grout consistency	10–14 cm
Bleeding rate	Less than 5% in 3 hours
Stone shrinkage rate	Less than 8% in 28 days
Grout specific gravity	1.8–2.0
Initial fluidity of grout	220–240 mm

can ensure that all scattered points of the test data are distributed on this curve as much as possible.

According to the results of the uniform test on proportioning of the grout, seven key grout performance indexes, including initial fluidity J, bleeding rate B in 3 hours, stone shrinkage rate S in 28 days, initial setting time T, and unconfined compression strength (C_1, C_7, and C_{28}) at different ages are finally obtained for multielement regression analysis. A quadratic multielement regression model is adopted to further reflect the influence of four independent variables—water-binder ratio, cement-sand ratio, bentonite-water ratio, and ash-cement ratio—on each property of the grout. The theoretical regression model is shown in the following formula:

$$y_i = \sum_{k=1}^{4} b_{ik}x_{ik} + \sum_{j=1}^{4}\sum_{k=j}^{4} b_{ijk}x_{ij}x_{ik} + \varepsilon_i \tag{8.8}$$

where y_i is each property level of the grout; x_{ij} and x_{ik} are test factors; ε_i, b_{ik}, and b_{ik} are regression coefficients; and i is respectively equal to 1, 2, 3, ... 7.

By analysis software the regression equations are obtained as follows:

$$\begin{cases} f_J = 110.363 - 138.383x_1 - 112.344x_2 - 2.104x_4^2 \\ \quad + 151.906x_1x_2 - 28.169x_3x_4 + 21.043x_1x_4 \\ f_B = 59.899 - 106.475x_1 - 44.939x_2 + 60.612x_1^2 \\ \quad + 169.458x_3^2 + 53.211x_1x_2 - 103.695x_1x_3 \\ f_T = 168.216 - 358.256x_1 - 19.815x_4 \\ \quad + 155.741x_1^2 + 43.271x_1x_2 - 193.668x_2x_3 \\ \quad + 156.641x_1x_3 + 28.668x_1x_4 \\ f_S = 52.163 - 51.91x_1 - 63.615x_2 - 4.132x_4 \\ \quad + 78.416x_1x_2 - 33.476x_1x_3 + 5.772x_1x_4 \\ f_{C1} = 5.664 - 7.708x_1 - 1.055x_4 + 3.142x_1^2 \\ \quad + 0.08x_4^2 + 0.607x_1x_3 + 0.501x_1x_4 \\ f_{C7} = 7.889 - 9.064x_1 - 0.404x_2 - 1.589x_4 \\ \quad + 3.25x_1^2 + 0.123x_4^2 + 0.748x_1x_4 \\ f_{C28} = 15.950 - 18.618x_1 - 2.761x_4 + 6.264x_1^2 \\ \quad + 0.143x_4^2 + 1.403x_1x_4 \end{cases} \tag{8.9}$$

where f_S is stone shrinkage rate (%) in 28 days; f_J is initial fluidity (cm); f_B is bleeding rate (%) in 3 hours; f_T is the initial setting time (%); and f_C is the unconfirmed compression strength (MPa) after maintenance.

Step 3: In combination with the grout performance of this project, objective functions and constraint conditions are obtained as follows (10):

$$\text{Objective functions:}\quad \begin{cases} \min & f_J \\ \min & f_B \\ \min & f_T \\ \max & f_{C_1} \\ \max & f_{C_7} \\ \max & f_{C_{28}} \\ \min & f_S \end{cases} \quad \text{Constraint conditions:}\quad \begin{cases} 220\,\text{mm} \le f_J \le 240\,\text{mm} \\ f_B \le 5\% \\ 10\,\text{h} \le f_T \le 14\,\text{h} \\ f_{C_1} \ge 0.2\,\text{MPa} \\ f_{C_7} \ge 0.6\,\text{MPa} \\ f_{C_{28}} \ge 3.0\,\text{MPa} \\ f_S \le 8\% \end{cases} \tag{8.10}$$

The recommended optimized proportioning parameters for the grout are a water-binder ratio of 0.745, a cement-sand ratio of 0.84, a bentonite-water ratio of 0.161 and a ash-cement ratio of 2.014. The actual optimized proportioning parameter of each material can be further obtained.

References

[1] Ye F, Mao J, Ji M, et al. Research status and development trend of back-filled grouting of shield tunnels. Tunn Constr 2015;35(08):739−52.

[2] Tang J, Huang J, Zhou Y, et al. Study progress of synchronous grouting material for shield tunneling method. Mod Urban Transit 2016;(04):89−92.

[3] Zhao T. Research on grout test and application technology of synchronous grouting for slurry shield. Tongji University; 2008.

[4] Mao J. Study on the grouts diffusion mechanism of shield tunnel back-filled grouts based on filtration. Chang'an University; 2016.

[5] Ye F, Mao J, Ji M, et al. Research status and development trend of back-filled grouting of shield tunnels. Tunn Constr 2015;35(08):739−52.

[6] Xu Q, Wang Y, Fan Y, et al. Development and application of grouting materilas. 21st Century Build Mater 2010;2(01):58−62.

[7] JGT/T70−2009. Test method for performance on building mortar. Ministry of Housing and Urban-Rural Construction of the People's Republic of China, Beijing, China 2009.

[8] GB/T25182-2010. Grouting admixture for prestressed structure. Standardization Administration of the People's Republic of China, Beijing, China. 2010.

[9] GB/T2419-2005. Test method for fluidity of cement mortar. Standardization Administration of the People's Republic of China, Beijing, China. 2005.

[10] GB/T1346-2001. Test methods for water requirement of normal consistency setting time and soundness of the portland cements. Standardization Administration of the People's Republic of China, Beijing, China 2001.

[11] Mao S, Zhou Z. Synchronous grouting construction technology of slurry shield in water-rich sand layer. West-China Explor Eng 2014;26(05):182−4.

[12] GB 50446-2017. Code for construction and acceptance of shield tunnelling method. Beijing: China Construction Industry Press; 2017.

[13] Huang H, Liu Y, Xie X. Application of GPR to grouting distribution behind segment in shield tunnel. Rock Soil Mech 2003;(s2):353−6.

[14] San B, Wu Y, Wei D, et al. Multi-objective optimization of membrane structures. China Civ Eng J 2008;(09):1–7.
[15] Fang K. Uniform design and uniform design tables. Beijing: Science Press; 1994.

Exercises

1. According to the relationship between backfill grouting and shield tunnelling, what kinds of backfill grouting can be classified in terms of timeliness?
2. Briefly introduce similarities and differences between shield tail grouting and segment grouting, and discuss their respective advantages and disadvantages.
3. What are the commonly used backfill grouting materials? How do the grouting materials match with the stratum?
4. When the water-binder ratio, the cement-sand ratio, the bentonite-water ratio, and the ash-cement ratio are 0.745, 0.84, 0.161, and 2.014, respectively. Please determine the material masses of water, cement, fly ash, and bentonite contained in each 1000 kg of grout.
5. If the inner diameter of the tail shield of a shield machine is 11.46 m, the thickness of the tail shield is 70 mm, the inner diameter of the segment is 10.3 m, the thickness of the segment is 0.5 m, the injection rate is set to 150%, and the shield tunnelling speed is 50 mm/min, what is the optimal injection rate during shield grouting?

CHAPTER 9

Muck conditioning for EPB shield tunnelling and muck recycling

9.1 Reasons for muck conditioning

The shield tunnelling method has become the primary construction method for underground projects such as urban tunnels, owing to the relatively safe, fast, and economical construction process. Common shield machines include EPB shields and slurry shields. The slurry shield has the disadvantages of large space occupation, not being environmentally friendly, and having a high cost, so it is used considered mainly for underwater tunnel construction (e.g., tunnels going under a river or sea). For land tunnels, the EPB shield is most often used. Choosing the appropriate shield is the key to performing shield tunnelling safely and efficiently. Inappropriate shield selection can bring fatal disasters to tunnelling construction. However, using the appropriate shield type cannot guarantee that there will be no issues at all during tunnelling, since the ground the tunnel goes through may be complicated, and it is hard to change the hardware once shield tunnelling has begun. Common problems that may occur in shield tunnelling include clogging on cutterhead and cutters, water spewing from screw conveyors, and abrasion of cutters, resulting from the complication of geological conditions, not enough recognition of soil-conditioning mechanisms, and inappropriate techniques for soil conditioning.

Clogging (Fig. 9.1) usually happens in soil or weathered rock with high content of clay minerals. It will lead to increases in the shield thrust and cutterhead torque and decreases in tunnelling efficiency and may even cause the clogging of cutterhead openings, eccentric abrasion of cutters, and disturbance of muck discharging out of the excavation chamber, requiring workers to stop the shield machine, open the chamber to remove the clogs, and replace the cutters. Engineers propose various ways to control clogging, such as optimizing the configuration of cutterhead openings and cutters, washing the clogs on the cutterhead with water under high pressure, or even using electro-osmosis to polarize the soil particles and present the soil from sticking to the cutterhead or cutters [1].

Shield Tunnel Engineering
DOI: https://doi.org/10.1016/B978-0-12-823992-6.00009-6

Figure 9.1 Clogging on cutterhead and cutters. (a) Clogging on the cutterhead. (b) Eccentric abrasion of cutters due to clogging on the cutter seat.

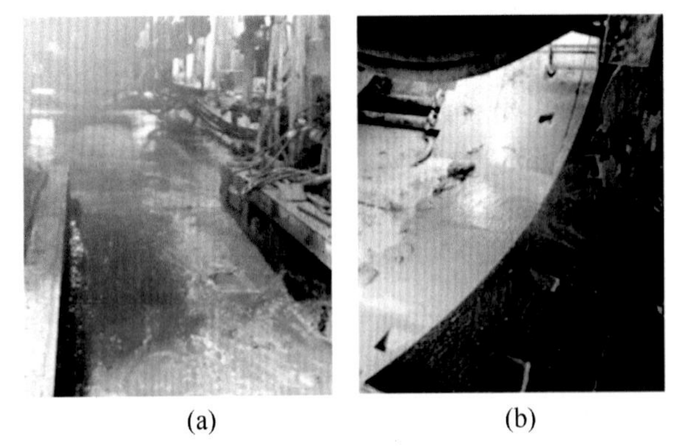

Figure 9.2 Water spewing from a screw conveyor. (a) Slurry on the bottom of the tunnel. (b) Slurry on the segments.

Although these methods do work in some situations, they cannot effectively solve the clogging problem for long-distance shield tunnelling in soils with high clay content.

Water spewing (Fig. 9.2) usually happens in highly permeable coarse-grained soils under high water pressure. Water spewing from screw conveyors will lead to the muck discharging ratio being out of control and the pressure inside the soil chamber being tremendously reduced, resulting in excessive settlement of the ground and even instability of the excavation face. Gentle water spewing from screw conveyors will not threaten the stability of the ground, but frequent water spewing could eject slurry inside the shield machine, staining the equipment and construction environment. To maintain a good water seal of the assembly joints, the slurry on the segments has to be washed away using clean water, significantly affecting the construction efficiency and presenting huge problems for construction workers. Attempting to modify the screw conveyor on the site could prevent the water spewing, such as by installing a pressure-retaining pump at the outlet of the screw conveyor, installing dual gates at the inlet of the screw conveyor, or using a secondary screw conveyor. These modifications relieve water spewing to some degree. However, the pressure pump will affect the soil discharging efficiency, and the dual gates must be closed together along with stopping of the shield tunnelling to prevent water spewing. These two measures are not appropriate for normal shield tunnelling but only for an emergency situation. A secondary screw conveyor can reduce the hydraulic gradient of the soil inside the primary screw conveyor. However, a secondary screw conveyor usually cannot be installed because of limited space inside the shield machine.

Abrasion (Fig. 9.3) usually happens in coarse-grained soils with high quartz content, especially for highly or completely weathered granite and sand pebble. Cutter abrasion will reduce the efficiency of shield tunnelling and may even result in having to stop the shield machine and replace the cutters. The replacement process needs at least one week and could last one to two months owing to dewatering or reinforcement of the ground, which significantly affects the construction progress. In addition, it very often causes ground instability accidents due to the replacement. Besides soil conditioning, two things can be done to reduce cutter abrasion. First

(a) (b) (c)

Figure 9.3 Cutter abrasion. (a) Granite with high quartz content. (b) Cutting drum damage. (c) Cutting circle damage.

highly abrasion-resistant materials can be used for the cutters, which are typically cemented carbide. An increase of carbon content in cemented carbide will increase its abrasion resistance. However, excessive carbon content will reduce the ductility of the cemented carbide, making the cutters more brittle. Second, using assistant air pressure and under-pressure modes for the excavation chamber can reduce the contact area between the cutters and soil and thus reduce the friction. However, pressure leakage inside the soil chamber causing serious risks such as ground instability could occur in using assistant air pressure and under-pressure modes for pretreated ground for solitary rock blasting or passing through adjacent buildings and other structures. Therefore these two measures have limited applications. However, muck conditioning could be an important way to reduce the cutter abrasion.

Besides clogging of cutterhead and cutters, water spewing from the screw conveyor, and cutters abrasion, inappropriate soil conditioning will also cause clogging of the soil chamber, difficulty in removing muck, large fluctuations in soil chamber pressure, and so on. The smooth process of shield tunnelling mostly relies on muck conditioning. By injecting conditioning agents in front of the cutterhead, inside the soil chamber, and in the screw conveyor to refine the state of the soil, problems such as clogging of clay soil, water spewing in water-rich and highly permeable soil, and cutter abrasion in soil with high quartz content can be avoided. Since the EPB shield was first used, researchers have conducted a significant number of studies on theories and applications of muck conditioning. These have reduced the construction risks of using the EPB shield, expanded the appropriate geological conditions for the application of the EPB shield, and improved the construction techniques of shield tunnelling.

9.2 Properties of the shield muck

Properties of the shield soil significantly influence the safety and smooth proceeding of shield tunnelling. They are closely related to the properties of the original rock and soil in the ground, as well as muck conditioning. Rock is an aggregate of minerals with certain types of structure (including crystalline and amorphous). Soil is a mixture of loosening particles generated in different natural environmental conditions from crust surface rocks, which are weathered, compressed, deposited and then corroded, transported, and accumulated via crust movement, water flow, glacier, and wind. Muck from EPB shields refers to the loosening particles cut from the strata by shield cutters. It can be categorized as natural soil but also has its unique physical and mechanical properties.

9.2.1 Composition of muck

The muck from EPB shields mainly consists of water, soil, air, and conditioning agents. It includes three phases, including solids, liquids, and air, and is categorized as unsaturated soil. The properties of each phase, their ratios, and their interaction between phases determine the physical and mechanical properties of the muck.

9.2.1.1 Solids

Solids of the muck consist of soil particles in the original ground (or soil particles cut from rock by cutters) and minor amount of solids from conditioning agents. Soil particles form the structure of soil and play a dominant role in its physical and mechanical properties. Since the muck properties vary significantly owing to the particle size differences, the grain gradation of soil greatly affects its physical and mechanical properties. To guarantee smooth removal, the shield muck typically needs to be well graded. When more coarse aggregates are present in the soil, fine aggregates such as bentonite should be injected to increase the muck plasticity.

9.2.1.2 Liquids

Liquids in the muck include water from the soil and conditioning agent, and can be classified as bound water or free water. The water attracted by the electrical fields and bonded to the surface of the soil particles is called bound water. Bound water can be classified as strongly bound water or weakly bound water, and the existence of weakly bound water is the reason that clay has plasticity. The water in the soil particles that is not affected by the attraction of electric fields is called free water. The ideal clayey muck mainly contains weakly bound water, which makes the soil plastic, while the free water between particles in the ideal coarse-grained soil leads to the fluidity of the muck.

9.2.1.3 Air

Air in the muck include the air in the soil skeleton, air in the foams used for muck conditioning, and air injected during the shield tunnelling process (non-full-chamber mode). The air in the muck can be classified as free air or enclosed air, depending on the connection to the atmosphere. Free air connects to the atmosphere and has an insignificant influence on the properties of the muck. The pressure of the enclosed gas is related to its volume. Foam is injected into the muck mainly to increase the volume of the enclosed air inside the muck, thus increasing the compressibility of the muck.

9.2.2 Physical and mechanical properties of muck

The ideal muck for EPB shield tunnelling should have appropriate physical and mechanical properties to support the tunnel face in the excavation chamber and to be removed smoothly by the screw conveyor. The muck needs appropriate plasticity for smooth removal. Currently, most researchers believe the slump value of the ideal muck state should be 10–20 cm. The cohesive index I_c for muck soil with high clay content should be 0.4–0.75, where I_c is calculated as follow:

$$I_c = \frac{w_L - w}{w_L - w_p} \tag{9.1}$$

where w_L is the liquid limit of the muck soil; w_P is the plastic limit of the muck soil; and w is the water content of the muck soil.

To reduce the torque of the EPB shield during tunnelling, as well as the abrasion of the cutters by the ground, the conditioned soil should possess relatively small shear strength, the undrained shear strength normally being 10–25 kPa[1] [2]. The soil with high clay content should also be less adhesive to prevent clogging during the shield tunnelling process. When the shield is tunnelling in water-rich soil, the soil should possess resistance to permeability to avoid water spewing, with a permeability coefficient less than 10^{-5} m/s [3]. In addition, when the compressibility of the soil is too small, the tunnelling speed of the shield and the rotating speed of the screw conveyor are both reduced, causing a large fluctuation in pressure in the excavation chamber. By contrast, when the compressibility of the soil is too high, resulting in too high fluidity of the soil, muck spewing from the screw conveyor is prone to happen, affecting the control of muck removal by screw conveyor. Therefore the muck should also have appropriate compressibility.

Because shield muck are loosened soil cut from the ground by the shield cutters, there is a difference in volume between the shield muck and the original soil. The ratio of the volume of shield muck against that of the original soil, called the loosening coefficient, is influenced by the original soil, the configuration of the shield cutters, the shield tunnelling parameters, and soil conditioning. Its theoretical equation is

$$K = K_1 \cdot K_2 \cdot K_3 \cdot K_4 \cdot K_5 \tag{9.2}$$

where K_1 is the loosening coefficient of the soil generated from being excavated to entering the excavation chamber; K_2 is the loosening coefficient of the soil generated by passing through the screw conveyor; K_3 is the loosening coefficient of the soil generated by falling freely from the outlets of the screw conveyor to the belt conveyor; K_4 is the loosening coefficient of the soil generated by passing through the belt conveyor (may be less than 1); and K_5 is the loosening coefficient of the soil generated by falling freely from the end of the belt conveyor to the muck carriage.

To determine the loosening coefficient, the unit weight of the dry soil is obtained first by oven drying a unit volume of soil (also removing the effect of added bentonite and foams). The loosening coefficient of the soil is then the ratio of the unit weight obtained from the test to the unit weight obtained from the geological survey report. Chinese *Code for Design of General Layout of Industrial Enterprises* (GB 50187-2012) [4] provides the reference values for the loosening coefficients of the soil, as shown in Table 9.1, where the soil grade is classified according to the soil classification method with 16 grades in total, the original loosening coefficient is used to calculate the volume of cut soil (in loosest state), and the final loosening coefficient is used to calculate the volume of filled soil (in densest state).

9.3 Types and technical parameters of soil-conditioning agents

9.3.1 Types of soil-conditioning agents

Soil-conditioning agents used in shield tunnelling can be classified as water, clay minerals, foaming agents, dispersants, flocculants, or water-absorbing agents on the basis of their functions. Clay minerals are mainly bentonite; foaming agents are mainly surfactants; and dispersants, water-absorbing agents, and flocculants are mainly polymers. These materials may be used alone or together. The applicable ground conditions are summarized as follows.

9.3.1.1 Water

The water content of the soil has a great influence on the soil properties, and its conditioning is mainly reflected in the following aspects: For coarse-grained soil and rock ground, by injecting water into the shield cutterhead and excavation chamber, the cutter abrasion can be reduced,

Table 9.1 The loosening coefficient of soils.

Soil category	Soil Grade	Soil name	Original loosening coefficient	Final loosening coefficient
Category I soil (loosened soil)	I	Sand with low clay content, powder humus soil and loosened planting soil; peat (silt) (exclude planting soil and peat)	1.08–1.17	1.01–1.03
		Vegetable soil, peat	1.20–1.30	1.03–1.04
Category II soil (normal soil)	II	Moisturized clay soil and loess; soft saline soil and alkaline soil; piled soil and planting soil containing construction materials debris, crushed stones, and pebbles	1.14–1.28	1.02–1.05
Category II soil (hard soil)	III	Medium dense clay soil and loess; clay soil and loess containing crushed stones, pebbles, and construction materials debris	1.24–1.30	1.04–1.07
Category IV soil (granular hard soil)	IV	Hard and highly dense clay soil and loess; medium dense clay soil and loess containing crushed stones and gravel (stones with volume within 10%–30% and weight below 25 kg); hardened heavy saline soil; soft marlstone (excluding marlstone and opal)	1.26–1.32	1.06–1.09
		Marlstone and opal	1.33–1.37	1.11–1.15
Category V soil (soft soil)	V–VI	Hard carboniferous clay; low cementation conglomerate; soft and joint-rich limestone and shell limestone; hard chalk; medium hard shale and marlstone.	1.30–1.45	1.10–1.20
Category VI soil (secondary-hard rock)	VI–IX	Hard argillaceous shale; hard marlstone; breccia granite; marl limestone; clay sandstone; micaceous shale and sandy shale; weathered granite, gneiss, and normal rock; talcum serpentinite; highly dense limestone; siliceous cementation conglomerate; sandstone; sandy calcareous shale.		
Category VII soil (hard rock)	X–XIII	Dolomite; marble; solid limestone, calcareous and quartz sandstone; hard sandy shale; serpentinite; coarse-grained syenite; andesite and basalt with traces of weathering; gneiss; trachyte; medium coarse granite; solid gneiss, trachyte; diabase; silt; medium coarse normal rock		
Category VIII soil (extra-hard rock)	XIV–XVI	Solid fine-grained granite; granite gneiss; diorite; solid porphyrite, amphibolite, gabbro, quartzite; andesite; basalt; the most solid diabase, limestone and diorite; olivine basalt; particularly solid gabbro; quartzite and porphyrite	1.45–1.50	1.20–1.30

Translated from China Metallurgical Construction Association, Code for design of general layout of industrial enterprises (GB50187-2012). Beijing: China Plan Press; 2012.

and the temperature of the cutter, cutterhead, and soil can also be reduced, as well as improving the fluidity of the soil. For soils with high clay content, injecting water into the shield cutterhead and excavation chamber can not only change the workability of the soil, facilitating shield soil removal, but also reduce its adhesion and prevent the soil from adhering to the cutterhead or soil chamber partitions; by injecting water into the cutterhead and the soil chamber, the soils possess an appropriate water content, which works with other conditioning agents jointly for soil conditioning to achieve the best conditioning state. For example, foam can be injected only when the soil contains an appropriate amount of water; otherwise, the foam dissipates very easily, and it is difficult to achieve the desired conditioning state.

9.3.1.2 Foaming agents

Foaming agents, also known as blowing agents, can reduce the surface tension of the liquid and produce a large amount of uniform and stable foam by mixing with pressurized air. Ingredients of a foaming agent include surfactants, foam stabilizers, and so on. The molecules of surfactants contain two parts, hydrophilic cluster and hydrophobic cluster, which tend to accumulate at the interface between liquid and gas in the solution, forming a thin molecular film to reduce the surface tension of the liquid and make the solution capable of foaming. The main function of the foam stabilizer is to slow down the dissipation of the foam and stabilize the foam. According to the effect, foaming agents can be categorized as general-purpose foaming agents or dispersing foaming agents. General-purpose foaming agents are used mainly for soils with low clay content, and dispersing foaming agents are used mainly for soils with high clay content.

The foaming agent is formulated into a solution based on a certain concentration, and a large amount of foam can be generated through foam generators. The generated foams, once mixed with the soil, can improve the performance of the soil for shield tunnelling. The conditioning effects of foams on the soil are mainly reflected in the following aspects: The foams work as lubricating agents after being injected into the soil, can significantly reduce the internal friction angle and improve the workability of the soil, facilitate shield soil removal, effectively establish the soil pressure in the excavation chamber to balance earth and water pressure on the excavation surface, and reduce the energy consumption of the shield machine. Because the foams fill the pores between the soil particles, the impermeability of the conditioned soil can be significantly improved. The foam-conditioned soil can form a cushion layer in the excavation chamber, which is similar to an impervious but compressible "sponge," resulting in improved compressibility of the soil. When the muck discharging rate by screw conveyor changes suddenly, the sensitivity of the soil pressure in excavation chamber is reduced, owing to the cushion effect, and it helps to maintain the stability of the excavation face [5–9]. A dispersive foaming agent can also reduce the adhesion between the particles, thereby preventing flocculation or agglomeration of the soil and reducing the possibility of muck clogging on the cutterhead in the soil chamber.

9.3.1.3 Dispersants

Dispersants are substances that disperse another substance in a medium such as water to form a colloidal solution. Their main function is to reduce the adhesion between particles and prevent flocculation or agglomeration. Dispersants are generally divided into two categories: inorganic dispersants and organic dispersants. Commonly used inorganic dispersants include silicates and alkali metal phosphates (e.g., sodium tripolyphosphate, sodium hexametaphosphate, and sodium pyrophosphate). Organic dispersants include cellulose derivatives, polycarboxylates, and guar gum. Currently, the commonly used dispersants for shield soil conditioning include cellulose derivatives and polycarboxylates. The dispersant can weaken the connection and release the bound water between the clay particles, thereby reducing the adhesion of soil with high clay content and reducing the possibility of muck clogging in the shield machine [10].

9.3.1.4 Clay minerals

Clay minerals are the main components in some conditioning agents when shield muck is coarse-grained soil. Their main function is to increase the content of fine particles, reduce the internal friction angle between coarse particles, and generate cohesive force, thus improving the continuity of the soil, increasing the fluidity of soil particles, and improving its impermeability. The clay mineral that is commonly used for shield soil conditioning is bentonite. Bentonite is a nonmetallic clay mineral with montmorillonite as its main component. Montmorillonite has a strong adsorption function, making bentonite capable of significant expansion. From the perspective of microstructure, bentonite particles are inorganic substances with a particle size of less than 2 μm. When the content of Na^{+} or Ca^{2+} in the bentonite accounts for more than 50% of the total exchangeable cations, it is called sodium-based bentonite or calcium-based bentonite, respectively. Sodium-based bentonite has greater water absorption and expansion ratio, higher cation exchange capacity, and better water dispersibility as well as better thixotropy, viscosity, lubricity, and thermal stability of its colloidal suspension. After hydration, bentonite forms an impermeable plastic colloid, which occupies the pores between the soil particles that are in contact with the bentonite, forming a dense impermeable layer, thereby reducing permeability [11].

The conditioning effects of bentonite on the soil are as followings: (1) The bentonite slurry that is injected into the excavation chamber and in front of the cutterhead will penetrate into the excavation face under pressure. The fine particles in the slurry will form a filter cake with a certain thickness in front of the excavation face during the penetration process, mainly composed of cemented and consolidated bentonite. This low-permeability filter cake can reduce water seepage force, ensures that the ground can maintain stability in front of the excavation face, and controls surface settlement. (2) The bentonite slurry is mixed with the excavated soil in the chamber, which increases the content of fine particles in the soil and improves the impermeability of the soil. (3) Owing to the cohesive property of the bentonite slurry, it will produce cohesiveness when mixed with the soil, which improves the workability of the muck and facilitates the removal of the muck. (4) The injection of bentonite slurry can play a role in suspending coarse particles in the soil chamber to facilitate the muck removal.

9.3.1.5 Flocculants

Flocculants are agents that make fine and subfine solids or colloids suspended in the solution form large loose flocs through bridging (Fig. 9.4),

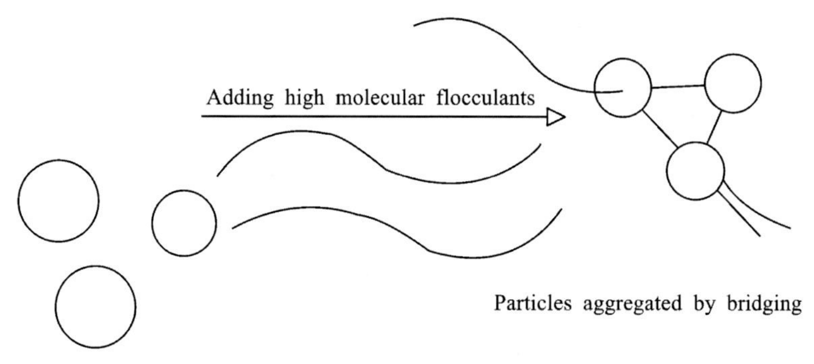

Figure 9.4 Schematic drawing of the mechanism of soil conditioning by flocculants.

thus achieving solid-liquid separation. The most commonly used flocculants for soil conditioning are polyacrylamide, carboxymethyl cellulose (CMC), and polyanionic cellulose (PAC). Soil conditioning by flocculants is used mainly for EPB shield tunnelling in water-rich ground. When the permeability of the soil is large, water spewing is prone to happen at the outlet of the screw conveyor. Flocculants injected into the soil chamber and screw conveyor can aggregate the particles in the soil and improve its plasticity, facilitating the formation of a soil plug in the screw conveyor and stopping water. When the pores in the soil are large, it is difficult to achieve the desired conditioning effect by using bentonite or flocculants alone. Adding flocculants (CMC or PAC) and bentonite together can cause the formation of larger flocs after the reaction to fill the pores in the soil and stop water [3].

9.3.2 Technical parameters and their determination for soil conditioning

9.3.2.1 Technical parameters of conditioning agents

Technical parameters of foaming agents

In shield construction the main parameters of a foaming agent are concentration, foam expansion ratio (*FER*), half-life, and foam injection ratio (*FIR*). The concentration of the foaming agent can significantly affect the *FER* and stability of the foam. The concentration is usually determined according to the manufacturer's recommendations or on-site tests and is generally 0.5%–5% [2].

The *FER* is a critical parameter that affects the behaviors of foams. Generally speaking, the larger the *FER*, the better the foam performance, but a *FER* that is too large will decrease the foam's stability. Therefore the *FER* is often controlled to 10–20 for soil conditioning. The calculation equation for *FER* is as follows:

$$FER = \frac{V_f}{V_L} \tag{9.3}$$

where V_f is the volume of foams under working pressure, *L*; and V_L is the volume of the foaming agent solution, *L*.

The foam dissipation ratio is the ratio of the mass of the dissipated foam to the mass of the total foam. This parameter reflects the change of the foam over time and is one of the important parameters to evaluate the stability of the foam. When the dissipation ratio *r* reaches 50%, the time required for half of the foam to dissipate is called the half-life [2,12].

According to engineering experience, the half-life of foam should exceed 5 minutes to meet the requirements of shield construction [13]. When the foam is mixed with the soil, the half-life of the foam will be greatly increased, owing to the slowdown of drainage, coarsening, and liquid film broken for the foam. When the foam is under certain pressure, its dissipation ratio (r) will also be greatly decreased [14]:

$$r = \frac{M_d}{M_0} \tag{9.4}$$

where M_0 is the mass of original foaming solution (g); and M_d is the mass of dissipated foams (g).

FER and half-life tests should be conducted at the shield construction site for determining the foaming agent concentration. When the *FER* is greater than 10 and the foam half-life is more than 5 minutes, the foaming agent concentration is appropriate for EPB shield tunnelling.

Technical parameters of clay minerals

The clay mineral agent used in the field is mainly bentonite slurry. The common types of bentonite in engineering can be categorized as calcium-based bentonite or sodium-based bentonite according to its composition. The technical parameters of bentonite slurry mainly include concentration and viscosity. When the slurry concentration is too high, its pumping performance is poor, and it is difficult to transport it to the soil chamber and the front of the cutterhead. When the slurry concentration is too low, it is difficult to effectively fill the pores between the soil particles. The concentration of bentonite slurry used on-site is generally 10%–20%. Kusakabe et al. [15] proposed the following equation for calculating the bentonite concentration:

$$D = a \times (30\text{-}p_{0.074}) \times \alpha + (40 - p_{0.25}) \times \beta + (60 - p_{2.0}) \times \gamma \tag{9.5}$$

where D is the slurry concentration; $p_{0.074}$ is the percentage of particles with particle size less than 0.074 mm; $p_{0.25}$ is the percentage of particles with particle size less than 0.25 mm; $p_{2.0}$ is the percentage of particles with particle size less than 2.0 mm; α, β, γ are three parameters, which $\alpha = 2.0$, $\beta = 0.5$, and $\gamma = 0.2$. When the value in parentheses in the equation is less than zero, it should be set to zero.

The value of coefficient a depends on the uneven coefficient U_c. When $U_c > 4$, $a = 1.0$; when $3 < U_c < 4$, $a = 1.1$; and when $1 < U_c < 3$, $a = 1.2$.

For shield construction, increasing the viscosity of the slurry can change the cohesion and internal angle values of the soil, facilitate filter cake formation on the excavation face, ensure the stability of the excavation face, prevent the deposition of gravel in the slurry tank, and facilitate soil removal. However, if the viscosity of the slurry is too large, its preparation and transportation are not easy, and the cost increases. For ground that is prone to water spewing at the outlet of screw conveyor, reasonably increasing the viscosity of the slurry can reduce the permeability of the muck and avoid water-spewing phenomenon. During construction the appropriate viscosity range can be determined by Marsh funnel based on specific ground conditions.

Technical parameters of dispersants and flocculants

The main technical parameter of dispersants and flocculants is concentration for muck conditioning. Concentration refers to the ratio of the volume of the conditioning agent to the volume of the solution before the conditioning agent is added. Since the chemical composition and concentration of the conditioning agents produced by different manufacturers are different, concentration is generally determined according to the parameters provided by the manufacturer and combined with the soil-conditioning effect on site.

9.3.2.2 Conditioning parameters and determination for soil

Conditioning parameters of foaming agent

One conditioning parameter of the foaming agent is the foam injection ratio (*FIR*), which is the ratio of the volume of the foam to the volume of the excavated soil:

$$FIR = \frac{V_F}{V_S} \times 100\% \tag{9.6}$$

where V_F is the volume of the added foam; and V_S is the volume of the excavated soil.

When the shield is tunnelling normally, the muck should first meet the workability requirements. Currently, the slump test is mainly used to evaluate the muck workability. When the shield is tunnelling in water-rich soil, the soil should also possess impermeability. In determining the FIR, the permeability test should also be conducted to ensure that water spewing will not happen during shield tunnelling. Kusakabe et al. [15] pointed out that the empirical value of the FIR for sand is

$$FIR = \frac{a}{2}\left[(60 - 4.0X^{0.8}) + (80 - 3.3Y^{0.8}) + (90 - 2.7Z^{0.8})\right] \tag{9.7}$$

where X is the percentage of particles with particle size less than 0.074 mm; Y is the percentage of particles with particle size less than 0.25 mm; and Z is the percentage of particles with particle size less than 2 mm.

When the value in brackets in Eq. 9.7 is less than zero, it should be set to zero. The value of coefficient a depends on the uneven coefficient U_c. When $U_c > 15$, $a = 1.0$; when $4 < U_c < 15$, $a = 1.2$; and when $U_c < 4$, $a = 1.6$.

The *FIR* of the sand given by Eq. 9.7 is for reference only. In field applications, slump tests and penetration tests should be considered to determine the final appropriate *FIR*.

Conditioning parameters of clay minerals

For EPB shield tunnelling, it is generally necessary to inject clay minerals into soil with low clay content to supplement the fine particles so that the muck will have suitable plasticity and impermeability. The field-conditioning parameter of clay minerals is the slurry injection ratio, which is the volume ratio of slurry to excavated soil. For different soils, this ratio varies greatly. The slurry injection ratio that is used for the field is generally determined by slump tests and permeability tests based on the specific geological conditions.

Conditioning parameters of dispersants and flocculants

The main conditioning parameter of dispersants and flocculants is the adding ratio. The adding ratio of conditioning agents refers to the volume ratio of the agent solution to the soil. Since the chemical composition and concentration of the conditioning agents produced by different manufacturers are different, the ratio is generally determined according to the parameters provided by the manufacturer and combined with the soil conditioning effect on site.

9.3.3 Soil adaptability of conditioning agents

In practical applications, the types and technical parameters of the conditioning agents should be appropriately determined according to the geological and hydrological conditions. The soil adaptability and conditioning characteristics of different conditioning agents are shown in Table 9.2. Note that in the actual soil-conditioning process, depending on the conditioning mechanisms of different agents, it is sometimes necessary to combine two or more agents to achieve the desired conditioning effect.

Table 9.2 Summary table of soil adaptability of conditioning agents.

Agent Type	Adaptive soil	Limitations	Precautions for on-site application
Water	All	Pay attention to the amount of injected water for water-rich soil.	The amount of injected water needs to be adjusted according to the soil state.
Foam agent	All	For formations lacking fine-grained particles and under high water pressure, it is difficult to ensure the good plasticity and impermeability of the soil by using a foaming agent alone.	Pay attention to the concentration of foaming agent and foaming pressure, and adjust the foam injection flow rate according to the soil removal situation.
Dispersants	With high clay content	The dispersion effect is better for soils with high clay content. Dispersion effect takes time, and the conditioning cost is high. Generally, it is combined with a foaming agent.	Pay attention to the selection of dispersant and the time required for effective dispersion of the soil.
Clay minerals	Lack of fine-grained particles	For soils with a large content of coarse-grained particles such as sand and pebbles, the plasticity of the conditioned soil is not desirable.	Pay attention to the bulging time, and adjust the injection amount in time to prevent clogging on the cutterhead and cutters.
Flocculants	Water-rich	When the fine-grained particles in the soil are insufficient, the plasticity and impermeability of the conditioned soil are not desirable.	Pay attention to the flocculant concentration and the conditioning ratio

9.4 Soil conditioning systems

The shield soil conditioning system mainly includes the foam injection system and clay mineral injection system. In addition, the foam and clay mineral systems are modified for adding dispersant and flocculant based on the ground conditions.

9.4.1 Foam injection system

The shield soil foam conditioning system usually adopts the single-pipe single-control method, in which each pipeline has a separate foam generator. Fig. 9.5 shows a schematic diagram of a common shield soil conditioning system. The concentration of foam mixture, *FER*, and the injection flow rate of foam for each pipeline are set on the operation panel. The control system controls each pipeline pump to deliver a certain proportion of the original foam liquid and water to mix in the mixed solution tank. Then the system calculates the mixed solution flow and air flow of each pipeline on the basis of the *FER* and foam injection flow rate. The flow meter of foam mixed solution and the air pressure control valve are controlled by the programmable logic controller system to adjust the mixed solution flow and air flow. The mixed solution and air are mixed in the foam generator to generate foam, which is injected at the front of the cutterhead and in the soil chamber through the pipelines.

The foam generator is the main device for generating foams. The generator squeezes the foaming agent solution under high air pressure

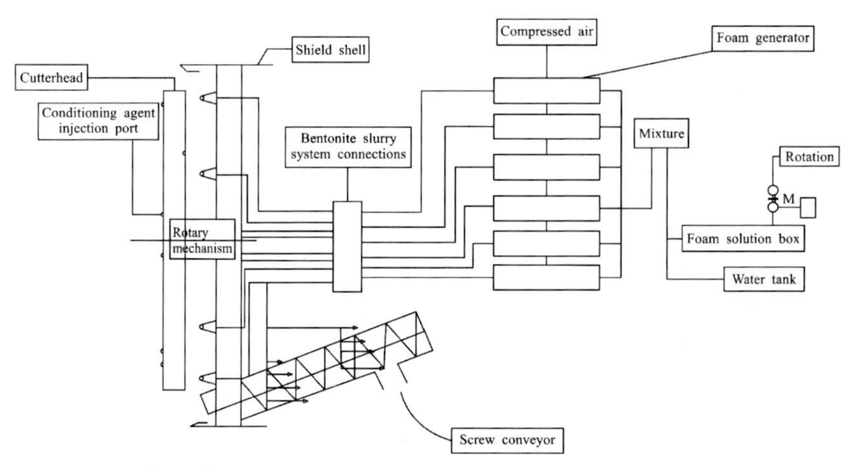

Figure 9.5 Soil-conditioning system.

through the fine porous structures to generate foams. The type of porous structure in the foam generator is classified as the grid type, consisting of multilayer grids; the film type, consisting of multilayer porous films; and the granular type, consisting of densely compacted granular materials (Fig. 9.6). A foam generator on site is shown in Fig. 9.7.

9.4.2 Clay mineral injection system

The clay mineral injection system (Fig. 9.8) is composed of the slurry tank, the squeeze pump, and the associated flow meters, sensors, and so on. The pipelines for the clay mineral injection system can be shared with the foam system and water nozzle at the front of the equipment bridge.

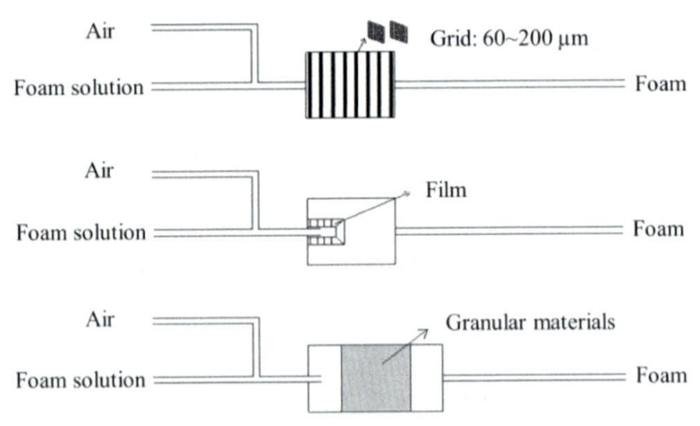

Figure 9.6 Types of foam generators.

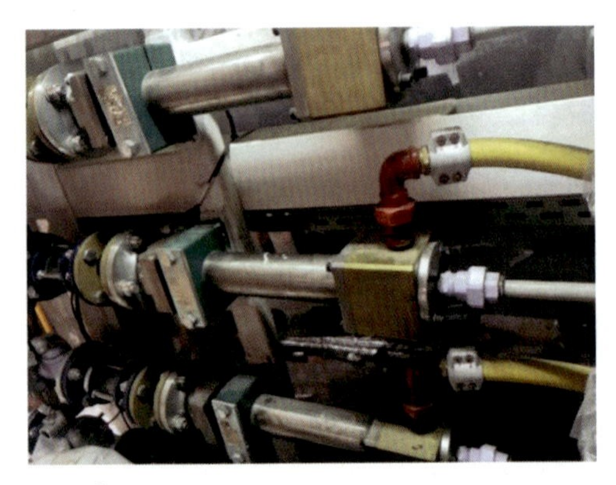

Figure 9.7 Foam generator.

9.4.3 Dispersant and flocculant injection system

Dispersant and flocculant injection systems are generally shared in the conditioning pipelines with clay mineral injection systems. When bentonite is required at the same time, because the synchronous grouting in China often uses a single-liquid grout, the dual-liquid stirring system is basically idle. Dispersant and flocculant solutions are often mixed in the liquid A (cement slurry) mixer and then injected through the foam pipeline or clay mineral pipeline. Fig. 9.9 shows the independent polymer injection system used in a project. After the system mixes the dispersant solution by a separate mixing tank, it is injected into the screw conveyor,

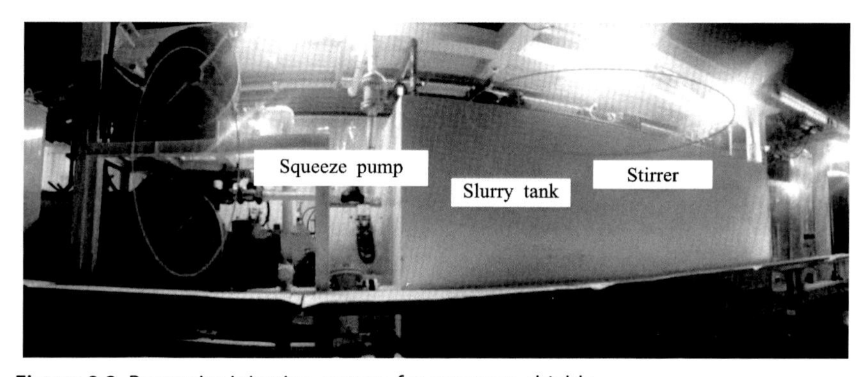

Figure 9.8 Bentonite injection system for common shields.

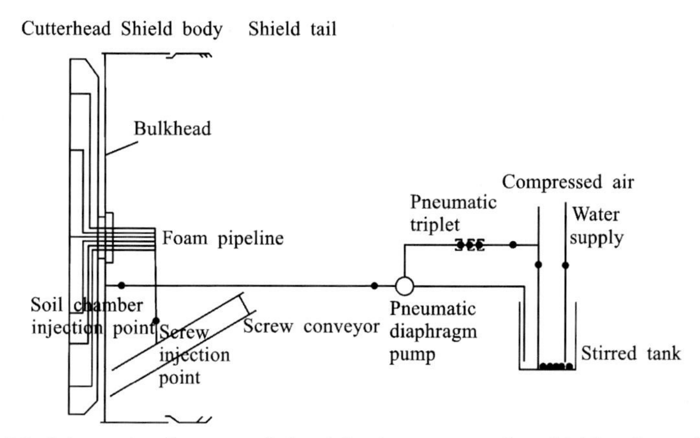

Figure 9.9 Schematic diagram of the injection system for shield soil-conditioning polymer.

soil chamber, front of the cutterhead, and other positions through the foam pressurized pump through pipelines.

9.5 Index properties and its determination method of conditioned soil

To improve the soil adaptability of the EPB shield and facilitate the muck discharging and pressure control of the excavation chamber, it is often necessary to perform soil conditioning. The ideal soil needs to have good fluidity, suitable plasticity, low shear strength and adhesion, small permeability coefficient, and certain compressibility.

9.5.1 Fluidity and plasticity

9.5.1.1 Slump test

The slump test was originally an important method to test the workability of freshly mixed concrete. Many researchers have introduced this method into the field of soil conditioning and use the slump value to evaluate the fluidity and plasticity of the conditioned soil. Table 9.3 shows the ideal slump values of the conditioned soils suggested by different researchers. The ideal one varies with soil type and tunnelling mode for EPB shield tunnelling.

Although the slump test has been widely used to evaluate the fluidity and plasticity of the excavated soil, the slump range of the ideal soil varies with relatively large differences. This is mainly because the ideal state of the soil is affected by the soil type, the chamber pressure, the tunnelling mode, and so on. For example, when the shield is tunnelling with full-chamber state, the state of the ideal soil may be completely different from the state with half-chamber state. In addition, the slump test has certain limitations, such as that it cannot directly reflect the adhesion of the soil

Table 9.3 Ideal slump ranges for soils.

Author	Slump (mm)	Author	Slump (mm)
Ye et al. [16]	170–200	Pena [17]	100-150
Quebaud et al. [12]	120	Zhang [18]	100-150
Qiao [8]	100-160	Vinai et al. [19]	150-200
Jancsecz et al. [20]	200-250	Peila et al. [21]	150-200
Leinala et al. [22]	50	Budach and Thewes [22]	100–200
Limited [17]	100-150	Qiu et al. [23]	195-210

with high clay content to metal materials such as cutters, and the maximum particle size of the test soil sample is required not to exceed 40 mm. In addition, the slump test is relatively subjective when used to evaluate the fluidity and plasticity of the soil. For example, it is necessary to observe whether the test soil is bleeding water or foam and whether the soil and the conditioning agents are mixed uniformly.

9.5.1.2 Fluidity test

A fluidity test for cement mortar can be used to evaluate the fluidity and plasticity of clayey soil with coarse and fine particles [24,25]. The cement mortar fluidity testing device consists of a cone-shaped cylinder and a glass base plate that can be shaken up and down. According to the Test Method for Fluidity of Cement Mortar, the soil sample is place in the cone in layers and the cone is then lifted up, and the glass plate is shaken up and down 25 times. The evaluation of the fluidity of the soil is based on the measured diameter of the upper surface of the sample.

9.5.1.3 Consistency test

The consistency test for cement mortar can also used to evaluate the fluidity and plasticity of soil [24,26]. The test instrument is mainly composed of a cone with a sliding rod (height: 145 mm, cone base diameter: 75 mm, weight: 300 ± 2 g) and a conical metal container. The evenly stirred soil is place in the container, the soil is tamped according to the test standards when it is about 1 cm from the top of the container, and the cone is allowed to sink freely from the center of the mortar surface. The consistency of the mortar, evaluated by penetration depth of the cone, is the reading of its falling distance after 10 seconds, which can be used to evaluate the fluidity and plasticity of the soil.

9.5.1.4 Mixing test

In the mixing test, a power meter is added to the indoor mortar mixer to measure the real-time energy consumed by mixing the soil when the mixer rotates and mixes the soil [10,27]. By recording the energy consumed by the small mixer when it runs and comparing the energy consumption for the soil with or without the addition of conditioning agents, the fluidity evaluation index of the soil, that is, the energy consumption reduction G_p, can be obtained to evaluate the fluidity and plasticity of the foam-conditioned soil.

9.5.2 Permeability

To avoid water spewing of the screw conveyor and instability of the excavation face, the soil in the shield excavation chamber needs to have low permeability. Foam and other conditioning agents injected into the coarse-grained soil can effectively fill the pores in the soil, cut off the water penetration channels, and reduce the permeability of the soil. A permeability test is the most direct method to determine the permeability of soil. To meet the needs of shield tunnelling, many researchers have proposed requirements for the permeability of conditioned soil, as shown in Table 9.4. The goal of soil conditioning is generally to limit the permeability coefficient of the soil below at least 10^{-5} m/s. Owing to differences in groundwater levels and shield types, there are certain differences in the value of permeability coefficient. In addition, considering the retention time of the soil in the excavation chamber under the circumstances of the shield machine shutdown and the like, the time for maintaining the permeability coefficient of soil below the limit value is at least 90 minutes.

9.5.3 Abrasion

When the shield passes through soil with a high content of high-hardness minerals (e.g., quartz), the soil tends to cause excessive abrasion to the shield cutters, which causes engineering problems such as low tunnelling efficiency and frequent cutter changes. Soil conditioning is an effective way to improve the abrasion resistance of shield cutters. Soil abrasion test (SAT), Laboratoire Central des Ponts et Chausees (LCPC), soft ground abrasion tester, and soil abrasion testing chamber (SATC) are the classic evaluation parameters for the current research on the abrasion of cutterheads and cutters by rocks [32,33], but the measurement of these parameters often requires large-scale test equipment. Therefore Barbero et al. [34] and

Table 9.4 Permeability coefficient range for appropriate soil conditioning.

Author	Permeability coefficient (m/s)
Zhu et al. [28]	$<(1.5-2.3)\times10^{-7}$
Ma [29]	$<(9.551-4.672)\times10^{-7}$
Quebaud et al. [10]	$<10^{-6}$
Budach and Thewes [22]	$<10^{-5}$
He et al. [30]	$<(4.456-5.601)\times10^{-7}$
Shen et al. [31]	$<10^{-7}$

Küpferle et al. [35] developed small instruments for measuring the abrasiveness of the soil as shown in Fig. 9.10, in which the PTFE an abbreviation of poly tetra fluoroethylene. The wear rate α of the rotating metal disk or cutter is used to characterize the abrasion of the conditioned soil and is calculated as follows:

$$\alpha = \frac{\Delta m}{m} \tag{9.8}$$

where Δm is the mass change of the cutterhead or cutters before and after abrading, g; and m is the original mass of the cutterhead or cutters, g.

The abrasion test uses related instruments to abrade metal scrapers on the soil at different speeds to evaluate the abrasion of the muck. The wear rate can directly demonstrate the abrasion capacity of the muck. However, the existing test equipment cannot measure the influence of temperature and water pressure on the abrasion of the muck.

9.5.4 Adhesion

In the process of shield tunnelling, it is often necessary to tunnel through soils with high clay content. The muck is prone to adhere to metal materials such as shield cutterheads and cutters, and it is easy for clogging to

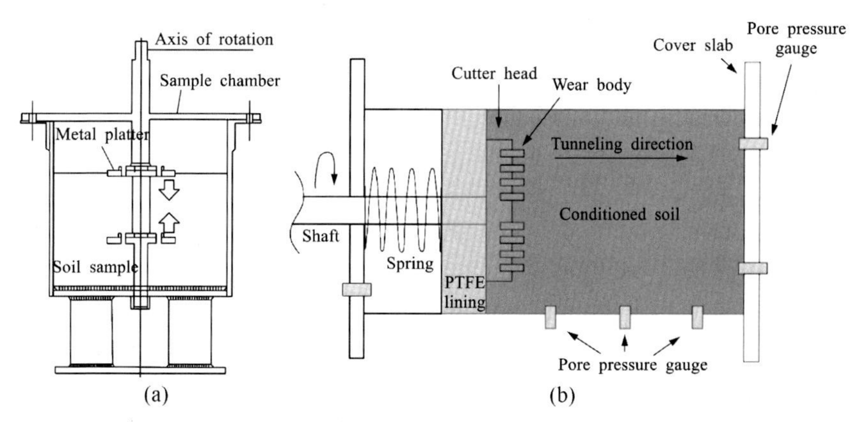

Figure 9.10 Small instrument for measuring abrasion of conditioned soils. (a) Barbero et al. instrument. (b) Küpferle et al. instrument. *From Barbero M, Peila D, Picchio A, et al. Test procedure for assessing the influence of soil conditioning for EPB tunnelling on the tool wear. Geoingegneria Ambientale e Mineraria, 2012;49(1):13–19. From Küpferle J, Zizka Z, Schoesser B, et al. Influence of the slurry-stabilized tunnel face on shield TBM tool wear regarding the soil mechanical changes—experimental evidence of changes in the tribological system. Tunnell Undergr Space Technol, 2018;74:206–216.*

occur on those metal materials under high temperature and pressure. In addition, this type of soil causes problems such as eccentric wear of the disc cutters and seriously affects tunnelling efficiency. However, the addition of conditioning agents can significantly reduce the adhesive strength of the interface between the muck and the metal, thereby avoiding soil clogging. There are several methods to evaluate the adhesion of soils.

9.5.4.1 Atterberg limit tests

The liquid limit characterizes the critical moisture content of soil transforming from the liquid state to the plastic state, and the plastic limit characterizes the critical moisture content of soil transforming from the plastic state to the solid state. Combined with the actual moisture content of the soil, the adhesion of the soil can be determined [36]. The most popular instrument for determining the Atterberg limits of soil is the liquid-plastic limit combined tester (fone cone), which is suitable for soils with a particle size of no more than 0.5 mm and an organic content of no more than 5% of the total mass of the sample.

Hollmann and Thewes [37] proposed a method for evaluating the risk of clogging in shield tunnelling through soil with high clay content based on the Atterberg limits and the actual water content of the soil. By measuring the liquid limit, plastic limit, and water content of the soil, the adhesiveness index can be calculated and compared with Fig. 9.11 to judge the risk of soil clogging. However, this method has certain limitations; for example, it cannot evaluate conditioned soil.

9.5.4.2 Sliding test

The sliding test involves putting a certain amount of soil sample on a rotatable metal plate and slowly raising the end of the metal plate until the soil sample starts to slide down the metal plate. The corresponding slope angle is the α angle for the soil. The larger the α angle, the stronger the adhesion of the soil to the metal surface [38].

9.5.4.3 Stirring adhesion test

The stirring adhesion test involves putting a certain amount of soil sample in a container, inserting the mixer into the sample, stirring for a certain period of time, and finally determining the mass of the soil adhering to the mixer (Fig. 9.12). The adhesion ratio λ of the soil is adopted to identify the soil adhesion. The larger the λ, the stronger the adhesion of the soil [39].

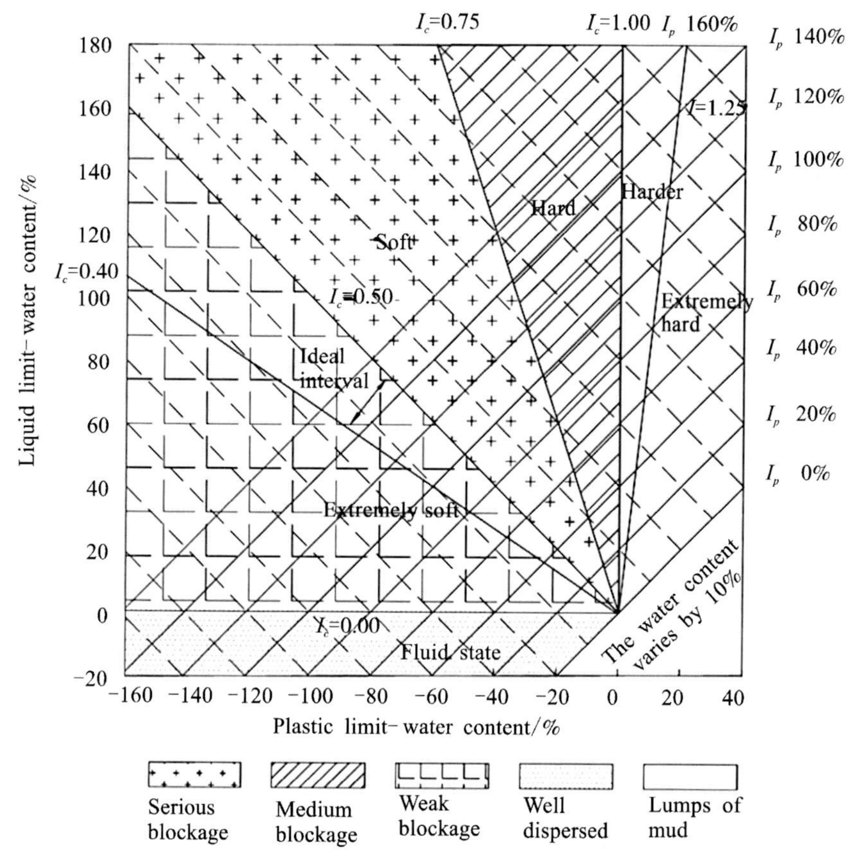

Figure 9.11 Chart for determining the degree of clogging for unconditioned shield soil with high clay content. *Modified from Hollmann F, Thewes M. Assessment method for clay clogging and disintegration of fines in mechanised tunnelling. Tunnell Undergr Space Technol, 2013;37(13):96-106.*

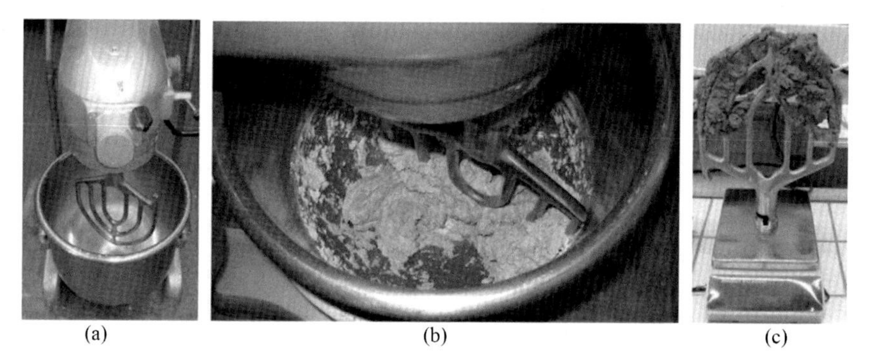

Figure 9.12 Test device for determining the adhesion ratio λ of conditioned soils. (a) Before stirring. (b) During stirring. (c) After stirring. *From Zumsteg R., Puzrin A., Stickiness and adhesion of conditioned clay pastes. Tunn Undergr Space Technol, 2012;86-96.*

$$\lambda = \frac{G_{MT}}{G_{TOT}} \tag{9.9}$$

where λ is the adhesion ratio; G_{MT} is the mass of the soil adhered to the mixer (g); and G_{TOT} is the mass of the total stirred soil (g).

9.5.4.4 Pull-out test

The pull-out test device consists of a conical metal block, a sample chamber, and a pull-out system. During the test, the sample chamber is filled with soil, the conical metal block is pressed into the soil sample for a period of time, and then the conical metal block is slowly lifted. The adhesion of the soil is evaluated on the basis of the tensile force subjected by the metal block and the mass of the adhered soil [40,41].

9.5.4.5 Rotational shear test

The rotational shear test device (Fig. 9.13) can measure the tangential adhesion strength of the soil with high clay content to the metal surface. The metal shear disk is buried in the soil sample. A certain air pressure is applied to the sample chamber to make the metal shear disk rotate and

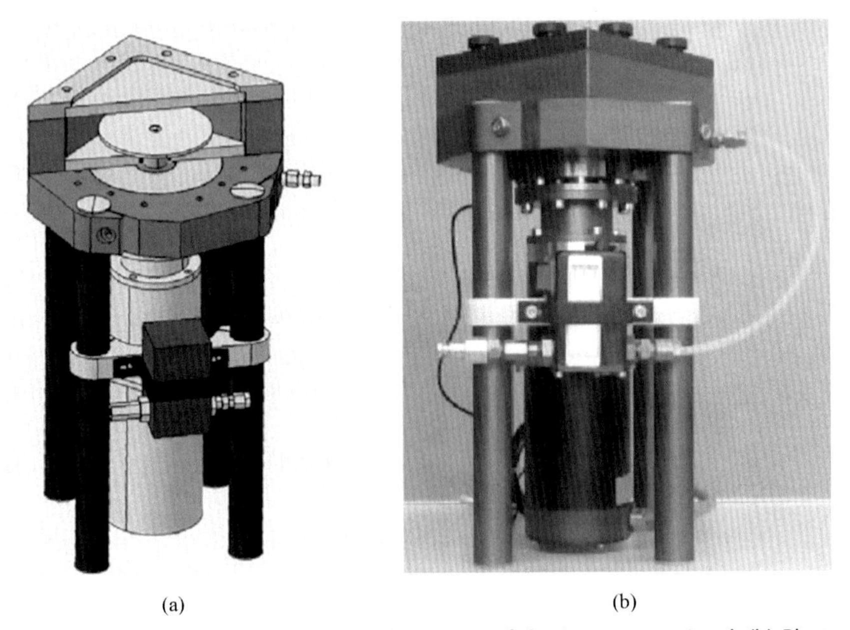

Figure 9.13 Rotational shear device. (a) Diagram of the instrument signal. (b) Photo of the instrument. *From Zumsteg R., Puzrin A., Stickiness and adhesion of conditioned clay pastes. Tunn Undergr Space Technol, 2012;86-96.*

shear under a certain soil pressure. The torque T required for shearing is recorded and can be converted to the adhesion strength of the soil-metal interface according to the following equation [39]:

$$a = \frac{6T}{\pi D^3} \tag{9.10}$$

where a is the adhesion strength of soil-metal interface (Pa); T is the rotational torque $(N \cdot m)$; and D is the diameter of the metal disk (m).

This device can determine the adhesion strength of the soil-metal interface under different pressure and rotational speeds during shield tunnelling. The effect of conditioning agents on the adhesion strength of the soil can be evaluated by comparing the adhesion strength of the soil before and after conditioning.

9.5.5 Shear strength

Soil conditioning can significantly reduce the internal friction angle of the soil and reduce the torque of the cutterhead and screw conveyor. Shield soils need to have small shear strength. European Federation of National Associations Representing Producers and Applicators of Specialist Building Products for Concrete (EFNARC) [2] proposed that the undrained shear strength should be 10–25 kPa. Current methods to determine the shear strength of soils include the direct shear test, the vane shear test, the large-scale penetration test, and the triaxial shear test.

9.5.5.1 Direct shear test

In the direct shear test, soils are put into the direct shear container, the normal load is applied, and shearing is initiated to obtain the shear strength of the soil. In the actual construction process of shield tunnelling, the muck in the excavation chamber and in front of the tunnel face is affected by conditioning agents such that it becomes weak in drainage performance. Because there is not enough time for the soil to drain during the shield tunnelling process, the undrained shear strength is measured by the quick shear test. The direct shear test instrument is simple in structure and convenient to operate, but there are some limitations. For example, the failure surface is not necessarily the surface with the weakest shear strength of the soil sample; the stress distribution on the shear surface is uneven, and the area of the shear surface is decreasing; the drainage conditions cannot be strictly controlled; and the change in pore water pressure during the shearing process cannot be measured.

9.5.5.2 Vane shear test

In the vane shear test, an vane shear instrument is used to shear the prepared soil sample, the shear torque T is recorded, and the undrained shear strength s of the soil under atmospheric pressure is obtained:

$$s = \frac{6T}{7\pi D^3} \tag{9.11}$$

where s is the shear strength of the soil sample (Pa); T is the shear torque ($N \cdot m$); and D is the diameter of the vane shear plate (m).

Since the conventional vane shear instrument can be operated only under atmospheric pressure, the stress state of the soil sample is not the same as the muck in the excavation chamber. Therefore researchers have independently developed a vane shear instrument that can measure the shear strength of soils under a certain pressure (Fig. 9.14). The conditioned soil sample is put into the sample chamber, the normal pressure is applied by vertically compressing the spring, and the soil sample is sheared by rotating the shear vane shaft. The maximum torque force is monitored with different soil-conditioning parameters, on which the calculation of shear strength of the conditioned soil is based [42]. Messerklinger et al. [43] independently developed a combined shear test apparatus that enables control of pressure on top of the soil sample and the rotational speed of the vane plate and the disk. The sealing performance of the sample chamber is good. It can measure the shear strength of soils under different pressure and rotational speed conditions, approximately simulating the stress state of the soil in the excavation chamber.

9.5.5.3 Large-scale penetration test

In the large-scale penetration test, a cone with a certain mass is dropped from the surface of the soil sample, and the penetration depth d of the cone in the soil sample is measured. This test is essentially measuring the undrained shear strength s of the soil sample, which can be obtained by Eq. 9.12:

$$s = \frac{K_\alpha W}{d^2} \tag{9.12}$$

where s is the undrained shear strength of the soil sample (kPa); K_α is the theoretical cone angle coefficient ($K_{30} = 0.85$, $K_{45} = 0.49$, $K_{60} = 0.29$, $K_{75} = 0.19$, where the subscripts represent the degree of the cone angle); W is the weight of the cone (N); and d is the penetration depth of the cone (mm).

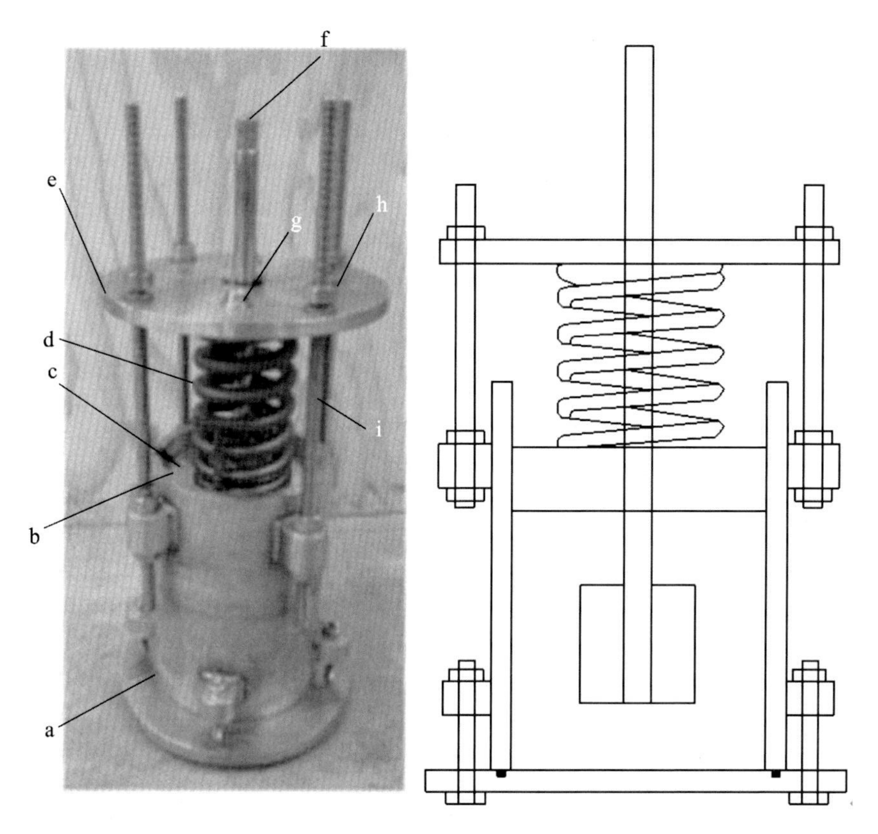

a. Chanber b. Top platen c. Air release d. Compression spring e. Reaction plate
f. Shear vane shaft g. Bulls eye level h. Bolts i. AMCE rods

Figure 9.14 Pressurized vane shear test device. *From Mori L., Mooney M., Cha M., Characterizing the influence of stress on foam conditioned sand for EPB tunnelling. Tunn Undergr Space Technol, 2018;454-465.*

The test results show that the undrained shear strength of the soil sample obtained by the large-scale penetration test is similar to that obtained by the vane shear test under the atmospheric pressure. Therefore the shear strength obtained by this test can be used to evaluate the change of shear strength of the muck before and after being conditioned [44].

9.5.5.4 Triaxial shear test

The triaxial shear test can be used to determine the shear strength of the conditioned soil. However, since the conditioned soil is a mixture of soil and conditioning agents, resulting in the soft state of the soil sample, it is very difficult to prepare a sample that meets the requirements of the triaxial shear test, especially for foam-conditioned sandy soil. In addition, the triaxial shear test takes a long time, but the properties of foam-conditioned soil are time-sensitive. The shear performance of the conditioned soil may change significantly during the long-term test. These disadvantages substantially limit the application of the triaxial shear test in evaluating the shear performance of conditioned soil.

9.5.6 Compressibility

The soil should have a certain compressibility so as to relatively suppress fluctuations in pressure in the excavation chamber when the rotational speed of the screw conveyor and the speed of tunnelling change. The Roche consolidator can measure the compression of the soil sample under a certain pressure, that is, the vertical displacement. By analyzing the relationship between the top pressure and the vertical displacement, the compressibility of the soil can be obtained, which can evaluate the compressibility of the soil before and after being conditioned [45]. The higher the compressibility, the easier it will be to control the pressure of the tunnel face during shield construction, and the smaller the pressure fluctuation in the excavation chamber. However, this instrument is suitable only for measuring the compressibility of fine-grained soils. For coarse-grained soils, considering the boundary effect caused by the increased particle size, the corresponding large-scale compression apparatus should be used for testing.

9.5.7 Shield tunnelling parameters

The final goal of soil conditioning is to ensure the safety and efficiency of shield tunnelling. Thus almost all the effects of soil conditioning techniques will eventually be tested by the shield tunnelling parameters. The success of the soil conditioning effect will directly affect the safety, efficiency, and engineering cost of the shield tunnelling. During the tunnelling process, the removed soil carried by the screw conveyor is required to have good workability, and the removal process should be continuous [46]. Foam, bentonite slurry, dispersants, flocculants, and their combinations are used to condition the soil generated during the shield tunnelling process. By monitoring the changes of cutterhead torque, shield thrust, tunnelling speed, chamber pressure, soil temperature, and the like before and after soil conditioning, combined with observation of the state of the soil being removed, the effect of soil conditioning during the shield tunnelling process can be reasonably evaluated. This evaluation method is relatively suitable for the project site and can directly reflect the conditioning effect. However, because of the numerous influencing factors on site, it is difficult to investigate the influence of muck conditioning on soil properties and tunnelling parameters.

9.5.8 Summary of evaluation methods for shield muck properties

According to the basic properties and testing methods currently used for evaluating the effect of shield soil conditioning, as summarized in Table 9.5, there are mainly two ways to evaluate the soil conditioning effect. One is to conduct indoor experiments to measure the physical and mechanical parameters of the conditioned soil; the other is to conduct field tests by comparing changes in shield tunnelling parameters. In dealing with the actual muck conditioning situation, the index properties and their determination methods for muck conditioning can be reasonably selected on the basis of the actual projects and existing muck conditions.

9.6 Numerical simulation of shield tunnelling under soil conditioning

Using foam, bentonite, and the like to condition soil in complex geological conditions can have a significant effect on the tunnelling parameters and further the ground response. The muck in EPB shield chamber needs to stay in a proper state. Low water content of the soil hinders soil discharging, increases cutterhead thrust and shield torque, reduces excavation speed, increases fluctuations in chamber pressure, and further causes uplift of ground surface. Excessive water content of the soil could cause spewing at the outlet of the screw conveyor and increase fluctuations in the shield thrust, cutterhead torque, excavation speed, and excavation chamber pressure, making it difficult to control the amount of excavated soil, resulting in relatively large ground deformation and even instability of the tunnel face. When the soil is conditioned by foam, bentonite, and the like to achieve an appropriate state, the thrust and torque of the shield are significantly reduced, the tunnelling speed is substantially increased, the fluctuation of the chamber pressure is reduced, and the abrasion of shield cutterheads and cutters by soils is also reduced, which facilitates the control of ground deformation during shield tunneling [47]. With the development of numerical methods and computer technology, ground deformation, changes in

Table 9.5 Summary of the index properties and their determination methods for muck conditioning.

Evaluation Index	Test	Overview	Advantage	Disadvantage
Fluidity and plasticity	Slump	Evaluate the fluidity and plasticity of soil based on the observation of slump value, water bleeding, etc.	Test is easy and provides reference for evaluation of the fluidity and plasticity of soil.	The ideal slump value is not consistently conclusive, and the maximum tested soil particle size cannot exceed 40 mm.
	Atterberg limits and water content	Evaluate the fluidity and plasticity of soil by using liquid and plastic limits and water content.	Easy to operate	Suitable only for soil with high clay content and particle size less than 0.5 mm.
	Fluidity	Measure the expansion ratio of the soil based on the method for cement mortar.	Measure the fluidity of the soil with high clay content better compared to the slump test.	Suitable only for fine-grained soil
	Consistency	Characterize the fluidity and plasticity of soil by measuring the settlement depth of the standard cone into the soil.	The fluidity of the soil is expressed by the penetration depth of the cone. The test is simple and intuitive.	The test cannot measure the cohesiveness of coarse-grained soils.
	Mixing	Measure the torque when stirring the conditioned soil. Evaluate the fluidity of the conditioned soil.	An electric machine is used to measure the fluidity of the soil. Reduce measurement errors.	The test equipment is complicated and needs to be customized.

Permeability	Permeability test	Measure the permeability coefficient of the conditioned soil and evaluate its permeability.	Can objectively reflect the permeability of soil.	The maximum particle size of the soil is limited by the diameter of the permeameter.
Abrasion	Abrasion	Abrade metal cutters on the soil at different rotational speeds to evaluate the abrasiveness of the soil.	Show the abrasiveness of rock and soil directly through the wear rate.	Cannot reflect the influence of temperature on abrasiveness.
Adhesion	Atterberg limits	Measure the liquid and plastic limits and plasticity index of the conditioned soil, and evaluate the adhesion of the soil with the adhesive index.	Test is easy and convenient.	Suitable only for soil with high clay content and particle size less than 0.5 mm.
	Sliding	Evaluate the adhesion of the soil by measuring the minimum inclination angle of the metal plate when the soil sample starts to slide.	Test method is easy to conduct.	The result is the inclination angle of the metal plate, which indirectly reflects the adhesion of the soil.
	Stirring adhesion	Measure the proportion of the soil adhered to the mixer by mixing the conditioned soil. Evaluate its adhesion.	Directly use the adhesion ratio of the soil as the adhesion evaluation index, which is simple and intuitive.	The adhesion amount is greatly affected by the shape of the mixer and the test operation, and the adhesion rate is related to the amount of soil in the container.

(Continued)

Table 9.5 (Continued)

Evaluation Index	Test	Overview	Advantage	Disadvantage
	Pull-out	Evaluate the adhesion of the soil by recording the tensile force when pulling the metal block out of the soil sample.	The operation is simple. Use tensile force to characterize the adhesion is intuitive.	The failure plane may be located inside the soil sample.
	Rotational shear	Rotate and shear the conditioned soil using a rotating shear disk. Evaluate the adhesion of the soil.	Reflect the adhesion of the soil and can shear the soil under pressure.	The test has high requirements for instruments.
Shear strength	Direct shear	Use direct shear test to determine the shear strength of soil.	Easy and convenient.	Difficult to perform direct shear test on soil with high water content.
	Vane shear	Use the vane shear test device to shear the prepared soil sample, and use the measured torque to obtain the undrained shear strength of the conditioned soil.	Truly reflect the undrained shear strength of the conditioned soil.	Applicable only to soft soils with a lateral earth pressure coefficient of approximately 1.
	Large-scale penetration	Drop the cone from the surface of the soil sample and measure its penetration depth in the soil, which is converted to the undrained shear strength of the conditioned soil.	Test instrument is simple.	Because it is a large-scale test, its test difficulty and cost are high.

	Triaxial shear	Use triaxial shear test to measure the shear strength of soil.	Simulate the stress state of conditioned soil in actual projects.	Very difficult to prepare samples of conditioned soil, and the test time is long.
Compressibility	Compressibility	Use the Roche consolidator to measure the compressibility of the soil sample and evaluate the compressibility of the conditioned soil.	The test method is simple and can measure the compressibility of the soil.	The sample size is limited by the size of the instrument.
Shield tunnelling parameters	Field tunnelling	Reasonably evaluate the effect of soil conditioning by monitoring the changes of the tunnelling parameters before and after the conditioning.	Suitable for the project site, directly reflect the conditioning effect.	Numerous influencing factors on site, and it is difficult to analyze the influence law.

tunnelling parameters, and muck movement characteristics during shield tunnelling can be analyzed by numerical simulation.

The current simulation methods of shield tunnelling mainly include the finite element method (FEM), the finite difference method (FDM), the discrete element method (DEM), and the computational fluid dynamics method (CFD). As an important construction procedure in shield tunnelling, muck conditioning can significantly change the mechanical parameters of soil. Therefore it is necessary to select appropriate parameters to simulate the mechanical behavior of the soil in a numerical simulation process.

Both the finite element method and the finite difference method divide the solid domain into finite elements before solving and are often used to calculate the force and deformation of rock and soil in geotechnical engineering. In using this method to simulate shield tunnelling, it is necessary to obtain the mechanical parameters of soil conditioned with different conditioning agents through laboratory tests and then select the appropriate constitutive model in the computation software, input the mechanical parameters of the soil, and analyze the mechanical characteristics of shield tunnelling in different soil-conditioning cases [48].

The discrete element method separates discontinuities into a collection of rigid particles with a certain mass and shape, makes each particle unit satisfy the equation of motion and contact constitutive equation, solves the movement and relative position of each particle unit using a time-step iteration method, and further obtains the deformation and evolution of the entire collection. Since the microparameters used by the discrete element are not the physical and mechanical parameters of the soil obtained through indoor experiments, in order to make the microparameters used by the discrete element to characterize the physical properties of the simulated soil, it is often necessary to use soil tests to calibrate the microparameters of the soil particles, and then the changes in the mechanical parameters of the conditioned soil can be characterized by changing the microparameters between soil particles [49]. Owing to the large size of the shield, if the discrete element method is used to simulate the shield tunnelling process, the amount of computation will be huge, and the basic discrete element particles cannot easily simulate the real particle size of the soil. Therefore some researchers use finite difference software to simulate the outer soil and use discrete element software to simulate the muck particle movement near the shield and inside the excavation chamber (Fig. 9.15a). The displacement and velocity transmission between the two regions can be determined by exchanging data through the coupling

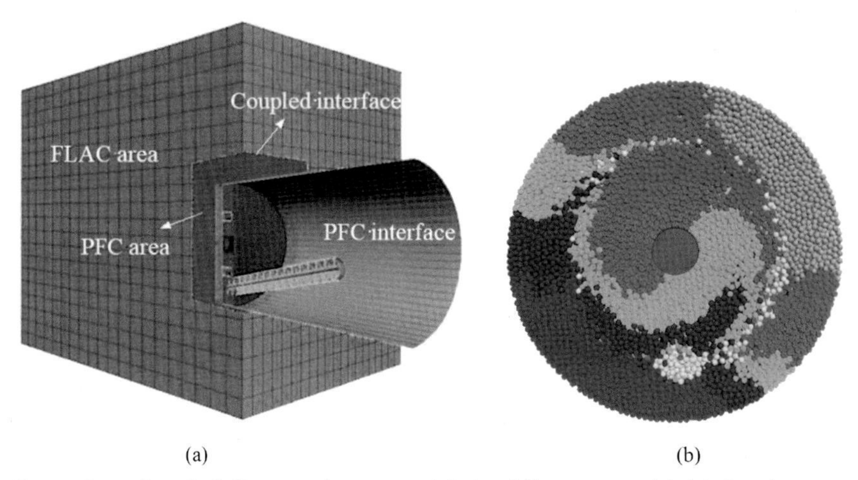

Figure 9.15 Coupled discrete element and finite difference model. (a) Coupling surface between FDM and DEM. (b) Soil movement inside the excavation chamber. *From Qu T., Wang S., Hu Q., Coupled discrete element–finite difference method for analysing effects of cohesionless soil conditioning on tunnelling behaviour of EPB shield. KSCE J Civ Eng, 2019;4538-4552.*

surface. Finally, the muck conditioning is reflected by reducing the friction coefficient between soil particles, which enables analysis of the flow characteristics of the muck in the excavation chamber (Fig. 9.15b), the pressure distribution of the excavation chamber, the average earth pressure distribution on the excavation face, the thrust of the shield, the torque of the cutterhead, and the difference of earth surface settlement under different muck conditioning cases, revealing the effect of muck conditioning on the muck state in the excavation chamber and the mechanical behavior of shield tunnelling.

The computational fluid dynamics method combines classical fluid dynamics and numerical computation methods; solves a series of partial differential equations composed of mass conservation, momentum conservation, and energy conservation through computers; and quantitatively describes the flow field in terms of time and space and also under specific boundary conditions. During the EPB shield tunnelling, the conveying process of the soil inside the screw conveyor has two failure modes: soil shear failure and plastic flow. The conditioned muck mainly exhibit cohesion and plasticity and thus can be considered as fluids, and a viscoplastic fluid constitutive relationship can be used to simulate the deformation of the muck.

9.7 Case study of muck conditioning for shield tunnelling

9.7.1 Project overview

The one-way total length of the shield tunnelling section between Zhongshan West Road Station and Zigu Road Station (SK13 + 016.072–SK13 + 681.556) in the fifth bid section of the first phase of Nanchang Metro Line 1 project is 1612.12 m. The ground in this section consists mainly of miscellaneous soil, silty clay, mucky silty clay, gravel sand, and argillaceous siltstone as well as a few pebble layers (see Fig. 9.16). The shield needed to pass through gravel sand, mud siltstone, and silty clay. Gravel sand is light yellow-brown in color, saturated; is mainly composed of quartz, feldspar, and siliceous rock with a small amount of pebbles; and has good roundness and a subround-to-round shape. The argillaceous siltstone is purple-red in color, with a medium-thick layered structure and an argillaceous structure, is soft, and belongs to soft rock. The rock is prone to soften in contact with water and to dry and crack when it loses water. Mucky silty clay is gray in color and is soft plastic to plastic. The mucky silty clay has high compressibility, is 1.6–10.2 m thick, and is mainly located in the Fu River section. The following subsections take the shield tunnelling through the argillaceous siltstone as an example to illustrate the method for determining the soil-conditioning parameters of muck with high clay content.

9.7.2 Necessities for soil conditioning

Mineral composition analysis found that the largest amount of mineral contents in argillaceous siltstone is quartz with 46.7%, and the content of clay minerals is also relatively high, including 27.8% of kaolinite, 8.3% of illite,

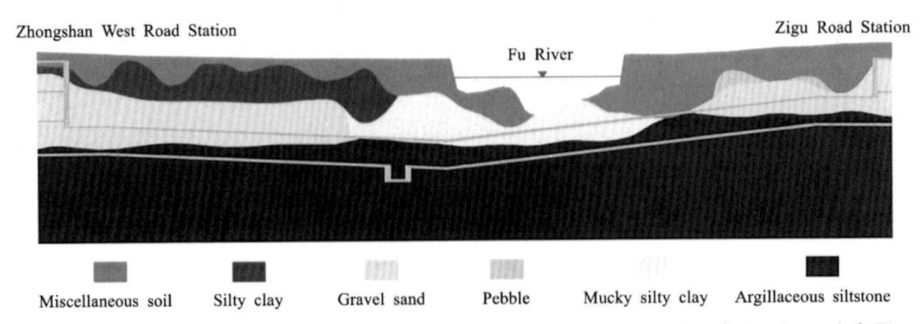

Figure 9.16 Vertical section view between Zhongshan West Road Station and Zigu Road Station.

and 4.4% of montmorillonite. Clay minerals such as montmorillonite, illite, and kaolinite are the main causes of clogging on the cutterhead, especially montmorillonite and illite. Because there is a possibility of clogging on the cutterhead for the shield to tunnel in this ground, it is necessary to perform soil conditioning in the argillaceous siltstone to prevent clogging.

9.7.3 Selection of conditioning agent

Dispersants and foaming agents are the main conditioning agents for ground with high clay content. Because this argillaceous siltstone contains quartz, resulting in less cohesion than clay, a foaming agent is mainly considered for conditioning the soil in this project. Three foaming agents—A, B, and C—were selected at the project site (Fig. 9.17). The foaming agents were chosen on the basis of the *FER* and half-life, combined with economic factors.

(a) (b) (c)

Figure 9.17 Foaming agents for the project. (a) Foaming agent A. (b) Foaming agent B. (c) Foaming agent C.

9.7.3.1 FIR test

According to the recommendations of the foaming agent manufacturer, the foaming agent was formulated into a solution with a volume concentration of 3% and then poured into the foam generator. The foaming pressure was set as 0.2–0.3 MPa for this project according to the chamber pressure. After the foam generator produced uniform foam, the foam was put into graduated cylinders, and the openings were sealed. The volume of the solution was read after the foam had completely collapsed. The test showed that the *FER*s for foaming agents A, B, and C were 16, 20, and 36, respectively. The foaming performance of the three foaming agents was as follows: The *FER* of foaming agent A was the highest with the best foaming performance, followed by foaming agent B, and the foaming effect of foaming agent C was the worst of the three. However, the *FER*s of these three foaming agents were all greater than 10, so all of them met the usage requirements.

9.7.3.2 Half-life test

The half-life test found that the half-lives of foaming agents A, B, and C were 5.3, 13.5, and greater than 18 minutes, respectively. Therefore the foam produced by foaming agent A had the worst stability, while foaming agent C had the best stability. However, all three foaming agents can did meet the requirement of having a half-life greater than 5 minutes.

Considering the *FER* and stability of the three foaming agents, foaming agent C had the best performance and foaming agent A had the worst, but all of them met the basic requirements of shield tunnelling. Considering the economic factors, foaming agent B was finally selected as the soil-conditioning agent for this project.

9.7.4 Determination of conditioning parameters

The slump, when used to evaluate the effect of muck conditioning, reflects the workability of the conditioned muck but also has certain defects. The slump of the conditioned muck may be relatively large during evaluation, and segregation or water bleeding may appear on the soil. This kind of conditioned muck obviously does not meet the conditioning requirements. It is reasonable to use the slump as well as the apparent state of the conditioned soil to evaluate the effect of soil conditioning. Therefore the evaluation of the muck conditioning test in this project combined the slump value and apparent state of the muck.

The water content of the muddy siltstone in the ground is 11.4%, so soil with this water content was obtained for testing. Water was further added on the basis of the test results, and then the foam was injected to achieve a good conditioning effect. Since it is difficult to control the amount of water in the actual construction process, a reasonable range of water content of the soil should be used in performing the slump test. The effect of muck conditioning can be evaluated on the basis of the amount of water added for the tests, which is calculated according to the different amounts of water injected during shield construction together with the different amounts of foam injected. Specifically, for every segment ring that the shield tunneled, the amounts of 6.9, 8.3, and 9.5 m^3 of water were injected, resulting in the water content of the soil being 19.7%, 21.2%, and 22.7%, respectively. The slump test was conducted on the basis of this water content. To obtain a more accurate injection ratio leading to a better muck-conditioning effect, conditioning tests with several different *FIR*s were conducted on the soil under the water content conditions. The detailed adding method was as follows: Muck conditioning tests were conducted by adding different amount of foam. The foam-adding ratio started from 0% with an increment of 5% each time until the slump reached the target value. The increment of the *FIR* should be adjusted on the basis of the test results of each group during the conditioning process. For example, the increment can be appropriately reduced if the soil is more sensitive to foam. Otherwise, it should be increased to obtain a more accurate foam injection amount. As shown in Table 9.6 and Fig. 9.18, when the water content (w) of the soil is 19.7%, the *FIR* for good soil conditioning effect is 30%–35%. When the water content is 21.5% and 22.7%, the corresponding *FIR*s are 20%–25% and 17.5%–20%, respectively. The slump of the ideal soil is 17–22 cm according to the laboratory tests.

9.7.5 Muck conditioning parameters for shield tunnelling

According to the laboratory tests, when the shield is tunnelling in this argillaceous siltstone, the water content of the soil should be within 19%–23%, the water injection amount is about 6–10 m^3 per segment, and the *FIR* is 15%–35%. These parameters can be adjusted according to the effect of soil conditioning on site. On the basis of the *FIR* obtained in laboratory tests, the flow velocity Q of each muck-conditioning pipeline of the shield on site can be obtained as follows:

$$Q = \kappa \frac{\pi D^2 v}{4n} FIR \tag{9.13}$$

Table 9.6 Slump test for the muck conditioning for shield tunnelling in argillaceous siltstone.

Water content (ω/%)	Foam injection ratio/%	Slump (cm)	Comments	Figure
19.7	5	1.0	Inappropriate, poor fluidity and plasticity, bleeding water	Fig. 9.18a
	10	7.8	Inappropriate, poor fluidity	Fig. 9.18b
	15	10.5	Inappropriate, poor fluidity and plasticity	Fig. 9.18c
	20	14.0	Inappropriate, poor fluidity and plasticity	Fig. 9.18d
	25	18.0	Inappropriate, poor fluidity and plasticity	Fig. 9.18e
	30	19.7	Appropriate	Fig. 9.18f
	35	21.5	Appropriate	Fig. 9.18g
21.2	5	7.0	Inappropriate, poor fluidity and plasticity, bleeding water	Fig. 9.18h
	10	12.0	Inappropriate, poor fluidity and plasticity, segregation	Fig. 9.18i
	12.5	14.3	Inappropriate, poor fluidity and plasticity, segregation	Fig. 9.18j
	15	15.0	Inappropriate, poor fluidity and plasticity	Fig. 9.18k
	20	17.0	Appropriate	Fig. 9.18l
	25	19.2	Appropriate	Fig. 9.18m
22.7	5	12.0	Inappropriate, poor fluidity and plasticity, segregation	Fig. 9.18n
	10	16.0	Inappropriate, poor fluidity and plasticity, segregation	Fig. 9.18o
	12.5	18.0	Inappropriate, poor fluidity and plasticity	Fig. 9.18p
	15	20.0	Inappropriate, poor fluidity and plasticity	Fig. 9.18q
	17.5	21.0	Appropriate	Fig. 9.18r
	20	21.0	Appropriate	Fig. 9.18s

where Q is the flow rate of each pipeline (L/min); κ is the loosening coefficient of muck; D is the shield cutting diameter (m); v is the shield tunnelling speed (mm/min); n is the number of modified shield pipelines; and FIR is the foam injection ratio.

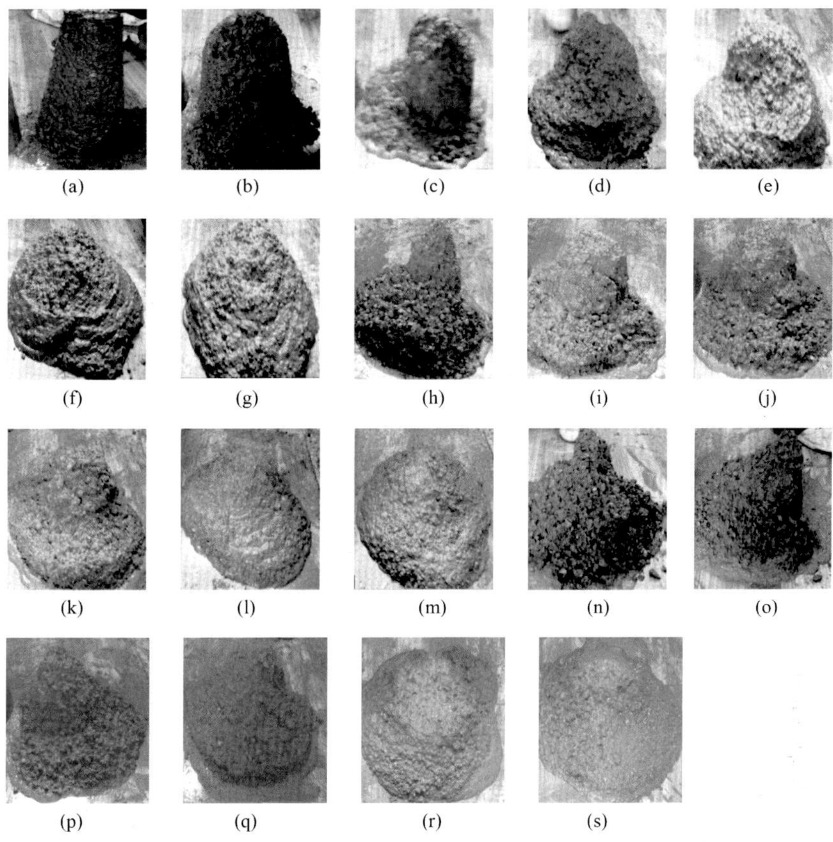

Figure 9.18 Results of slump tests. (a) w = 19.7% and *FIR* = 5%. (b) w = 19.7% and *FIR* = 10%. (c) w = 19.7% and *FIR* = 15%. (d) w = 19.7% and *FIR* = 20%. (e) w = 19.7% and *FIR* = 25%. (f) w = 19.7% and *FIR* = 30%. (g) w = 19.7% and *FIR* = 35%. (h) w = 21.2% and *FIR* = 5%. (i) w = 21.2% and *FIR* = 10%. (j) w = 21.2% and *FIR* = 12.5%. (k) w = 21.2% and *FIR* = 15%. (l) w = 21.2% and *FIR* = 20%. (m) w = 21.2% and *FIR* = 25%. (n) w = 22.7% and *FIR* = 5%. (o) w = 22.7% and *FIR* = 10%. (p) w = 22.7% and *FIR* = 12.5%. (q) w = 22.7% and *FIR* = 15%. (r) w = 22.7% and *FIR* = 17.5%. (s) w = 22.7% and *FIR* = 20%.

9.7.6 Soil conditioning effect analysis

During shield tunnelling, after the muck-conditioning parameters were adjusted to the target values for 5 minutes, the slump tests were performed on the excavated soil. The tunnelling parameters (shield thrust and cutter-head torque) were compared and analyzed, and the ranges of tunnelling parameters that ensured good soil conditioning status were obtained, which justified the muck conditioning effect.

As shown in Fig. 9.19, the slump of the soil was inversely proportional to the shield thrust in sampling the shield tunnelling muck (11 samples) of the argillaceous siltstone. When the slump was large, the thrust force of the shield was low accordingly. This was because the larger slump resulted from thinner and more fluid soil, which reduced the resistance from the soil when the shield was tunnelling.When the slump of the soil exceeded 20 cm, the fluidity of the shield soil was excessive, and there was a risk of the soil being ejected at the screw conveyor outlets. Therefore the maximum slump of the soil should not exceed 20 cm. When the soil slump was less than 17 cm (for the samples of No. 9 - 11), the shield thrust showed an increase with an decrease of slump, so the slump of the soil should not below 17 cm. When the slump of the conditioned soil was 17–20 cm, the corresponding shield thrust was 16500–1800 kN.

The cutterhead torque during shield tunnelling can also be used as a factor in evaluating the soil conditioning. As shown in Fig. 9.20, the relationship between the torque of the cutterhead and the slump of the soil in the excavation chamber was similar to the relationship between the thrust and slump. When the slump increased, indicating that the soil is thinner and more fluid, the torque required for the cutterhead to stir the soil decreased accordingly. This changing trend was also consistent with the actual situation. When the soil stayed in an ideal state (slump = 17–20 cm), the cutterhead torque was 3300–3500 kN · m.

Considering the shield tunnelling parameters combined with the continuity of the muck-conditioning parameters and so on, the parameters when the shield tunnels 160 segment rings normally were selected for

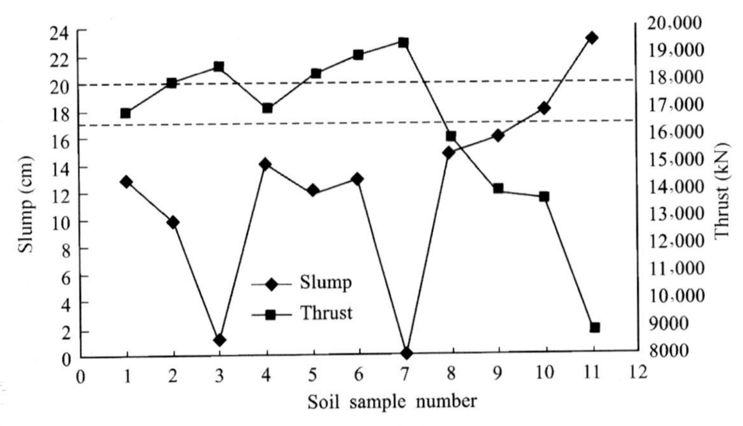

Figure 9.19 Relationship between slump and thrust.

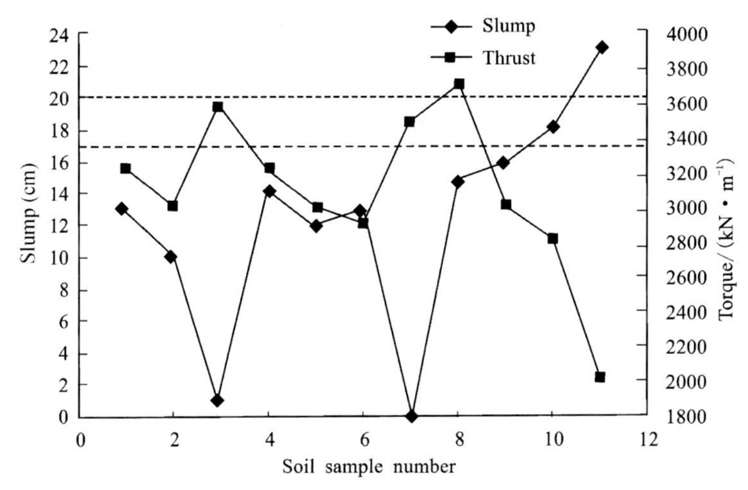

Figure 9.20 Relationship between slump and cutterhead torque.

analysis; that is, the soil-conditioning parameters for No. 220–300 segment rings and No. 380–460 segment rings were collected and compared with the parameters obtained from the laboratory tests. The analysis results are shown in Table 9.7.

The statistical results in the table show that the muck-conditioning parameters were not significantly different from those obtained from laboratory experiments. Specifically, the amount of foaming agent was between 56–76 L, and the average amount per segment was 61.2 L, which was consistent with the results from the laboratory experiments. Comparably, the laboratory results shows that when the foaming agent dosage was 56–65 L, the water injection amount was 6.9 m^3 per segment, while the actual water injection amount was 5.6 m^3. The reason for the difference may be that the natural water content of the argillaceous siltstone formation varies slightly, resulting in a decrease in water injection amount.

In summary, when the shield is tunnelling in an argillaceous siltstone formation, foam can be used for soil conditioning. When the volume concentration of the foaming agent is 3%, injection of 56–76 L of foaming agent and 4.8–6.8 m^3 of water per segment ring tunnelling can obtain the ideal state soil with a slump of 17–20 cm, the corresponding shield thrust is 16500–1800 kN, and the torque is 3300–3500 kN · m.

Table 9.7 Statistics of thesoil conditioning parameters for shield tunnelling inargillaceous siltstone formation.

Conditioning parameters	Argillaceous siltstone formation Foaming agent injection amount (L)	Water injection amount (m^3)
Mean	61.2	5.6
Observations	160	80
Median	62.5	5.6
Standard deviation	14.4	2.0
Skewness	−0.19	0.27
Kurtosis	0.31	1.05
Minimum value	24.4	0.8
Maximum value	108.5	10.8
Distribution histogram figure		

9.8 Recycling of shield muck as resources

9.8.1 Significance of muck recycling

In urban subway construction, the shield tunnelling method has become one of the main options for tunnel construction. The large amount of waste soil produced by shield tunnel construction has brought many challenges to the urban environment. The processes of dehydration, transportation, and landfilling of excavated soil will have a large adverse effect on the environment [50]. With the development of large-scale construction the soil landfill is seriously insufficient. Taking Changsha City in China as an example, the total storage capacity of the existing waste yards is only around 800×10^4 m^3, while the amount of muck processed in Changsha in 2015 was 1800×10^4 m^3, and it was increasing at a rate of 15% every year. The construction of Changsha Metro Line 2 produced about 400×10^4 m^3 of muck. The shield soil had a high water content and poor stability, resulting in potential danger if it was landfilled without any treatment [51]. On December 20, 2015, a landslide accident occurred at the Hongao waste soil yard in Guangming New District, Shenzhen, China, resulting in 73 people killed, four people missing, 17 people injured, 33 buildings destroyed and buried, and the direct economic loss of 881 million yuan.

Therefore recycling and reusing the excavated soil from the shield can not only reduce the development of other natural resources, but also protect the urban environment and meet the government's requirements for developing a recycled economy, protecting the environment, and building a resource-saving and environmentally friendly society. The most common ways to recycle excavated soil have been in the fields such as roadbed fillings, grouting materials, cement admixtures, haydite, and brick making. If the shield muck can be treated efficiently, economically, and in an environmentally friendly way through methods such as green recycling and comprehensive utilization, this will have very important practical significance and engineering application value [52].

9.8.2 Methods and case studies of muck recycling

9.8.2.1 Synchronous grouting

The prices of raw materials such as cement, bentonite, sand, and gravel are getting higher and higher, which in turn increases the cost of synchronous grouting for shield tunnelling. The geological conditions in China are complex and diverse, and they are significantly different in different cities. For example, the ground in Changsha is mainly soft on top and hard underneath, the ground in Chengdu is dominated by sand and pebble, and the ground in Zhengzhou is dominated by sandy silty clay and fine sand. General soil contains some which can be used as synchronous grouting materials. For example, sandy silty clay soil contains clay and sand. After the soil is sieved, clay can partially replace bentonite [53], and sand can be used directly as synchronous grouting material. If the shield soil can be refined and replace the existing bentonite or sand for the grout preparation, it will not only reduce the cost of purchasing large amounts of bentonite and sand, but also reduce the export of some of the muck, resulting in less environmental pollution [54]. Shield mud sand can replace bentonite and river sand, and by adjusting and optimizing the mix ratio through the orthogonal test method, it can produce economical and environmentally friendly synchronous grouting materials that meet the performance requirements [55].

Case 1

Project overview The ground between Huanghe Road Station and Jinshui Road Station of the Zhengzhou Rail Transit Line 3 Phase I Project is dominated by Quaternary loose sediments. The underlying bed-rock is buried deeply. The main soil layer through which the tunnel roof passed was clayey silt intercalated with sandy silt, with partial soil layers of silty clay and clayey silt intercalated with sandy silt. After correction the classification of surrounding ground was all grade VI, and the stability of surrounding ground was poor. The ground in the test region consisted of miscellaneous soil, silty sand, sandy silt, clayey silt, fine sand, and silty clay. The main ground consisted of fine sand and silty clay.

Field application To ensure the recycling of shield muck as the synchronous grouting material, special equipment was required to stir soil with high silty clay content. The special mixing equipment, based on the needs of on-site mixing, was mainly composed of crushers, mixers, silos, shelves, and so on. The synchronous grouting ratio used for field tests was the optimal ratio obtained from laboratory tests, as shown in Table 9.8. The fine sand in the soil replaced the sand in the original synchronous grouting materials, the silty clay in the soil replaced the bentonite in the original synchronous grouting materials, and other materials were the same as in the original synchronous grouting materials.

The detailed process is as follows: The mixer was used to mix the silty clay from soil with a certain amount of water to form slurry; then the slurry pump was used to pump the slurry into the mixing tank in the shield mixing station, mix with other materials synchronously, and finally formed the synchronous grouting slurry, which was transported into the tunnel by a mortar truck.

Application effect

1. When part of the soil was used as the synchronous grouting material, during the grouting process the average grouting pressure was 1.68 bar, there was no problem such as clogging of the grouting pipe, and

Table 9.8 Grouting ratio of shield soil.

Category	Cement	Fly ash	Silty clay from soil	Fine sand from soil	Water
Mass	166 kg	456 kg	53 kg	740 kg	462 kg

there was no difference between the current grouting efficiency and the original synchronous grouting efficiency. In addition, the filling effect of the slurry was better.

2. Compared with using conventional synchronous grouting material, the settlement of the segment after escaping from the shield was relatively large when part of the soil was used as the synchronous grouting material, but it satisfied the code requirements for the control of the ground settlement of the shield. The code limited the ground settlement of shields to a maximum allowable value of ± 30 mm. When the shield passed through structures that needed special requirements for settlement, the use of soil as part of the grouting material requires careful consideration.
3. When part of the muck is used as the synchronous grouting material, it can reduce the environmental pollution of the soil and save on the material cost of the synchronous grout material. Taking Zhengzhou Metro as an example, the cost of grout material that could be saved was 596,000 yuan per kilometer, and the cost of waste soil exportation that could be saved was 208,000 yuan per kilometer, leading to a total cost saving of 804,000 yuan per kilometer.

9.8.2.2 Unfired bricks

The soil that is excavated from the shield tunnelling can be used as the base material for making unfired bricks, which can have better working performance by the addition of corresponding curing agents and graded reinforcing materials. Seco et al. used a mixture of concrete, ceramic waste, and marl clay to make unfired bricks [56]. Espuelas et al. [57] added magnesium oxide during the production of unfired bricks. The study found that their mechanical properties and water absorption are optimized to a certain degree. The production of unfired soil bricks not can only solve the problem of disposing some construction waste, but also allows use the produced unfired bricks for building construction, demonstrating green disposal and recycling of solid waste [58].

Case 2

Project overview The main ground passed by the shield section between North Bus Station and China National Exhibition Center Station of Hangzhou Metro Line 10 consists of silty clay, silty clay intercalated with silt, and silty clay. The ground along the shield tunnelling line is mainly hard on top and soft at the bottom, with a partial region that is soft on top and hard at the bottom, resulting in nonuniform ground along the line [59]. In addition, the ground in this section is mainly composed of clay, so the excavated soil from foundation pits and shield tunnels can be used as the base material for making unfired bricks.

Field application The test process was divided into the raw material preparation stage and the brick-making stage. In the raw material preparation stage, the muck was dried in an oven at 105°C for more than 24 hours; then a grinder was used to crush it to less than 0.25 mm in diameter. In the brick-making stage, the process of making unfired bricks can be divided into the raw material mixing stage, the compression molding stage, and the curing state. In the raw material mixing stage, the various raw materials were weighed according to the test conditions (see Table 9.9 for the material proportion of soil bricks) and mixed in a wheel-mill forced mixer for 10 minutes. In the compression molding stage, the mixed material was put into a 240 × 115 mm mold and placed on a digital pressure-testing machine for compression. In the curing stage, the prepared brick samples were labeled according to the test conditions and arranged upright on a shelf in a transparent rain shed, where the natural drying method and spraying water for curing were used in the first three days.

Application effect The 28d compressive strength of the soil bricks reached up to 10.2 MPa, and the softening coefficient was 0.80, which met the MU10 requirements of JC/T 422-2007, Unfired Rubbish Gangue Bricks" [60]. An analysis of the ratio of materials used in various working conditions showed that ordinary Portland cement was the key material for making unfired bricks by using silty clay. The hydration reaction of cement generated calcium silicate hydrate (C-S-H) gel, and so on, which could better wrap sand and clay and cement them together, enhancing the integrity of the unfired brick with higher strength and better engineering properties.

The actual production cost of a single brick was about 0.144 yuan per brick by using the muck as basic brick material. The market price of

Table 9.9 Proportion of soil bricks.

Item	Cement	Sand	Fly ash	Straw fiber	Soil
Value	20%	15%	7.5%	0.2%	57.3%

nonfired rubbish gangue bricks with the same performance is 0.4 yuan per brick, so the savings on a single brick produced using muck as the basic material was 0.256 yuan per brick. The daily total profit for the standard production of 45,000 bricks in a single day was thus 11,520 yuan.

9.8.2.3 Cement admixtures

Cement admixtures are minerals added to cement during production to improve cement performance and adjust cement grades. Their main functions are to improve certain properties of the cement, adjust cement grades, reduce heat of hydration, and reduce costs. Currently, many kinds of admixture materials are used in cement; the most widely used are slag powder, fly ash, limestone, and burnt clay. However, with the rapid growth of cement production, the amount of granulated blast furnace slag and fly ash is insufficient to meet the cement industry's demand for admixtures, and the current prices of fly ash and slag are even higher than the price of cement clinkers. Therefore the application of shield muck as an admixture to cement production is promising [61].

Case 3

Project overview The shield soils in a certain section of Zhengzhou Metro Line 4 and a certain section of Line 14 are mainly silty clay and clay silt, the particle diameter being below 0.15 mm. The water content of the two kinds of soil is relatively large: 23.43% and 14.38%, respectively. The muck was prone to cluster during the drying process, leading to difficulty in evaporating water inside the soil. The dried soil had a certain strength. After crushing and sieving, the contents of particles smaller than 0.08 mm in the two types of soil are 60% and 30%, respectively.

Field test The changing trend of the compressive strength to the calcination temperature showed that below 700°C the compressive strength did not change significantly with temperature, increasing by only 3.4%–10.3%; When the calcination temperature was greater than 700°C, especially greater than 800°C, the compressive strength increased significantly with the increase in temperature.

An appropriate amount of calcined soil mixed with cement powder can reduce the calcium hydroxide content of the system; increase the possibility of ettringite, C-S-H, and so on accumulating and growing near the interface; and improve the interface structure. The fine particles in the calcined soil can provide a large number of crystals for the crystallization of calcium hydroxide, improving its microstructure and thereby increasing the strength and durability of cement mortar.

When CaO and $CaSO_4 \cdot 2H_2O$ were used as activators, the compressive strength showed a trend of first increasing and then decreasing. Excessive addition not only failed to further improve the activity of the soil, but also caused serious strength reduction. There were two main reasons based on experimental observation and related resources. First, the increase of admixture content increased the water demand of the test block, reduced the fluidity of the muck, and reduced the compactness of the test block during molding, which affected strength of the test block. Second, excessive CaO and $CaSO_4 \cdot 2H_2O$ expanded in volume during the hydration process, resulting in poor stability and further reduction in the strength of the test block.

Application effect

1. The activity of the muck was increased after calcination at 400°C–800°C, and the unconfined compressive strength of the test block was increased by 64%–85.7%. When the temperature was below 700°C, the compressive strength was only increased by 3.4%–10.3%. Above 700°C, the compressive strength increased rapidly, reaching 85.7% at 800°C, which met the requirements of active admixtures for cement. The calcination temperature was an important factor affecting the activity of the soil.
2. Both CaO and $CaSO_4 \cdot 2H_2O$ as activators had certain effects that improved the activity of shield soil, but the effects were limited. The optimal dosages of the two activators were 2% and 1%, respectively. Beyond the optimal dosages, the unconfined compressive strength of the test block was severely reduced.
3. Owing to the uniqueness of metro construction, the compositions of muck are dramatically different in different areas and regions. It significantly hinders the large-scale resource utilization and forming a unified standard. The utilization needs to be based on the local conditions.
4. Although it has been proved that the shield soil can be used as a cement admixture through calcination, it still has many limitations and requires more theoretical support and experimental research. In addition, the instability of its composition and market recognition are factors that restrict its development.

9.9 Technical issues of muck conditioning

As an assistant procedure for shield tunnel construction, muck conditioning has achieved rapid development in recent years. Conditioning agents are diversified, and their effects are getting better. The level of recognition of the physical and mechanical properties of the conditioned soil has been gradually improved, which promotes the safe shield construction under complex geological conditions. However, the use of traditional trial-and-error empirical methods of muck conditioning is not uncommon in field.

With the rapid development of tunnel engineering, the geological and hydrological conditions of shield tunnel projects are becoming more and more complex, and the surrounding environment is more sensitive. The muck-conditioning technology urgently needs to bypass the traditional trial-and-error method. The theory and application of shield soil conditioning need improvement to better enable safe, efficient shield tunnel engineering. The recommendations are as follows [47]:

1. Currently, the commonly used conditioning agents mainly include water, foam, dispersants, clay minerals, and flocculants. The proper water content of the soil is a prerequisite for other agents to have good effects. Water is the first priority for muck conditioning. The muck must stay within the optimal range of water content; otherwise, other conditioning agents will not be able to play their roles effectively. Therefore the function of water should be emphasized for muck conditioning. In addition, high temperature will promote soil clogging, accelerate the dissipation of bubbles, and reduce the effect of the conditioning agents. The pH value and chemical composition of groundwater will have different levels of influence on the conditioning effect. Current research on the influence of soil temperature and the chemical composition of groundwater on the effect of conditioning agents is limited. Research and development of conditioning agents that are resistant to high temperature and corresponding groundwater conditions are needed.
2. When the shield passes through rock with high clay mineral content, clogging is a major problem. Currently, adding foam and adding dispersants are the main methods for soil conditioning, but the effects should be strengthened. Some researchers have proposed that the soil-conditioning agents for ground with high clay content should be able to effectively seal the soil surface and prevent soil with high clay content from decomposing when exposed to water. Muck conditioning

can not only reduce the adhesive strength of the soil, but also maintain the shear strength of the soil, thus effectively preventing clogging. However, it is necessary to develop an environmentally friendly soil-conditioning agent that can achieve the purpose of conditioning.

3. It is difficult to perform conditioning of water-rich coarse-grained soil, especially pebbles. On one hand, the permeability of soil is strong. On the other hand, the soil generates heat through friction. Excessively high temperature causes the rapid collapse of the foam, leaving only the foam surfactant to play a lubricating effect. If bentonite slurry is injected to reduce the permeability of the soil and lubricate the cutterhead and cutters, the high temperature of the soil due to friction causes the clogging of bentonite wrapped with coarse-grained soil, which not only reduces the muck-conditioning effect, but also causes the secondary hazard of clogging on the cutterhead and cutters. Therefore the conditioning of coarse-grained soil is not only difficult, but also prone to cause secondary risks. There is an urgent need to study the muck-conditioning techniques applied to coarse-grained soil with strong permeability and ease of heat generation due to friction.
4. Conditioning agents can significantly increase the fluidity and compressibility of the soil; reduce the shear strength, adhesion strength, and permeability coefficient of the soil; and reduce the abrasion of the soil to the cutters. However, owing to the lack of unified evaluation standards, different researchers use different test instruments and evaluation factors. Therefore the establishment of a unified evaluation system to effectively guide the shield tunnel construction is an urgent problem to be solved to promote muck conditioning.
5. Muck conditioning affects the fluidity of the soil and the control of soil removal, which in turn affects the pressure control inside the excavation chamber, ground deformation, and stability. Although a few researchers have tried to establish coupled numerical models of formation, excavation chambers, screw conveyors, and the soil based on coupling technologies such as discrete element and finite element methods, due to the limitation of the number of discrete element particles, the simulated mechanical behavior of the soil is still somewhat different from the reality. It is limited to qualitative soil removal analysis and cannot be effectively applied to engineering practice. Therefore it is urgent to adopt new numerical simulation technology to conduct studies on shield soil discharging and ground response, explore the effect of soil conditioning, and provide theoretical support for refining soil-conditioning technology.

6. Owing to the ground changes, there is a lag in on-site guidance-based muck-conditioning laboratory experiments. Research on smart muck-conditioning technologies should be conducted, that is, research on quickly adjusting the soil-conditioning plan and parameters on the basis of the tunnelling parameters and the state of soil discharging in order to adapt to changes in ground conditions.

References

[1] Heuser M, Spagnoli G, Leroy P, et al. Electro-osmotic flow in clays and its potential for reducing clogging in mechanical tunnel driving. Bull Eng Geol Environ 2012;71:721–33.
[2] EFNARC. Specification and guidelines for the use of specialist products for soft ground tunnelling. Surry: European Federation for Specialist Construction Chemicals and Concrete Systems; 2005.
[3] Milligan G. Lubrication and soil conditioning in tunnelling, pipe jacking and micro-tunnelling: a state-of-the-art review. London: Geotechnical Consulting Group; 2000.
[4] China Metallurgical Construction Association. Code for design of general layout of industrial enterprises (GB50187-2012). Beijing: China Plan Press; 2012.
[5] Zumsteg R, PlÖtze M, Puzrin A. Effect of soil conditioners on the pressure and rate-dependent shear strength of different clays. J Geotech Geoenviron Eng 2012;138(9):1138–46.
[6] Qiao G. Development of new foam agent for EPB shield machine and foamed-soil modification. Beijing: China University of Mining and Technology-Beijing; 2009.
[7] Zumsteg R, Langmaack L. Mechanized tunnelling in soft soils: choice of excavation mode and application of soil-conditioning additives in glacial deposits. Engineering 2017;863–70.
[8] Liu P, Wang S, Ge L, et al. Changes of atterberg limits and electrochemical behaviors of clays with dispersants as conditioning agents for EPB shield tunnelling. Tunn Undergr Space Technol 2018;73:244–51.
[9] Wei K. Microscopic mechanism analysis of foam and bentonite improved soil in earth pressure balance shield construction. Mod Tunn Technol 2007; 44(1):73–7.
[10] Quebaud S, Sibai M, Henry J. Use of chemical foam for improvements in drilling by earth-pressure balanced shields in granular soils. Tunn Undergr Space Technol 1998;13(2):173–80.
[11] Yan X, Gong Q, Jiang H. Soil conditioning for earth pressure balanced shield excavation in sand layer. Chin J Undergr Space Eng 2010;6(3):449–53.
[12] Wu Y, Mooney M, Cha M. An experimental examination of foam stability under pressure for EPB TBM tunnelling. Tunn Undergr Space Technol 2018;77:80–93.
[13] Kusakabe O, Nomoto T, Imamura S. Geotechnical criteria for selecting mechanized tunnel system and DMM for tunnelling (Panel discussion). In: Proceedings of 14th international conference on soil mechanics and foundation engineering, Rotterdam; 1999, p. 2439–40.
[14] Ye X, Wang S, Yang J, et al. Soil conditioning for EPB shield tunnelling in argillaceous siltstone with high content of clay minerals: case study. Int J Geomech 2016;17(4):05016002.

[15] Jancsecz S, Krause R, Langmaack L. Advantages of soil conditioning in shield tunnelling: experiences of LRTS Izmir. Congress Challenges for the 21st Century. Balkema: WTC; 1999, p. 865–75.
[16] Leinala T, Grabinsky M, Klein K. A review of soil conditioning agents for EPBM tunnelling. Rotterdam: NATC; 2000.
[17] Limited PM. Easing the way-soil conditioning. Tunn Tunn Int 2003;35(6):48–50.
[18] Peña M. Soil conditioning for sands. Tunn Tunn Int 2003;35(7):40–2.
[19] Zhang F. Shield tunnel. Beijing: People's Communications Press; 2004.
[20] Vinai R, Oggeri C, Peila D. Soil conditioning of sand for EPB applications: a laboratory research. Tunn Undergr Space Technol 2008;23(3):308–17.
[21] Peila D, Oggeri C, Borio L. Using the slump test to assess the behavior of conditioned soil for EPB tunnelling. Environ Eng Geosci 2009;15(3):167–74.
[22] Budach C, Thewes M. Application ranges of EPB shields in coarse ground based on laboratory research. Tunn Undergr Space Technol 2015;50:296–304.
[23] Qiu Y, Yang X, Tang Z, et al. Soil improvement for earth pressure balance shield construction in watered sandy stratum. J Tongji Univ (Nat Sci) 2015;43(11) 1703–8.
[24] Li P, Huang D, Huang J, et al. Experiment study on soil conditioning of shield construction in hard-plastic high-viscosity layer. J Tongji Univ (Nat Sci) 2016;44 (1):59–66.
[25] Oliveira D, Thewes M, Diederichs M, et al. Consistency index and its correlation with EPB excavation of mixed clay–sand soils. Geotech Geol Eng 2018;2:1–19.
[26] Langmaack L. Advanced technology of soil conditioning in EPB shield tunnelling. In: Proceedings of the North American tunnelling conference. Rotterdam: NATC; 2000. p. 525-542.
[27] Liu D. The appraisal of improving the soil condition of epbs with new foam. Beijing: China University of Geosciences; 2012.
[28] Zhu W, Qin J, Wei K. Research on the mechanism of the spewing in the EPB shield tunnelling. Chin J Geotech Eng 2004;26(5):589–93.
[29] Ma L. Study on ground conditioning for EPB shield In water rich cobble ground. Tunn Constr 2010;30(4):411–15.
[30] He S, Zhang S, Li C, et al. Blowout control during EPB shield tunnelling in sandy pebble stratum with high groundwater pressure. Chin J Geotech Eng 2017;39 (9):1583–90.
[31] Shen X, Gao F, Wang F, et al. Test research on sediment improvement in EPB shield tunnelling through gravelly sand stratum. Railw Stand Des 2017;61(4): 121–5.
[32] Jakobsen P, Langmaack L, Dahl F, et al. Development of the soft ground abrasion tester (SGAT) to predict TBM tool wear, torque and thrust. Tunn Undergr Space Technol 2013;38(9):398–408.
[33] Barzegari G, Uromeihy A, Zhao J. Parametric study of soil abrasivity for predicting wear issue in TBM tunnelling projects. Tunn Undergr Space Technol 2015;48:43–57.
[34] Barbero M, Peila D, Picchio A, et al. Test procedure for assessing the influence of soil conditioning for EPB tunnelling on the tool wear. Geoing Ambientale e Mineraria 2012;49(1):13–19.
[35] Küpferle J, Zizka Z, Schoesser B, et al. Influence of the slurry-stabilized tunnel face on shield TBM tool wear regarding the soil mechanical changes—experimental evidence of changes in the tribological system. Tunn Undergr Space Technol 2018;74:206–16.
[36] Liu P, Wang S, Yang J, et al. Effect of soil conditioner on atterberg iimits of clays and its mechanism. J Harbin Inst Technol 2018;50(6):91–6.

[37] Hollmann F, Thewes M. Assessment method for clay clogging and disintegration of fines in mechanised tunnelling. Tunn Undergr Space Technol 2013;37(13):96–106.
[38] Peila D, Picchio A, Martinelli D, et al. Laboratory tests on soil conditioning of clayey soil. Acta Geotechnica 2015;11(5):1061–74.
[39] Zumsteg R, Puzrin A. Stickiness and adhesion of conditioned clay pastes. Tunn Undergr Space Technol 2012;31:86–96.
[40] Spagnoli G. Electro-chemo-mechanical manipulations of clays regarding the clogging during EPB-tunnel driving. Aachen: RWTH Aachen University; 2011.
[41] Sass I, Burbaum U. A method for assessing adhesion of clays to tunnelling machines. Bull Eng Geol Environ 2009;68(1):27–34.
[42] Mori L, Mooney M, Cha M. Characterizing the influence of stress on foam conditioned sand for EPB tunnelling. Tunn Undergr Space Technol 2018;71:454–65.
[43] Messerklinger S, Zumsteg R, Puzrin AM. A new pressurized vane shear apparatus. Geotech Test J 2011;34(2):1.
[44] Merritt S. Conditioning of clay soils for tunnelling machine screw conveyors. Cambridge: University of Cambridge; 2005.
[45] Houlsby G, Psomas S. Soil conditioning for pipejacking and tunnelling: properties of sand/foam mixtures. London: UCS; 2001, p. 128–38.
[46] Guo C, Kong H, Wang M. Study on muck improvement of EPB shield tunnelling in waterless sandy-cobble-boulder stratum. China Civ Eng J 2015;48(S1):201–5 (in Chinese).
[47] Wang S, Liu P, Hu Q, et al. State-of-the-art on theories and technologies of soil conditioning for shield tunnelling. China J Highw Transp 2020;33(5):8–34.
[48] Hu C, Zhang Y, Tan B. Soil conditioning experiments for EPB shield tunnelling in water-rich sandy cobble strata. Mod Tunn Technol 2017;54(6):45–55.
[49] Qu T, Wang S, Hu Q. Coupled discrete element–finite difference method for analysing effects of cohesionless soil conditioning on tunnelling behaviour of EPB shield. KSCE J Civ Eng 2019;23(10):4538–52.
[50] Yang D, Tan L, Li S. Physical properties and utilization of debris produced by earth pressure balance shield tunnelling. Geotech Invest Surv 2019;47(11) 17-22 + 34.
[51] Wu Z, Xiong S, Yao W, et al. Optimized design of the platforms at the slope bottom of construction solid waste landfill. Geotech Invest Surv 2018;46(06):7–12.
[52] Zhu K, Zhang Y, Xue Z, et al. Environmental issues and green treatment of shield residues. Urbanism Architecture 2018;29:108–10.
[53] Dai Y, Yang J, Zhang C, et al. Research on recycling of discarded soil produced from slurry shield in synchronous grouting materials. J Huazhong Univ Sci Technol (Natural Sci Ed) 2019;47(10):40–5.
[54] Hao T, Li X, Leng F, et al. Synchronous grouting materials for shield slag in silty clay of Zhengzhou metro. J Chang'an Univ (Natural Sci Ed) 2020; 40(03):53–62.
[55] Xu K. The research and application on high-performance grouting made by shield sediment. Wuhan: Wuhan University of Technology; 2011.
[56] Seco A, Omer J, Marcelino S, Espuelas S, Prieto E. Sustainable unfired bricks manufacturing from construction and demolition wastes. Constr Build Mater 2018;154.
[57] Espuelas S, Omer J, Marcelino S, et al. Magnesium oxide as alternative binder for unfired clay bricks manufacturing. Appl Clay Sci 2017;146:23.
[58] Jiang J, Yin B. Research on development of new wall materials by shield muck. Brick-Tile 2019;03:45–8.
[59] Yao Q, Cai K, Liu C, et al. Study on proportioning and mechanical properties of unfired bricks based on foundation pit muck in silty clay stratum. Tunn Constr 2020;40(S1):145–51.

[60] National Development and. Reform Commission of China. Non-fried rubblish gangue brick. (JC/T 422)-2007;2007.
[61] Hao T, Wang S, Li X, et al. Feasibility study on preparation of cement mixture by shield muck. Bull Chin Ceram Soc 2019;38(04):1018–23.

Exercises

1. Briefly describe the effects of muck conditioning on shield tunnelling.
2. List several common types of muck-conditioning agents, and briefly describe their working mechanisms and adaptable ground.
3. Briefly describe the common types of foam-generating systems and their working mechanisms.
4. Why does water content have an important influence on the effect of conditioning agents on the muck?
5. Analyze common methods and their characteristics for evaluating muck adhesion.
6. Briefly describe common methods and their characteristics for determining the shear strength of muck.
7. Discuss how to evaluate the effect of soil conditioning, and propose ideas for soil-conditioning evaluation based on field application and other perspectives.
8. Collect a typical soil conditioning case for EPB shield tunnelling, evaluate its technology, and propose ideas for optimization.

CHAPTER 10

Slurry treatment for shield tunnelling and waste slurry recycling

10.1 General

The slurry shield stabilizes the excavation face with pressurized slurry to form a filter cake in the excavation chamber. The slurry made by mixing water, bentonite, and additives is pressed into the slurry pool through a delivery pipeline. A soil material is used as a medium to stabilize the excavation face in slurry shield tunnelling. A mud chamber is formed between the excavation face and the bulkhead behind the cutterhead. When the cutterhead rotates to excavate the soil, discharge soil will be transported out through discharging pipelines [1–3].

The excavated soil continuously enters the excavation chamber during the slurry shield tunnelling, so the composition and properties of the slurry in excavation chamber constantly change. This change will cause the engineering properties of the slurry to deteriorate, so they will not be able to meet the engineering needs of excavation face stabilization, transportation, discharging cutterhead cooling, and smooth construction. Therefore the slurry should be handled or discarded in a shorter time. The slurry circulation process of the shield on site is shown in Fig. 10.1. The discharge soil enters the mud-water separation equipment with the slurry, and the sand sludge with larger particle size is first separated through screening, primary swirling, secondary swirling, and parameter detection of relative density and particle size. It is then mixed with the remaining slurry. If the slurry conforms to the requirements, it will be properly mixed and then be pumped onto the excavation chamber with slurry feeding pipelines to form a filter cake on excavation face. If not, it will be sent to the waste slurry pool for tertiary treatment, and then discharge soil with a smaller particle size but still not meeting the requirements will be screened out for dehydration [4,5].

The discharged soil are mixed and transported to the separation plant outside the tunnel by the slurry pump, and the slurry is recycled after separation.

Shield Tunnel Engineering
DOI: https://doi.org/10.1016/B978-0-12-823992-6.00010-2

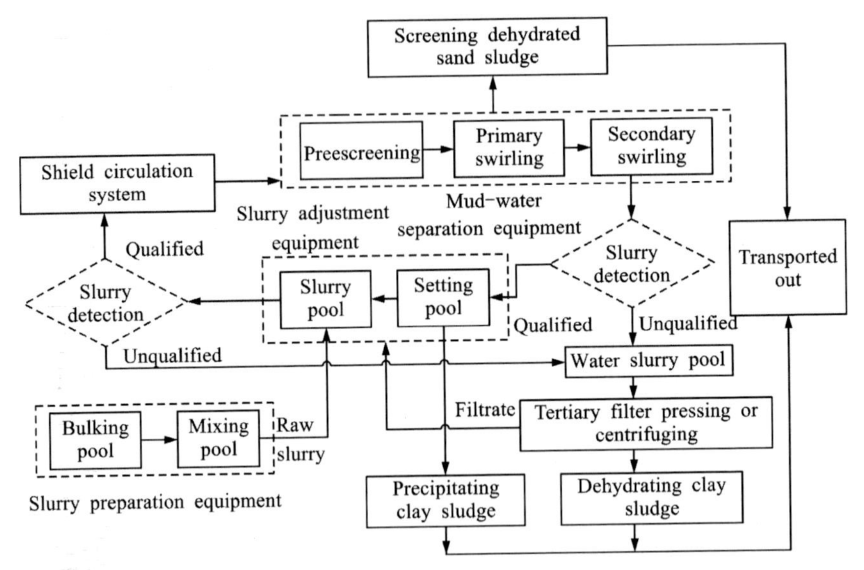

Figure 10.1 Slurry processing flowchart.

10.2 Shield slurry

10.2.1 Functions of slurry

10.2.1.1 Filter cake formation and excavation face stabilization

The filter cake is an impermeable or slightly permeable compact material layer formed by slurry under the action of pressure on the excavation face. It can reduce seepage forces in the strata ahead of excavation face and avoid large ground deformation and collapse, further ensure the safety and stability of the stratum and surrounding buildings (structures). The formation mechanism of the filter cake is shown in Fig. 10.2.

When the stratum in front of the excavation face is contacted by the slurry, because the slurry pressure on the excavation face is larger than the groundwater pressure of the stratum ahead, fine particles and water in the slurry flow into the stratum through stratum pores and fissures. The fine particles fill the stratum pores and fissures, resulting in a smaller permeability coefficient of the stratum. Owing to the clogging effect of aggregation in the soil and the bridging effect of coarse particles, the permeability coefficient reduces, the amount of water in the slurry flowing through the voids (the dehydration amount) becomes less, the increase in excess groundwater pressure (the water pressure in the stratum voids rises due to dehydration, and the increased part of the water pressure in the stratum void is the excess groundwater pressure) is smaller, the excess groundwater

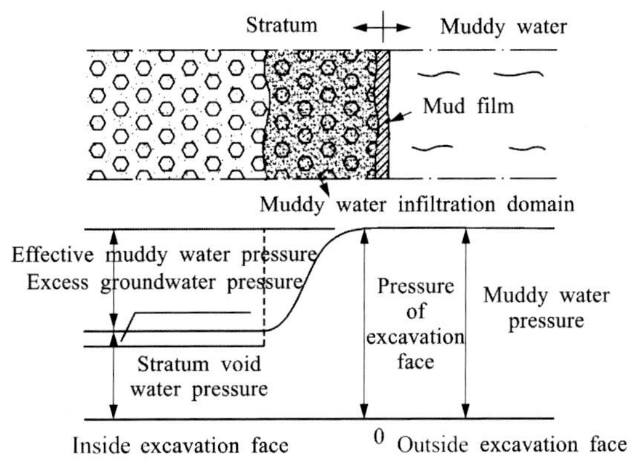

Figure 10.2 Schematic diagram of mud film formation mechanism.

pressure is finally stabilized at a certain value, the impermeability of the excavation face is improved, and negatively charged clay particles in the slurry are absorbed and aggregated onto oppositely charged soil particles on the excavation face under the action of electrostatic attraction, forming the filter cake. With the formation of the filter cake, the slurry in the excavation chamber is prevented from entering the stratum, and the groundwater in the stratum is prevented from entering the excavation chamber, thus achieving a bidirectional isolation effect and ensuring stabilization of the excavation face.

10.2.1.2 Discharging excavation soil with slurry

The soil excavated by shield cutterhead needs to be transported to the ground in the form of slurry. In the actual construction process, whether the excavated soil can be successfully removed depends on whether the slurry flow rate is larger than the minimum slurry flow rate of fine particles without deposition and the minimum starting slurry flow rate of coarse particles such as blocks. Besides, the slurry needs to have good viscosity, which not only can ensure that the stratum soil or rock debris is suspended in the slurry without deposition, but also can prevent pipeline blockage and reduce the wear of the pipeline due to particle deposition during transportation.

10.2.1.3 Cutterhead washing, lubrication, and cooling

When the cutterhead cuts the strata, the acting force (thrust and torque) between the cutterhead and the strata is very large, easily generating a large amount of heat and even having a rick of fatiguing and deforming the cutterhead. By washing of the slurry, a large amount ofclayed soil sticking to the cutterhead due to cutting can be removed, as can the heat, thereby cleaning the cutterhead, reducing the load on the cutterhead, and effectively reducing the thrust of the shield. In addition, the slurry can play a role in lubricating the tunnel face and the cutter and reducing the torque of the cutterhead to a certain extent.

10.2.2 Composition of shield slurry

The slurry is a suspension composed of water, granular materials (bentonite, pottery clay, limestone, silt, and fine sand), and additives (chemical agents), in which the water accounts for 70%–80% and the solid particles account for 20%–30% [6–9].

10.2.2.1 Clay (bentonite)

Clay is composed mainly of various types of clay minerals and other mineral impurities. The clay minerals include montmorillonite, kaolinite, illite, and the like, and the mineral impurities include unweathered rock debris, quartz sand, feldspar, mica, calcite, pyrite, organic impurities, and the like. Bentonite is a kind of clay. It is the most commonly used material in slurry preparation because montmorillonite is its main composition and it has good hydration and dispersion properties and high slurry-making ability. Two types of bentonite are used: sodium-based bentonite and calcium-based bentonite. Because clay is the main component of the slurry, the clay that is excavated and discharged in the muddy water should be used to the greatest extent possible to reduce the cost of raw materials. To obtain better engineering performance in the construction process, it is necessary to screen recycled clay; clay particles with particle diameter less than 0.075 mm are reserved.

10.2.2.2 Carboxymethyl cellulose

Carboxymethyl cellulose (CMC) is a high molecular polymer compound remaining after the chemical treatment of wood and bark. CMC has extremely high viscosity when dissolved in water, and it can be used as a thickening agent in gravel layers. It can reduce the water filtration amount, prevent mud escape, and resist cationic pollution.

10.2.2.3 Water

The main liquid phase material of the slurry should be clean tap water on site. If groundwater or river water is used as a water source for conditioning the slurry, a water quality inspection and a muddy water blending test should be conducted, impurities must be removed, and the pH must be adjusted.

10.2.2.4 Sand

When the shield advances in a loose strata with large voids, a certain amount of sand needs to be added to the slurry to fill voids in the strata ahead of excavation face, and the particle size of the sand is confirmed according to the permeability ratio ($n = 14-16$).

10.2.2.5 Other additives

In the slurry preparation process, appropriate additives can be added to adjust the performance of the slurry according to the site conditions. Polyacrylic acid (PAA) (acrylic acid modified resin) can facilitate the penetration of the slurry into the soil particle voids in the excavation face, cement the soil on the excavation face, and further form a filter cake layer with a certain thickness, thereby reducing mud escape and preventing groundwater from surging into the excavation chamber. Dispersing agents such as sodium humate and sodium carbonate can effectively reduce the density and viscosity of the slurry, and acid agents such as dilute sulfuric acid and phosphoric acid can neutralize the alkaline components in the slurry to prevent deterioration of the slurry performance.

In actual engineering, the additives are used only in traditional slurry because of its high cost, poor environmental protection, and inability to form a filter cake. The additives are usually added to the base slurry in the form of an aqueous solution. Polymer additives that are difficult to dissolve in water need to be dissolved in water to form a uniform solution by stirring and other means and then be added to the base slurry.

10.2.3 Shield slurry performance and index

10.2.3.1 Physical stability

Physical stability is the ability to maintain the clay particles in the physical state of floating and dispersing after the slurry stands for a long time. The slurry stability can be indicated by the interface height. The interface height is the height between the clear water precipitated at the top and the slurry interface after a certain amount of slurry, placed in a measuring cylinder, stands for some time. The smaller the change in interface height with time, the better the physical stability of the slurry.

10.2.3.2 Chemical stability

Chemical stability refers to the chemical deterioration of the filter cake function after impurities with positive ions have been mixed in the slurry. According to research, the main reason for the deterioration is that the negatively charged slurry particles will be converted from a suspension state to an agglomeration state by the impurities with positive ions in the slurry, thereby increasing the viscosity of the slurry and making the formation of a filter cake more difficult.

10.2.3.3 Relative density

Generally speaking, the larger the relative density of the slurry, the better the filter cake formation, the lower the groundwater pressure in strata ahead of excavation face, and the smaller the deformation of the excavation face. Moreover, the buoyancy of the excavated soil in the slurry with higher relative density is also greater, which can ensure the smooth delivery of the excavated soil in the discharging pipelines. Certainly, slurry with a large relative density is high in flowing friction and poor in fluidity, which will easily make the slurry transport pump run in an overload manner and increase the difficulty of slurry separation. Slurry with a small relative density has the characteristics of low flowing friction and good fluidity, but there are still drawbacks, such as slow filter cake formation speed, which is not conducive to the stability of the excavation face. Therefore slurry with a reasonable relative density should be determined in the actual construction process to ensure normal operations.

10.2.3.4 Viscosity

In the construction process, the slurry should be kept within a certain viscosity range to prevent particle components of the slurry from being deposited in the cabin, to prevent mud from escaping, and to ensure smooth transportation of the excavated soil. The viscosity values of the slurry commonly used in engineering are shown in Table 10.1.

10.2.3.5 Water loss

Filter loss is the amount of water in theslurry flowing into the strata through the voids in the filter cake formation process. The groundwater pressure (pore water pressure) of the stratum will increase by water filtration, and the increased part of the pore water pressure in the strata is called excess pore pressure. Besides, too much water filtration will cause

Table 10.1 Viscosity values (S) of slurry for stabilizing excavation face.

Excavated soil	Viscosity measured by funnel viscometer (500 mL)	
	Situation in which groundwater has little impact	Situation in which groundwater has a large impact
Silt with sand	25–30	28–34
Sandy clay	25–30	28–37
Sandy silt	27–34	30–40
Sand	30–38	33–40
Gravel	35–44	50–60

the excess groundwater pressurein the strata to increase; that is, the effective pressure of the muddy water will decrease.

10.2.3.6 Permeability ratio

Because of particle blocking factors, the ratio n of the diameter L of the stratum pore to the effective diameter G of clay particle in the slurry is commonly used to determine whether the slurry can form the filter cake on the excavation face. If $n<2$, the slurry particles cannot permeate the stratum; if $2<n<4$, the slurry particles can permeate the stratum; if $n>4$, the slurry particles will flow out through pores.

Under conditions of sand and gravel formation, it is necessary to replace L and D with $0.2D_{15}$ and D_{85}, respectively. D_{15} is the grain diameter (mm) corresponding to 15% passing, by weight, and D_{85} is the grain diameter (mm) corresponding to 85% passing, by weight.

10.2.3.7 pH

Excessively high concentration of positive ion impurities in the slurry will corrode equipment and pipelines. Studies have shown that the pH of slurry that has not been polluted by positive ions ranges from 7 to 10, while the pH of slurry that has been polluted by positive ions far exceeds 10. Therefore the degradation degree caused by positive ions can be judged by measuring the pH on the construction site to identify the chemical stability of the slurry.

10.2.4 Shield slurry performance requirements

Generally speaking, the related properties of the slurry need to be evaluated before the slurry enters the shield machine, and whether the slurry is qualified or not is evaluated according to relative density, viscosity, and water loss in projects. Furthermore, the engineering properties that the slurry needs to meet are quite different under different stratum conditions, so the stratum properties should be comprehensively considered in the slurry configuration. The following is a detailed description of the slurry preparation steps:

1. Calculating D_{15} of the tunnelling stratum through a particle size test in the geological survey.
2. Selecting the type of bentonite, and determining the particle size gradation accumulation curve of the bentonite.
3. Selecting two or three kinds of granular additives to mix with the selected bentonite.
4. Adding a thickening agent and a dispersing agent to the mixed liquid of the selected bentonite and the granular additives. According to the standard quality with a relative density of 1.2, a funnel viscosity of 25–30 seconds, and an n value of 14–16, the excess groundwater pressure output, corresponding to the muddy water proportioning determined by the method, of the excavation stratum is the minimum, and the slurry characteristic is the best.

A silty clay stratum has a small permeability coefficient and small pores, and a filter cake forms easily on the excavation film. The content of silty clay in the stratum particles is high, and the self-slurry-making ability of the stratum is strong, but the relative density parameter of the slurry is difficult to control. Furthermore, excessive sludge particles in the slurry are small in size and cannot be easily separated by the slurry separation equipment, so the slurry will have a high relative density. In a silty–fine sand stratum, the stability of the excavation face is particularly important if the stratum has a single particle size and poor gradation. To ensure the stability of the excavation face, a compact filter cake must be formed on the excavation face. However, the silty–fine sand stratum has a small particle size, and the slurry cannot easily penetrate the stratum to form a permeability zone; it is stuck on the excavation face, where it can form a filter cake. Therefore it must be ensured that the filter cake with high compactness and strong antidisturbance ability is formed on the excavation face. A gravelly sand or gravel stratum has high permeability, and the content of fine particles is low. Therefore the self-slurry-making ability of the stratum is poor, and it is necessary to add fine particles such as clay to the muddy water system to ensure normal construction of the project. Meanwhile, because of the large permeability coefficient of the strata, a certain amount

Table 10.2 Slurry proportion for typical projects and their properties.

Projects	Stratum condition	Material proportioning	Relative density ($g \cdot cm^{-3}$)	Viscosity (s)	Loss water (mL)
Lanzhou Metro Line 1	Sandy cobble stratum	Water:clay:bentonite:foaming agent = 800:(80−120):(80−120):120:70	1.15−1.2	25−35	15−25
Underground Diameter Line from Beijing Railway Station to Beijing West Railway Station	Sandy cobble stratum	Red modified sodium bentonite: PAA:positive gel: clay = 4:0.0038:0.0028:5.940	1.08	30	18
Shenyang Metro Line 10	Water-rich grit (cobble) stratum	Water:bentonite:NaOH: CMC = 1000:100:1.1:2.2	1.05	26.5	13.4
Yangtze River crossing piping engineering in Tianxing Zhou, Wuhan	Artificial fill layer, new lacustrine formation, clay layer, silty soil, sandy soil, and silty-fine sand	Qingshan field soil:sodium carbonate:CMC:LG: XC = 700:400:40:40:3	1.09	50−70	11.5
Power central line tunnel in Hiroshima, Japan	Boulder bed	Water:general clay:bentonite: CMC = 875:270:35:0.5	1.14−1.25	25−35	5

of fine sand must be contained in the slurry to help plug the pores of the stratum to prevent instability of the excavation face caused by a large amount of slurry loss to the stratum.

The proportioning of the slurry needs to be verified by laboratory and field tests. Past engineering experience has important significance for reference in slurry preparation. The stratum conditions, the slurry proportioning, and the corresponding slurry properties of typical engineering projects are summarized in Table 10.2.

10.3 Slurry treatment site layout and equipments

10.3.1 Slurry treatment site layout

As one of the core components of the slurry shield tunnel, the importance of the slurry treatment system becomes evident. Reasonable determination of the layout of the slurry treatment system has a very significant effect on the quality and efficiency of on-site construction [10].

The slurry treatment system is composed of five parts: a mud-water separation system, a waste slurry system, a raw slurry mixing system, a water taking system, and a slurry treatment control system. It includes a waste slurry storage tank, a waste slurry treatment tank, a raw slurry storage tank, an adjustment tank, a transition tank, a sedimentation tank, a clean water tank, a mud-water separation equipment foundation, a raw slurry material warehouse, a bentonite mixing tank, and an equipment pit. The sedimentation tank, slurry mixing tank, raw slurry storage tank, bentonite mixing tank, and clean water tank are arranged in as concentrated a form as possible, and the waste treatment area is arranged according to the principle of convenient mechanical transportation. The area of each region should be calculated according to the principle of facility capacity. The slurry treatment system should be kept as far away as possible from the living area, office area, and external housing to avoid noise pollution and should be close to the large yard to prevent a large amount of mud from being poured inside and affecting efficiency. Furthermore, loud slurry treatment equipment should be concentrated at corners to minimize the impact of the noise. Fig. 10.3 shows a typical layout.

The slurry treatment process can generally be subdivided into primary treatment, secondary treatment, and tertiary treatment. The primary treatment separates coarse particles with particle size larger than 74 μm, such as gravel, sand, and clay agglomerates discharged from the slurry, and its treatment equipment may include a vibration screen, a centrifuge, or a

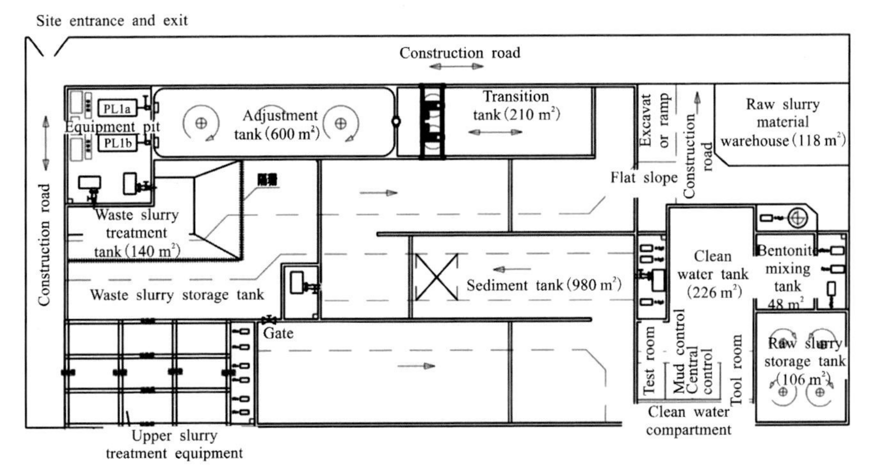

Figure 10.3 Layout plan of a slurry treatment site. Source: *Modified from Tang J. Design of slurry treatment system and layout of slurry treatment yard for slurry balance shield tunnel. Constr Mater Decor 2019(13):232-233.*

combination of the two. The secondary treatment further separates soil (fine-grained components) and water (aggregation and dehydration) from the remaining slurry after primary treatment, and its treatment equipment includes aggregation separation equipment, dehydration equipment, and so on. The tertiary treatment will process the water with high pH after the secondary treatment, such as drainage in the pit into water that meets the discharge standard for discharge, and treatment usually neutralizes the wastewater with sulfuric acid to reduce its pH. The working process of the slurry treatment equipment is shown in Fig. 10.4.

10.3.2 Slurry treatment equipments

The slurry treatment equipments can be divided into two parts. One part includes a screening unit, a cyclone separation unit, and a slurry mixing unit, which is used for improving the composition and characteristics of the slurry so that the related indexes of the slurry meet the engineering requirements again. In the other part, the remaining slurry is separated through the dehydration unit to further make it harmless and to reduce the remaining slurry. In actual engineering, there are big differences in the application of the slurry treatment equipment. Geological conditions, treatment flow, and economic cost are all deciding factors in the selection of the slurry treatment equipment. For example, for slurry generated

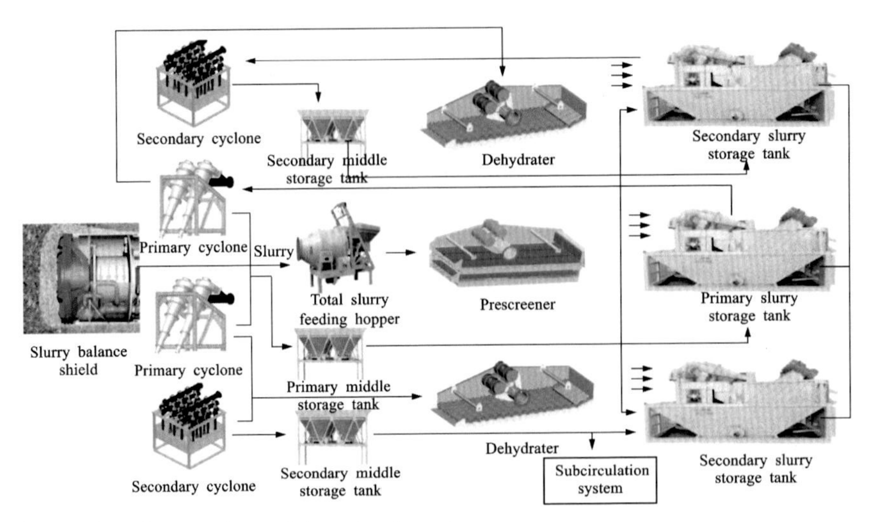

Figure 10.4 Process flow of slurry treatment.

under stratum conditions dominated by fine sand, siltstone, shale, and so on, vibration screens or rolling screens, cyclones, and other separation equipment are often used to separate coarse particles in the slurry, while fine particles in the slurry enter the equipment for processing. In a weak saturated water-bearing layer dominated by clay, silt, silty clay, sludgy clay, and other fine particles (0.1–0.0035 mm), the use of the above equipment alone usually has little effect on mud separation, and it is necessary to use other mud separation methods such as sedimentation in slurry treatment to achieve the necessary effect.

10.3.2.1 Screening unit

Screening is a common equipment for slurry treatment (Fig. 10.5). The slurry discharged from the shield machine is first screened by the prescreener to remove slurry particles with a diameter greater than 2 mm.

In the prescreener, the slurry is vibrated through a single-layer or multilayer screen with uniformly distributed holes so that the slurry particles with particle size larger than the sieve are trapped on the drying surface while the slurry particles with particle size smaller than the sieve pass through the screen. Under the action of the vibration of the screen box, the structure of the slurry particles on the screen becomes looser, increasing the possibility of the fine particles in the upper layer of the screen being transferred to the lower layer through the gap so that the originally

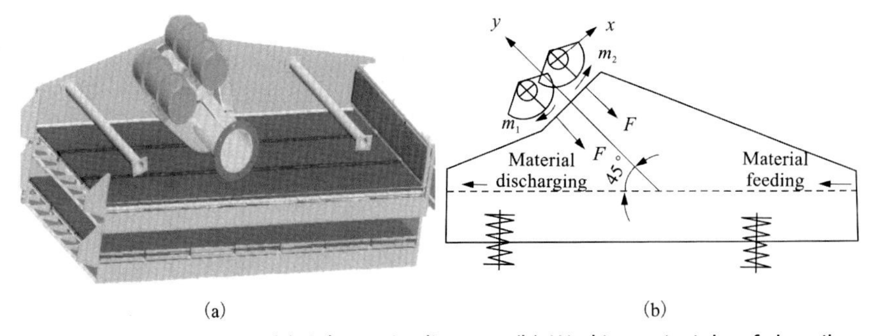

(a) (b)

Figure 10.5 Prescreener. (a) Schematic diagram. (b) Working principle of the vibration screen.

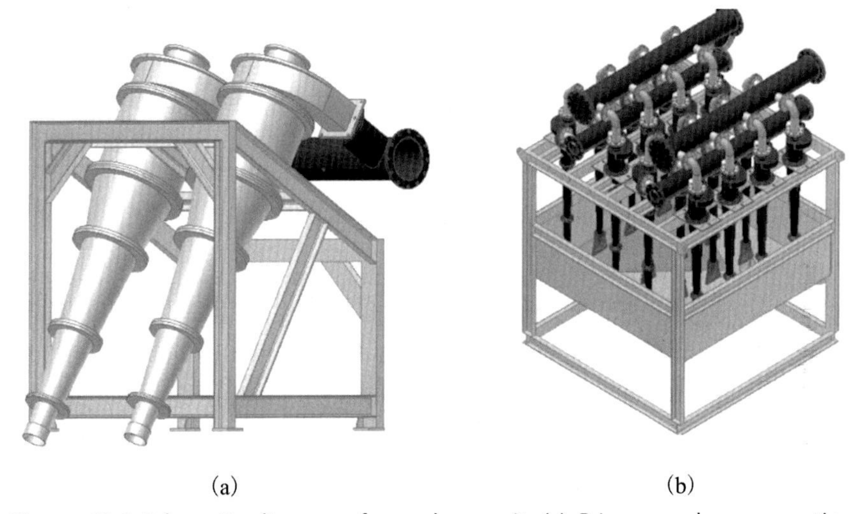

(a) (b)

Figure 10.6 Schematic diagram of a cyclone unit. (a) Primary cyclone separation unit. (b) Secondary cyclone separation unit.

slurry particles are distributed such that the fine particles are below and the coarse particles are above.

10.3.2.2 Cyclone separation unit

The cyclone separation unit (as shown in Fig. 10.6) is divided into a primary cyclone separation unit and a secondary cyclone separation unit. The function of the primary cyclone separator is to separate the particles with particle size of 74 μm to 2 mm in the prescreened slurry, and the function of the second stage cyclone separator is to separate the particles with particle size of 4/9—4 μm in the slurry separated by the primary cyclone. The working principle of the cyclone separator is mainly to use the different centrifugal force, resistance, and centripetal buoyancy between the coarse and fine particles of the slurry. The result of centrifugal sedimentation is that the coarse particles will overcome the hydraulic resistance and move down the wall to form an external eddy current, which is then discharged through the bottom outlet of the cyclone separator, while most of the fine particles will flow out of the overflow pipe with the liquid, thus separating the fine particles from the coarse particles.

10.3.2.3 Slurry mixing unit

The slurry mixing unit is a key part of slurry shield construction and is divided into slurry mixing when the shield originates and slurry mixing during normal shield advancing. During originating, clay or bentonite is generally used to mix raw slurry in a mixing tank. The slurry should conform to the requirements of stratum film formation, and quality control can be conducted according to the properties of relative density, viscosity, stability, and particle gradation of the slurry. During shield advancing, the coarse particles in the slurry are filtered after passing through the whole circulation system of the slurry shield. The remaining slurry can be recycled, but it must be tested to ensure that the properties of the slurry meet the requirement of stratum film formation. If not, the slurry needs to be adjusted. During the adjustment the slurry can be pumped from the slurry mixing tank to the slurry mixing tank, and the slurry mixing is done to meet the requirements of formation film formation.

The main precautions for slurry mixing during slurry shield tunnel construction are as follows:

1. The properties of the slurry should meet the requirement of stratum film formation.
2. The properties of the slurry in each type of stratum after passing through the circulation system and entering the slurry mixing tank again are preliminarily estimated according to the principles of particle gradation and mass conservation so as to further determine how much sludge needs to be discharged and how much bentonite, clay, sodium CMC, and other slurry-making materials need to be used and when the remaining slurry should be abandoned.
3. Before slurry mixing, various stratum film formation tests that may be encountered in shield advancing should be carried out, and the slurry properties corresponding to the formation of stable and dense filter cake in various strata are proposed. Especially for strata (such as coarse

sand, gravel, and cobble strata) with high permeability, a more systematic film formation test should be carried out, and greater economic efficiency can also be achieved.

4. The waste slurry should be recycled as much as possible to reduce consumption of more slurry mixing materials and reduce the waste slurry transportation cost. For example, if the shield-originating stratum is silty clay, the clay solid particles in the slurry cannot be easily removed by the filtering system, resulting in a large amount of slurry waste, which is actually a good slurry-making material. In a coarse sand and gravel stratum, a large amount of slurry-making agents and bentonite needs to be added, owing to lack of fine particles in the stratum and poor self-slurry-making ability. At this time, if the slurry that was abandoned previously can be supplemented, a large amount of slurry-making materials can be saved, and considerable economic benefits are obtained.

10.3.2.4 Mechanical dehydration unit

Mechanical dehydration uses pressure difference or density difference between the two sides of a filter medium as a driving force to separate the solid particles from water in the slurry. The water content of the treated slurry is generally lower than 60%, which meets the requirement of loading and transportation. Commonly used forms of mechanical dehydration include filter pressing dehydration, centrifugal dehydration, and vacuum filtration. In addition, the particle size of the slurry processed by the screening unit and the cyclone separation unit is generally less than 20–45 μm, causing difficulty in mechanical dehydration and often needing an agent-dosing unit.

Dehydration of filter press

A filter press (shown in Fig. 10.7) is a commonly used dehydration machine. The filter press can be divided into a plate and frame type press filter, a box type filter press, a vertical filter press and a belt press filter. The plate and frame type press filter has the advantages of simple operation, low cost, good separation effect, and high material adaptability, especially for processing of waste slurry with high clay content, and is widely used in waste slurry filter pressing. This method also has the drawbacks of a large area and intermittent operation. It is worth noting that different types of filter presses have their own advantages and limitations in terms of processing efficiency, processing effect, economy, and material

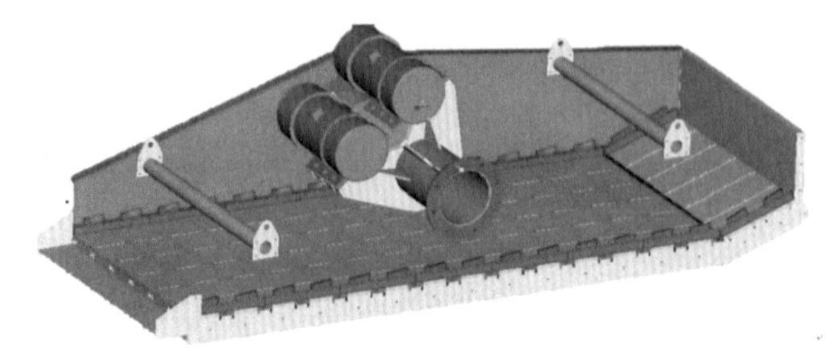

Figure 10.7 Schematic diagram of a dehydration unit.

adaptability. Therefore technicians should consider the technical economy of filter press models in combination with the actual situation in selecting the most suitable filter press model.

Centrifugal hydration

According to Stokes' law, the particle sedimentation velocity is proportional to the density difference between the solid phase and the liquid phase, inversely proportional to the viscosity of the liquid, and proportional to the square of particle diameter. For a heterogeneous system with a small density difference between the two phases, large viscosity, and small particle size, it is more difficult to centrifugally separate the particles, and it is often necessary to extend the centrifugal separation time to achieve better separation. The separation efficiency can be effectively improved by adding coagulants to increase the rotation speed of the drum and reduce the viscosity, and the slurry particles are deposited on the surface of the drum as early as possible.

In terms of the separation mechanism, centrifuges may have a centrifugal sedimentation form or a centrifugal filtration form. Centrifugal sedimentation generally refers to the process of centrifugal separation of waste slurry on a nonporous drum, and its separation process can be divided into the three steps of solid-phase sedimentation, sediment compaction, and partial removal of liquid from sediment pores. The process of centrifugal separation of the slurry with a perforated drum is called centrifugal filtration, and its separation process can be divided into the four steps of solid-phase sedimentation, liquid filtration and filter residue formation, filter residue compaction, and removal of the liquid retained by the molecular force in the filter residue. Generally speaking, centrifugal separation has the characteristics of high processing efficiency, small occupancy area, and high material adaptability. The efficiency of centrifugal separation can be effectively improved by adjusting operation parameters such as differential speed, rotation number, and processing flow of the centrifuge in combination with related properties of the slurry.

Vacuum filter dehydration

Vacuum filter dehydration involves the use of the pressure difference formed between atmospheric pressure and the generated vacuum to overcome the resistance of a filter material layer and trap the solid phase in the filter material layer while the solid liquid phases with small pore diameters pass through the filter material layer. In terms of solid-liquid separation, vacuum filtration and filter pressing have high similarity. The difference between the two treatment methods lies in the way the pressure difference of the filter material layer is applied. Since the vacuum filter needs to be maintained frequently during operation and has poor adaptability to the fluctuation of solid concentration and particle size distribution of the materials, it is rarely used in actual slurry treatment projects.

10.3.2.5 Coagulation pretreatment

Slurry has the characteristics of high stability, small particle size, and high water content. These characteristics make it difficult for mechanical dehydration to achieve the desired effect. To improve dehydration of the slurry, it is necessary to add coagulants to the slurry in advance before the slurry is mechanically dehydrated. Coagulation is the aggregation process of colloidal particles and tiny suspended matters in water. This aggregation process aims to break the long-term dispersion and suspension characteristics of colloidal particles in water.

Coagulants

Coagulants have three kinds of coagulation effects on colloidal particles in water: electric neutralization, adsorption bridging, and sweeping net capture. Under different water quality conditions, adding different types and dosages of coagulants will produce different mechanisms of action. Table 10.3 lists the commonly used inorganic coagulants.

Coagulant aids

When the coagulant alone cannot achieve the desired effect, a coagulant may be used as an auxiliary to improve the structure of the coagulant. Commonly used coagulants in engineering are bone glue, polyacrylamide and its hydrolysate, active silicic acid, and sodium alginate.

Table 10.3 Commonly used inorganic coagulants.

Title		Chemical formula	Applicable conditions
Aluminum series	Aluminum sulfate	$Al_2(SO4)_3 \cdot 18H_2O$ $Al_2(SO4)_3 \cdot 14H_2O$	Relatively narrow optimal pH range: 6–8 Temperature: 20–40°C
	Alum	$KAl(SO4)_2 \cdot 12H_2O$ $NH_4Al(SO4)_2 \cdot 12H_2O$	
	Poly aluminum chloride	$[Al_2(OH)_nCl_{6-n}]_m$	Wide optimal pH range: 5–9 Low-temperature and low-turbidity or high-turbidity slurry environment
	Poly aluminum sulfate	$[Al_2(OH)_n(SO_4)_{3-n/2}]_m$	
Iron series	Ferric trichloride	$FeCl_3 \cdot 6H_2O$	Optimal pH range: 8.5–11 High-turbidity slurry environment
	Ferrous sulfate	$FeSO_4 \cdot 7H_2O$	Optimal pH range: 8.5–11 Slurry environment with high alkalinity and hardness
	Poly ferric sulfate	$[Fe_2(OH)_n(SO4)_{3-n/2}]_m$	Wide optimal pH range: 5–11 Low-temperature and low-turbidity or high-turbidity slurry environment
	Poly ferric chloride	$[Fe_2(OH)_nCl_{6-n}]_m$	Appropriate pH: 6–9
Organic coagulant	Polyacrylamide	$(C_3H_5NO)_n$	Appropriate pH: 6–9
	Polyoxyethylene	H-(-O-CH2-CH2-)n-OH	Wide applicable range, high polymerization and microtoxicity

The coagulant is usually a high molecular polymer, which can absorb polymer chains onto several attachment points on the surface of slurry particles, and most of the remaining polymer chains are projected into the surrounding solution and adhere to other slurry particles. Broadly speaking, all agents that can enhance or improve the flocculation effect of coagulants are considered to be coagulants. In engineering practice, lime-based alkaline substances are often added before filter pressing of the slurry, which can promote the hydrolysis reaction of the coagulant and can also increase the permeability of the filter cake and promote the filtration of the water in the filter cake through the filter cloth.

10.4 Case study on shield slurry treatment

10.4.1 Project profile

The river-crossing tunnel on Wangjiang Road in Hangzhou, a large-diameter river-crossing tunnel projects, is located between the Qianjiang Third Bridge (Xixing Bridge) and Qianjiang Fourth Bridge (Fuxing Bridge), with an upstream distance of 2.4 km from the Qianjiang Fourth Bridge and two banks connecting to Wangjiang East Road in Shangcheng District and Jianghui Road in Binjiang District. The tunnel has a total length of 1837 m and a buried depth of about 12—22 m. It was constructed by two large-diameter slurry shield machines with a segment length of 2.0 m. The excavation diameter of the shield machine was 11.75 m, the maximum tunnelling speed reached 45 mm/minute, and the slurry output per ring was about 1900 m^3. A large amount of slurry needed to be treated during excavation, and the maximum slurry output during the construction period was about 2081 m^3. In addition, the water in this area belongs to the secondary protection zone for drinking water resources. Improper slurry treatment would have endangered the safety of drinking water for the population. According to relevant government regulations, the slurry had to be treated in a concentrated manner before loading and transportation, and the treated slurry and wastewater could not cause secondary pollution to the environment. Therefore this project was equipped with a centrifuge and a filter pressing system to form a complete slurry treatment system, which greatly improved working efficiency (Fig. 10.8).

10.4.2 Engineering geology

10.4.2.1 Hydrological situation

The tunnel is perpendicular to Qiantang River. The cross-section width of this river is about 1300 m, and the ground elevation of coastal area is

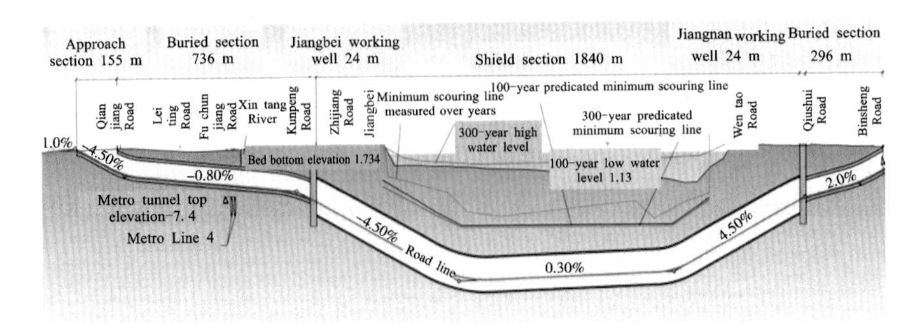

Figure 10.8 Longitudinal profile of the river-crossing tunnel on Wangjiang Road in Hangzhou.

generally 5.0–7.0 m. The terrain is generally open and flat. The riverbed elevation of Qiantang River is generally −2.30–0.85 m. Owing to erosion of channels and main streams in some sections, the water depth is 10–13 m, and the relative elevation is about −8 to −12 m. The buried depth of the tunnel in river is about 12–22 m, and the maximum water pressure is 0.42 MPa.

10.4.2.2 Stratum geology

The shield section is mainly located under Qiantang River. The main excavation layers are silty clay intercalated with silt sand, silty clay, sandy silt intercalated with silty clay, silty clay silt, and gravels. The soil particles in the soft and hard layers of the shield crossing vary greatly in shape, getting larger as the soil layers go down.

Fig. 10.9 shows the proportion of each layer along the shield tunnel. The shield tunnelling stratum is dominated by silty clay, which is a kind of soft soil with the characteristics of low shear strength, high compressibility, low permeability, and high natural water content. Under such geological conditions the circulation slurry from the slurry shield can better meet the requirements of film formation of the excavation face without adding material or by adding only a small amount of bentonite and polymer materials.

10.4.3 Slurry disposal

This river-crossing tunnel adopted ZXS II-2500/20 slurry separation equipment. This equipment consists of a screening unit, a primary desanding unit, a secondary desanding unit, a vibration screening, and a dehydration unit (i.e., a slurry storage tank flushing unit). In addition, considering the large slurry treatment capacity of the project and the high content of

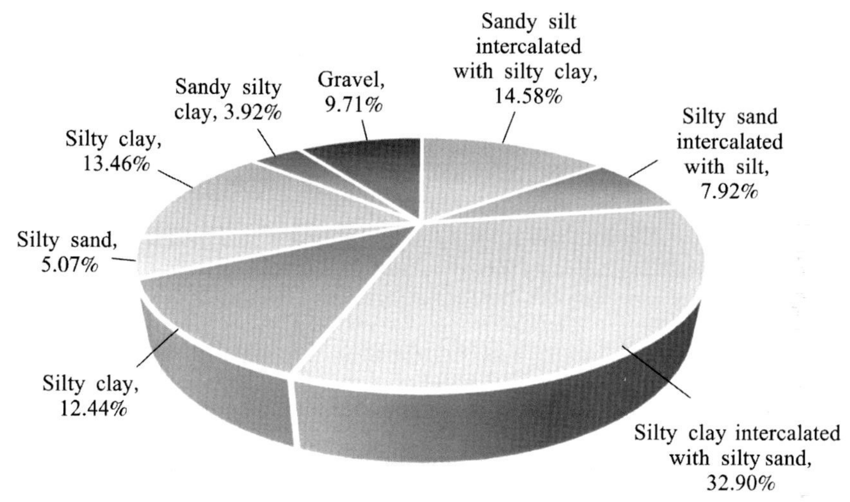

Figure 10.9 Proportion of layers passed by the shield along the line.

fine slurry particles in specific strata, six sets of ZXYL-60 filter pressing systems and a set of Centrisys centrifuge system were applied as a mechanical dehydration unit in the slurry treatment equipment unit.

10.4.3.1 Screening unit

To ensure the separation effect of coarse sand, a double-layer vibration screen with a downslope angle was installed. The upper layer of the vibrating screen was installed with a stainless steel tensile sieve plate measuring 1830 × 4270 × 6 mm, the rear one and lower one being installed with stainless steel slotted sieves measuring 1830 × 370 × 3 mm and 1830 × 4270 × 3 mm, respectively. The sieve plate adopted a unique tensioning method, and its produced secondary vibration effectively prevented screen blocking and pasting. It had a significant effect on separation of clay blocks, gravel sand, and slurry and did not easily block meshes. The treatment capacity of a single machine can reach 13,000 m^3/h, and the peak value of coarse sand can reach 96 T/h. In addition, the upper sieve plate of the prescreener adopted a tensioned stainless steel screen, which was welded by orthogonal stainless steel bars, with φ10 steel bars in a tension direction at a spacing of 40 mm and φ6 steel bars in a flow direction at a spacing of 12 mm, to form a good elastomer (Fig. 10.10).

10.4.3.2 Cyclone separation unit

The cyclone separation unit had two stages: a primary cyclone unit and a secondary cyclone unit. Both were provided with vacuum-controlled

(a)

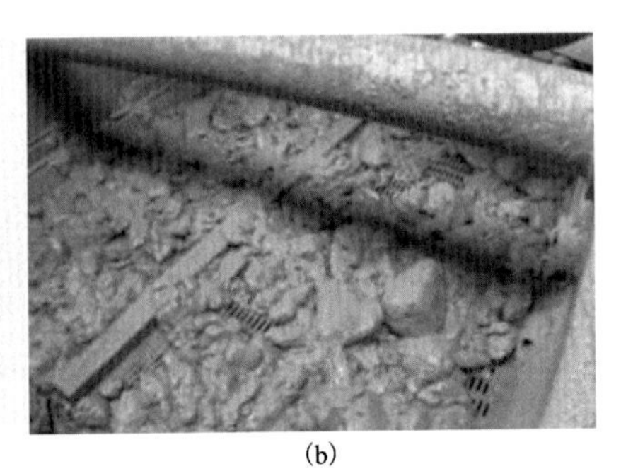

(b)

Figure 10.10 Schematic diagram of the screening unit. (a) Double-layer vibration screen. (b) Stainless steel tensile sieve plate.

underflow discharge ports, whose function was to facilitate reasonable adjustment of the underflow density, accelerate formation of a cushion layer on a dehydration screen surface, and provide suitable slurry for the downstream dehydration screen to speed up the formation of the cushion layer on the dehydration screen surface.

In the primary cyclone unit treatment, the prescreen slurry entered the primary slurry storage pool and then was transported to six sets of $2 \times \varphi 500$ cyclone groups for desanding by two stock pumps. The primary cyclone adopted a $\varphi 500$ variable-tape-angle cyclone for separation, and the primary slurry inlet pressure was controlled to 0.15–0.18 MPa. At the same time, a fishtail device for underflow concentration adjustment and a vacuum adjustment device were arranged.

In the secondary cyclone unit treatment, the slurry after primary desanding entered the secondary slurry storage pool and then was transported to the secondary cyclone for desilting by two stock pumps. The secondary cyclone adopted an imported 150 mm small-diameter, multicone cyclone for separation. At the same time, a fishtail device used for underflow concentration adjustment and a vacuum adjustment device were arranged as shown in Fig. 10.11.

10.4.3.3 Mechanical dehydration unit

In the centrifugal dehydration unit an American Centrisys centrifuge system was used. The system included a screw pusher, a rotating drum, a cover, a hydraulic differential, main and auxiliary motors, a shock absorber, and an engine base. With the increase of advancing distance, the specific gravity of the slurry will continuously increase, affecting the normal circulated slurry dreg carrying capacity. The maximum slurry

(a)

(b)

Figure 10.11 Schematic diagrams of primary and secondary cyclones and vacuum adjustment device. (a) Primary and secondary cyclones. (b) Vacuum adjustment device.

processing capacity of the centrifuge could reach 200 m^3/h, separating the slurry into sludge with a smaller water content and wastewater that met the discharge standard. In addition, before the slurry entered the centrifuge, the inorganic coagulants or organic coagulants (i.e., anionic coagulants, cationic coagulants) were added to improve the processing performance of the centrifuge according to the construction situation.

In the filter pressing dehydration unit, the filter press adopted a ZXYL-60 filter pressing system. This system mainly includes a box type filter press, a filter press pump, a slurry tank, an agitator, an air compressor, an air tank, a slurry pipeline, a pneumatic pipeline, and various control valves. When the shield machine advanced through the clay stratum, the separation index of the cyclone decreased, owing to a large amount of fine clay particles in the slurry. The fine clay particles in the slurry after secondary cyclone treatment were gradually enriched. If they are not removed in time, the specific gravity and the viscosity of the slurry will increase, directly resulting in reducing the slurry dreg carrying capacity and the pumping capacity of the circulation system to further reduce the advancing efficiency of the shield machine. The function of the filter press system is to carry out a thorough solid-liquid separation when the clay layer cyclone screening equipment cannot separate enough solid phase and cannot reduce the specific gravity of the slurry to the low value of about 1.05–1.10 g/cm^3 at the beginning of advancing. By separating the dry soil with sufficient low water content (below 23%–27%) and recovering a filtrate with sufficient low solid content (below 50 mg/L), the specific gravity of the filtrate was restored to the required value at the beginning of advancing. In addition, to improve the filter pressing

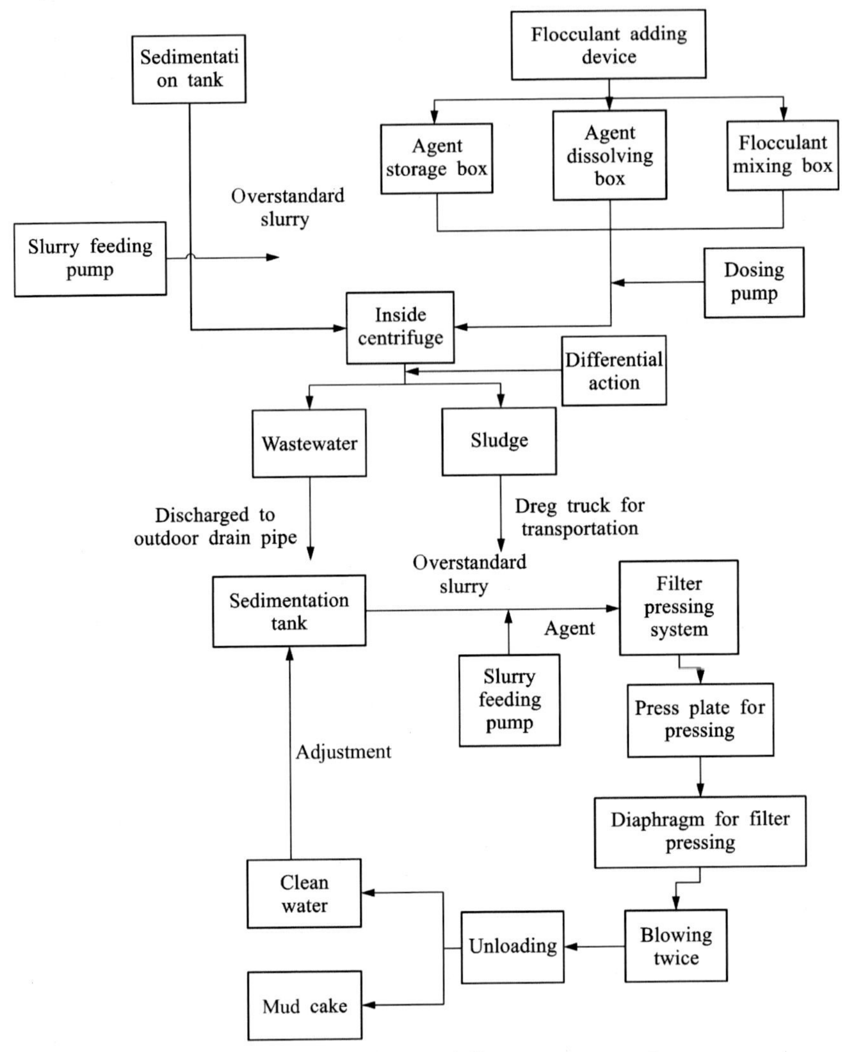

Figure 10.12 Flowchart of the centrifuge and filter press processing.

efficiency and reduce the clogging of the filter cloth in the filter pressing process, the slurry needed to be improved before filter pressing. To accomplish that, lime in an amount of 20–50 kg/set was added to the slurry and fully stirred for about 10 minutes (Fig. 10.12).

10.4.3.4 Layout of slurry disposal yard

The slurry disposal yard of the river-crossing shield tunnel was a rectangular area 90 m long by 45 m wide, including a waste slurry storage pool, an abandoned slurry treatment pool, a raw slurry storage pool, an adjustment pool, a transition pool, a sedimentation pool, two clean water pools, a mud-water separation equipment foundation, a raw slurry material warehouse, a bentonite mixing pool, and five equipment pits. The equipment that was loud during operation was concentrated in the southwest corner of the yard to keep the noise from the dense residential area. According to relevant engineering experience [10], the slurry treatment system of this project needed to be repeatedly calculated according to its own yard and working conditions to determine the layout plan most favorable for construction.

10.4.4 Site slurry treatment effect

The river-crossing tunnel project adopted a large-diameter slurry shield for construction, and a large amount of slurry was discharged in the construction-advancing process. The waste slurry is treated with slurry separation equipment. The sludge was filtered and screened, treated by the primary cyclone separator and the secondary cyclone separator, and then discharged to obtain mud with a specific gravity that met the construction requirements. Excess slurry was transported to the filter press and the centrifuge for processing, which greatly improved the working efficiency. The clean water that was processed in the filter press with a low solid content was transported to the slurry sedimentation tank for slurry configuration. After being processed in the centrifuge, the sludge was transported by the sludge track, and the wastewater that reached the discharge standard was discharged into the drain. The slurry separation treatment system had the advantages of energy saving, high efficiency, and environmental protection. The slurry treatment results were introduced by taking the sandy silt intercalated with silty clay stratum for an example, as shown in Table 10.4.

10.5 Recycling of waste slurry and case studies

In many cities in China, slurry-balanced shields are commonly used in the excavation of large-section tunnels in high water content (or underwater) sand or clay layers. In these conditions, the performance of slurry shield tunnelling is better than that of conventional drilling and blasting methods and of EPB shield tunnelling methods. It is worth noting that several million tons of waste slurry is produced by the slurry balance shield methods every year in China. Almost all of this waste slurry is treated as waste.

Table 10.4 Gradation and slurry treatment results of sandy silt intercalated with silty clay.

<table>
<tr><th colspan="2">Stratum name</th><th colspan="4">Sandy silt intercalated with silty clay</th><th colspan="2">Stratum equivalent ratio</th><th>3.07%</th></tr>
<tr><td rowspan="2">Particle size distribution</td><td>mm</td><td>>2</td><td>2–0.075</td><td colspan="2">0.075–0.045</td><td colspan="2">0.045–0.02</td><td><0.02</td></tr>
<tr><td>%</td><td>0</td><td>7.10</td><td colspan="2">37.63</td><td colspan="2">31.36</td><td>23.91</td></tr>
<tr><td colspan="2">Sludge outlet proportion</td><td>Prescreened mud pie 20%</td><td colspan="3">Primary cyclone 20.73%</td><td colspan="2">Secondary cyclone 40.14%</td><td>Tertiary filter pressing 19.13%</td></tr>
<tr><td colspan="2" rowspan="3">Sludge grade/t</td><td rowspan="2">Total quantity of materials</td><td rowspan="2">screening</td><td colspan="2">Primary cyclone</td><td colspan="2">Secondary cyclone</td><td>Tertiary filter pressing</td></tr>
<tr><td>Underflow</td><td>Overflow</td><td>Underflow</td><td>Overflow</td><td>Filter pressing sludge outlet</td></tr>
<tr><td>425.06</td><td>85.01</td><td>88.13</td><td>251.92</td><td>170.61</td><td>81.31</td><td>81.31</td></tr>
</table>

Since it needs to be processed, transportation and disposal plants, equipment, and personnel are required, and the construction cost of projects is increased. Besides, the waste slurry is not effectively treated during transportation and landfill, resulting in serious ecological damage to the surrounding environment (including groundwater and air pollution). Traditional waste slurry treatment methods have difficulty meeting the stipulation of the environmental protection agency, aggravate the dreg stacking pressure on the waste dreg yard, and even increase the risk of instability of the waste dreg yard. Therefore it is of great significance that research on green treatment technology of shield dregs be carried out to improve engineering economics and urban environmental protection.

Mass recycling has always been a hot issue in the construction field. For quite some time, many developed countries have regarded the reduction and the reuse of construction waste to be a strategic goal of sustainable development, and the related technology research is relatively mature. The management of construction waste in China started late, but in recent years, attention to related issues has gradually increased. A series of policies, systems, and standards have been published with the goal of reducing the emission of construction waste, supporting research on construction waste, promoting industrialization and resource development of construction waste, and promoting healthy development of the urban environment. For example, it is important to vigorously research and develop solid waste treatment and disposal technologies and resource utilization technologies and to encourage and support the adoption of measures that are conducive to environmental protection for concentrated solid waste disposal. Therefore treating shield waste slurry from the perspective of green recycling and comprehensive utilization not only is in line with current national conditions and international development trends, but also can solve such engineering issues.

10.5.1 Recycling of waste slurry in backfill grouting

The river-crossing shield tunnel on Wangjiang Road in Hangzhou has a total length of 1.837 km and is a double-line tunnel. The shield section is mainly located under Qiantang River, and the lines are perpendicular to Qiantang River, with a buried depth of about 12–22 m. When the slurry shield passes through the fine-grained stratum, more abandoned slurry and clay dregs will be produced because the mass ratio and viscosity of the slurry need to be controlled. This significantly increases the construction cost and causes some environmental pollution during the treatment process. For the large-diameter slurry shield of this project, a huge amount of waste slurry is produced, owing to the particularity of the stratum. Therefore if the waste slurry can be recycled as backfill grouting material, the purchase cost of an enormous amount of required grouting materials can be reduced, and the waste slurry disposal cost and urban pollution can also be reduced.

On the basis of orthogonal test design, the optimal on-site proportioning that is obtained is 0.745 for the water-binder ratio, 0.84 for the cement-sand ratio, 0.161 for the bentonite-water ratio, and 2.014 for the ash-cement ratio. The actual optimized proportioning parameters (calculated by 1000 kg) are 108.405 kg for cement, 218.274 kg for fly ash, 388.904 kg for river sand, and 284.416 kg for waste slurry, wherein the mass ratio of the clay waste slurry is 1.08. Indoor slurry preparation and related performance tests are carried out according to optimized proportioning parameters. The performance data of the obtained synchronous grouting is shown in Table 10.5. As Table 10.5 shows, the actual measured values of various performance indexes of the slurry after proportioning optimization are all within the required value range of the performance indexes of the synchronous grouting material, indicating that the performance of the synchronous grouting after proportioning optimization can meet the rapid tunnelling requirements of the large-diameter slurry shield on site.

The single line of this project can save about RBM42,200 of synchronous grouting material cost and transportation and waste disposal cost,

Table 10.5 Performance checklists of optimized proportioning.

Performance index	Experiment value	Required upper limit	Required lower limit
Consistency (cm)	13.4	14	10
Initial fluidity (cm)	23.1	25	22
Stone shrinkage ratio in 28 days (%)	5.8	2.0	8.0
Bleeding ratio in 3 h (/%)	2.0	5.0	2.0
Initial setting time (h)	10.82	14	10
Unconfined compressive strength in 1 day (MPa)	0.6	5.0	0.2
Unconfined compressive strength in 7 days (MPa)	1.0	2.0	0.6
Unconfined compressive strength in 28 days (MPa)	2.6	5.0	2.0

save nearly 5% of engineering site resources, reduce pollution on the construction site, prevent high-frequency transportation of dreg trucks on urban roads, protect the urban road environment, reduce the degree of stacking pressure on the waste dreg yard, limit the amount of urban waste slurry discharge, and protect the urban ecological environment to a certain extent.

10.5.2 Recycling of waste slurry in subgrade engineering

When the slurry shield passes through the sandy stratum, a large amount of waste sand will be produced through slurry treatment, and conventional outward transport can easily cause resource waste. Analysis of particle gradation, water content, and extremely fine particle content of waste residue finds that the slurry can be used for subgrade materials.

The Weser River Tunnel in Germany used a 60-m-long slurry shield machine with a cutterhead diameter of 11.3 m for advancing. Soil excavated from the tunnel was pumped to the separation equipment on the ground with the supporting liquid to separate mud and water. The crusher that was installed behind the cutterhead was specially designed to crush boulders with a diameter of less than 0.9 m into transportable materials. The engineer continuously monitored the separated silt and measured the content of clay (particle size less than 0.063 mm) by using the water content after it had been determined. The sieving curve and the extremely fine particle content were used as control indexes, and the sedimentation analysis of individual cases was conducted. The conventional test needed to screen only the dried sample. The summarized typical characteristic curve shows than the separated silt was all fine sand or medium sand. The weight percentage of water was about 20%, and the water content was quite stable.

On the basis of research, the project proposed the following reuse scheme: The sand with a proportion of clay less than 5% could be used to prevent frost and be applied to highway subgrade; when the Weser River Tunnel was excavated, this type of soil was used as a load-bearing layer for connecting road sections; and the sand with a proportion of fine particles more than 5% was used to build soundproof walls on both sides of the road. Compared with stacking and purchasing related materials, a preliminary calculation shows that the cost could be decreased by 10%–20% if water was sprayed onto the dehydration screen for cleaning in a targeted manner.

10.5.3 Recycling of waste slurry in reclamation materials

Waste slurry can be solidified into soil materials with a certain strength after a certain treatment, which can be used for various types of engineering fillers. The Yamatogawa route of the Hanshin Expressway is a 9.7-kilometer-long expressway connecting the cities of Sakai and Matsubara City in Osaka Prefecture in Japan. Most of the route is underground, with a 3.9 km tunnel section that was constructed by using the slurry pressure shield method. It is estimated that the amount of soil produced by the shield machine with a diameter of about 12.5 m was about 1 million m^3. The large amount of excavated soil produced by engineering has become a problem, and it is expected that the soil can be reasonably and effectively reused. Because the tunnel was closely connected to the reclamation project, it was planned to reclaim land to fill an area of 8.3 hectares, and the produced silt could be used as reclamation materials after solidification treatment. To avoid unfavorable settlement of land and facilities near the reclamation site, some countermeasures were taken, such as improvement of reclamation land and the original ground and consideration of reclamation procedures. Improving the excavated soil enables sufficient mechanical properties. Furthermore, in this project, an earth terminal complex (ETC) manifest system was developed and operated to ensure the traceability of large amounts of excavated soil related to transportation. The system improved the efficiency of manifest issuing, reduced traffic congestion, and avoided vehicle overload.

References

[1] Shi Z, Xue D, Peng M, et al. Experiment on modified-curing and strength properties of waste mud from slurry shield. J Eng Geol 2018;20(01):103−11.
[2] Zhang R. Regeneration of the engineering slurry and its waste disposal. Railw Eng 2003;(03):42−3.
[3] Liu Y, Wang H. Slurry treatment and separation for a slurry shield. Mod Tunelling Technol 2007;(02):56−60 + 71.
[4] Chang G, Li C, Ding G, et al. Coagulation and separation of waste slurry from qianjiang river tunnel shield. Tech Equip Environ Pollut Control 2016;6(10):3752−6.
[5] Li X. Flocculation and dehydration experimental study of waste mud in slurry shield. Railw Eng 2018;58(05):144−7.
[6] He C, Feng K. Review and prospect of structure research of underwater shield tunnel with large cross-section. J Southwest Jiaotong Univ 2011;46(01):1−11.
[7] He C, Wang B. Research progress and development trends of highway tunnels in China. J Mod Transportation 2013;21(4):209−23.
[8] Lin C, Zhang Z, Wu S, et al. Key techniques and important issues for slurry shield under-passing embankments: a case study of Hangzhou Qiantang river tunnel. Tunn Undergr Space Technol 2013;38:306−25.

[9] Zhai N, Wang W, Zheng B. Properties of mud slurry for slurry shield. Chin J Undergr Space Eng 2017;13(S1):58–64.
[10] Tang J. Design of slurry treatment system and layout of slurry treatment yard for slurry balance shield tunnel. Constr Mater Decor 2019;13:232–3.

Exercises

1. Please explain the functions of filter cake at excavation face for slurry shield tunnelling.
2. What influence does the relative density of slurry have on its engineering performance for slurry shield tunnelling?
3. In a sandy cobble stratum, what is the criterion for judging whether the slurry can form a stable filter cake on the excavation face?
4. What are the advantages and disadvantages of the inorganic coagulants and organic coagulants that are commonly used in slurry?
5. What are the commonly used coagulants, and what is their action mechanism?
6. In addition to the additives in the tunnelling slurry components mentioned in this chapter, please introduce at least three other slurry additives through investigation in details.
7. Propose a plan for reusing the waste slurry that is discharged from the slurry treatment system to reduce the site waste sludge treatment pressure.

CHAPTER 11

Ground deformation and its effects on the environment

The construction of shield tunnels in cities in China has developed rapidly, and more and more urban traffic tunnels, subway tunnels, and intercity railway tunnels have been constructed by using safer and faster shield tunnelling methods in recent years. Most of these tunnels need to pass under dense surface buildings in cities, inevitably resulting in adverse impacts on the usability and safety of adjacent buildings in the surrounding environment. Ground displacement caused by construction of urban tunnels is the main cause of safety problems to adjacent buildings. Disturbances caused by shield tunnel construction in the surrounding strata will inevitably cause ground deformation. The deformation will be transmitted to adjacent buildings, which will have to be rebalanced. Damage and accidents will be caused if the final deformation of the building exceeds the ultimate bearing capacity. Therefore accurate prediction of the ground displacement induced by shield tunnel construction and its impact on surrounding environments and determination of targeted reinforcement control methods are of great significance to ensure safety in the construction and operation of shield tunnels [1].

11.1 Shield tunnelling induced ground deformation

11.1.1 Main reasons and mechanisms of ground deformation

During shield tunnel construction, ground deformation is closely related to construction conditions and geological characteristics, and there are many reasons and impact factors. Table 11.1 summarizes the main reasons and mechanisms of ground settlement caused by shield tunnel construction [2,3].

11.1.2 Effects of shield tunnelling induced groundwater loss

In the shield machine tunnelling process, the synchronous grouting in the tunnel vault area generally can not be fully dense, resulting in hydraulic

Shield Tunnel Engineering
DOI: https://doi.org/10.1016/B978-0-12-823992-6.00011-4

Table 11.1 Main reasons and mechanisms of ground deformation caused by shield tunnel construction.

Types	Main reasons		Stress disturbance	Deformation mechanism
Initial settlement	Groundwater level lowered, soil compacted		Pore water pressure reduced, effective stress increased	Porosity reduced, consolidation
Deformation in front of excavation face	Swelling	Too large shield thrust	Reverse earth pressure increased	Elastoplastic deformation produced by compression
	Settlement	Too small thrust or excessive soil excavated	Stress release, disturbance	Deformation due to unloading
Settlement when shield passes	Construction disturbance		Disturbance, stress release	Compression
Settlement of shield tail void	Insufficient supporting such as small grouting pressure or overexcavation		Ground stress release	Elastoplastic deformation
Consolidation settlement	Residual effect		Stress relaxation	Creep compression

passages at the vault along the tunnel longitudinal direction. When the shield machine is stopped for a long time, groundwater will easily flow from a place behind the shield machine to the excavation face, leading to a large loss of groundwater. When the stratum fluctuates largely or the sealing quality of geological drilling are not good, hydraulic passages easily form in the upper stratum, which may directly penetrate the impermeable layer and cause the groundwater level to drop. Large groundwater loss from the excavation face will also be caused by shutdown of the shield machine in a water-rich stratum. For example, in a shield tunnel construction in Shenzhen Metro, because the shallow upper overlaying soil was loose, with some geological boreholes that were not well blocked, vertically hydraulic passages formed, and the groundwater dropped by more than 2 m during the shield crossing, causing a surface settlement of up to 120 mm. When the shield tunnelling parameters were adjusted and the grouting amount was increased, the grout oozed out from the surface.

11.1.3 Effects of shield tunnel construction on ground deformation

Shield tunnel construction disturbance is the main cause of ground displacement. Construction disturbance causes soil stress changes and leads to primary consolidation compression, elastoplastic shear, and viscous aging creep of the ground. Studies have shown that the larger the range of ground disturbance, the larger the maximum settlement of the surface above the tunnel center, and the settlement during the shield tunnel excavation is roughly linear with the range of ground disturbance. The range of ground disturbance is related to the jacking force during advancing, the backfill grouting time, and the ratio of overburden thickness (H) to tunnel excavation diameter (D). The larger the jacking force during shield tunnelling, the higher the extrusion degree of the soil mass in front of the shield excavation face and the greater the degree of ground disturbance, indicating that the surface settlement at the trough center is also larger. The longer the time interval between shielding passing and backfill grouting, the higher the degree of soil stress release; that is, total surface settlement is highly related to the degree of ground stress disturbance. A greater H/D ratio will lead to a smaller disturbance to the ground as well as smaller ground displacements.

Owing to construction factors such as the squeezing effect of the shield during advancing and the grouting effect of the shield tail, a positive excess pore water pressure is formed in the surrounding stratum. The excess pore water pressure is dissipated and restored within a period of time after shield tunnel construction. In this process, the stratum is formed by drainage and consolidation, and the surface settlement of the stratum caused by the change in the pore water pressure is called the main consolidation settlement. After the stratum has been disturbed, the skeleton of the stratum will undergo compression and deformation for a long time. The surface settlement that is produced during the creep of the stratum is called the secondary consolidation settlement. In soft plastic and fluidal plastic clay with large porosity and sensitivity, the secondary consolidation settlement often lasts for several layers or even more, and its proportion can be as high as 35% or more of the total settlement.

11.2 Ground loss in shield tunnel construction

11.2.1 Concept and composition of ground loss

11.2.1.1 Concept of ground loss

In the excavation process of the shield tunnels, the stratum will be inevitably deformed and spread to the surface to form a surface settlement trough due to unloading and construction loads. The calculation of the surface displacement field is often determined according to the change of the boundary. From simple empirical methods and analytical solutions to advanced finite element analysis, the ground deformation caused by tunnel excavation is often described by ground loss as a characteristic parameter and is usually represented by taking ground loss rate as a parameter, that is, the percentage of the stratum volume loss per unit length to final tunnel volume. In a narrow sense, the ground loss rate refers to the percentage of the difference between actually excavated soil volume in shield tunnel construction and the theoretically calculated soil discharge volume to the theoretical excavation volume, while in a generalized sense, the ground loss rate can be expressed as the percentage of the surface settlement trough volume caused by shield tunnel construction to the theoretical excavated volume of the tunnel [3].

11.2.1.2 Composition of ground loss

For modern closed shield tunnelling, the source of ground loss during tunnelling may consist of the following (Fig. 11.1) [4]:

1. Component a: axial displacement in the tunnel excavation face due to unbalanced pressure on the excavation face.
2. Component b: radial ground displacements around the tunnel shield due to excessive cutting, shield taper, and self-weight or yaw and pitch motions of shield steering.
3. Component c: radial ground displacements entering the voids at the tail due to a relatively small gap between the excavation profile and the diameter of the lining. In normal cases, the annular gap should be grouted, but before the lining is installed, the ground loss will occur due to insufficient grouting.
4. Component d: radial ground displacement caused by deformation of the tunnel lining.
5. Component e: deformation of the stratum in the direction of the tunnel due to consolidation and creep.

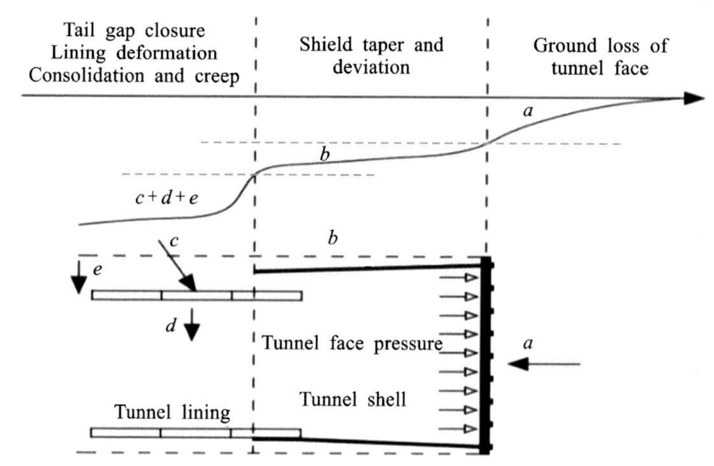

Figure 11.1 Composition of ground deformation volume loss caused by shield tunnelling.

11.2.2 Calculation method of ground loss

11.2.2.1 Empirical method

Generally speaking, ground volume loss is the percentage of volume loss of the final tunnel volume per unit length. This definition has been widely used in the existing prediction methods for ground displacement caused by tunnel excavation. The value of the ground volume loss parameter is often based on the experience of similar tunnel projects in the past. By using a large number of cases, the empirical value of the ground volume loss rate under different stratum conditions and shield excavation methods can be obtained, as shown in Table 11.2.

11.2.2.2 Empirical formula method

Since the ground loss is directly related to the ground deformation caused by tunnel excavation, the ground volume loss can be estimated according to the stability coefficient (N) of the tunnel excavation face, where the stability coefficient of the excavation face is defined as follows:

$$N = \frac{\gamma z_0 + P_l - P_s}{S_u} \tag{11.1}$$

where z_0 is the depth to the axis of the tunnel; γ is the soil bulk density; P_l is the load acted on excavation face; P_s is the support pressure applied to the tunnel face; and S_u is the undrained shear strength.

Table 11.2 Summary of ground loss values.

Stratum condition	Excavation method	Volume loss V_1	Explanation
Hard clay	Open excavation face	1%–2%	Supported by mass case data
Tight sand or pebble	Closed excavation face (EPB or slurry shield)	0.2%–1%	Deeply buried
		0.8%–1.3%	Shallowly buried
Soft clay	Closed excavation face (EPB or slurry shield)	1%–2%	Reinforcement excluded
Mixed conditional sand covered above hard clay	Closed excavation face (EPB or slurry shield)	0.03%–1%	Ratio of overburden thickness to diameter greater than 0.6
		2%–4%	Small ratio of overburden thickness to diameter

When N is greater than or equal to 1 but less than or equal to 2, the ground deformation is elastic, the ground loss is small, and the tunnel face is stable; when N is greater than or equal to 2 but less than or equal to 4, a local plastic zone may form near the excavation face; when N is greater than or equal to 4 but less than or equal to 6, larger ground deformation probably occurring should be evaluated in consideration of larger ground loss at the excavation face caused by plastic ground deformation; when N is greater than 6, shear failure may occur on the tunnel excavation face and the ground surface, and the tunnel excavation face is not stable.

Table 11.3 summarizes the empirical relationships between the stability coefficient of the tunnel excavation face and the short-term ground loss. However, in applying them to design purposes, it is recommended that the ground volume loss value obtained through the empirical formulas be matched with the empirical values in Table 11.2.

11.2.2.3 Equivalent formula method

The ground loss parameter obtained from relevance of case experience lack a rigorous theoretical basis, and the influence of ground surface conditions, tunnel excavation methods, and tunnel structures cannot be

Table 11.3 Empirical relation expressions between stability coefficient of tunnel excavation face and short-term ground loss.

No.	Bibliography	Expressions	Explanation
1	[5]	$V_i = m \times \exp(N-1)$, $(N \geq 1)$; $V_i = m \times N$, $(N < 1)$	Calculation assumes that $E_u/S_u = 500-1500$, where $m = 0.002 \sim 0.006$. Data points of overconsolidated soil and normally consolidated soil are all scattered.
2	[6]	$V_i = \left(\frac{S_u}{E_u}\right) \times \exp\left(\frac{N}{2}\right)$	The Eu/Su ratio is usually 200–700, and V_1 should be increased to three times for inferior processes.
3	[6]	$V_i = 1.33 \times N - 1.4$, $(1.5 < N < 4)$	75% of case data are within the scope of the design guidelines.
4	[7,8]	$V_i = 0.23e^{4.4\left(\frac{N}{N_c}\right)}$, $\left(\frac{N}{N_c} \geq 0.2\right)$	N_c represents a critical stability coefficient. The formula is based on field data of overconsolidated clay within upper and lower design limits.

considered. Generally speaking, the tunnel can be analyzed according to plane strain problems, and the ground volume loss is defined according to the cross section of the tunnel. Therefore some equivalent parameters can be used to express the volume loss proportional to the tunnel volume. For shield tunnels, traditional ground loss parameters can be defined as equivalent ground loss parameters:

$$V_l = \frac{\pi\left(R + \frac{g}{2}\right)^2}{\pi R^2} \times 100\% = \frac{g}{R} \times 100\% \tag{11.2}$$

where R is the radius of tunnel; and g is the estimated equivalent gap parameter at the vault.

The second-order gap (g^2) is ignored because its influence on the ground loss value can be negligible; that is, the second-order component of ground loss 1% is about 0.01% (the ground loss estimation error is only 1%). The ground deformation caused by shield tunnel construction is

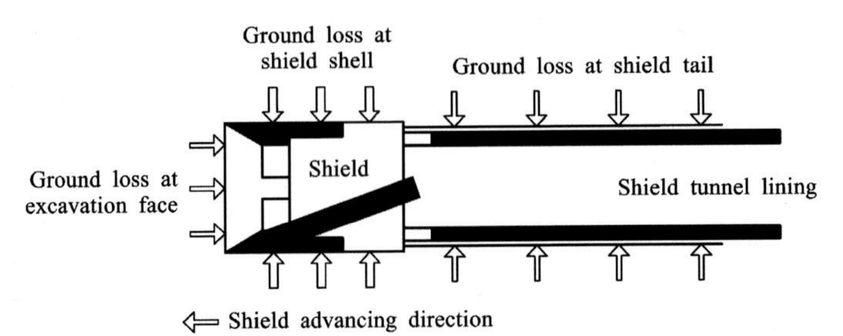

Figure 11.2 Components of ground loss in shield tunnel construction. *Modified from Loganathan N., An Innovative Method for Assessing Tunnelling-Induced Risks to Adjacent Structures. One Penn Plaza: Parsons Brinckerhoff Inc; 2011.*

greatly affected by the ground loss parameter. The ground loss includes the sum of the excavation face loss (V_f), shield loss (V_s), and shield tail loss (V_t) (as seen in Fig. 11.2) [8]. Each aspect is represented and calculated by the following variables.

Excavation face loss (V_f)

The stratum intruding into the tunnel excavation face due to stress release of the tunnel excavation face is eventually excavated, and the ground loss volume of the excavation face is equal to the volume of the overexcavated soil material around the excavation face. Therefore provided that the soil in front of the tunnel excavation face moves to the back of the excavation face to cause the radial settlement deformation of the stratum, the convergent deformation of the stratum is expressed by the radial equivalent gap parameter g_f, and the ground loss of the tunnel excavation face can then be expressed as V_f:

$$V_f = \frac{g_f}{R} \times 100\% \tag{11.3}$$

where R is the radius of the tunnel; and g_f is the equivalent gap at the tunnel vault caused by ground loss of the excavation face, which can be obtained by using the following formula:

$$g_f = \frac{k}{2}\frac{\Omega R P_0}{E} \tag{11.4}$$

where k represents the friction coefficient between intruding soil and the skin of the shield shell; Ω is the dimensionless axial displacement ahead of the tunnel face; R is the radius of the tunnel; P_0 is the total stress release amount at tunnel excavation face; and E is the elastic modulus at tunnel spring line (usually the undrained Young's modulus).

The relevant variables are as follows:

1. Parameter k

Owing to the squeezing effect of shield advancing, the friction between the skin of the shield shell and the surrounding ground will generate longitudinal tensile stress, leading to failure and plastic flow into the tunnel face and the annular void between the tail skin. Based on a large amount of three-dimensional elastoplastic finite element analysis results [9]. The expression of the friction coefficient k is obtained as follows:

$$k=\begin{cases} 0.7 & \text{stiff strata } (q_u > 100\text{ kPa } or\ N_{30} > 10) \\ 0.9 & \text{Soft strata } (25\text{ kPa} < q_u < 100\text{ kPa } or\ 3 < N_{30} < 10) \\ 1.0 & \text{Very soft strata } (q_u < 25\text{ kPa } or\ N_{30} < 3) \end{cases} \tag{11.5}$$

where N_{30} is the 300 mm standard penetration test (SPT) hammering number; and q_u is the unconfined compression strength which equal to $2S_u$(S_u is the undrained shear strength).

2. Parameter Ω

The determination of coefficient Ω is related to the stability coefficient N of the excavation face, as shown in Eq. 11.6, that is

$$\Omega=\begin{cases} 1.12, & N<3 \\ 0.63N-0.77, & 3<N<5 \\ 1.07N-2.55, & 5<N \end{cases} \tag{11.6}$$

where P_i is the pressure of shield excavation face; $N=(\gamma H-P_i)/S_u$; H is the depth to the tunnel spring line; and S_u is the undrained shear strength.

3. Parameter P_0

The total amount of stress release at the tunnel face due to excavation can be estimated by the following equation:

$$P_0=k_0P_v+P_w-P_i \tag{11.7}$$

where k_0 is the lateral earth pressure coefficient; P_v is the effective stratum pressure at position of spring line of tunnel; P_w is the water pressure; and P_i is the pressure at tunnel excavation face (calculated by Terzaghi earth pressure theory or a wedge sliding model).

Shield loss (V_s)

The shield shell is composed of a cutterhead and a shield shell. The cutterhead is slightly larger than the shield shell to minimize the friction between the shield and the surrounding ground. The edge of the cutterhead is provided with cutter beads to make the excavation contour slightly larger than the cutterhead body. The shield shell is usually tapered, with a slightly smaller diameter at the tail. Some shields have both cutter bead and a tapered shield shell. The overexcavation thickness of the cutter bead is shown as t_b, and the taper of the shield shell is shown as t_t. The values of the two components are usually in the ranges of 5 ~ 15 mm for t_b and 30 ~ 60 mm for t_t although they may vary according to project requirements (Fig. 11.3).

As a result of overexcavation of the cutterhead and the design of the tapered shield shell, a gap U_i will be formed between the shield body and the surrounding strata. Usually, the gap will be filled with slurry or groundwater when the shield excavation chamber is pressurized; in a stable stratum when the excavation chamber is not pressurized, the gap may be deformed without being filled and supported until segment lining and shield tail grouting is completed. The deformation of the stratum into the shield gap can be obtained by using the following formula:

$$U_i = R(1+\nu)\frac{(\gamma H + p_w - p_i)}{E} \tag{11.8}$$

where ν is the Poisson's ratio of the stratum, and E is the elastic modulus of the stratum.

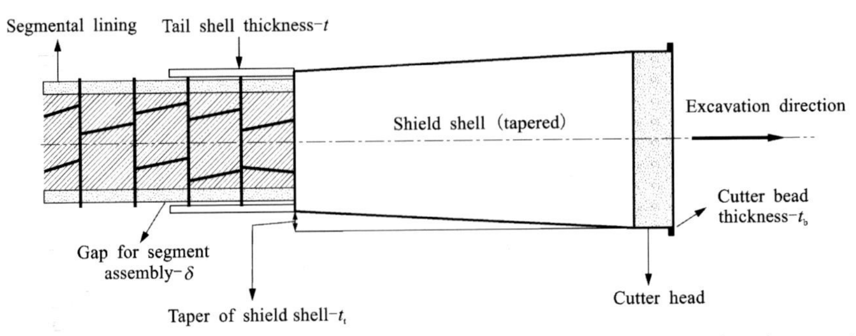

Figure 11.3 Schematic diagram of shield configuration with cutter bead and tapered shield shell. *Modified from Loganathan N., An Innovative Method for Assessing Tunnelling-Induced Risks to Adjacent Structures. One Penn Plaza: Parsons Brinckerhoff Inc; 2011.*

The other parameters are the same as above.

The ground loss of the shield shell can be estimated through the following equation:

$$V_s = \frac{g_s}{R} \times 100\% \tag{11.9}$$

where R is the radius of tunnel, and g_s is the equivalent gap parameter of ground loss of shield shell. Its volume and the deformation of the stratum entering the shield gap can be deduced as follows: If $U_i > t_t + t_b$, $g_s = (t_t + t_b)$; if $U_i \leq t_t + t_b$, $g_s = 0.5U_i$.

Shield tail loss (V_t)

A physical gap is formed at the tail as a result of the thickness t of the shield tail skin and the gap δ reserved for mounting the erection of the segment lining. The gap shall be filled with synchronous grouting immediately after the lining is assembled to minimize the ground loss. However, in practice, there will be a time dependent shrinkage in the hardening process of cement-soil slurry mixture, owing to hydration of the cement. Related experimental studies have shown that the volume change (shrinkage) of the cement slurry with a water-cement ratio of 0.4 is about 7%–8% [10]. Similarly, laboratory tests of the cement slurry mixture show that the thickness of the cement slurry mixture sample is reduced by 7%–10% [3]. Therefore if synchronous grouting is used to fill the physical gap, the final shield tail loss gap value is assumed to be 7%–10% of the total tail gap. In consideration of the possible gap in the grouting due to poor workmanship, it is generally assumed that about 10% shrinkage will be generated in tunnel construction, including any possible volume reduction due to incomplete synchronous grouting. The equivalent gap parameter formed by shield tail can be expressed as

$$g_t = 0.1(t + \delta) \tag{11.10}$$

The ground loss component caused by shield tail can be estimated as follows:

$$V_t = \frac{g_s}{R} \times 100\% \tag{11.11}$$

The value of total ground loss during shield tunnel excavation can be obtained by adding excavation face loss, shield loss and the shield tail loss together.

11.2.3 Calculation cases

[Case 11.1] A shield tunnel has a buried depth of 30 m and an excavation diameter of 6 m. The tunnel is driven by an earth pressure balance shield machine. The parameters of the soil layer are shown in Table 11.4, and the configuration information of the tunnel boring machine (TBM) is shown in Table 11.5. Try to determine the following:

1. Provided that the pressure P_i of the TBM excavation face is equal to 100 kPa and the total stress release amount P_0 is 292.8 kPa, please estimate the ground loss caused by shield excavation.
2. Determine the actual effective pressure P_F of the TBM excavation face based on Terzaghi earth pressure theory.

   [Solution]:

11.2.3.1 Calculation of ground loss

1. Calculating the ground loss V_f of the excavation face

   From formula (11.5), $3 < n = 5 < 10$, so the friction coefficient $k = 0.9$.

   From formula (11.6) the stability coefficient of the excavation face is $N_R = 5.9 > 5$, so the coefficient is

$$\Omega = 1.07 \times N\text{-}2.55 = 3.7$$

   From formula (11.4),the equivalent gap at the tunnel vault is

$$g_f = \frac{k\Omega RP_0}{2\ \ E} = \frac{0.9 \times 3.7 \times 3000 \times 292.8}{2 \times 75000} = 19.6 \text{ mm}$$

   From formula (Eq. 11.3), the ground loss of the excavation face is

$$V_f = \frac{g_f}{R} \times 100\% = 0.65\%$$

2. Calculating the ground loss V_s of shield shell

   The deformation of the shield gap is determined by formula (11.8) if $U_i > t_t + t_b$, $g_s = (t_t + t_b)$; if $U_i \leq t_t + t_b$, $g_s = 0.5U_i$

$$U_i = R(1 + \nu)\frac{(\gamma H + p_w - p_i)}{E} = 29.4 < t_b + t_t = 30 \text{ mm}$$

   Thus $g_s = 0.5U_i = 14.7$ mm

   The ground loss of the shield is

$$V_s = \frac{g_s}{R} \times 100\% = 0.49\%$$

Table 11.4 Basic information of soil layer.

Poisson's ratio	Young's modulus (kPa)	Shear strength (kPa)	Earth pressure coefficient k_0	Unit weight (KN/m^3)	Depth from tunnel to groundwater level (/m)	Stability coefficient of excavation face	Cohesive force (kPa)	Internal friction angle	Hammering number N of SPT
0.5	75000	150	0.8	18	25	5.9	75	1	50

Table 11.5 Configuration information of the shield machine.

Length of shield shell (m)	Tail thickness t (mm)	Lining assembly gap δ (mm)	Shield taper t_t (mm)	Overcutting thickness t_b
9.14	15	25	30	0

3. Calculating the ground loss V_t caused by shield tail separation

The equivalent gas parameter formed by shield tail separation is

$$g_t = 0.1(t+\delta) = 4\ \text{mm}$$

The ground loss of the shield tail is

$$V_t = \frac{g_t}{R} \times 100\% = 0.13\%$$

In summary, the total ground loss is

$$V_L = V_f + V_s + V_t = 1.27\%$$

11.2.3.2 Calculating the pressure of the shield excavation face

1. Based on Terzaghi theory, the loading width B can be expressed as

$$B = R\tan\left(45° - \frac{\varphi}{2}\right) + \frac{R}{\cos\left(45° - \frac{\varphi}{2}\right)} = 7.15\ \text{m}$$

2. The vertical earth pressure at the top of tunnel can be expressed as

$$\sigma_v = \frac{\gamma B - C}{k_0 \tan\varphi}\left(1 - e^{-k_0 \tan\varphi H/B}\right) + P_0 e^{-k_0 \tan\varphi H/B} = 197.54\ \text{kPa}$$

3. Calculating the effective pressure of the excavation face

Arch height $h_a = 10.97$ m

The water level below ground is

$$h_w = 30 - 25 = 5\ \text{m} < h_a$$

Thus the effective pressure of the excavation face is

$$P_F = K_1\left[\gamma(H_a - H_w) + \gamma'\left(H_w + \frac{D}{2}\right)\right] = 138.45\ \text{kPa}$$

11.3 Prediction of ground deformation

The prediction of ground displacement caused by shield tunnel construction is done by using the empirical estimation method based on sorting of data observed in actual engineering. The stochastic medium theoretical method using ground settlement calculation method in mining engineering, and the semiempirical formula method based on analytical or numerical calculation. The empirical formula method is a fitting formula based on statistical analysis of a large number of actually measured data and fails to consider the stratum characteristics and construction factors of the shield tunnel. The stochastic medium theoretical method and the analytical and numerical semiempirical method are more rigorous in their theoretical derivation process but can conduct analysis only for simple boundary condition problem owing to the simplified processing of calculation conditions. With the advances in finite element analysis and computer technology, the calculation of the stratum displacement caused by shield tunnel construction can be analyzed by a large-scale numerical simulation model, which can consider the nonlinear mechanical characteristic of geotechnical materials and segment lining structures as well as complicated factors such as stratum characteristics, shield tail gaps, backfill grouting, shield cutterhead propulsion, and adjacent buildings (structures).

11.3.1 Empirical formula method

11.3.1.1 Surface settlement

The empirical prediction method based on statistical analysis of data observed in actual engineering is a primary choice for the prediction of stratum displacement caused by tunnel excavation. What is most commonly used is the empirical formula proposed by Professor Peck at the International Conference for Soil Mechanics in 1969 through analysis of a large amount of surface settlement data from engineering cases [11]. He stated that the volume of the transverse surface settlement trough should be equal to the volume of the ground loss and that the transverse surface settlement curve roughly conforms to normal distribution. The corresponding surface settlement estimation formula is

$$S_{(y,z)} = \frac{V_L A}{\sqrt{2\pi} i_z} \exp\left[\frac{-y^2}{2i_z^2}\right] \tag{11.12}$$

where V_L is the ground loss rate in tunnel construction; A is the theoretical excavation area of the tunnel $\left(A = \pi D^2/4\right.$, and D is the diameter of the tunnel); y is the distance from the central line of the tunnel along the transverse direction; i_z is the width coefficient of the settlement trough caused by tunnel construction, and it vary with depth according to the formula proposed by Mair et al. [12]. This width coefficient of the settlement trough at the depth z of the stratum is:

$$i_z = K_z(z_0 - z) \tag{11.13}$$

where K_z is a dimensionless coefficient related to depth z. The value of K_z is related to the type of stratum. Generally, the empirical value is 0.4 for a hard stratum to 0.7 for a soft silt stratum. K_z is 0.2–0.3 for a gravel stratum below groundwater. These empirical values were calculated by Rankin [13] based on fitting of a large number of actually measured data. Rankin also put forward that for practical engineering applications, K_z is usually 0.5 for soft or hard clay strata and 0.24–0.45 for sand and gravel stratum (Figs. 11.4–11.6).

As a general rule, the width of surface settlement trough excavated in the clayey stratum is three times the buried depth of the tunnel, and the value of K does not change with the buried depth of the tunnel in calculating the surface settlement. The ground loss rate V_L is defined as the

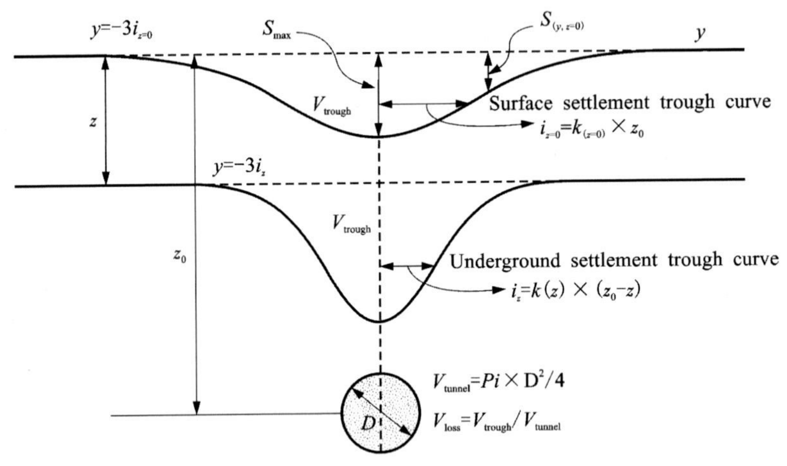

Figure 11.4 Transverse surface settlement curve caused by tunnel excavation.

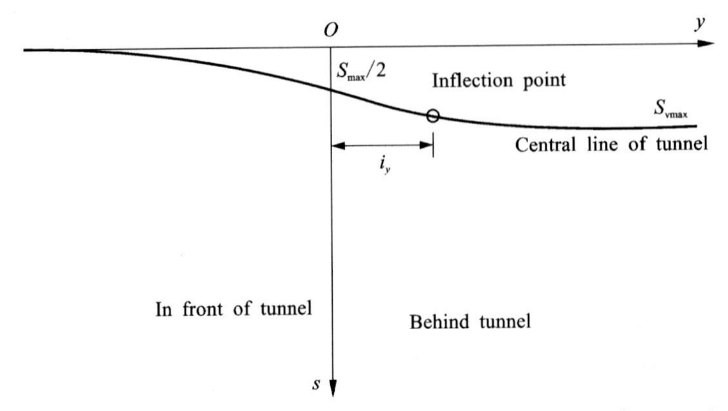

Figure 11.5 Longitudinal surface settlement curve caused by tunnel excavation.

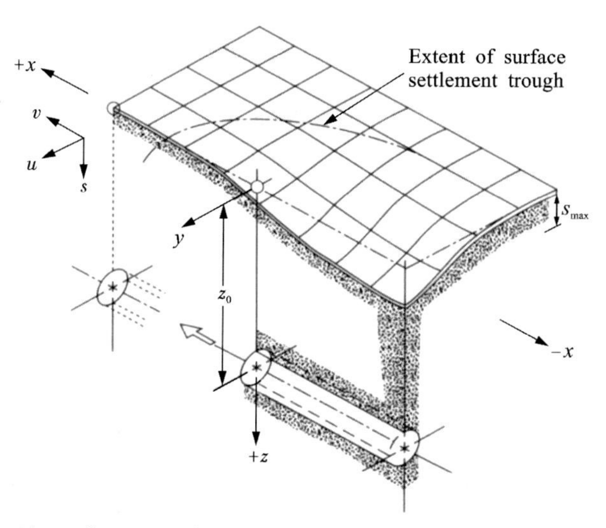

Figure 11.6 Three-dimensional tunnelling-induced surface settlement trough. *Modified from Attewell P., Yeates J., Selby A.R., Soil Movements Induced by Tunnelling and their Effects on Pipelines and Structures Chapman and Hall Ltd. New York 1986. From Leca E., New B., Settlements induced by tunneling in soft ground Tunn Undergr Space Technol 2007 119–149.*

percentage of the volume V_s of the settlement trough per unit length to the excavation volume per unit length of the tunnel. Under undrained conditions, the following formula can be obtained according to integral of Peck's formula:

$$V_S = \sqrt{2\pi} i S_{\max} \tag{11.14}$$

For a tunnel where the excavation area is A:

$$V_L = \frac{\sqrt{2\pi} i S_{\max}}{A} \tag{11.15}$$

For a circular tunnel with a radius of d:

$$V_L = \frac{0.798 i S_{\max}}{d^2} \tag{11.16}$$

On the basis of the above formulas, the surface settlement of any point at a distance of y from the central line of the tunnel along the transverse direction of the tunnel can be obtained as follows:

$$S_{(y,z)} = \frac{V_L A}{\sqrt{2\pi} K z_0} \exp\left[\frac{-y^2}{2K^2 z_0^2}\right] \tag{11.17}$$

For a circular tunnel with a radius of d:

$$S_{(y,z)} = \frac{1.253 V_L d^2}{K z_0} \exp\left[\frac{-y^2}{2K^2 z_0^2}\right] \tag{11.18}$$

11.3.1.2 Horizontal displacement

Horizontal displacement will also cause damage to adjacent buildings, so it is necessary to predict the horizontal displacement caused by tunnel excavation. However, there are very few studies on horizontal displacement and little actually measured data based on case studies. The method proposed by O'Reilly et al. [14] to predict the surface horizontal displacement caused by tunnel excavation is generally accepted; and the method provided that the stratum displacement vectors all point to the axial of the tunnel opening. The horizontal displacement S_h and the settlement S_v are in the following relation.

For surface horizontal displacement:

$$h_{(y,z)} = \frac{S_{(y,z)} \cdot y}{z_0} \tag{11.19}$$

For horizontal displacement at depth z below the surface:

$$h_{(y,z)} = \frac{S_{(y,z)} \cdot y}{(z_0 - z)} \tag{11.20}$$

where z_0 is the axial depth of tunnel (Fig. 11.7).

Horizontal displacements at different ground conditions are easily given by using the above formula combined with Peck's formula. Meanwhile, the horizontal strain ε_h at any point can be obtained through differentiation:

$$\varepsilon_h = \frac{dS_h}{dy} = \frac{S_{\max}}{z_0}\left(1 - \frac{y^2}{i^2}\right)\exp\left(-\frac{y^2}{2i^2}\right) \tag{11.21}$$

The maximum horizontal displacement appears at the inflection point, while the maximum horizontal strain appears at the position where $y = 0$ (compression) and $y = \sqrt{3}i$ (stretching).

Fig. 11.8 shows the relationships among the surface settlement trough curve, horizontal displacement, and horizontal strain. In the region $i > y > -i$, the horizontal strain is compressive strain, and the minimum horizontal strain is $\varepsilon_h = 0$ at the inflection point; in the region $i < y$ or $y < -i$, the horizontal strain is tensile strain.

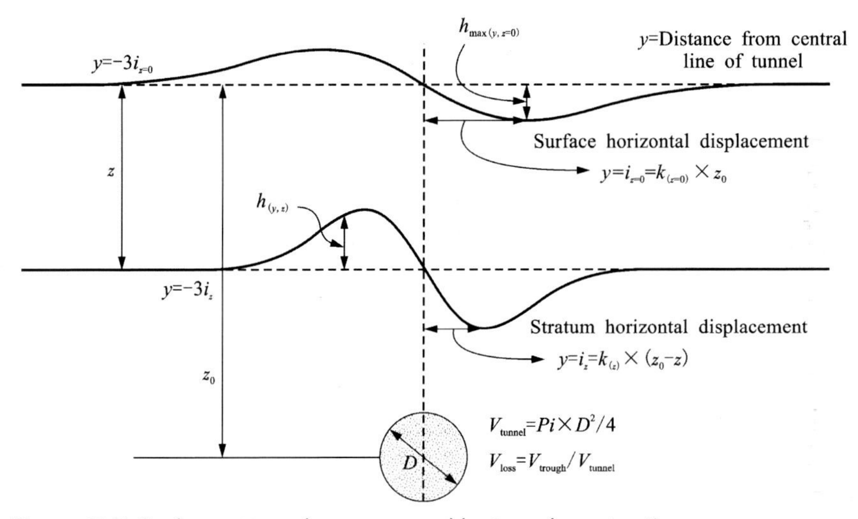

Figure 11.7 Settlement trough curve caused by tunnel construction.

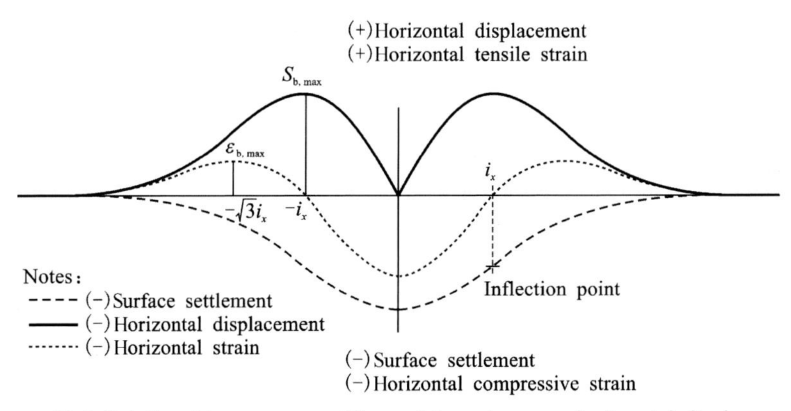

Figure 11.8 Relationships among settlement trough curve, horizontal displacement, and horizontal strain [14].

In summary, the prediction of stratum displacement through the Peck's formula and its development results has been proved to be an effective method for estimating settlement, owing to its simple principles, basis on field and laboratory test results, great simplification in mathematics, decades of engineering practices, and similarity of the obtained normal distribution probability curve to the shape of the actually measured curve. This effective method has become a classical formula in the field and has been enriched in practical applications. Because the empirical method is generally summarized on the basis of actually measured data analysis, more realistic results can often be obtained. At the same time, such methods are widely used in engineering practices because of their simple calculation, fewer calculation parameters, and strong pertinence.

11.3.2 Method based on stochastic medium theory

Because the rock-soil mass is very complicated in structure, the rock-soil excavated from the tunnel are usually composed of various rocks or soils with different compositions and organizations. As a result, the deformation characteristic of the ground are very complicated, resulting in limitations on the study and application of traditional continuum mechanics in deformation performance of the ground mass. Under such circumstances, Polish scholar J. Litwiniszyn proposed the stochastic medium theory in the late 1950s [15–17]. Through further development and improvement by Chinese scholars Liu Banchen and Yang Junsheng et al. [16], the stochastic medium theory has come to be regarded as an effective method for predicting surface settlement caused by underground excavation and urban subway tunnel excavation. The theory of this method regards the ground mass as a kind of stochastic medium and considers the surface subsidence caused by tunnel excavation as a stochastic process. Starting with unit excavation, the entire tunnel is decomposed into an infinite number of small units for excavation, and the stratum displacement and deformation caused by tunnel excavation are equal to the sum of the upper stratum displacement and deformation caused by the excavation of each infinitely small unit. On this basis, the distribution of surface subsidence, horizontal displacement, surface inclination, horizontal strain, and surface curvature can be analyzed to further predict their impact on adjacent structures.

11.3.2.1 Surface movement by unit excavation

From a statistical point of view, the entire excavation can be decomposed into an infinite number of infinitely small excavations. The impact of the entire excavation on the surface should be equal to the sum of the effects of many infinitely small excavations on the surface.

Thickness, length, and width, all being for infinitely small excavations, are defined as unit excavation $d\xi d\zeta d\eta$ (Fig. 11.9), and the depth from the tunnel center to the surface is H. On any horizontal plane Z ($Z \leq H$) above the excavation level, the surface subsidence basin caused by the unit excavation is called a unit subsidence basin. The unit subsidence is expressed as $W_e(X, Y, Z, t)$ in a four-dimensional coordinate system, and the expression of the unit subsidence of the ground mass at the Z level is

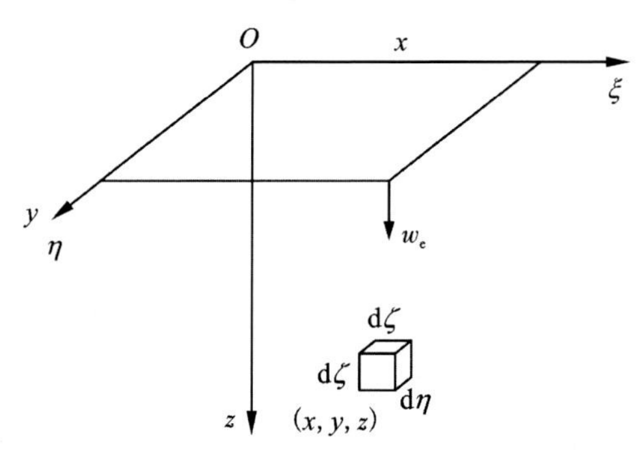

Figure 11.9 Unit excavation.

$$W_e(X, Y, Z, t) = \frac{1}{r^2(Z)}\left[1 - \exp(-Ct)\right]\exp\left[-\frac{\pi}{r^2(Z)}(X^2 + Y^2)\right]d\xi d\zeta d\eta \tag{11.22}$$

Considering the plane problem, the unit excavation is infinitely long along the Y-axis and is integrated by formula (11.22):

$$\begin{aligned} W_e(X, Z, t) &= \int_{-\infty}^{+\infty}\frac{1}{r^2(Z)}\left[1 - \exp(-Ct)\right]\exp\left\{-\frac{\pi}{r^2(Z)}\left[X^2 + (Y-\zeta)^2\right]\right\}d\xi d\zeta d\eta \\ &= \frac{1}{r(Z)}\left[1 - \exp(-Ct)\right]\exp\left[-\frac{\pi}{r^2(Z)}X^2\right]d\xi d\eta \end{aligned} \tag{11.23}$$

where $r(Z)$ is the main influence range of unit excavation at the Z level. This range depends on the stratum condition where the excavation is located and can be in a linear or nonlinear relation with Z. If the main influence angle β of the stratum is introduced, then the $r(Z)$ is considered to be linearly related to Z.

$$r(Z) = \frac{Z}{\tan\beta} \tag{11.24}$$

The value $\tan\beta$ of in the formula depends on the stratum conditions where the excavation is located. For the ground surface, the main influence range $r(H)$ is equal to $H/\tan\beta$.

After a long time, maximum surface subsidence of the unit excavation is reached. In view of the plane strain condition, the final unit subsidence value is

$$W_e(X) = \frac{1}{r(Z)}\exp\left[-\frac{\pi}{r^2(Z)}X^2\right]d\xi d\eta \tag{11.25}$$

Formula (11.25) is a basic formula for studying the subsidence of each point on the surface under the influence of arbitrary excavation in plane strain conditions.

To study the horizontal movement of each point on the surface caused by tunnel excavation, the deformation of the ground caused by excavation can be regarded as incompressible process, that is, the volume deformation of the ground mass tends to be zero. For three-dimensional problems,

$$\varepsilon_{eX} + \varepsilon_{eY} + \varepsilon_{eZ} = 0 \tag{11.26}$$

where ε_{eX}, ε_{eY}, and ε_{eZ} are strains of the unit rock-soil mass along X, Y, and Z directions, respectively.

For the two-dimensional plane strain problem, $\varepsilon_{eY} = 0$. At the same time, the movement of overlying ground mass caused by unit excavation can be regarded as macroscopic continuous. Under plane strain conditions, the unit excavation causes final surface horizontal displacement $U_e(X)$:

$$U_e(X) = \frac{X}{r(Z)}\cdot\frac{1}{Z}\exp\left[-\frac{\pi}{r^2(Z)}X^2\right]d\xi d\eta \tag{11.27}$$

Substituting formula (11.24) into formula (11.27) gives

$$U_e(X) = \frac{X\tan\beta}{Z^2}\exp\left[-\frac{\pi\tan^2\beta}{Z^2}X^2\right]d\xi d\eta \tag{11.28}$$

Formula (11.28) is a basic formula for studying the subsidence of each point on the surface under the influence of arbitrary excavation in plane strain conditions.

11.3.2.2 Surface movement and deformation induced by tunnel excavation

Excavation of a tunnel with arbitrary section at a certain depth is obviously a plane strain problem. As shown in Fig. 11.10, the distance between the center of the underground excavation section and the surface is H, coordinate system $\xi O\eta$ is used for the unit excavation of ground

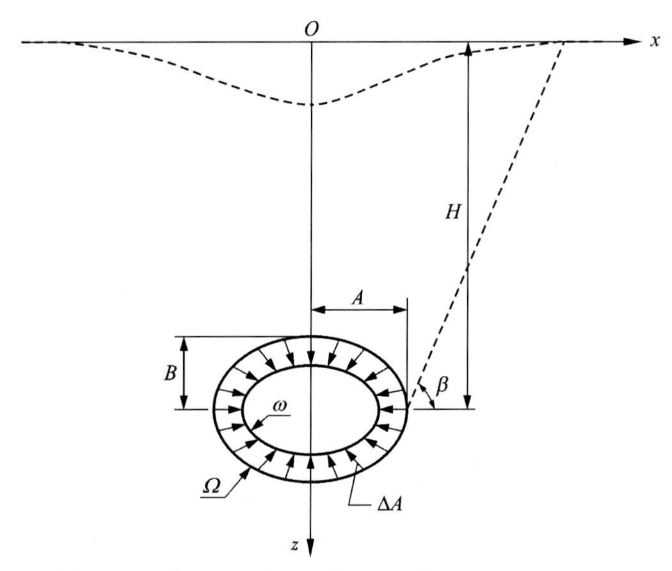

Figure 11.10 Schematic diagram of tunnel excavation.

mass, and coordinate system XOY is used for the ground surface. If the tunnel completely collapses, maximum subsidence of the surface will be caused after a long time. The entire excavation range is divided into an infinite number of unit excavations. Under the influence of unit excavation $d\xi d\eta$, it can be seen from formula (11.25) that the final subsidence value of the ground at a distance of X from the center of the unit is

$$W_e(X) = \frac{1}{r(\eta)} \exp\left[-\frac{\pi}{r^2(\eta)} X^2\right] d\xi d\eta \tag{11.29}$$

Provided that each excavation unit completely collapses in the whole excavation range Ω, formula (11.24) is substituted into formula (11.29) according to the superposition principle, then the surface subsidence value at the moment can be obtained by:

$$W(X) = \iint_{\Omega} \frac{\tan\beta}{\eta} \exp\left[-\frac{\pi \tan^2\beta}{\eta^2}(X-\xi)^2\right] d\xi d\eta \tag{11.30}$$

In fact, the surface settlement caused by tunnel construction is only the contraction of the tunnel excavation section caused by the movement of the ground mass around the tunnel to the excavation space. If the initial section of tunnel excavation is Ω, the excavation section contracts from Ω to ω after the tunnel has been completed, and the surface subsidence should be equal to the difference between the subsidence caused by the excavation range Ω and the surface subsidence caused by the excavation range ω, that is,

$$W(X) = W_{\Omega}(X) - W_{\omega}(X) = \iint_{\Omega-\omega} \frac{\tan\beta}{\eta} \exp\left[-\frac{\pi \tan^2\beta}{\eta^2}(X-\xi)^2\right] d\xi d\eta \tag{11.31}$$

Similarly, according to the superposition principle, the surface horizontal displacement $U(X)$ caused by tunnel construction should be equal to the difference between the horizontal displacement $U_{\Omega}(X)$ on the surface caused by the excavation range Ω and the horizontal displacement $U_{\omega}(X)$ caused by the excavation range ω. The formula $U(X)$ can be obtained from formula (11.28):

$$U(X) = U_{\Omega}(X) - U_{\omega}(X) = \iint_{\Omega-\omega} \frac{(X-\xi)\tan\beta}{\eta^2} \exp\left[-\frac{\pi \tan^2\beta}{\eta^2}(X-\xi)^2\right] d\xi d\eta \tag{11.32}$$

The surface deformation caused by tunnel construction mainly refers to inclination $T(X)$ of a surface point caused by differential settlement and horizontal strain $E(X)$ of the surface point caused by differential horizontal displacement:

$$T(X) = \frac{dW(X)}{dX} = \iint_{\Omega-\omega} \frac{-2\pi \tan^3\beta}{\eta^3}(X-\xi)\exp\left[-\frac{\pi \tan^2\beta}{\eta^2}(X-\xi)^2\right] d\xi d\eta \tag{11.33}$$

$$E(X) = \frac{dE(X)}{dX} = \iint_{\Omega-\omega} \frac{\tan\beta}{\eta^2}\left[1-\frac{2\pi \tan^2\beta}{\eta^2}(X-\xi)^2\right] \exp\left[-\frac{\pi \tan^2\beta}{\eta^2}(X-\xi)^2\right] d\xi d\eta \tag{11.34}$$

The curvature $K(X)$ of the surface subsidence curve $W(X)$ is approximately expressed by the following formula:

$$K(X) = \frac{d^2W(X)}{dX^2} = \iint_{\Omega-\omega} \frac{2\pi \tan^3\beta}{\eta^3}\left[\frac{2\pi \tan^2\beta}{\eta^2}(X-\xi)^2-1\right] \exp\left[-\frac{\pi \tan^2\beta}{\eta^2}(X-\xi)^2\right] d\xi d\eta \tag{11.35}$$

For subway tunnels, water supply pipe tunnels, electric power and communication optical cable tunnels, and other tunnels constructed by shield tunnelling methods, it is common for the construction section to be circular or quasicircular. For a circular section tunnel as shown in Fig. 11.10, the depth of the tunnel center from the surface is H, and the initial radius of excavation is $A = B$. Provided that the tunnel section is uniformly deformed, the radius of the section uniformly shrinks by ΔA after completion of the tunnel, and the subsidence value $W(X)$ of the surface can be obtained from formula (11.31):

$$W(X) = \int_a^b \int_c^d \frac{\tan\beta}{\eta} \exp\left[-\frac{\pi \tan^2\beta}{\eta^2}(X-\xi)^2 \right] d\xi d\eta - \int_e^f \int_g^h \frac{\tan\beta}{\eta} \exp\left[-\frac{\pi \tan^2\beta}{\eta^2}(X-\xi)^2 \right] d\xi d\eta \tag{11.36}$$

The surface horizontal displacement $U(X)$ can be obtained from formula (11.32):

$$U(X) = \int_a^b \int_c^d \frac{(X-\xi)\tan\beta}{\eta^2} \exp\left[-\frac{\pi \tan^2\beta}{\eta^2}(X-\xi)^2 \right] d\xi d\eta - \int_e^f \int_g^h \frac{(X-\xi)\tan\beta}{\eta^2} \exp\left[-\frac{\pi \tan^2\beta}{\eta^2}(X-\xi)^2 \right] d\xi d\eta \tag{11.37}$$

From formulas (11.36)–(11.37), the upper and lower limit values of a, b, c, d, e, f, g, and h of the double integral are, respectively, as follows

$$a = H - A, \quad b = H + A, \quad c = -\sqrt{A^2 - (H-\eta)^2}, \quad d = -c$$
$$e = H - (A - \Delta A), \quad f = H + (A - \Delta A), \quad g = -\sqrt{(A-\Delta A)^2 - (H-\eta)^2}, \quad h = -g$$

Because it is difficult to write an integral formula for an original function in the above-mentioned stochastic medium theory prediction formula, the integration can be calculated by compiling a computer program.

The above two prediction methods described in this chapter are widely used but also have shortcomings. In fact, it can be considered that the Peck's formula method is actually a method similar to the stochastic medium theory method for the tunnel cases with a large buried depth and small-section size ($H/R > 5$). For shallow-buried tunnels with large sections, more accurate results can be obtained by applying the stochastic medium theory. These two simplified calculation methods require some empirical parameter values, and the prediction can be improved if there are a large amount of application experiences in engineering practice. However, it is undeniable that these methods have the following limitations in practical application: (1) The complexity of geological conditions, such as influence of groundwater and soil strength variety, cannot be considered. (2) The surface displacement and deformation under special boundary conditions cannot be considered, e.g. surface load. (3) The influence of tunnel shapes and construction technologies cannot be considered.

11.3.3 Analytical method

The influence of many parameters should be considered in surface deformation prediction. These parameters include construction methods, tunnel face advancing details, tunnel depths and diameters, groundwater conditions, initial stress states, and stress-strain-strength characteristics of the ground mass before and after tunnel excavation. The analytical method for estimating stratum deformation caused by tunnel construction is generally based on the relevance between some variables and the displacement values observed in actual tunnels. The analytical method is to consider the ground mass in the tunnel excavation region as elastic, elastoplastic, or viscoelastic medium to analyze the problems by mathematical methods combined with mechanical theory to obtain the theoretical calculation formulas.

11.3.3.1 Closed-form analytical solution method

Sagaseta [18] proposed that in initially isotropic and uniform incompressible soil, an analytical solution of the strain field can be obtained due to near-surface ground loss caused by tunnel excavation. Verruijt and Booker [19] used the approximation method, proposed by Sagaseta [18], about the ground loss and considered the influence of long-term tunnel ellipse (tunnel lining formation) to give an analytical solution of the tunnel in the uniform and elastic half-space under incompressible conditions and arbitrary Poisson's ratio. Loganathan [20] corrected the analytical solution of Verruijt and Booker [19] by considering the influence of the actual deformation boundary conditions of the tunnel, and its specific calculation expression is as follows:

1. Surface settlement

$$U_{z=0} = \varepsilon_0 R^2 \frac{4H(1-\nu)}{H^2+x^2} \exp\left(-\frac{1.38x^2}{(H\cot\beta+R)^2}\right) \tag{11.38}$$

2. Vertical stratum displacement

$$U_z = \varepsilon_0 R^2 \left(-\frac{z-H}{x^2+(z-H)^2} + (3-4\nu)\frac{z+H}{x^2+(z+H)^2} - \frac{2z\left[x^2-\left(z+H^2\right)\right]}{\left[x^2+(z+H^2)\right]^2}\right) \cdot \exp\left(-\left[\frac{1.38x^2}{(H\cot\beta+R)^2}+\frac{0.69z^2}{H^2}\right]\right) \tag{11.39}$$

3. Horizontal stratum displacement

$$U_x = -\varepsilon_0 R^2 x \left(\frac{1}{x^2 + (H-z)^2} + \frac{3-4\nu}{x^2 + (z+H)^2} - \frac{4z[z+H]}{[x^2+(z+H^2)]^2} \right) \cdot \exp\left(-\left[\frac{1.38x^2}{(H\cot\beta + R)^2} + \frac{0.69z^2}{H^2} \right] \right) \tag{11.40}$$

where ε_0 is the average ground loss rate; β is the critical angle equal to $45° + \phi/2$; z is the buried depth under the surface; R and H are the radius of the tunnel and the buried depth of the axis, respectively; x is the horizontal distance from the central line of the tunnel; and ν is the Poisson's ratio of stratum.

This method can quickly determine the stratum displacements caused by shield tunnel construction, and the influence of the lateral pressure coefficient of the stratum is considered through Poisson's ratio. In this formula the value k_0 of the lateral pressure coefficient of the stratum is

$$K_0 = \frac{\nu}{(1-\nu)} \tag{11.41}$$

In estimating the ground loss value, the strength, stiffness and elasto-plastic properties of the stratum medium are considered. In most cases, tunnel excavation is carried out within the elastic strain range of the stratum. The tunnel-induced strain around the excavation face can be controlled by applying appropriate working face pressure, installing tunnel supporting systems in time, or improving the ground around the tunnel.

11.3.3.2 Analytical solution of a complex variable function

Problem statement

This problem involves a circular tunnel in an isotropic uniform elastic half-plane (z-plane, $y<0$) shown in Fig. 11.11. The upper boundary of the half-plane is regarded as unstressed, and the boundary of the tunnel is assigned with a given radial displacement. The radius of the tunnel and the depth of the tunnel axis are expressed by r and h, respectively.

Basic equations and boundary conditions

In the complex variable method, the solutions of stress and displacement can be expressed by two functions $\varphi(x)$ and $\psi(x)$. These two functions

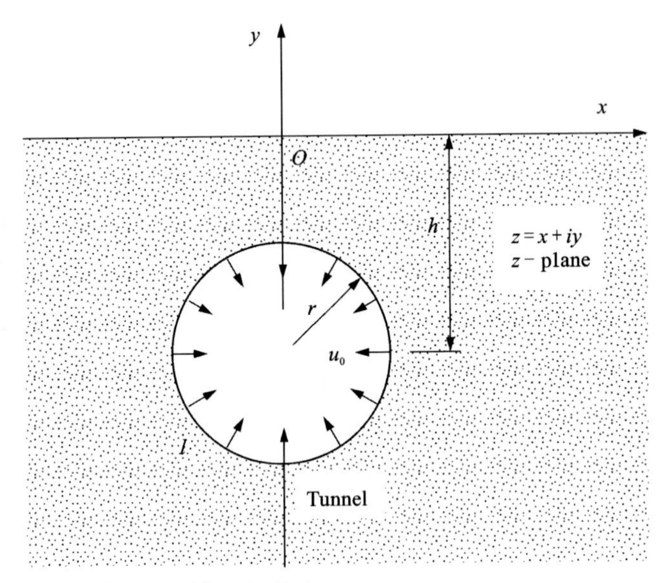

Figure 11.11 Circular tunnel in a half-plane.

need to be analyzed in the z-plane, excluding a circular cavity. The stress can be expressed by the following equations:

$$\sigma_{xx} + \sigma_{yy} = 2\{\varphi'(z) + \overline{\varphi'(z)}\} \tag{11.42}$$

$$\sigma_{yy} + \sigma_{yy} + 2i\sigma_{xy} = 2\{\overline{z}\varphi''(z) + \psi'(z)\} \tag{11.43}$$

Displacement can be expressed as follows:

$$2G(u_x + iu_y) = \kappa\varphi(z) - \overline{z\varphi'(z)} - \overline{\psi(z)} \tag{11.44}$$

where g is the shear modulus of an elastic material on the half-plane, and κ is related to Poisson's ratio ν. In this study, it is assumed that the problem is a plane strain condition, so

$$\kappa = 3 - 4\nu \tag{11.45}$$

The upper boundary of the z-plane is assumed to be stress free, $z = \overline{z}$, and the boundary conditions around the tunnel can be expressed as $|z + ih| = r$. Therefore the boundary conditions can be expressed by the following equations:

$$z = \overline{z}{:}\varphi(z) + \overline{z\varphi'(z)} + \overline{\psi(z)} = 0 \tag{11.46}$$

$$|z + ih| = r{:}2G(u_x + iu_y) = \kappa\varphi(z) - \overline{z\varphi'(z)} - \overline{\psi(z)} \tag{11.47}$$

These expressions are applicable to a single circular tunnel in an elastic half-plane. These expressions can be used separately for each of the two parallel tunnels.

Conformal mapping

The region in the z-plane occupied by the elastic material can be conformally mapped to the ring on the plane (region γ) and defined by circles $|\zeta| = 1$ and $|\zeta| = \alpha$ (Fig. 11.12). The circle $|\zeta| = 1$ corresponds to the upper boundary of the half-plane, and the circle $|\zeta| = \alpha$ corresponds to the tunnel boundary. The value of α can be determined by the following equation:

$$\frac{r}{h} = \frac{2\alpha}{1+\alpha^2} \tag{11.48}$$

The closer the ratio r/h tends to 1, the shallower the buried depth of the tunnel. The closer the ratio r/h tends to 0, the greater the buried depth of the tunnel. The conformal transformation is

$$z = \omega(\zeta) = -ih\frac{1-\alpha^2}{1+\alpha^2}\frac{1+\zeta}{1+\zeta} \tag{11.49}$$

Verruijt solution for stratum displacement

Provided that there is only one tunnel on the elastic region in the z-plane, $\varphi(z)$ and $\psi(z)$ are analyzed in this region, and $\omega(z)$ is in region γ.

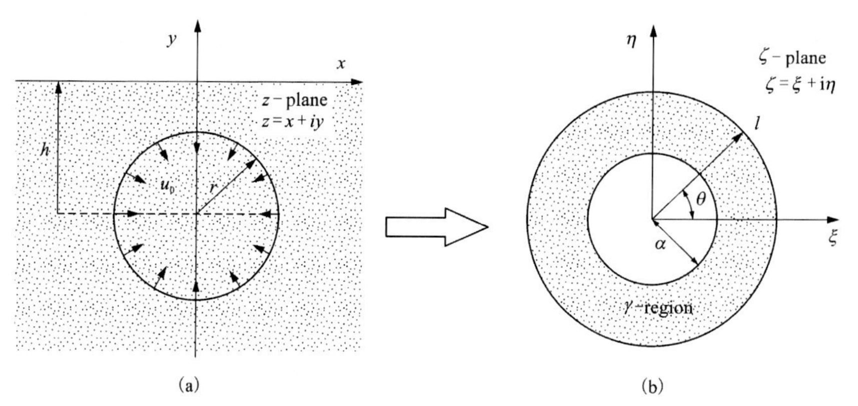

Figure 11.12 Conformal mapping from the z-plane to the ζ-plane; (a) half plane; (b) γ region.

Therefore functions $\varphi(z)$ and $\psi(z)$ will be analyzed in region γ. These two formulas can be extended to the form of Laurent series:

$$\varphi(z) = \varphi(\omega(\zeta)) = \varphi(\zeta) = a_0 + \sum_{k=1}^{\infty} a_k\zeta^k + \sum_{k=1}^{\infty} b_k\zeta^{-k} \tag{11.50}$$

$$\psi(z) = \psi(\omega(\zeta)) = \psi(\zeta) = c_0 + \sum_{k=1}^{\infty} c_k\zeta^k + \sum_{k=1}^{\infty} d_k\zeta^{-k} \tag{11.51}$$

The iterative calculation of these Laurent series in the ring domain γ is convergent. Therefore the coefficients a_k, b_k, c_k, and d_k can be determined according to the stipulated boundary conditions. The circle with radius $\rho = 1$ in the ζ-plane corresponds to a free face in the half-plane, and ζ can be expressed as $\zeta = \rho\sigma$ in the polar coordinate system, where $\sigma = \exp(i\theta)$. At the same time, considering

$$\varphi'(\zeta) = \frac{d\varphi}{dz}\frac{dz}{d\zeta} = \varphi'(z)\omega'(\zeta) \tag{11.52}$$

the boundary condition (11.46) can be rewritten as

$$|\zeta| = 1{:}\varphi(\zeta) + \frac{\omega(\zeta)}{\overline{\omega'(\zeta)}}\overline{\varphi'(\zeta)} + \overline{\psi(\zeta)} = 0 \tag{11.53}$$

where

$$\frac{\omega(\zeta)}{\overline{\omega'(\zeta)}} = -\frac{1}{2}\frac{(1+\rho\sigma)(\sigma-\rho)^2}{\sigma^2(1-\rho\sigma)} \tag{11.54}$$

Eqs. (11.49)–(11.52) and (11.54) are substituted into Eq. (11.53), and the coefficients of all powers of σ are set to 0, and the following results are obtained:

$$c_0 = -\overline{a_0} - \frac{1}{2}a_1 - \frac{1}{2}b_1 \tag{11.55}$$

$$c_k = -\overline{b_k} + \frac{1}{2}(k-1)a_{k-1} - \frac{1}{2}(k+1)a_{k+1}, k = 1, 2, 3, \ldots \tag{11.56}$$

$$d_k = -\overline{a_k} + \frac{1}{2}(k-1)b_{k-1} - \frac{1}{2}(k+1)b_{k+1}, k = 1, 2, 3, \ldots \tag{11.57}$$

The circle $|\zeta| = \alpha$ in Fig. 11.12 can be expressed as $\zeta = \alpha\sigma$, where $\sigma = \exp(i\theta)$. The boundary condition in Eq. (11.53) can be written as

$$|\zeta| = \alpha{:}2G(u_x + iu_y) = \kappa\varphi(\zeta) - \frac{\omega(\zeta)}{\overline{\omega'(\zeta)}} - \overline{\psi(\zeta)} = U(\zeta) = U(\alpha\sigma) \quad (11.58)$$

To simplify the form of Eq. (11.57), Verruijt [21] ingeniously introduce a function $U'(\zeta)$ which can be extended to a Fourier series:

$$U'(\zeta) = U'(\alpha\sigma) = (1 - \alpha\sigma)U(\alpha\sigma) = \sum_{k=-\infty}^{+\infty} A_k\sigma^k \quad (11.59)$$

By substituting Eqs. (11.50)–(11.52) and (11.58) into Eq. (11.59), and the final coefficients in Eqs. (11.50) and (11.51) must conform to the following equations (i.e., all powers of σ on both sides of the equation are equal):

$$\begin{aligned}&(1-\alpha^2)(k+1)\overline{a_{k+1}} - (\alpha^2 + \kappa\alpha^{-2k})b_{k+1} = \\ &(1-\alpha^2)k\overline{a_k} - (1 + \kappa\alpha^{-2k})b_k + A_{-k}\alpha^{-k}, k = 1,2,3,\ldots\end{aligned} \quad (11.60)$$

$$\begin{aligned}&(1 + \kappa\alpha^{2k+2})\overline{a_{k+1}} + (1-\alpha^2)(k+1)b_{k+1} = \\ &\alpha^2(1 + \kappa\alpha^{2k})\overline{a_k} + (1-\alpha^2)kb_k + \overline{A_{k+1}}\alpha^{k+1}, k = 1,2,3,\ldots\end{aligned} \quad (11.61)$$

$$(1-\alpha^2)\overline{a_1} - (\kappa + \alpha^2)b_1 = A_0 - (\kappa + 1)a_0 \quad (11.62)$$

$$(1 + \kappa\alpha^2)\overline{a_1} + (1-\alpha^2)b_1 = \overline{A_1}\alpha + (\kappa + 1)a^2\overline{a_0} \quad (11.63)$$

The coefficients a_1 and b_1 can be determined by using the value of the coefficient a_0 according to the values of Eqs. (11.62) and (11.63), and all other coefficients can be obtained recursively by using Eqs. (11.60) and (11.61). Verruijt [21] suggested calculating the coefficient a_0 through repeated numerical examples. If $k \to \infty$, the coefficients a_k and b_k tend to zero.

Boundary condition related to ground loss

1. Uniform radial displacement

Verruijt [21] provides the solution for a single tunnel problem with uniform radial displacement u_0 at the tunnel boundary. For the given ground loss parameter V_l, it is related to u_0 through the following relation:

$$u_0 = r\left(1 - \sqrt{1 - V_l}\right) \quad (11.64)$$

If u_0 is considered to be positive inwards, the boundary condition can be expressed as

$$2G\left(u_x + u_y\right) = -2Gu_0\frac{z + ih}{r} \tag{11.65}$$

On the basis of formulas (11.48), (11.49), and (11.59) the coefficient of Fourier series expansion in the formula (11.59) can be expressed as

$$\begin{cases} A_0 = -2Gu_0\alpha i \\ A_1 = 2Gu_0 i \\ A_k = 0, k \geq 2 \\ A_k = 0, k \leq -1 \end{cases} \tag{11.66}$$

All coefficients in Eqs. (11.50) and (11.51) can be determined by Eqs. (11.55)–(11.57). Because Laurent series expansions are all unique, correct solutions can be obtained.

2. Differential radial displacement

In actual engineering, the deformation around the tunnel is often differential. The general expression of such differential deformation is as follows:

$$u_r = -u_0(1 - \beta\cos 2\theta' + \eta\sin\theta') \tag{11.67}$$

where the coefficient β represents the degree of elliptical deformation of the tunnel, which is manifested as vertical compression and horizontal expansion, or vice versa, without causing changes in tunnel volume. When $\beta = 0$, the horizontal deformation and vertical deformation around the tunnel are consistent, and no elliptical deformation is caused. When $0 < \beta < 1$, the convergence rate of the tunnel along the horizontal direction is smaller than the vertical deformation. When $\beta = 1$, the vertical deformation rate of the tunnel is doubled, while the horizontal convergence is zero. When $\beta > 1$, the convergent deformation in the horizontal direction is negative, while the vertical deformation gradually increases. The tunnel will float integrally owing to the difference in stratum stiffness between the top and the bottom of the tunnel and the difference in gravity between the tunnel itself and the excavated soil. η is a coefficient that describes the overall vertical displacement of the tunnel.

Therefore the boundary condition around the tunnel can be written as follows:

$$2G\left(u_x + iu_y\right) = -2Gu_0(1 - \beta\cos 2\theta' + \eta\sin\theta') = U(\alpha\sigma) \tag{11.68}$$

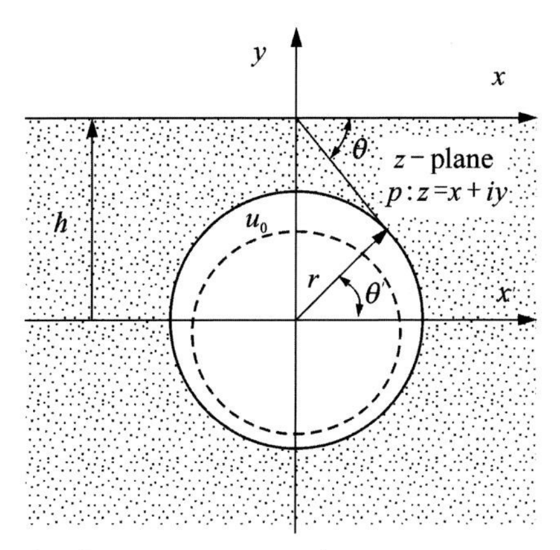

Figure 11.13 Relationship between two coordinate systems.

If point P at the boundary of the tunnel in the xOy coordinate system is $z = x + iy$, as shown in Fig. 11.13, it can be expressed in the $x'Oy$ coordinate system as follows:

$$z' = z + ih \tag{11.69}$$

However, on the basis of trigonometric functions, the following relation can be obtained:

$$\sin\theta' = \frac{z'_p - \overline{z'_p}}{2ir}; \cos\theta' = \frac{z'_p + \overline{z'_p}}{2r} \tag{11.70}$$

According to Eq. (11.49), Eq. (11.70) can be rewritten as follows:

$$\begin{cases} \sin\theta' = \dfrac{1+\alpha^2}{2\alpha} - \dfrac{\left(1-\alpha^2\right)^2}{2\alpha}\dfrac{1}{(1-\alpha\sigma)(1-\alpha\sigma^{-1})} \\ \cos\theta' = \dfrac{i\left(1-\alpha^2\right)}{2\alpha}\dfrac{\left(\sigma^{-1}-\sigma\right)}{(1-\alpha\theta)(1-\alpha\sigma^{-1})} \end{cases} \tag{11.71}$$

By substituting Eqs. (11.68) and (11.71) into Eq. (11.59) and using results given by Verruijt and Strack [22], the coefficient in the Fourier series can be obtained as follows:

$$\begin{cases} A_0 = -Gu_0 i\left[2(1-\eta)\alpha + 2\beta\alpha + \eta(1+\alpha)^2\right] \\ A_1 = Gu_0 i\left[2(1-\eta) + \eta\left(2+3\alpha+a^3\right) + 2\beta\alpha^2(2-\alpha)^2\right] \\ A_k = -Gu_0 i\left[\eta\left(1-\alpha^2\right)^2\alpha^{k-2} + \beta\left[3+(k+1)\left(\alpha^2-1\right)\left(1-\alpha^2\right)\right]\alpha^{k-3}\right](k \geq 2) \\ A_k = Gu_0 i\left[\beta\left(1-\alpha^2\right)^2\alpha^{-k-1}(k \leq -1)\right] \end{cases} \tag{11.72}$$

For all values of β and η, it should be possible to conduct the Fourier series expansion by using the coefficients given by Eq. (11.72). This also means that the coefficients of the two functions $\varphi(z)$ and $\psi(z)$ can be obtained by the above method, and the stress and displacement fields obtained therefrom can be expressed analytically.

Other analytical solution methods such as that proposed by Mindlin [23] have considered the tunnel to be a cylindrical cavity in a half-space elastic solid infinite medium under the action of gravity and have found an accurate analytical solution meeting the free boundary condition of the upper part surface and the free boundary condition of the cavity. Timoshenko and Goodier [24] used the Airy stress function to obtain a general solution for the deformation of the soil around the tunnel. Bobet [25] proposed an analytical method that is applicable only to shallow tunnels and ground deformation around the shallow tunnel where the soil is saturated. Park [26] proposed an elastic analytical solution for stratum movement induced by tunnel excavation in soft soil. However, owing to the limitation of calculation conditions, these methods can solve only some simple questions for a simplified strata material model. They are not suitable for considering the influence of complex geological conditions, construction methods, and other factors on stratum displacement, so their application is greatly restricted.

11.3.4 Numerical analysis

Owing to limitations of the empirical formula method and analytical solution method, various factors in practice cannot be considered comprehensively, so there is a great need for more advanced calculation method. However, the numerical analysis method can comprehensively consider the influence of various factors and can more accurately simulate the whole process of the actual excavation. Therefore this method has been popularized among various engineering researchers. Numerical analysis methods used for shield tunnel construction methods are mainly include the finite differential method, the finite element method, the discrete element method, and the boundary element method, etc.

The research on the finite differential method and the finite element method is relatively mature. In the excavation process of the shield tunnelling method, the shield machine excavation, groundwater level change and other related external effects may disturb the surrounding soil, resulting in displacement and deformation of the strata around the tunnel excavation face.

Current numerical analysis methods mainly focus on the research of factors that cause stratum settlement during shield tunnel construction. Although the focus of each research factor is different, it can be seen that the stratum deformation value caused by the shield tunnel excavation process is mainly related to the characteristics of the rock-soil mass through which the tunnel passes, shield construction technologies, buried depth and section sizes of the tunnel, and the grouting situation. Moreover, the magnitude of deformation shown by each actual project is also different, owing to changes in complicated environment factors.

In recent years, with the development of large general-purpose numerical calculation software, powerful tools have been provided for numerical analysis in geotechnical engineering. Compared with aforementioned methods, numerical analysis is widely used in engineering fields because it can conveniently handle various nonlinear problems and can flexibly simulate complicated construction and mechanical processes of tunnel construction and underground geotechnical engineering. With the help of computers and the adoption of numerical analysis methods, many influence factors can be analyzed together and the characteristics of material nonlinearity and geometric nonlinearity can be considered. Some assumptions in analysis of classical elastoplastic theory and the main influence parameter affecting stratum displacement can be more comprehensively considered to obtain the displacement field and the stress field of the surrounding strata caused by tunnel construction. Numerical analysis can also be used to analyze the influence of the construction process, excavation section sizes, construction time sequence, supporting parameters, and the like on stratum displacement, so it has become the most common and mature analysis method.

However, owing to inconsistency between constitutive relation of the rock-soil mass and the spatial variety of rock-soil parameters, it is difficult for the results of numerical simulation to be completely consistent with the actual situation. The numerical simulation software that is commonly used in analysis of geotechnical engineering problems includes ABAQUS, ADINA, FLAC, and ANSYS. Valuable results can be obtained through conducting sophisticated analysis on many engineering problems by using those softwares.

11.3.5 Model experiment

The model experiment method is an important research means, mainly including similar material model experiments and centrifuge model experiments. It is often difficult to select similar materials for the rock-soil mass. At present, the centrifuge experiment is most commonly used. In the centrifuge model test, a scaled tunnel model is put in a high-speed rotating centrifuge and subjected to a centrifugal acceleration greater than that of gravity to compensate for the loss of weigh of the geotechnical structure caused by the reduction of the model size. This method can obtain stress state, displacement changes, and deformation failure mechanism of the models that are the same as those for the prototype.

11.3.6 Other methods

Other analysis methods are based on the analysis of actually measured data of surrounding stratum displacement around the tunnel excavation. By using these actually measured data from the excavated field and adopting some new analysis method, the deformation of the excavated stratum can be predicted very well by establishing a data model.

11.3.6.1 Prediction method based on grey system theory and time series analysis

The grey system theory posits that a system containing both known information and unknown or nondeterministic information can predict the grey process that changes in a certain direction and is related to time. Although the phenomena displayed in the process are stochastic and disorganized, they are orderly and bounded after all. Therefore this data set has a potential law. Grey prediction uses this law to establish a grey model to predict the grey system.

The time series analysis method uses a set of numerical sequences in chronological order and applies mathematical statistics to predict the development of things in the future. Such method recognizes the continuity of the development of things, uses existing data to speculate on the development trend of things, and also considers the stochasticity of things to deal with accidental influence factors.

The deformation of the surrounding ground mass caused by tunnel excavation is a dynamic and continuous process, but it is difficult to establish a clear functional relationship between this kind of deformation and other factors. It is a new research direction to approximate, simulate, and reveal the deformation law and dynamic characteristics of the ground mass through establishing a mathematical model by the time series analysis method and grey system theory. A discrete and stochastic time series can be formed according to the actually measured data of stratum and surface settlement. Therefore the time series analysis theory and method based on the actually measured data can be used to analyze and predict the trends and laws of ground deformation and settlement.

11.3.6.2 Method based on artificial neural network analysis

Neural network theory is a nonlinear discipline that has developed rapidly in recent decades, which tries to simulate the basic characteristics of the human brain, such as self-organization, self-adaptation, and fault tolerance. The method has the characteristics of information memory, autonomous learning, knowledge reasoning, and calculation optimization in handling problems involving very complicated information, unclear background knowledge, and uncertain inference rules. In the shield tunnel construction process, the deformation of the stratum around the tunnel is a complicated nonlinear system with complex geological conditions and diverse influence factors. Traditional methods and techniques have difficulty revealing its inherent laws, while the high nonlinear mapping ability of the neural network shows high objectivity and adaptability in modeling the highly complicated and nonlinear deformation based on actually measured data. In practical applications, the sample database is expanded through continuous addition of actually measured data, and the network is trained, which can gradually improve the accuracy of estimation and the ability to predict.

Therefore, the simulation of stratum deformation caused by tunnel excavation is a complicated, nonlinear, and imprecise system. The use of artificial neural network can avoid traditional complicated mathematical models and the constitutive relation of stratum materials, as a result, this method is regarded as an effective method for predicting the surface movement and deformation caused by tunnel excavation. However, this method is still in the initial stage, and it is worth discussing how to consider and pretreat accumulation of samples and physical factors (geological conditions and construction processed).

11.3.6.3 Statistical regression analysis method

The statistical regression analysis method is a widely used deformation analysis method. The method conducts regression analysis to establish a functional relation between measured influence factors and deformation values based on actually measured data. With this functional relation, the deformation can be predicted, and can also be physically explained. The more abundant and high quality of the observation data, the more reliable the result. However, in regression analysis the choice of which factory system and the choice of which expression are sometimes just speculative, and the regression analysis is also limited in some cases, owing to diversity of the deformation influence factors and unpredictability of some factors. At the same time, this method does not accurately reflect the discreteness and stochastic volatility of observation values because it uses an average curve for fitting and forecasting. The regression model is a static model, which reflects only the relevance between the deformation value and the independent variables at the same moment and does not reflect the time sequence of deformation observation sequence, interdependence, and deformation continuity. If a single variable is used as an independent variable to establish a deformation prediction model (i.e. trend analysis), the first step of quantitative prediction is usually to find out the main change trend and predict the final settlement amount.

11.4 Assessment of the effects of shield tunnelling on existing buildings

Owing to the complexity of construction technologies, surrounding environments, and geotechnical materials, it is impossible to completely eliminate stratum deformation caused by tunnel construction even if the most advanced shield tunnelling method is adopted. When stratum movement and surface deformation exceed a certain limit, excessive deformation will cause damage to the use function and bearing capacity of surrounding structures, leading to serious economic loss and adverse social impacts, thereby affecting the normal use and operation safety of the tunnel and surface buildings.

The deformation of the stratum and adjacent structures is a complicated soil-structure interaction problem as a result of the effect of tunnel construction. In traditional methods of evaluating building damage, provided that the building follows the free surface settlement and the self-weight of the building is not considered, the deformation index of the building is calculated for evaluation. In fact, because of factors such as relative stiffness, relative position, building weight and structural characteristics, not only does the settlement caused by tunnel affect the existing adjacent buildings, but the existing buildings also affect the ground deformation characteristics caused by the tunnel. Therefore to evaluate the impact of shield tunnel construction on adjacent existing buildings (structures), the interaction among the tunnel, the stratum, and the buildings should be fully considered.

11.4.1 Buildings damage forms

The construction of urban shield tunnels will inevitably cause disturbance to the surrounding stratum, causing surface settlement or uplift to trigger settlement, differential settlement, and inclination of buildings (structures). Excessive deformation of the buildings (structures) can cause damage to the use function and the bearing capacity, leading to safety accidents. The main control indexed for the deformation and damage of buildings (structures) caused by shield tunnel construction mainly include settlement amount, differential settlement amount, inclination amount (angle variable), crack width, and so on [27].

Fig. 11.14 lists the main forms of building deformation. Generally, the following main factors cause building damage [28–30].

11.4.1.1 Settlement damage

The uniform settlement of the surface can cause the building (structure) to have overall uniform settlement, which has little effect on its upper structure and will not have much impact on stability and use conditions. However, when the settlement amount is too large, the indoor floor may be lower than the outdoor surface, causing rainwater backflow, pipe breakage, and so on. Excessive settlement often produces uneven settlement. Differential settlement of the building (structure) often causes structural members to be damaged by shear distortion. Frame structures are particularly sensitive to settlement differences.

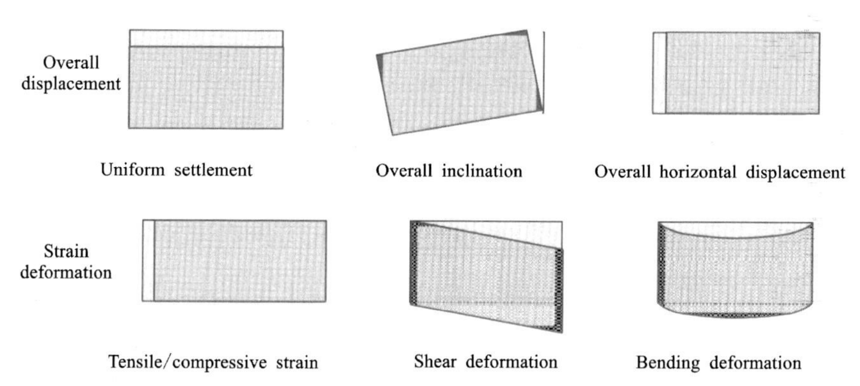

Figure 11.14 Manifestation of building damage.

11.4.1.2 Inclination damage

The differential settlement of the stratum causes the surface inclination to have a great impact on the building with a small bottom area and a large height, causing the center of gravity to deviate and leading to changes in structure stress to bring damage. Even if the structure of an ordinary building is not severely damage, excessive inclination will damage its use function. Precision instruments in certain buildings are more sensitive to inclination. For some buildings with small stiffness and deformation modulus, angular deformation of the building will be caused as a result of differential inclination between the base and the upper structure, resulting in cracking and damage to the buildings.

11.4.1.3 Curvature damage

The surface is deformed to form a curved surface settlement trough. When the building (structure) is in a relatively concave negative curvature of the surface and the foundation looks like a beam supported at both ends, suspended in the middle, compressed at the upper part, and tensioned at the lower part, the wall is more likely to produce cracks in splayed pattern and horizontal cracks, and the bottom of the building (structure) when its length is too large may be broken under the action of gravity. When the surface is in a relatively convex positive curvature and the foundation is partially suspended on both ends, tensioned at the upper part, and compressed at the lower part, the wall can easily suffer cracks in inverted-splayed pattern, and the end of the roof truss or the beam may be pulled out from the wall or columns, causing collapse.

11.4.1.4 Damage resulting from ground horizontal deformation

The ground horizontal deformation refers to stretching and compression of the ground. The foundation bottom of the building located in the stretched region is subjected to external friction, and a small tensile deformation is enough to make the building crack, especially if it is a masonry building. General buildings have relatively high resistance to compression, but too large a compression deformation can cause crushing damage to weak points of the structure, and the damage may be more serious than that caused by stretching.

11.4.2 Assessing methods for building deformation and damages

The deformation and damage control parameters of the building (structure) generally include settlement, differential settlement, inclination (angle variable), crack, deflection, strain, horizontal deformation, and so on. The specific values of the control index are affected by construction age, foundation forms, structure types, geological conditions, and other factors. Therefore it is generally difficult to adopt a uniform deformation index to quantify the building damage owing to the complexity of the interactions between the shield tunnel construction, soil, and structure. The specific structure forms of the building, shield construction parameters, load situation, historical deformation, and subsequent deformation within the service life should be considered in the use of various control indexes in engineering evaluation [16,31,32].

Evaluation methods for deformation and damage of buildings (structures) generally include statistical analysis of monitoring cases, engineering structural mechanics calculation, numerical simulation analysis, and so on. Owing to the complexity of research problems, the actual deformation of the building (structure) can be reflected through survey statistics based on actual projects, but predictive evaluation cannot be provided. Although methods such as structural mechanics calculation and numerical simulation analysis can provide predictive evaluation results, skilled professional technicians and analysis software are required to carry out detailed evaluation. In fact, according to the simplified method suggested by Mair [33], the building damage assessment can be carried out in three stages to simplify the assessment procedure. To focus the assessment on the buildings that is most vulnerable to be damaged, preliminary assessment of the first stage is conducted to screen out buildings with low expected damage risks. The assessment of the second stage and the third stage will be conducted on the buildings that are predicted to have high potential damage risks (medium and severe damage classification) at the first stage.

11.4.2.1 First stage: preliminary assessment

In the first step, a settlement contour is drawn along the alignment direction of the tunnel project, and the occupied area of all existing buildings is drawn on the settlement contour. The maximum settlement and the angular deformation are estimated for each structure. The assessment is based on estimation of the surface settlement values and surface slope in free field condition. Six risk categories ranging from 0 (negligible) to 5 (very severe) are used to define the degree of possible damage. Table 11.5 shows the standards for ranking building damage level of preliminary assessment results. For building damage level lower than slight only needs further observation and generally not need to take reinforcement measures.

11.4.2.2 Second stage: assessment at the second stage

At the second stage, the deformation indexes such as horizontal and shear strain of the building are estimated and calculated again in view of the interaction between the ground and the building. The criteria based on the critical strain of building proposed by Boscardin and Cording [27] and Burland [34] are adopted (seen in Table 11.6). The assessment at the second stage is conservative because it assumes that the building has no stiffness and is deflected to conform to the settlement trough in the free field condition. However, in practice, the actual degree of damage may be lower than the assessment result owing to the presence of structural stiffness of the building.

Fig. 11.15 shows the alternative method of evaluation based on angular distortion and horizontal strain proposed by Boscardin and Cording [27]. This method adopts the horizontal strain and angular deformation of the building to classify the damage level of the building, and has been applied to the early research on excavated and backfilled tunnels. Both methods proposed by Boscardin and Cording [27] and Burland [34] provide a consistent damage classification.

In addition, for buildings with different foundation forms, the assessment at the second stage requires different calculation procedures, formulas and related parameters. For shallow foundation buildings, the ultimate goal of the second assessment is to calculate critical strain and compare it with the standard. Therefore it is first necessary to estimate the characteristics of the deformation curve of the building and figure out the convex region and the concave region of settlement deformation of the building caused by shield tunnel construction (Fig. 11.16).

Then the bending strain ε_b, the angular strain ε_d, and the horizontal strain ε_d of the building need to be estimated according to the following formulas:

$$\frac{\Delta}{L} = \left(\frac{L}{12t} + \frac{3IE}{2tLHG}\right)\varepsilon_{b\max} \tag{11.73}$$

$$\frac{\Delta}{L} = \left(\frac{HL^2G}{18IE} + 1\right)\varepsilon_{d\max} \tag{11.74}$$

$$\varepsilon_{\mathrm{h}} = \frac{\Delta L}{L} \tag{11.75}$$

where H is the height of building; E/G is the ratio of Young's modulus to shear modulus of building; L is the length of building in selected span

Table 11.6 Damage assessment criteria at first stage and second stage.

Building damage classification [34,35]					Equivalent surface settlement and slope [13]	
1	2	3	4	5	6	7
Damage level	Damage level description	Description of typical masonry building for restoration	Crack width (mm)	Maximum tensile strain (%)	Maximum surface slope	Maximum building settlement (mm)
0	Negligible	Small cracks	< 0.1	Smaller than 0.05	—	—
1	Very slight	Small cracks easily appear during normal decoration. They may be isolated minor cracks in the building. Appearance cracks are visible after a careful check.	0.1 ~ 1.0	0.05–0.075	Smaller than 1/500	Smaller than 10
2	Slight	Cracks can be filled easily. Redecoration may be needed. There are a few slight cracks inside the building. Cracks are visible outside and may need to be repainted for weather protection. Doors and windows may stick slightly.	1 ~ 5	0.075–0.15	1/500–1/200	10-50

(*Continued*)

Table 11.6 (Continued)

Building damage classification [34,35]					Equivalent surface settlement and slope [13]	
1	**2**	**3**	**4**	**5**	**6**	**7**
3	Moderate	Cracks may need to be repaired. A suitable liner can cover recurring cracks. A small number of external bricks may need to be replaced. Doors and windows are stuck. Utility services may be interrupted. Airtightness is often impaired.	5 ~ 15 or crack number greater than 3	0.15–0.3	1/200–1/50	50-75
4	Serious	Extensive repair is needed, including wall, door, and window removal and replacement. Window and door frames are deformed. The floor has an obvious slope. The wall obviously inclines or protrudes. Some bearings of the beam are damaged. Utility services are interrupted.	15–25, but it also depends on crack number	Greater than 0.3	1/200–1/50	Greater than 75
5	Very serious	Partial repair or full construction is needed. The beam cannot bear the weight. The wall inclines severely and needs to be supported. Windows are broken due to deformation. There are dangers of instability.	Usually greater than 25, but it also depends on crack number	—	Greater than 1/50	Greater than 75

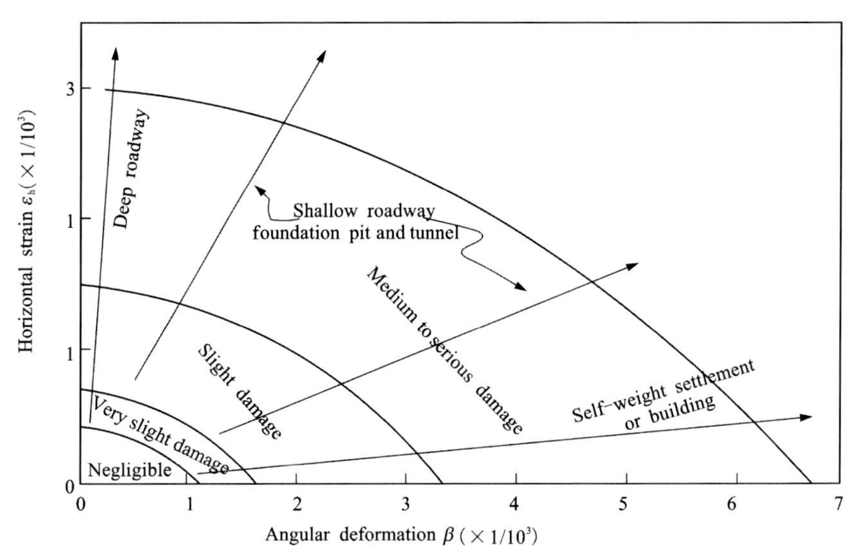

Figure 11.15 Relation between angular deformation damage and horizontal strain [27].

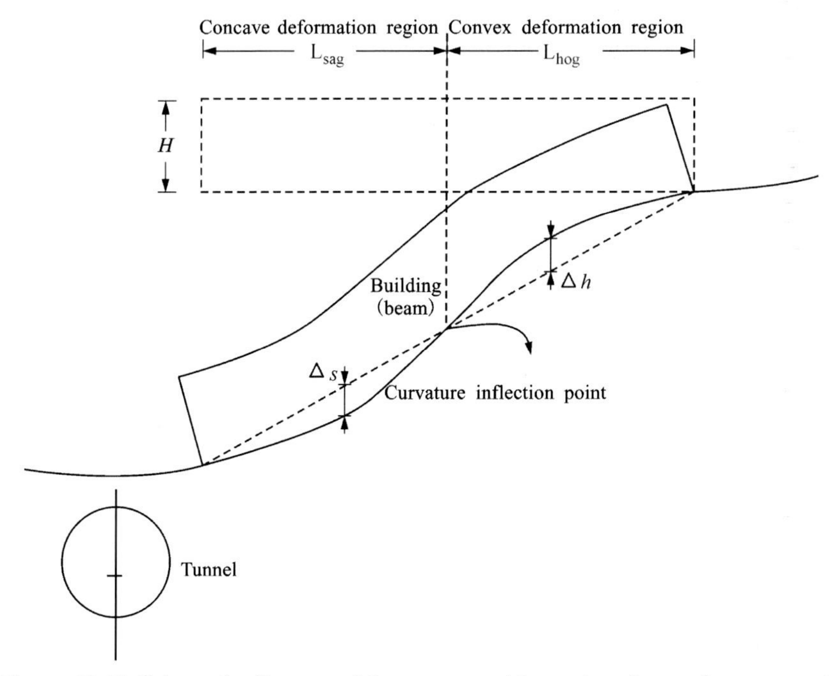

Figure 11.16 Schematic diagram of the convex settlement region and concave settlement region of a building.

range; I is the bending moment of the equivalent beam of the building in the respective region (calculated on the basis of the position of the neutral axis); t is the minimum distance from the neutral axis to the edge of the equivalent beam; Δ is the maximum settlement selected in calculation of the span range; and Δ/L is the ratio of maximum settlement and span length in calculation of the span range.

The total bending strain and the angular strain are further estimated to finally obtain the critical strain value. Comparison with Table 11.6 is made to update the damage level of the building.

Total bending strain:

$$\varepsilon_{bs} = \varepsilon_{b,\max} + \varepsilon_h \tag{11.76}$$

Angular strain:

$$\varepsilon_{ds} = \varepsilon_h \left(\frac{1-\nu}{2}\right) + \sqrt{\varepsilon_h^2 \left(\frac{1-\nu}{2}\right)^2 + \varepsilon_{d\max}^2} \tag{11.77}$$

Critical strain:

$$\varepsilon_{\text{critical value}} = \max(\varepsilon_{bs}, \varepsilon_{ds}) \tag{11.78}$$

In evaluating the piled building, it is necessary to evaluate the damage level by checking and calculating the combined stress of the pile foundation. The formulas for calculating the combined stress of the pile are as follows:

$$\sigma_{\max} = \frac{(M + \Delta M)}{Z} + \frac{(P + \Delta P)}{Z} \tag{11.79}$$

$$\sigma_{\min} = \frac{(M + \Delta M)}{Z} - \frac{(P + \Delta P)}{Z} \tag{11.80}$$

where M is the designed bending moment; P is the designed axial force; ΔM is the induced bending moment; ΔP is the induced axial force; and Z is the regional section modulus.

For concrete piles under working loads, a 25% extra compressive stress is suggested adding to the 28-day cubic strength of concrete when calculating allowable stress based on total cross-section area of the pile foundation. It is not recommended that the above-mentioned overstress assumption be applied to the calculation of the old piled building. If $\sigma_{\max} < \sigma_{\text{permit}}$, the pile foundation will not fail.

11.4.2.3 Third stage: detailed assessment

In the assessment at the third stage, a detailed assessment is conducted on buildings with a risk level of 3 or above (medium damage or worse). At the third stage, not only the consequences of building damage caused by shield tunnelling are considered, but also the types of control measures of damage. The assessment at the third stage starts with a field investigation, in which visual crack inspection are conducted, the stiffness of buildings is evaluated, and existing conditions and potential damage consequences are estimated. On the basis of the field investigation and with reference to existing data, the following factors can be considered in revising the damage risk level of the buildings:

1. Geological conditions, stratum layer and groundwater conditions
2. Stiffness and types of buildings (wood, masonry, or framed structure)
3. Building foundation types
4. Building history and detailed information about the age of buildings
5. Sensitivity and use of the buildings, such as offices, private houses, public buildings, sport facilities, etc.

Any buildings that have been designated to medium damage or worse damage level after field inspection will be analyzed in detail according to the methods proposed by Addenbrook [36] by taking into account the relative stiffness of the building and the stratum, and the monitoring frequency of the monitored items should be submitted.

For piled buildings, numerical simulation methods are generally used directly for detailed evaluation (stage 3). At present, various numerical methods are used to estimate the response of pile groups under combined external loads. The computer programs used for this analysis vary in types of used methods and complexity of handling with different aspects of pile groups, and elaborated models should be established for detailed analysis.

11.4.3 Calculation engineering cases

[Case 11.2] The details of the size and other parameters of a building are shown in Table 11.7, and the stratum settlement curve caused by nearby tunnel construction is shown in Fig. 11.17. The specific parameters of the

Table 11.7 Details of the building.

Length L (m)	Height H (m)	Width W (m)	Poisson's ratio v	Description	Young's modulus E/shear modulus G
18	2.5	19	0.3	Single-layer brick structure	2.6

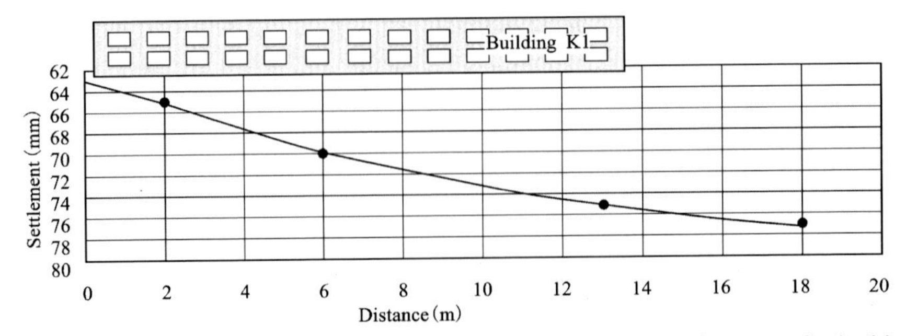

Figure 11.17 Schematic diagram of the settlement curve at the bottom of a building. *Modified from Loganathan N., An Innovative Method for Assessing Tunnelling-Induced Risks to Adjacent Structures. One Penn Plaza: Parsons Brinckerhoff Inc; 2011.*

Table 11.8 Calculation parameters of concave range and convex section of settlement curve.

Concave range		Convex range	
Length L	14 m	Length L	4 m
Deflection Δ	2 mm	Deflection Δs	0.8 mm
$t = H/2$	1.25 m	$t = H/2$	2.5 m
$I = H^3/12$	1.3 m^3	$I = H^3/12$	5.2 m^3
Length B	14 m	Length B	4 m
Horizontal displacement Δhs	10 mm	Horizontal displacement Δhs	10 mm

calculated settlement curve are shown in Table 11.8. Please determine the second-stage deformation damage assessment of the building [3].

[Solution]: Regarding the building as a beam structure, from Table 11.8:

$t =$ distance from neutral axis to the edge of the beam

$I =$ cross-section moment area of the beam

$B =$ length of building related to horizontal displacement

Evaluation can be conducted on the basis of the second-stage evaluation method for shallow foundation buildings given in Section 11.4.2.

11.4.3.1 Calculation on concave section of settlement curve

The local bending strain is obtained from formula (11.73):

$$\varepsilon_b = \left(\frac{L}{12t} + \frac{3IE}{2tLHG} \right) \frac{B}{\Delta} = 0.000136$$

The local angular strain is obtained from formula (11.74):

$$\varepsilon_d = \left(\frac{HL^2G}{18IE} + 1 \right) \frac{B}{\Delta} = 0.000016$$

The horizontal strain is obtained from formula (11.75):

$$\varepsilon_h = \frac{\Delta_h}{B} = 0.000714$$

The total bending strain is obtained from formula (11.76):

$$\varepsilon_{bs} = \varepsilon_b + \varepsilon_h = 0.000850$$

The total angular strain is obtained from formula (11.77):

$$\varepsilon_{ds} = \varepsilon_h\left(\frac{1-\nu}{2}\right) + \sqrt{\varepsilon_h^2\left(\frac{1-\nu}{2}\right)^2 + \varepsilon_d^2} = 0.000500$$

The critical tensile strain is obtained from formula (11.78):

$$\varepsilon_{\text{critical value}} = \max(\varepsilon_{bs}, \varepsilon_{ds}) = 0.00085$$

11.4.3.2 Calculation on convex section of settlement curve

In the same way, the strain values in the convex section of the settlement curve can be obtained as:

$$\varepsilon_b = 0.0002115; \quad \varepsilon_d = 0.000172; \quad \varepsilon_h = 0.002500;$$

$$\varepsilon_{bs} = 0.002711; \varepsilon_{ds} = 0.001767; \varepsilon_{\text{critical value}} = 0.002711$$

In summary, the control strain is

$$\varepsilon_{\text{critical value,max}} = 0.002711 = 0.2711\%$$

As can be seen from Table 11.6, the damage level is moderate when the maximum tensile strain is in the range 0.15%–0.3%. The specific description is as follows: cracks may need to be repaired, a suitable liner can cover recurring cracks, a small number of external bricks may need to be replaced, and the doors and windows of buildings are more likely to get stuck.

11.5 Common methods for ground deformation control

11.5.1 Common methods for ground deformation control

11.5.1.1 Initial settlement control in shield tunnel construction

The key to settlement control in the early stage of shield tunnel construction is to maintain groundwater pressure and the corresponding measures are:

1. Reasonably setting the earth pressure (slurry pressure) control value and maintaining it as constant during the advancing process to balance the earth pressure and the water pressure of the excavation face.

2. Prerequisites for maintaining stability of the earth pressure (slurry pressure) of the excavation face: For earth pressure shields, the prerequisite is that the plastic fluidization improvement effect of soil conditioning which requires appropriate selection of additives and injection parameters according to the stratum conditions; For slurry shields, the prerequisite is that the slurry performance which requires appropriate selection of slurry materials and their proportioning according to the stratum conditions.
3. Preventing groundwater infiltrating from cutterhead spindle seals, hinged seals, shield tail, and assembled lining structures. For this reason, it is necessary to keep the cutterhead drive, hinges, shield tail, and other parts sealed in good condition; ensure the injection pressure and injection volume of sealing oil for the shield tail; and allow the sealing and assembly quality of the segment to conform to the specification requirements.
4. When the earth pressure shield is advancing in the stratum with high groundwater level and good permeability, effective antispouting measures should be taken to prevent the gushing of groundwater from the screw conveyor.

11.5.1.2 Settlement (uplift) control ahead of excavation face

Earth pressure (slurry pressure) management is the main measure to control settlement (uplift) ahead of excavation face, so earth pressure (slurry pressure) balance must be truly achieved during shield tunnelling process. The usual measures are as follows:

1. Reasonably setting the earth pressure (slurry pressure) control value and keeping it constant in the advancing process to balance the earth pressure and the water pressure of the excavation face.
2. Maintaining the stability of the excavation face.
3. Strengthening control of soil discharge.
4. For earth pressure shield machine, control of shield tunnelling parameters such as shield thrust, advancing speeds, and cutterhead torques when necessary.

11.5.1.3 Settlement (uplift) control when shield machine passing

There are two main kinds of control measures for settlement (uplift) when shield machine passing:

1. The moving posture of the shield machine is controlled to avoid unnecessary correction. Deviation correction should be carried out

frequently in a moderate way but avoid too large correction at once. In the case of tunnelling in a hard stratum, overexcavation is required during deviation correction or curved alignment section tunnelling, and the overexcavation radius and range should be determined reasonably to minimize the volume of overexcavation.

2. When the earth pressure shield machine is tunnelling in a soft or loose stratum, grouting measures should be conducted to reduce the viscous resistance or friction between the periphery of the shield machine and the surrounding stratum.

11.5.1.4 Settlement (uplift) control for shield tail voids

The key of settlement (uplift) control for shield tail voids is to adopt appropriate grouting measures behind the lining, including the following:

1. Using synchronous grouting to fill shield tail voids in time.
2. Reasonably selecting single cement slurry grouting or cement-silicate mixed slurry grouting according to geological conditions, engineering conditions, and other factors and correctly selecting grouting materials and the proportioning to stabilize the assembled lining structure in a timely manner.
3. Strengthening the control of grouting volume and grouting pressure.
4. Carrying out secondary compensation grouting in time.

11.5.1.5 Subsequent settlement control

Subsequent settlement occurs during construction mainly in soft and viscous stratum, and the main control measures are as follows:

1. The disturbance to the stratum should be minimized as much as possible during shield advancing, deviation correction, and grouting.
2. If subsequent settlement is too large and does not meet the stratum settlement requirement, grouting in the specific location of the stratum can be taken.

11.5.2 Control of shield tunnelling parameters

11.5.2.1 Optimal matching of shield tunnelling parameters

Optimal shield tunnelling refers to minimal impact on the surrounding stratum and surface, small stratum strength reduction, little disturbance to surrounding environment, small excess pore water pressure, small surface uplift, and small abrupt subsidence during shield tail leaving the segment ring. All of these are primary conditions and fundamental methods for controlling surface settlement and protecting environments in shield tunnel construction.

To achieve the optimal tunnelling state, reasonable parameters must be selected to guide construction during shield tunnelling according to the buried depth of the tunnel, geological conditions, ground loads, design slope, alignment curve radius, axis deviation, and shield machine attitude. However, each parameter is independent, and there are also problems of mutual matching and optimal combination, showing control effects of surface deformation on a macro level. Therefore it is necessary to monitor the surface deformation along the tunnel and continue to optimize accordingly to ensure that the optimal construction parameters are truly achieved.

11.5.2.2 Trial tunnelling to determine parameters for guiding construction

A discrete and anisotropic three-phase body is the typical feature of the ground medium. The main working medium faced by shield tunnel construction is the rock and soil mass, coupled with interaction between the structures and the rock and soil mass, so it is difficult for physical and mechanical properties, calculation models and theoretical analysis results to achieve the accuracy of general continuum mechanics. For the above reasons, it is necessary to use a certain tunnelling section as a test section according to the stratum conditions and building (structure) conditions along the tunnel alignment.

Generally speaking, the first 100 m section of initial tunnel excavation is regarded as a trial section. In the actual tunnelling process, the 100 m trial tunnelling section can be divided into three divisions. The first division is 15 m long and is called an initial tunnelling, a total of three groups of tunnelling parameters are set, and the law of stratum changes and axis control can be explored based on monitoring results of surface settlement. The second section is 35 m long in which the three groups of parameters set at the first stage are adjusted to obtain optimal parameters according to ground conditions, adjacent buildings, and underground pipelines. The third section is 50 m long and is the preparation stage for formal tunnelling, in which the control measures are determined for surface settlement, tunnel axis control, lining installation quality, and so on. Thus the shield tunnelling parameters are basically determined, and information is fed back to guide the construction. The exploration of the tunnelling parameters and the ground deformation law of the 100 m trial section lays a good reference for determination of construction parameters during the entire tunnelling process.

11.5.2.3 Setting of pressure soil chamber

During the entire tunnelling process, the setting of the soil chamber pressure is a critical parameter, so its value should be considered separately here. If the earth pressure setting value is too small, the surface settlement amount will increase, whereas if the earth pressure setting value is too large, the ground surface will uplift. Details of theoretical basis are shown in Fig. 11.18.

Generally speaking, there are three conventional methods for theoretical calculation of water pressure and earth pressure on the tunnelling working face. The first one is calculation based on the traditional Rankine-Coulomb earth pressure theory, which is generally used in situations when the buried depth is not large. The second one is calculation based on Terzaghi theory, which is applicable to situations when the buried depth is large and a self-supporting arch can be formed above the

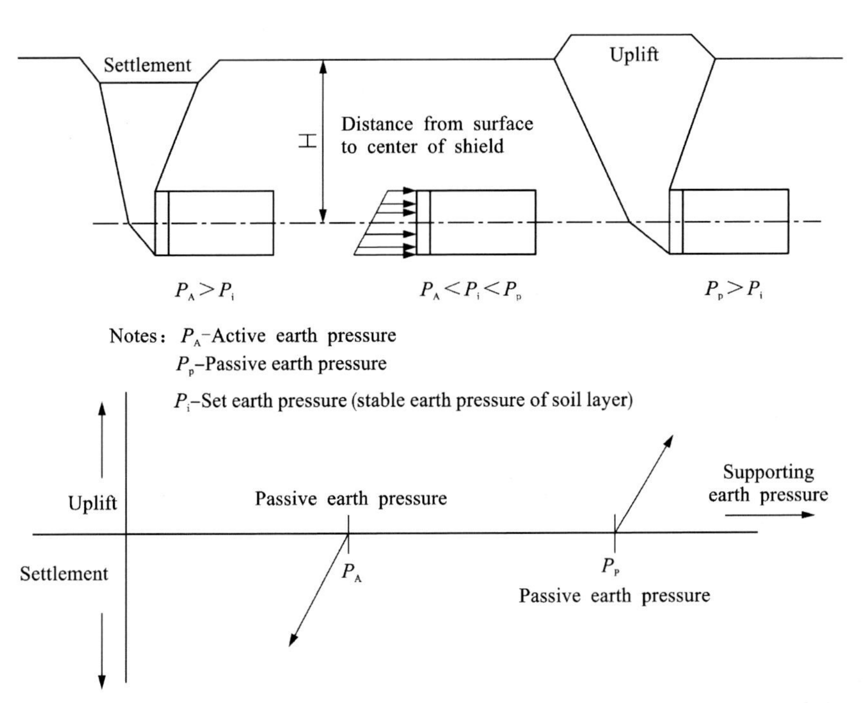

Figure 11.18 Schematic diagram of the stratum supporting earth pressure of the excavation face and consequent ground deformation.

tunnel excavation. The third one is the Murayama calculation method, which is an improvement on Terzaghi theory.

11.5.2.4 Determination of synchronous grouting parameters at shield tail

In the shield tunnelling process, synchronous grouting is carried out on an annular building gap on the back of the lining released from the shield tail with proper grouting pressure, grouting volume, reasonably proportioned grouting materials, and so on. This is the key measure to control or reduce the surface deformation.

The key parameter control of the synchronous grouting process at the shield tail mainly includes the following points:

1. Reasonably proportioned grout: The consistency value is controlled to 10.5 ~ 11.0, and the density approximates that of the originally undisturbed soil.
2. Grouting pressure: The appropriate grouting pressure is 0.5 ~ 0.6 MPa, and over-volume grout must be appropriately higher than the calculated grouting pressure before it can be injected into the shield tail voids, because the actual grouting volume is greater than the calculated grouting volume.
3. Grouting time: The injection time of the shield tail grouting has a significant effect on the grouting construction effect. If grouting is not timely, the expected grouting effect will not be achieved by grouting after the ground deformation has occurred. Therefore the injection time of the grout should be synchronized with the moment when the segment is pulled apart from the shield tail, and the time for uniform grout injection should be the same as the time for the segment assembly to advance by one ring.
4. Grouting volume: The control of synchronous grouting volume at the shield tail can be obtained according to the calculation of the shield tail void. However, in the actual grouting process, the soil at the shield tail is not compact and voids may exist in the soil at the shield tail, coupled with the disturbance of shield tunnel construction to surrounding stratum, so the actual synchronous grouting volume at the shield tail is much larger than the theoretical calculation. In construction of the No.17 section of Beijing Metro Line 5, the appropriate grouting volume in the sandy gravel stratum was 160%–220% of the theoretical grouting volume, and the appropriate grouting volume in

the silty soil and clay soil was 140%–180% of the theoretical grouting volume.

5. Distribution of grouting positions: Grouting pipes distributed in the shield tail shell at equal angles are purposely selected for grouting. The grouting pressure and grouting volume of each grouting pipe are determined according to different geological conditions and control standards, which can make the tail end of the tunnel lining floating in the grout producing a controllable displacement. Such arrangement not only can improve the original deviation of the tunnel axis, but also can effectively improve the jamming of the segment and the shield tail.

11.5.3 Ground reinforcement

11.5.3.1 Reinforcement method by grouting

The grouting reinforcement effect is mainly to enhance the cohesion of the soil and has little effect on other factors. As far as constructability, construction can be conducted from the ground and in the shaft; that is, the constructability is good and economical. However, the reliability of stratum reinforcement is low, and the reinforcement strength is limited, so it is used mostly to improve the waterstop property of the strata. In the grouting method, there are many kinds of materials and construction techniques, which need to be determined on the basis of groundwater, geology, construction environment, and so on. The expected reinforcement effect also needs to be considered, including countermeasures for stratum uplift caused by excessive grouting.

11.5.3.2 High-pressure jet grouting method

The high-pressure jet grouting method can be used as a construction method for general temporary structures such as concrete walls. The reinforcement thickness can be determined through calculation of soil quality and structural mechanics. The minimum requirement for reinforcement thickness is shown in Table 11.9. The reinforcement effect is good with

Table 11.9 Minimum reinforcement thickness (columnar jet grouting method).

D	$0.0 \le D < 1.0$	$1.0 \le D < 3.0$	$3.0 \le D < 5.0$	$5.0 \le D < 8.0$
B	1.0	1.0	1.5	2.0 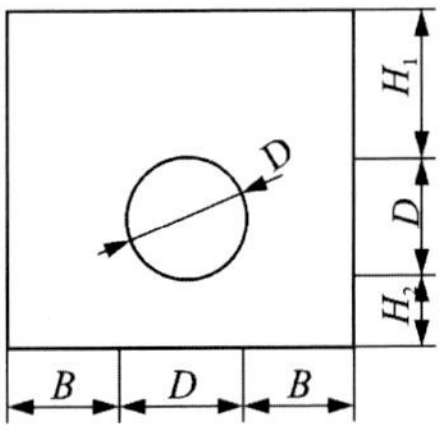
H_1	1.0	1.5	2.0	2.5
H_2	1.0	1.0	1.0	1.5

high reliability. During construction a sludge discharge trench for processing excavated soft mud by high-pressure jet must be arranged.

11.5.3.3 Freezing method

In the freezing method, the nonuniform ground in a natural state is frozen into ground with uniform mechanical properties. The advantage of this method is that the frozen soil wall can be controlled with temperature to ensure a long-term stable state with good reinforcement effects. The determination of the reinforcement range is calculated in accordance with methods used for general temporary structures.

The method of vertical freezing from the ground surface is mostly adopted for stratum reinforcement while the method of horizontal freezing is mostly adopted in the launching and receiving shaft section. The freezing method can reach high strength and stable reinforcement effects. In the presence of flowing groundwater, the advancing speed of shield machine cannot reach 1 ~ 2 m/day. The time required to form frozen the stratum varies with conditions, which generally 40 ~ 60 days. In addition, frozen stratum will produce frost heave and thawing settlement, especially in cohesive soil ground, so necessary measures should be taken to avoid such consequence.

References

[1] Fu J. Analysis on Impact of Shallow Excavation Subway Construction on Adjacent Buildings in Water-Rich Composite Stratum. Changsha: Central South University; 2010.
[2] Zhang F, Hehua Z, Deming F. Shield Tunnel. Beijing: China Architecture & Industry Press; 2004.
[3] Loganathan N. An Innovative Method for Assessing Tunnelling-Induced Risks to Adjacent Structures. One Penn Plaza: Parsons Brinckerhoff Inc; 2011.
[4] Fu J, Yang J, Klapperich H, Wang SY. Analytical prediction of ground movements due to a nonuniform deforming tunnel. Int J Geomech 2016;16(4).
[5] Wayne Clough G, Schmidt B. Chapter 8-Design and performance of excavations and tunnels in soft clay. Developments in Geotechnical Engineering. Elsevier B.V; 1981.
[6] Attewell P, Yeates J, Selby AR. Soil Movements Induced by Tunnelling and their Effects on Pipelines and Structures. New York: Chapman and Hall Ltd.; 1986.
[7] Macklin S. The prediction of volume loss due to tunnelling in overconsolidated clay based on heading geometry and stability number. Ground Eng 1999;32(4):30–3.
[8] Mair RJ, Gunn MJ, O'Reilly MP. Ground movements around shallow tunnel in soft clay. In: Proceedings of 10th International Conference on Soil Mechanics and Foundation Engineering, vol. 1. Rotterdam, The Netherlands: Balkema; 1981. p. 323–328.
[9] Lee K, Rowe R, Lo K. Subsidence owing to tunnelling. I. Estimating the gap parameter. Can Geotech J 1992;29:929–40.

[10] Lagerblad B, Fjällberg L, Vogt C. Shrinkage and Durability of Shotcrete. Swedish Cement & Concrete Research Institute; 2010.
[11] Peck RB. Deep excavation and tunnelling in soft ground. In: Proceedings of the seventh International Engineering. Mexico; 1969. p. 225–290.
[12] Mair R, Taylor RN, Bracegirdle A. Subsurface settlement profiles above tunnels in clays. Geotechnique 1993;43(2):361–2.
[13] Rankin W. Ground movements resulting from urban tunnelling: predictions and effects. Geol Soc Lond Eng Geol Spec Publ 1988;5(1):79–92.
[14] O'Reilly MP, New BM. Settlement above tunnels in the United Kingdom—their magnitude and prediction. In: Tunnelling 82, Proceedings of the third International Symposium, Brighton. London: IMM; 7-11 June, 1982. p. 173–181. International Journal of Rock Mechanics & Mining Sciences & Geomechanics Abstracts, 1983.
[15] Litwiniszyn J. The theories and model research of movements of ground masses. Colliery enginering 1958;(1):1125–36.
[16] Yang J, Li J, Fu J. Assessment system of the effects of tunneling on adjacent structures. Chin J Undergr Space Eng 2011;7(1):168–73.
[17] Yang J, Liu B. Surface Movement and Deformation Caused by Urban Tunnel Construction. Beijing: China Railway Publishing House; 2002.
[18] Sagaseta C. Analysis of undrained soil deformation due to ground loss. Géotechnique 1987;37(3):301–20.
[19] Verruijt A, Booker J. Surface settlements due to deformation of a tunnel in an elastic half plane. Géotechnique 1998;46(5):753–6.
[20] Loganathan N. Analytical prediction for tunneling-induced ground movements in clays. J Geotech Geoenviron Eng 1998;124(9):846–56.
[21] Verruijt A. A complex variable solution for a deforming circular tunnel in an elastic half-plane. Int J Numer Anal Methods Geomech 1997.
[22] Verruijt A, Strack OE. Buoyancy of tunnels in soft soils. Géotechnique 2008; 58(6):513–15.
[23] Mindlin R. Stress Distribution Around a Tunnel, 105. American Society of Civil Engineers; 1940. p. 1117–40.
[24] Timoshenko S, Goodier J. Theory of Elasticity, 3rd ed. New York: McGraw-Hill Book Company; 1970.
[25] Bobet A. Analytical solutions for shallow tunnels in saturated ground. J Eng Mech 2001;127(12):1258–66.
[26] Park K. Analytical solution for tunnelling-induced ground movement in clays. Tunn Undergr Space Technol 2005;20(3):249–61.
[27] Boscardin M, Cording E. Building response to excavation-induced settlement. J Geotech Eng 1989;115(1).
[28] Attewell P. Ground movements caused by tunnelling in soil, large ground movements and structures. In: Proceedings of Conference at University of Wales Institute of Science and Technology. New York: Wiley. 1977; p. 812–948.
[29] Bezuijen A. Bentonite and grout flow around a TBM. Tunn Tunn Int 2007;(6): 39–43.
[30] Burland JB, Wroth CP. Settlement of buildings and associated damage. In: Proceedings of conference on the Settlement of Structures. Pentech Press. 1974; p. 611–654.
[31] Fu J, Yu Z, Wang S, Yang J. Numerical analysis of framed building response to tunnelling induced ground movements. Eng Struct 2018;158(3):43–66.
[32] Wu F. Deformation controlling indices of buildings and structures. Rock Soil Mech 2010;31(S2):308–16.
[33] Mair RJ. Settlement effects of bored tunnels. In: Proceedings of International Symposium on Geotechnical Aspects of Underground Construction in Soft Ground. London: [s.n.]. 1966; p. 43–53.

[34] Burland JB. Assessment of risk of damage to buildings due to tunnelling and excavation. In: Kogakkai J, editors. Proceedings of first International Conference on Earthquake Geotechnical Engineering. A.A. Balkema: Tokyo, Japan. 1995; p. 495–546.
[35] Mair RJ, Taylor RN, Burland JB. Prediction of ground movements and assessment of risk building damage due to bored tunneling. In: Mair RJ, Taylor RN, editors. Proceedings of Geotechnical Aspect of Underground Construction in Soft Ground. Rotterdam: A. A. Balkema. 1996; p. 713–718.
[36] Addenbrooke T, Potts D, Puzrin A. The influence of pre-failure soil stiffness on the numerical analysis of tunnel construction. Géotechnique 1997;47(3):693–712.

Exercises

1. What are the reasons for ground deformation caused by shield tunnel construction?
2. What are components of ground loss during shield tunnel construction?
3. Which shield machine configurations will cause ground loss?
4. What are the types of prediction methods for ground displacement caused by shield tunnel construction? What are their characteristics?
5. Please introduce the evaluation process for determining the impact of shield tunnelling on adjacent buildings?
6. Please briefly introduce the control measures for ground deformation and their characteristics during shield tunnel construction.

CHAPTER 12

Defects of shield tunnel lining and their treatments

Under the combined action of various factors such as loads and natural and usage environment variations during construction and operation, the shield tunnel lining undergoes segment cracking, deformation, water leakage, and gradual deterioration with the extension of service life. Especially in the construction stage of shield tunnels in some developing countries, structural defects of segment lining are significant, owing to the fast construction speed, tight schedule, shortage of technicians and on-site workers, segment assembly problems, and effects of the complex surrounding environment after construction. In recent years a large number of shield tunnels have been put into operation, and the health diagnosis and defect prevention and control of the shield tunnel structure have presented long-term problems that need to be faced and solved.

12.1 Common defect causes and control measures

Defects in the shield tunnel lining structure exist across the tunnel's lifetime. Some tunnels have defects before they are put into usage. The causes of lining defects are complicated, and their effects on the tunnel's service life vary [1]. According to a considerable amount of statistical data surveys, a large proportion of shield tunnels experience cracking and water leakage after completion of construction and at the beginning of operation. Some tunnels even have defects such as segment lining cracks during construction. After the tunnel has been completed and checked for acceptance, there will inevitably be defects resulting from its structural characteristics and maintenance during its operation period. The defects of the segment lining structure in the shield tunnel are the most prominent.

12.1.1 Uneven circumferential joint

An uneven circumferential joint means that the segment surface facing the jack at one side of the same ring is not on the same plane after being assembled (Fig. 12.1). Unevenness occurs between different segments and

Shield Tunnel Engineering
DOI: https://doi.org/10.1016/B978-0-12-823992-6.00012-6

Figure 12.1 Uneven circumferential joint of shield tunnel lining.

will affect the assembly of a next ring. As a result, it is difficult to penetrate circumferential bolts and easy to cause segment breakage and so on.

The reasons for uneven circumferential joint of the circular segment lining include: (1) accumulation of size errors in production of segments; (2) inclusion of debris between two ring segments during assembly; (3) uneven jacking force, causing the compression amounts of waterproof strips between circumferential joints to be different; (4) lack of firm adherence of the waterproof strips and being turned out of the trough during assembly, so they are not closely attached to the ring surface of the previous ring, causing the segment to protrude; (5) bolts of ringed segments not being tightened and retightened in time; and (6) longitudinal uneven settlement of the segment rings during the operation period, causing the circumferential joint of the tunnel to be differentially deformed, resulting in opening of circumferential joints, damage of waterproof filling materials, and leakage.

The measures that can be taken to avoid uneven circumferential joints of the segment lining include: (1) checking the ring surface of the previous ring segments before assembly and determining the amount of deviation correction necessary and correction measures during ring assembly; (2) removing all sundries on the ring surface and at the shield tail; (3) controlling the jacking force of the jack so that it is uniform; (4) checking the adhesion of the waterproof strips to ensure reliable sticking; and (5) using joint straddling jacks during shield advancing to ensure a flat ring surface.

12.1.2 Uneven longitudinal joint

Uneven longitudinal joints can result from misalignment of adjacent segments in the same ring, causing quality problems in longitudinal joints

Figure 12.2 Uneven longitudinal joint of segment tunnel lining.

such as horn shape and angle opening, stepping on inner arc surfaces, excessively wide longitudinal joints, and relative rotation between two segments (Fig. 12.2). Such problems will lead to serious harm to both waterproofing and stress performance of the segment tunnel lining.

The reasons for uneven longitudinal joint may include the followings: (1) The segments are not aligned when assembled, and there are sundries in the shield shell so that the segment at the bottom of the ring is not positioned in place or is turned upward or downward. The sundries clamped in the circumferential segment surface will also cause horn shape joints; (2) The segment fails to form a perfect circle during assembly, resulting in internal and external angle opening of the circumferential joint; (3) The benchmark of the previous segment ring is not accurate, causing inaccurate positioning of a newly assembled ring; (4) The tunnel axis is inconsistent with the actual central line of the shield. As a result, the segment collides with the shield shell and cannot be assembled into a perfect circle but can only be assembled into an ellipse, and the quality of the longitudinal joint cannot be guaranteed; (5) During the servicing period, the segment radially and unevenly converges and deforms, causing the horizontal diameter of the tunnel to increase, the vertical height diameter to decrease, and the longitudinal joint of the segment ring to continuously change. As a result, the lateral elastic seals lose efficacy, leading to defects such as jointing material fall-off and water leakage.

In view of the reasons for uneven longitudinal joint, the following measures that can be taken include (1) cleaning up the shield shell and each side of the segment before assembly to prevent sundries from being intercalated between segments; (2) frequently correcting deviation while the shield is advancing to minimize the deviation between the axis of the shield and the design axis as much as possible so that the segment can be assembled in the center, and enough construction gap is reserved between the segments and the shell to make the segments being assembled form a perfect circle; (3) timely correction of the deviation of the circumferential ring surface to ensure that the error between the central line of the assembled segment ring and the design axis is minimized and the segments can always be assembled around the shield tail; (4) positioning the segment in place, applying uniform force to the jacks when getting close to the segments, and arranging at least two jacks to support each segment expect the capping block.

12.1.3 Segment cracks

For various reasons, cracks may occur in segments during production, tunnel construction, and operational service. Cracks in the segment will affect the safety and durability of the concrete lining structure. In severe cases, cracks will affect the use of the tunnel and the safety of other facilities, causing great harm as a result of improper handling.

According to different locations and directions of cracks and their relationships with the tunnel longitudinal direction, the cracks can be divided into three types: longitudinal cracks, circumferential cracks, and local cracks at edges and corners (Fig. 12.3). Longitudinal cracks, which are parallel to the axis of the tunnel, are the most harmful and can cause the tunnel crown to fall-off, the side wall to break, and even the entire tunnel to collapse. Circumferential cracks are mainly caused by longitudinal uneven loads and geological changes in surrounding strata and mostly occur at the

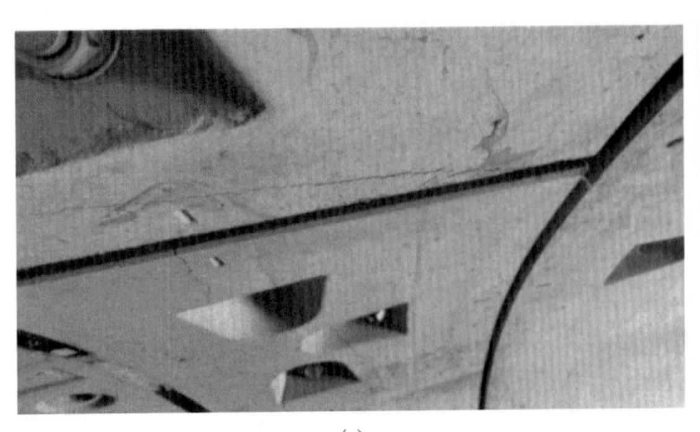

(a)

(b)

Figure 12.3 Shield tunnel segment cracks. (a) Cracks close to the corner and edge of segment. (b) Cracks crossing the segment.

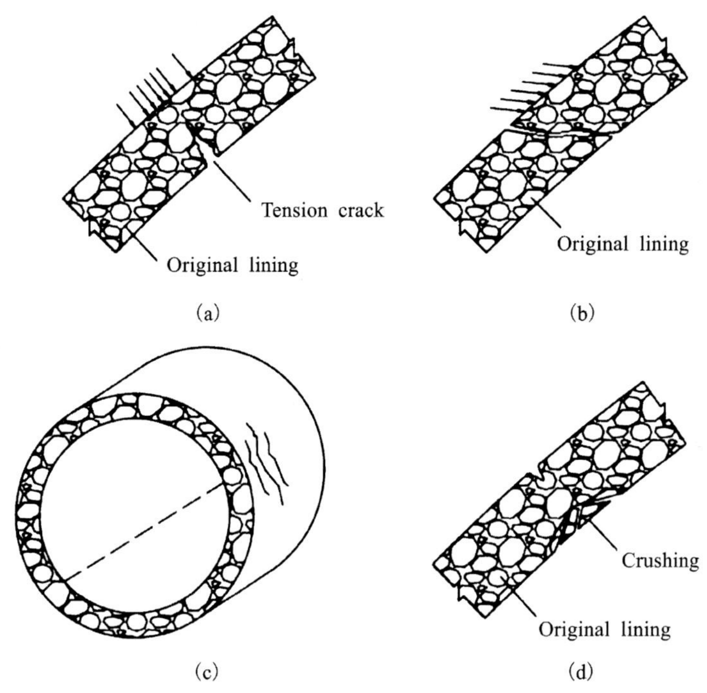

Figure 12.4 Cracks of the segment lining caused by different stress characteristics. (a) Bending and tension crack. (b) Shear crack. (c) Torsion crack. (d) Compression shear crack. *Source: Modified from W. Gao Structural damage of shield tunnel and its causes and treatment. Spec Struct 2012;29(3):70–73.*

entrance of the tunnel or between the junction of unfavorable geological zones and intact rock formation. Local cracks at edges and corners are often caused by tensile stress, local compressive crushing, and concentrated stress at bolt holes due to complex stress performance of the concrete lining. The harmfulness of local cracks is only somewhat less than that of longitudinal cracks, and they need to be carefully reinforced.

Segment cracks can be divided into four types—bending and tension cracks, shear cracks, torsion cracks, and compression shear cracks—according to their stress characteristics, as shown in Fig. 12.4.

12.1.3.1 Cracks caused by segment production

Segment cracks in the production process are mainly caused in two stages. The first stage is the curing stage after the segment has been demolded, and cracks of the segment are mainly on the surface and can be found through visual observation. The second stage is the stage after 28 days, when microcracks are occurred in the process of segment delivery, lifting, unloading, and assembling. Such cracks are not easily observed during factory inspection and are easy to ignore. However, they will quickly expand to form larger cracks once they are subjected to concentrated stress. The main causes of segment cracks in the production process include unreasonable and unscientific production technology and unreasonable concrete mix proportion. The corresponding controlling measures are as follows.

Adjusting the mix proportion according to local conditions

Depending on the geographical location and climatic conditions of the construction site, the selected concrete materials, such as stone, sand, cement, additives, and fly ash, have different compositions, contents, and properties. Even if the segments use the same concrete grade (C50), their mix proportion is not the same and cannot be directly applied. A series of tests should be carried out to determine the appropriate local concrete mix proportion, which should be adjusted according to climatic conditions such as seasonal temperature and humidity changes.

Improving construction technology

The construction procedures for segment production sequentially include concrete mixing, concrete pouring, vibrating, in-mold self-curing or steam-curing, demolding, water storage or spray curing, and so on. The vibrating and curing processes (including steam-curing) have the greatest impact on segment quality control, especially concrete density.

1. Vibrating: Integral vibrating and manual vibrating in a steel mold are the methods commonly used by segment production plants. In integral vibrating, the energy consumption is large, and the operation is simple. The concrete slurry on the same horizontal plane vibrates uniformly, but it is not easy to homogenize on the vertical plane, and it is easy to form floating slurry on the upper arc surface or outer arc surface. Furthermore, overvibration segregation will occur at special locations such as bolt holes and hoisting holes, owing to dense structural ribs or steel components. Manual vibration is controlled by people, so workers must master the vibrating process, and each step must be completed carefully. It is easy to control the vibrating quality at special locations where the vibrating energy is small. For upper floating slurry produced by integral vibrating, additional concrete must be added to compensate; otherwise, the upper arc surface or the outer arc surface will produce many surface shrinkage cracks, seriously affecting the quality of the protective layer.

2. Curing: Concrete segment curing can be divided into curing before demolding and curing after demolding. There are two methods of curing before demolding: natural curing and steam-curing. Steam-curing can accelerate the turnover speed of steel molds and has been widely used, but the maximum temperature of steam-curing, the temperature difference between inner side and outer side, and temperature rising and dropping gradient must be strictly controlled. At present, the experience of segment production in China shows that the steam-curing time should be controlled to 6–8 hours, the maximum temperature at the constant temperature stage should not exceed 60°C (90°C in some specifications), the temperature difference between inner side and outer side should be less than 20°C, and the temperature rising and dropping gradient should be less than 20°C/hour. Curing after demolding usually includes spray curing and water storage curing, with 7 days as a curing cycle. A comparison showed that when the total thrust force reached 12,000–15,000 kN under the same conditions of geological characteristics, segment reinforcement, concrete mix proportioning, and construction parameters, some spray-cured segments began to crack, while the water storage–cured segments were intact, and no cracks were found in the latter even at 28,000 kN thrust force. These cases show that at present in China, by water storage curing of most of the C50 concrete segments with cement content of more than 400 kg/m^3 for more than 7 days, hydration can be fully carried out, and the compactness of the concrete can be enhanced, thereby more effectively preventing cracks at the curing stage.

12.1.3.2 Segment cracking during shield tunnel construction

Many factors can cause segment cracking during shield tunnel construction. The main influence factors and corresponding countermeasures are as follows.

Excessive thrust force

The force acting on the segment is the most basic factor causing segment cracking, and excessive total thrust force during shield tunnelling is the most direct cause of segment cracking. At present, in the construction of subway shield tunnels in China (with inner diameters of 5.4–5.5 m) the common total thrust is between 5000 and 15,000 kN. The total thrust will exceed 15,000 kN when the earth pressure balance is established or a mud cake accumulates on the cutterhead. When the total thrust exceeds

15,000 kN, the segments that have not been subjected to water curing or that are poorly cured and have a thicknesses of 30 ~ 35 cm and reinforcements below 150 kg/m^3 are probably cracked. The total thrust is increased, and the cracking frequency is also increased. Control measures that can be taken include (1) injecting conditioning agents into the excavation chamber to avoid soil clogging and reducing the tunnelling torque and total thrust force and (2) carefully setting the earth pressure balancing state in the ground with a relatively stable excavation face to avoid excessive thrust force.

Gravity center deviation of the jack support boot during segment assembly

When the capping block is installed, the roundness of the first installed segment may be not enough, the gap between adjacent blocks may be too small, and lubricant is not daubed on both sides of the block as required. These imperfections causes the joint between the capping block and adjacent blocks to be broken, in which broken joint occurs at the upper parts of adjacent blocks and both sides of the capping block. Owing to uneven circumferential contact surface of the segment in the assembly process, local concentrated stress is formed under the action of the jack, causing the segment to be broken; The thrust center of the jack support boot does not coincide with the position of the central line of the segment, causing eccentric stress or local concentrated stress of the segment to break the segment. According to theoretical calculation, even a uneven difference of 0.5 ~ 1.0 mm will cause the maximum splitting torque of the next ring to be 1241 kN · m. Although shield segments are all precision prefabricated, a misalignment of 0.5 ~ 1.0 mm may occur as a result of imprecision and staggered assembly (in terms of structural stiffness, staggered assembly is superior to straight-joint assembly). To prevent such cracking, the installation accuracy of the segments should be improved as much as possible to reduce the design deviation of the jack support boot.

Mismatch between shield machine attitude control and the curved alignment section

If the attitude control of the shield machine does not match the curved tunnel alignment, the shield shell squeezes the segment and the rounder squeezes the segment, causing segment cracking (see Fig. 12.5). However, wider segments (1.5 m) are also a disadvantage for a small curve radius of

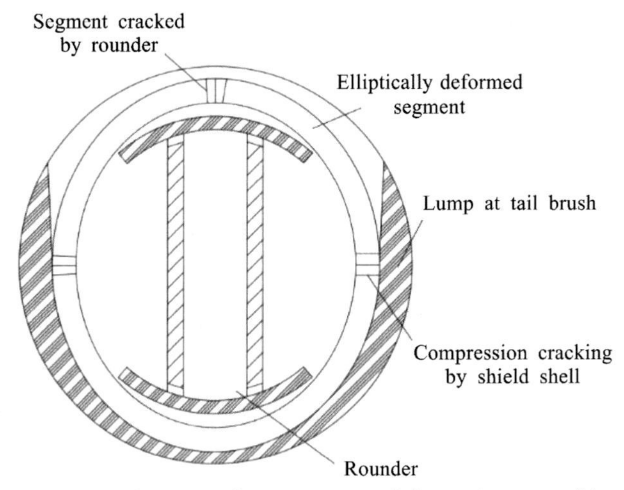

Figure 12.5 Schematic diagram of segment ring deformation caused by agglomeration and hardening of the tail brush.

tunnel alignment. The reason why the shield shell squeezes the segment is also related to agglomeration and hardening of the shield tail brush, elliptical deformation of the shield tail shell, tunnel rotation, and untightened segment connecting bolts (which making it easy for the segment to deform). This inference can be confirmed in the shield tunnelling practice; the positions of the segment cracking basically correspond to the positions where the tail brush is seriously damaged and the inner shell of the shield tail is polished. Corresponding countermeasures include correctly controlling the shield attitude at the tunnel turning section, slowly advancing with careful deviation correction, and replacing all damaged seal brushes as much as possible when passing the shaft or deep excavation section (e.g., station section in metro line).

Deformation and defect of the segment ring in coming out of the shield tail

The deformation and defect of the segment ring in coming out of the shield tail are caused by influential factors such as unreasonable shield attitude, improper segment selection, shield tail filler and filling technology, segment ring attitude in the shield shell, the size of shield tail gap, asymmetric earth pressure of the stratum, and so on (Fig. 12.6). These factors may cause the segment ring seriously deformed or even cracked, among which the shield tail filler and the filling technology have the greatest impact on the deformation of the segment ring. Segment cracking after

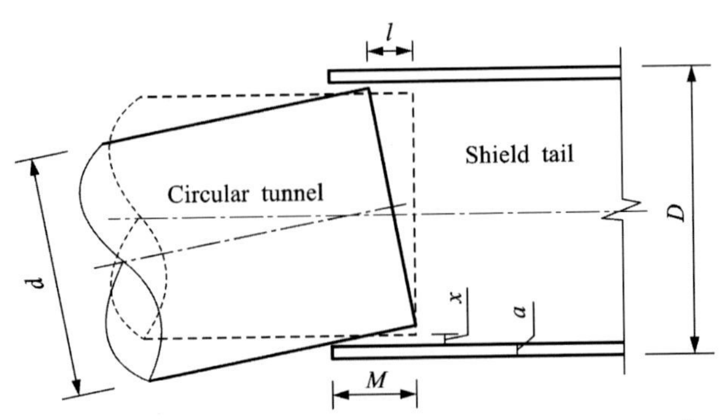

Figure 12.6 Schematic diagram of deformation of a segment ring when comes out of the shield tail.

the segment ring comes out of the shield tail is rare, and measures such as improving the proportioning of the filler and shortening the solidification time can be taken to control segment cracking.

12.1.3.3 Cracking during service life of the shield tunnel

Cracking during the use of the shield tunnel has two main causes: (1) the changes of earth pressure and groundwater level around the tunnel and (2) liquefaction of the surrounding sand layer caused by vibration of the train in the tunnel or earthquake. The measures to be taken include the following:

1. Reinforcing the surrounding strata around the tunnel and improving the properties of the soil to prevent changes in earth pressure and groundwater level.
2. Grouting in the tunnel for reinforcement, strengthening the bearing capacity of the tunnel foundation, and improving the stability of the tunnel itself.

12.1.4 Segment dislocation

Segment dislocation refers to unevenness of the inner arc surface between two ring segments after they have been assembled, i.e., a height difference between the segment rings, or an excessive height difference between segment blocks in one ring. Segment staggering not only affects the tunnel clearance, but also causes cracking and defects in the segment lining structure and brings hidden dangers to waterproofing of the tunnel (Fig. 12.7).

The reasons for segment dislocation may include the followings: (1) The center of the segment assembly is not concentric with the center of

Figure 12.7 Segment dislocation and local cracking of a metro shield tunnel.

the shield tail, the segment collides with the shield tail, and the segment is radially moved to assemble it in the shield tail, resulting in excessive ring height difference; (2) The ellipticity of the segment assembly is large, causing an excessive ring height difference; (3) The shield attitude is improperly controlled, and the ring face of the segment is not perpendicular to the tunnel axis, as a result, the shield tail will collide with the segment if assembling is being done in the direction of the previous ring by moving the segment in an opposite direction, causing excessive high ring difference; (4) The construction gap is not filled in time or is improperly grouted after the segment comes out from the shield tail, and the segment sinks under its own weight, causing an excessive ring height difference.

Segment ring height differences are caused mainly in the construction process, so during that process, it is important to pay attention to (1) installing the segment in the center of the shield to prevent the segment from colliding with the shield tail; (2) ensuring the ellipticity of the segment assembly; (3) correcting any nonperpendicularity between the ring face of the segment and the tunnel axis; and (4) carrying out adequate synchronous grouting in a timely manner and supporting the segment by synchronous grouting using grout to reduce the ring height difference.

12.1.5 Segment joint damage

A segment joint damage is a breakdown or cracking in joint parts that occurs during segment production and installation or a damage in the joint concrete during service life (Fig. 12.8). The deviation of the segment size and quality in the prefabrication segment production stage will cause hidden dangers in the segment assembly and tunnel use stages. A damage

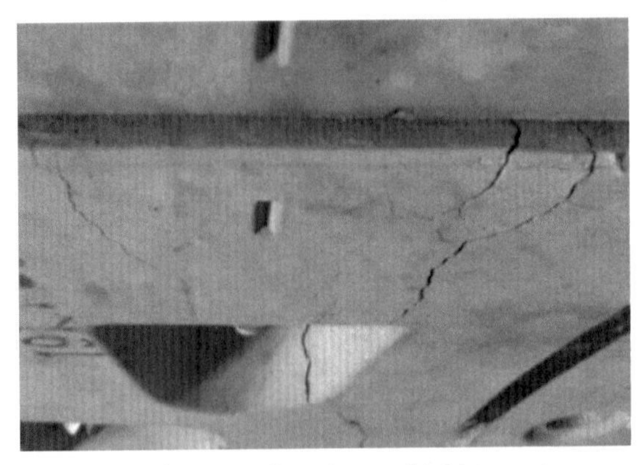

Figure 12.8 Segment joint damage of a subway shield tunnel.

in the segment joint in the assembly stage has a great impact on subsequent maintenance and repair costs.

For a fabricated concrete lining tunnel, joints generally play a dual role in segments internal force transfer and in tunnel waterproofing. But the joints are also the weak link in the tunnel lining structure. Because of the influence of longitudinal deformation of the tunnel structure and the convergence deformation of the lining ring, dislocation between rings and within rings frequently will be induced. If the segment joints opened too large, joint cracks and leakage may occur. Furthermore, as the waterproof materials of the tunnel joints aging, their waterstop effect gradually degrades, making the lining joints the main passage for leakage.

In general, segment joint connecting bolts bear large shear loads and tensile stress for a long time, are prone to distortion or even fracture, and often suffer more serious stress corrosion effects. Their aging degree and lifetime will directly affect the safety and service performance of the entire tunnel structure. The connecting bolts generally do not have special anticorrosion measures and rust easily. Bolt holes tend to become leakage passages. Therefore, the problems of the durability of the connecting bolts and their structural mechanical behavior should be fully considered.

12.1.6 Water leakage of a segment

Water leakage is the most obvious manifestation of a defect to the shield tunnel segment lining. Such leakage of groundwater from joints of the assembled segments into the tunnel causes many adverse effects, and even threatens the stability of the tunnel, facilities in the cavity, traffic safety, surface buildings, and the water environment around the tunnel. Water

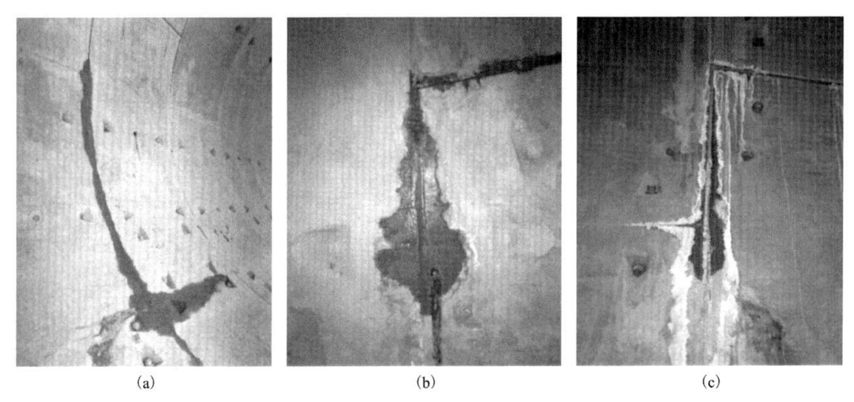

Figure 12.9 Water leakage of a shield tunnel segment. (a) Water leakage from joints and bolt holes. (b) Water leakage from joints. (c) Water leakage with calcium carbonate.

leakage of the segment lining occurs mainly in circumferential joints, longitudinal joints, grouting holes, deformation joints, bolt holes, and other parts, reflecting design level, construction quality, and service state to a certain extent (Fig. 12.9). In a sandy stratum the leakage of groundwater into the tunnel will wash way soil behind the lining, forming a cavity and reducing the strength and stability of the surrounding strata as well as causing greater water pressure. Long-term water leakage is also likely to reduce the reliability of the lining structure. If corrosive materials are contained in the leaking water, the lining will deteriorate, and the bearing capacity of the lining will be reduced.

Reasons for leakage of the segment joints include the followings: (1) Owing to poor quality of segment assembly, sundries in the joints, angled or horn opening of the longitudinal segment joints, uneven gap between the segments, and too large local gap, the waterstop strips cannot meet the sealing requirement, and the surrounding groundwater will leak into the tunnel; (2) When the segment is broken and the damage range reaches a point at which the grooves of the waterstop strips are attached, the waterstop strips and the segments cannot be hermetically attached, leading water leakage into the tunnel from the damaged positions; (3) The sticking quality of the waterstop strip is not good, and its adhesion is not strong, which causes the waterstop strip to become loose or deformed during assembly, and the waterstop effect cannot be achieved; (4) The segment that has been attached with the waterstop strip is not protected very well, so the waterstop strip swells with water before assembly, as a result, the segment is difficult to assemble, and the waterstop ability is reduced; (5) The segment fabrication had accuracy and construction problems; (6) Vibration from vehicles and construction activities in the safety protection zone aggravates the settlement deformation of the tunnel structure, resulting in gaps between the segments and water leakage; (7) The bolt hole asphalt gasket is incomplete or was crushed during construction, and the grouting of bent bolt holes is not dense enough to effectively prevent water from entering the tunnel; (8) The grouting material at the joint is not dense or hardened to fall off; (9) The aging of rubber waterstop strips and gasket materials constitutes a hidden danger that may cause water leakage in the tunnel.

The measures that can be taken to prevent the segment joint leakage include: (1) improving the assembly quality of the segment joints, correcting the ring face in time, ensuring the roundness of the segment and the normal working condition of the waterstop strip during assembly; (2) repairing the damaged segment in time and repairing the defect caused during transportation before the waterstop strip is attached, and repairing the damaged segments during the advancing or assembly process in order to protect the waterstop strip; (3) operating strictly according to the procedures for applying the waterstop strip, cleaning the waterstop trough, and sticking the waterstop strip only after glue does not flow; (4) arranging protective facilities such as a rain shed on the construction site to strengthen the protection of the segment. Expansion retardants can also be applied to expanded waterstop strips to ensure the construction quality.

12.1.7 Segment corrosion

Harmful substances such as chloride and sulfate in the ground corrode steel bars, causing soluble corrosion and expansive corrosion of concrete. Carbon dioxide and carbon monoxide in the air and in soft soil in the tunnel dissolve in water and penetrate into the concrete, causing neutralization of the concrete [3].

12.1.8 Frost damage in a segment

Frost damage in a segment refers to frost-heaving, cracking, chipping, flaking, water accumulation, hanging ice, freezing, icicles, and other defects of the tunnel in a cold or severely cold area caused by the freeze-thaw action of a cold environment and affecting the normal use of the tunnel [4]. In cold area tunnel, frost damage is the main reason for deterioration of the segment lining. If drainage equipment of the tunnel is in the freezing circle of the tunnel, ice jams are likely to occur in winter. If the lithology of the surrounding rock in the freezing circle is non frost heaving soil, frost-heaving damages will not occur.

Frozen earth areas are widely distributed in China, and permafrost regions account for one-fifth of the land area of the country. In cold places, water leakage in the tunnel will further cause frost damage, and frost damage will also aggravate leakage in the tunnel and worsen the service environment of the tunnel. A temperature that is too low will cause moisture inside the concrete, between the concrete and the lining, and inside the surrounding ground to freeze, resulting in a huge frost-heaving force. The increase in the surrounding earth pressure will cause damages in the lining structure and at the same time will cause the volume of the surrounding strata and the concrete to expand. As a result, the surrounding strata and the concrete would be crushed, and their strength to be reduced, leading to reduction in the bearing capacity of the lining structure.

12.2 Investigation of defects in the shield tunnel segments

The investigation of defects in the shield segment lining structure is of great significance in maintenance and repair of the shield tunnel. A comprehensive and detailed investigation can find existing defects and their extent in the tunnel to provide a basis for maintenance and repair of the tunnel, prolong the service life of the tunnel, and avoid premature construction and reconstruction work.

12.2.1 Defects investigation methods

Methods for investigating defects in the segmental lining structure include the ultrasonic-rebound combined method, the ground-penetrating radar method, the tunnel laser cross section scanning method, and the infrared field photographic method [5]. The ultrasonic-rebound combined method can detect the dynamic elastic modulus and strength of the concrete, thickness and defects in segments, and so on [6]. This technical method is relatively mature. Usually, on the basis of detection results of the ultrasonic-rebound method, it is also necessary to use a core-drilling method to verify the thickness and strength of the tunnel segment. The ground-penetrating radar method can detect changes in construction material properties; structural abnormalities (with voids); and changes in lining thickness, lining composition materials (such as reinforcement steel bars), and surrounding strata conditions (voids between lining and surrounding strata, spring holes, changes in composition materials, and abnormalities). This detection method has been widely used worldwide. The tunnel laser profiler is based on laser ranging technology without cooperative target and precise digital angle measurement technology. With polar coordinate measurement, computer and image-processing technology, the tunnel sectional view can be quickly obtained and compared with the design contour in order to quickly judge any deformation of the tunnel lining. The infrared field photographic method can measure the flow of water between the lining and the surrounding strata at different temperatures, changes in geological conditions behind the lining, defects in the lining, and voids behind the lining. Since the method relies on temperature gradient measurement, it is best to use it in winter when there is a large temperature difference. When detection requirements are high and dry and fine cracks need to be detected, a multispectral analysis method can be used. As more and more tunnels are constructed and the demands for maintenance are required, the use of nondestructive inspection techniques such as ground-penetrating radar, infrared testing, and spectrum analysis to inspect the tunnel lining has been of increasing interest to tunnel engineers.

12.2.2 Main investigation content

12.2.2.1 Crack detection

The detection of cracks in lining structures usually includes the quality of the concrete around the cracks; location, number, and direction of the cracks; length, depth, and width of the cracks; the presence or absence of foreign objects and water in the cracks; load conditions; and surrounding environmental conditions, including changes in temperature and humidity; cracking time, changes in the cracking process, and stability or instability.

1. The location, number, and direction of the cracks are generally recorded in the form of photos and crack expansion diagrams.
2. The length of the crack is measured with a ruler and a tape measure, and its width can be measured with a crack width comparison card, a crack width gauge, and a feeler gauge. The crack width detection is also usually conducted by using a scale magnifying lens, also called a crack microscope. The objective lens of the microscope is aligned to

the to-be-observed crack, the image is focused by rotating the knob on the side of the microscope, and the width of the crack can be read from the eyepiece.

3. The borehole coring method and the ultrasonic method are often used for crack depth detection. In the borehole coring method, a core sample is drilled at the crack in the concrete structure, and the quality inside the concrete and depth penetration of the crack are directly observed. In the ultrasonic rebound method the travel time curve is obtained according to the ultrasonic propagation speed in the lining concrete. By fixing the position of the ultrasonic transmitter and allowing the receiver to move along a certain direction of the lining, the crack depth can be calculated according to changes in the ultrasonic propagation time at the crack position, such as delay time. The ultrasonic rebound detection method is simple and easy to use and does no harm to the detection structure, which should be popularized in structural inspection.

Finally, on the basis of the inspection results, the expansion map of the tunnel lining is drawn, taking the vault as the central line, expanding at two sides, and observing from the tunnel in a perspective mode. Mileage numbers are marked along the axial direction of the tunnel. The tunnel lining is horizontally divided so that the detected defects can be correctly displayed on the expansion map, which can be used to evaluate the severity and cause of the tunnel defect.

12.2.2.2 Water leakage detection

The most important part of water leakage detection is judging the water leakage state of the tunnel, distinguishing types of water leakage defects, and determining the locations, scope, and characteristics of the leaks. The current water leakage of the tunnel is usually divided into the following three types: wet stain, water leakage, and water seepage. The detection and judgment methods are as follows:

1. A wet stain generally disappears under artificial ventilation, that is, a state in which the evaporation amount is larger than the infiltration amount. There is no feeling of moisture when the wet stain is touched with dry hands during detection. Paper will not change color when blotting paper or newspaper is attached to the wet stain. During detection the range of the wet stain is drawn with chalk, the height and width are measured with a steel ruler, and the area is calculated to be marked on the expansion map.

2. Water leakage will not disappear under conditions of enhanced artificial ventilation, that is, the leakage amount is greater than the evaporation amount. There is a feeling of moisture when the leakage area is touched with dry hands during detection, and moisture will stick to the hands. Paper will be soaked and change color when blotting paper or newspaper is attached to the leakage area. During detection the range of water leakage is drawn with chalk, the height and the width are measured with a steel ruler, and the area is calculated to be marked on the expansion map.
3. The inspection method for water seepage is the same as that for water leakage except that the amount of water flow is larger, and sometimes mud and sand are entrained. During detection the dripping speed of the flow of seeping water can be observed.

12.2.2.3 Lining concrete strength detection

Aside from appearance defects in the tunnel lining, the strength of the lining concrete is an important aspect in understanding the durability and deterioration probability of the tunnel service environment [7]. Common methods of measuring strength include the rebound method, an ultrasonic detection method, the ultrasonic-rebound method and the borehole coring method.

1. When the rebound method is used to test the concrete strength, detection needs to be conducted in accordance with specific requirements of the related technical regulations. Measuring points should be evenly arranged within the measurement area but should not be arranged on pores or exposed stones. The distance between two adjacent measurement points should not be less than 30 mm, and the same measurement point may to be bounced in only one time.
2. Ultrasonic detection include longitudinal wave velocity, surface wave velocity, and spectral characteristics, which are processed separately according to the relevant research needs. The longitudinal wave velocity is calculated on the basis of the travel time difference of a first break recorded between two points and point distance. The surface wave velocity is obtained by correlation analysis. If this velocity is lower than the normal longitudinal wave velocity index corresponding to the concrete mark, the quality of the concrete itself is defective. The spectral characteristics are used to determine the integrity of the lining. Through normalizing the actually measured main frequencies at the far and near points, the result is the structural integrity coefficient. When

the structural integrity coefficient approximates to 1, the structure is complete; when the coefficient is less than 1, the structure is broken with cracks; and when the coefficient is much greater than 1, the lining has voids with acoustic resonance. In the ultrasonic-rebound combined method the sound velocity and the rebound value of the structure or the component are tested in the same measurement area by using the ultrasonic instrument and rebound instrument. The ultrasonic sound velocity reflects the internal strength information of the concrete, and the rebound value reflects the surface strength of the concrete. By using the two methods for mutual complementation, the true quality of the concrete can be thoroughly reflected.

3. In the borehole coring method, core samples are drilled from structural concrete in order to detect the concrete strength or observe the internal quality of the concrete. Because this method causes local defects in the structural concrete, it is a semidestructive field detection means. In measuring the strength, it is necessary to follow the specific requirements in the related technical regulations to detect the concrete strength by the borehole coring method and combine the results with the actual conditions of the tunnel structure.

12.2.2.4 Other defect detection

1. Preliminary judgment of segment dislocation is carried out by visual inspection. The suspected locations can be confirmed by touching, or a searchlight can be placed flat on the segment to illuminate the suspected dislocation parts. If there is a dislocation phenomenon, there will be a clear contrast between light and dark, and the height of dislocation can be measured with a steel ruler.
2. Preliminary judgment of the opening of the segment joint is carried out by visual inspection. For locations that are opened largely, bolts can be found after lighting. The opening size of the joint needs to be measured on the spot by using a ladder truck.

The process of investigation of defects in the shield tunnel lining should include detailed written records of the location, scope, and characteristics of the defects, and comprehensive photos and video data for future reference. With the development of artificial intelligence, machine vision methods are used to photograph, correct, splice, and reconstruct appearance defect such as cracks, dislocation, and water leakage (Fig. 12.10) [8–10]. Then artificial intelligence methods are used to identify and detect the defects based on a large amount of data photos. This is

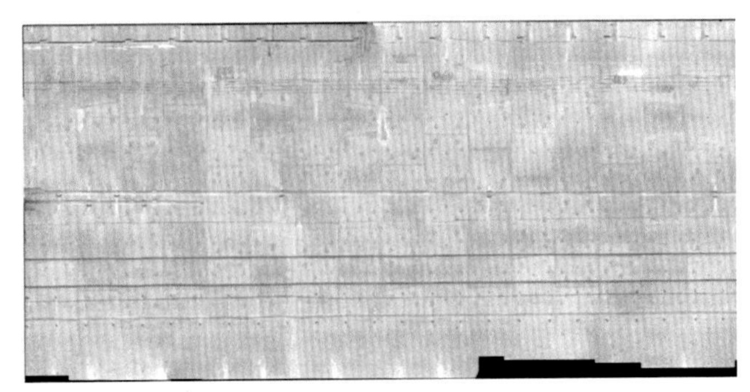

Figure 12.10 Panorama photo for detection of defects in the segmental lining based on machine vision.

a new development direction for intelligent detection technology for tunnel lining inspection.

12.3 Safety evaluation of the shield tunnel structure

Defects such as water leakage, segment breakage, segment dislocation, and tunnel uneven settlement will occur during construction and service life of the shield tunnel segment lining, and its structural behavior will continue to deteriorate, affecting the normal use of the shield tunnel, even causing accidents. Therefore to reduce the structural safety risks of the shield tunnel due to various defects, it is necessary to carry out field detection and monitoring, conduct safety assessments of the lining structure, and formulate a reasonable maintenance or repair plan. In general, the safety assessment of the shield tunnel lining structure is based on field detection and monitoring results. On the basis of mechanics theory and comprehensive evaluation theory, the safety status of the shield tunnel structure and the important factors affecting the structural safety are analyzed to provide the basis and guidance for maintenance and rehabilitation of defects in the shield tunnel.

12.3.1 Safety monitoring of the shield tunnel lining structure

12.3.1.1 Deformation monitoring

The deformation of the shield tunnel can be divided into two classifications: longitudinal deformation and cross-section deformation [11]. The longitudinal deformation reflects the overall mechanical characteristics of the shield tunnel in the longitudinal direction and can be used to evaluate the longitudinal stability and longitudinal seismic performance of the tunnel structure. The cross-section deformation can be used to judge the mechanical properties of the tunnel structure at the section, providing a basis for safety evaluation of the lining structure.

In normal circumstances the section deformation of the weak part and the key section of the tunnel are determined. Generally, the stress and the cross-section deformation performance of the shield tunnel lining structure can be evaluated by monitoring the convergent deformation of the shield tunnel (diameter change amount ΔD, where D is the inner diameter of the tunnel). In observing the convergent deformation of the section, a spring tension type convergent meter is a simple and practical tool. Its basic principle is the use of a mechanical method to transfer the relative displacement between two measurement points and convert the relative displacement to the difference between two readings of the dial indicator. Such method has the characteristics of high test accuracy, reasonable structure, good stability, and convenient to use, and it meets the requirements of field monitoring of projects [12].

12.3.1.2 Internal force monitoring

The internal force monitoring of the tunnel lining is usually done by pre-embedding stress gauges or strain gauges in the concrete segments or installing measurement strain gauges on the assembled lining surface. These are used to calculate the internal force level or the bearing capacity of the lining structure based on the relation between the stress and the strain, and the overall stress and the deformation state of the segment lining structure are analyzed. The basic principle that when the resistance strain gauge has been installed on the inner side surface of the structure member, the slight deformation of the structure member after being loaded will make the sensitive grating of the strain gauge deform accordingly, and the resistance of the strain gauge will change. The rate of change is proportional to the strain of the structure member where the strain gauge is installed. By measuring the change of the resistance, the strain at the corresponding position of the structure member and the corresponding stress can be calculated. Commonly used stress or strain gauges include the vibrating wire and fiber grating types.

12.3.1.3 Crack monitoring

Crack monitoring includes monitoring of crack width, depth, and length. The crack length can be monitored and recorded at the end of the crack at regular time intervals. A tape measure is used to measure the change in the crack length with time. The crack width can be measured by coating the straddle joint with gypsum and using a graduation magnifying glass, or a crack width gauge can be used to observe the same location regularly. In addition, methods such as installing the strain gauges across joints can be used to measure the change of the width with the time.

12.3.1.4 Visual monitoring

Visual monitoring is a simple and practical method and is the most commonly used method in tunnel maintenance. Observing the safety of the lining structure by visual monitoring is usually done by a competent maintenance person on foot or on a rail car under sufficient lighting conditions. By visually observing the appearance of the tunnel, it is easy to find any defects that has occurred in the tunnel. The disadvantage is that it is difficult to find hidden defects inside the tunnel. Visual observation and measurement are used to investigate changes in appearance of the tunnel, and any variation is recorded in a notebook. Visual monitoring generally includes locations and types of lining leakage, locations (ranges) of lining cracks, deterioration of joint materials, lining crushing or peeling, and corrosion.

12.3.2 Safety evaluation content of the shield tunnel lining structure

Safety assessment of the shield tunnel segment lining structure is a multifactor and multilevel hierarchical analysis problem [13]. The judgment of the safety status of the segment lining structure should consider not only the local behavior reflected by a single project, but also the overall behavior reflected by multiple projects. With reference to existing experience in detecting common defects in railway tunnels, highway tunnels, and subway shield tunnels, the safety evaluation of the shield tunnel can be summarized in the following three aspects:

1. The appearance evaluation content of the tunnel structure, including cracks, breakage, leakage, and joint opening.
2. The safety evaluation of the tunnel structure, including lining concrete strength, lining reinforcing bars' strength, segment bolt strength, joint strength, and lining bearing capacity as well as the elastic modulus of reinforcing bars and the lining.
3. The durability evaluation content of the tunnel structure, including water seepage quality (including content of harmful ions such as sulfate, pH value, etc.), air smoke quality (including content of carbon

dioxide and sulfur dioxide in the air), concrete carbonization depth, concrete protective layer thickness, content of chloride and sulfate ions in concrete, reinforcement corrosion, concrete penetration, and material waterproofing.

12.3.3 Classification standard for typical defects of the shield tunnel lining structure

12.3.3.1 Evaluation of segment lining deformation

Tunnel lining deformation refers mainly to convergent deformation of the lining cross section, resulting in reduction of the tunnel clearance or occupancy of the reserved space. The two main types of deformation are transverse deformation and longitudinal deformation, of which transverse deformation is the dominant form. Transverse deformation is the change of cross section shape caused by stress of the lining structure. Deformation of the lining is described by the deformation value and the deformation speed. Table 12.1 lists deformation evaluation standards for railway tunnels in China [14], and the deformation classification standards of straight-joint assembled shield tunnels in Shanghai [15].

12.3.3.2 Evaluation of segment lining joint opening and dislocation

Excessive transverse deformation of the tunnel may cause the longitudinal joint of the tunnel to open. The development of uneven longitudinal settlement will increase opening of the circumferential ring joint, and joint opening will cause secondary defects such as yielding of the tunnel joint under tension and tunnel leakage.

Dislocation is an important manifestation of deformation between tunnel segmental rings. Under the influence of uneven longitudinal settlement of the tunnel, dislocation deformation occurs between the tunnel segmental rings. The development of dislocation will cause the longitudinal connecting bolt to tend to yield, shear failure occurs in the tenon, and the elastic waterstop gasket to be deformed and fail, as well as other secondary defects to structures.

Table 12.2 shows the safety classification criteria for segment dislocation and joint opening of the straight-joint assembled shield tunnel in Shanghai [16].

12.3.3.3 Classification of cracks in segment lining

The causes of cracks in the segment lining structure are different, and the geometric conditions are complicated and changeable. The feature

Table 12.1 Deformation classification criteria of shield tunnel lining structure.

Evaluation rating	Deformation (railway tunnel)	Longitudinal deformation curvature radius/km (straight joint)	Transverse diameter change quantity in percentage of the diameter (straight joint)
I	Deformed, but stable now, and no effect on use	>15	<0.5
II	Deformed, with a deformation speed less than 3 mm/year	15～4.5	0.5～1
III	Deformation or displacement speed is 3～10 mm/year, and new deformation occurs	4.5～3	1～1.6
IV	Deformation or displacement speed larger than 10 mm/year	<3	>1.6
V	Deformation, displacement and sinking of the lining develop rapidly, threatening driving safety	Affecting normal use	Affecting the bearing capacity of structure

Table 12.2 Classification criteria for joint opening and dislocation of segment lining.

Evaluation rating	Joint opening (mm)	Segment dislocation height (mm)
I	<0.35	<0.033
II	0.35～2	0.033～4
III	2～4	4～10
IV	>4	>10
V	Joint waterproofing failure and joint damage	Joint waterproofing failure and intrusion into tunnel clearance

description parameters of the lining crack include the initial position of the crack, the spreading shape, the opening state, the width, the depth, the length, penetrating or not, and the relative dislocation distance. The first two parameters are unified feature parameters of the tunnel, which

Table 12.3 Classification criteria for cracks in tunnel lining.

Rating	Evaluation index and standard		
	Crack length L (m)	Crack width δ (mm)	Crack depth ratio K
I	$L <$ minimum distance from bolt hole to edge	$\delta < 0.05$	Smaller than thickness of protective layer
II	Minimum distance from bolt hole to edge $\leq L < 0.5$	$0.05 \leq \delta < 0.2$	$K < 1/4$
III	$0.5 \leq L < 1$	$0.2 \leq \delta \leq 0.5$	$1/4 \leq K < 1/3$
IV	$L \geq 1$	$0.5 < \delta$	$1/3 \leq K < 1/2$
V	Penetrating through the whole segment	Mud and sand particle leakage	$K \geq 1/2$

can be represented by coordinates of several key points. The last five parameters can be unified features of the crack or can correspond to state representations of different key points, depending on the state uniformity and accuracy requirements of the crack. If the depth and the width of the crack similar to each other at different positions of the crack, the parameter can be regarded as the unified feature of the crack; otherwise, the parameters of different key points of the crack need to be recorded and described separately. If analysis of the development of the crack is needed, it is necessary to record the changes of the quantitative parameters at key points over time.

Table 12.3 shows the safety evaluation criteria for main evaluation indexes, including width δ, length L and joint depth, of the cracks, wherein K represents the percentage of the crack depth accounting for the structure thickness.

12.3.3.4 Classification for water leakage effects

Defects in the shield tunnel affected by water leakage are significant, not only affecting normal driving of vehicles, but also aggravating the development of structural corrosion and settlement. In general, water leakage factors can be divided into six manifestations: wetting, water seepage, dripping, water leakage, water jetting, and water gushing. Mud and sand leakage is a special form of water leakage that will more quickly cause tunnel settlement, so it needs special attention. In addition, since water leakage in the tunnel usually occurs at the joints, not only is the impact of

Table 12.4 Classification criteria for cracks in tunnel lining.

Evaluation rating	Water leakage location and phenomenon
I	Meeting the requirement of national secondary waterproofing technology; slight falling and damage without seeing the reinforcing bar; and there is a possibility of water seepage
II	Joint deformation is not obvious; slight water seepage at the tunnel side wall; without mud and sand; and there is a possibility of obvious water leakage
III	Obvious water seepage at the tunnel side wall; bolts can be seen in the segment joint; water seepage in the bypass channel; there is a possibility of water leakage, and mud and sand particle leakage
IV	Water gushing; accumulated mud and sand particle; continuous seepage with visible water film on the surface; or linear water flow; seepage at the top joint
V	Joint waterproofing failure and joint damage

the water leakage on the tunnel affected by the leakage rate, but also the leakage location has a significant impact on the lining structure considering the feature of joint deformation. Therefore in classifying the water leakage rating, both the speed and the location of the water leakage in the tunnel should be considered (Table 12.4).

12.4 Defect treatment in the shield tunnel lining

12.4.1 Treatment principle for defects in the segment lining

Since the causes of defects in the shield tunnel segment lining are complicated, the treatment for defects is first to strengthen the structure and stabilize the surrounding strata and then to eliminate damage inducement factors and to prevent deterioration of the defects. A comprehensive treatment plan combining multiple treatment measures should be adopted. The principles of preventing lining damages are as follows:

1. Improving segment precast technology such as vibrating and curing in the segment production, optimizing the proportioning, and minimizing defects or cracks in the production process of the segment.
2. Strengthening the observation and investigation, determining the deformation of the tunnel surrounding strata, the deformation of the segment lining crack, geological data, and occurrence time, determining

causes, adopting different engineering measures for different causes and locations of cracks. It is necessary to ensure that the treatment does not affect the long-term monitoring of the tunnel deformation and the deformation and cracks do not develop further after treatment.

3. Minimizing the adverse impact on traffic for an operating tunnel while ensuring that normal use of equipment and pipelines in the tunnel is not affected.
4. Comprehensively treating defects such as water leakage and corrosion, and following a principle of thorough treatment.
5. Taking reinforcement measures to stabilize the tunnel floor or base.
6. Carefully measurement to ensure that the clearance of the reinforced tunnel meets the tunnel space requirements and reasonably arranging a low speed traffic plan during construction to minimize interference to normal operation.
7. Ensuring the construction quality of crack grouting, backfill grouting, partial lining replacement, and so on.

12.4.2 Treatment for lining cracks

The prevention and control of lining crack defects involves first eliminating all defects in the lining cracking zone that affect structures and operation and then preventing cracks from expanding. The tunnel lining model test proves that the cracked lining may still have a certain bearing capacity. Even if the lining is severely damaged and dislocated or is partially intrusion into tunnel clearance. For scenarios like this, the intrusive lining part can be chiseled or removed under the protection of a temporary support arch, and backfill grouting can be used restore and improve the lining bearing capacity. Only when the lining is seriously deformed without bearing capacity and most of cross-section invades the clearance space, the lining should be replaced. In general, appropriate methods should be selected for comprehensive treatment according to the tunnel working environments, geological conditions, construction records and technical data, crack information (length, width and depth) to ensure the structural safety of segmental lining and the driving comfort in the tunnel.

1. In a section where cracks have a slight influence, crystalline and permeable mixed materials can be applied to the surface of the cracked area layer by layer if there are many cracks. Crystalline and permeable mixed materials can be applied along cracks to close the cracks if there are only a few cracks.
2. In a section where the cracks have a small influence, the cracked area can be lined with reinforced concrete, or the surface can be coated

with crystalline and permeable mixed materials layer by layer for reinforcement if there are many cracks. Inverted trapezoidal grooves can be chiseled along the cracks and high-strength binding agents are embedded for reinforcement if there are only a few cracks.

3. In a section where cracks have a great influence, the cracked area can be lined with reinforced concrete for reinforcement if there are many cracks. Inverted trapezoidal grooves can be chiseled along the cracks and high-strength binding agents are embedded for reinforcement if there are only a few cracks.
4. In a section where cracks have a serious influence, the cracked area should be demolished and rebuilt, or the cracked area should be lined with strong reinforced concrete for reinforcement if there are many cracks and the structure loses its function. Inverted trapezoidal grooves can be chiseled along the cracks and high-strength binding agents can be embedded for reinforcement if there are only a few cracks.

12.4.3 Treatment for water leakage of segment lining

12.4.3.1 Surface painting and closing

For esthetic reasons the surface should be painted and sealed after leakage treatment. The surface painting and sealing are done mainly after treatment of cracks within the allowable range of the structure without slight water leakage, or after treatment of cracks by measures such as in-seam grouting and backfill grouting.

For surfaces without water leakage, waterproof mortar plastering and waterproof coating can be implemented after necessary treatment and protection measures. For surfaces with slight water leakage, the crack is cut into a V-shaped groove with a certain depth and width as required, and the groove is filled and compacted with a rapid-curing material.

12.4.3.2 In-seam grouting

In-seam grouting is generally used in the following situations: (1) The crack is large, and there is slight leakage or relatively serious leakage; (2) cracks are seriously leaking after treatment such as backfill grouting; or (3) cracks are still leaking although they have been repaired.

12.4.3.3 Backfill grouting

Backfill grouting refers to the use of a pressure pump to inject waterstop grout into the back of the concrete segment through a special injection pipe to form a protection layer on the back to completely block water leakage in the crack.

12.4.3.4 Joint repair

Joint repair involves comprehensive treatment measures, i.e., the combination of grouting with caulking and surface plastering. The original material in the joint is removed at a certain depth, then the joint with small leakage is blocked by rapid-curing material, and a sealing waterproof material is embedded, and finally holes are drilled for back grouting to block water.

The choice of leak-stopping and waterstopping materials has a great impact on the effect of defects treatment. Different leak-stopping and waterstopping materials are used to deal with different leakage situations. The leak-stopping and waterstopping materials are divided into inorganic leak-stopping materials, organic leak-stopping materials, and rapid-curing leak-stopping agents. Table 12.5 shows the classification of tunnel lining leakage situations and the corresponding treatment measures using different materials.

12.4.4 Treatment for corrosion of segment lining

Three cause elements for the corrosion of the segment lining are the presence of a corrosive medium, the presence of corrosive substances, and the presence of mobility in the groundwater. In view of causes and conditions of erosion, the main measures taken to prevent tunnel erosion are as follows:

1. Improving the compactness and corrosion resistance of segment concrete, such as using anticorrosion concrete.
2. Improving corrosion resistance by using additives.
3. Reasonably selecting corresponding cement with better corrosion resistance in view of different environmental water corrosion media (low alkaline and high sulfate resistance cement and quick-setting and rapid-hardening cement are most suitable).
4. Improving and strengthening the construction of the waterproofing and drainage systems of the tunnel, improving the waterproofing performance of the segment joint in terms of materials and structures, and using dense materials that do not react chemically with concrete as an isolation waterproofing layer on the outer surface of the lining.
5. Using anticorrosion grout for synchronous grouting and backfill grouting of segment lining.

Table 12.5 Classification of segmental tunnel lining water leakage and the corresponding treatment measures.

Description of leakage	Water leakage treatment
Small-area leakage at the segment assembly joints	Drilling for grouting, caulking, and surface painting
Large-area leakage at the segment assembly joints	Through hole is drilled on segments, and materials such as water glass cement slurry and polyurethane are injected into the positions behind the segment to block water leakage passages, then rapid curing cement can be used to close the holes and the peripheral seam, and the surface should be painted.
Leakage in lifting holes and bolt holes	Cleaning up dirt in the holes, tightening the plug after the hole is blocked by super rapid-curing cement slurry; If the leakage is serious, tightening the plug after the hole is injected by pressurized ultrafine cement slurry through buried aluminum pipes.
Leakage at the connection section between the cross passage and the segment ring	Backfill grouting (using ultrafine cement slurry, water glass slurry, modified epoxy resin chemical slurry, etc.) and waterproof caulking.
Leakage at the segment joints	Cracks with a width greater than 0.2 mm should be blocked by grouting and then coated with neoprene latex, acrylic emulsion, or the like material for sealing; microcracks with a width smaller than 0.2 mm are brushed with coating material for concrete wall, cement sealing materials, or the like for sealing.
Leakage at a damaged segment	The interface is painted with cement base material, embedded with wire mesh, and backfilled with epoxy mortar.

12.4.5 Prospect for durability guarantee technology in the segment lining structure

At present, the prefabricated segment lining structure is the main structural type for shield tunnels. Under the influence of a corrosive environment, extreme conditions, disaster, and etc., the long-term safety of the single-layer segment lining structure needs to be verified. The lining structure is often cracked and damaged by construction loads during construction, and the strength of the structure cannot be guaranteed by simple repair after completion of construction, so risks to the safety of the tunnel structure in the service life increase. Therefore more and more attention is being paid to research new types of shield tunnel lining segments. Breakthroughs in reinforced concrete materials, the development of the new segment lining type with both high performance and high durability are current goals. At the same time, establishing the whole life design of the shield tunnel lining structure must be done.

Furthermore, effective inspection and maintenance are necessary during shield tunnel service life, and the development of new detection methods has become an important trend. For example, tunnel inspection robots and pipeline robots for water delivery and drainage tunnels can be used to detect defects that cannot be recognized by the naked eye in extremely difficult situations and find out defects such as cracks, rust stains, corrosion, and steel bar exposure; Detection method such as ground penetrating radar, elastic wave computed tomography (CT), X-Ray detection and the others can be used to find out interior defects like voids behind the lining concrete, crack depth, and so on.

References

[1] Zhou W. Shield Tunnelling Technology. Beijing, China: Construction Industry Press; 2004.
[2] Gao W. Structural damage of shield tunnel and its causes and treatment. Spec Struct 2012;29(03):70−3.
[3] Zhang M, Zhang N. Analysis on common disease and its influence on shield tunnel. Urban Roads Bridges Flood Control 2009;09 182-187 + 237-238.
[4] Yang X, Huang H. Tunnel Defect Prevention and Treatment. Shanghai: Tongji University Press; 2003.
[5] Edited by Basic Department of Ministry of Railways Transportation. Technical Manual for Railway Tunnel Inspection. Beijing: China Railway Press, 2007
[6] TB 10233-2004. Process for Nondestructive Test of Railway Tunnel Lining Quality. Beijing, China: China Railway Press; 2004.
[7] Ye Y. Tunnel Operation Management and Maintenance Guide. Beijing: People's Communications Press; 2013.
[8] Zhu Z, Fu J, Yang J, et al. Panoramic image stitching for arbitrarily shaped tunnel lining inspection. Comput Aided Civ Infrastruct Eng 2016;31(12):936−53.
[9] Peng B, Zhu Z, Yang J, et al. Research on digital identification method of tunnel lining leakage based on panoramic image expansion. Mod Tunn Technol 2019;56 (03):31−37 + 44.
[10] Huang H, Li Q. Image recognition for water leakage in shield tunnel based on deep learning. Chin J Rock Mech Eng 2017;36(12):2861−71.
[11] Yan B, Yang C, Chen L. Quality and safety evaluation system for operation and maintenance of metro shield tunnel. Prestress Technol 2012;01:25−7.
[12] TB 1003-2016. Code for Design of Railway Tunnel. Beijing, China: China Railway Press; 2016.

[13] He C, Feng K, Sun Q, Wang S. Consideration on issues about structural durability of shield tunnels. Tunn Constr (CN-EN) 2017;37(11):1351–63.
[14] TB/T2820.2-1997. Evaluation Standard for Deterioration of Railway Bridges and Tunnels. Beijing, China: China Railway Press; 1999.
[15] Gu L, Zhang D. Study on evaluation index of stress and deformation of shield tunnel structure. In: The Eighth National Civil Engineering Forum for Graduate Students. 2011.
[16] Lin P, Zhang D, Yan J. Study on structural safety assessment method of operating shield tunnel. Tunn Constr 2015;35(S2):43–9.

Exercises

1. What are common defects in shield tunnel lining structure? How are they related to each other?
2. Please explain the reasons for shield tunnel lining cracking?
3. How to carry out a defects investigation for shield tunnel lining structure?
4. How to evaluate the safety of the shield tunnel lining structure?
5. How are the common defects of shield tunnels classified?

Index

Note: Page numbers followed by "*f*" and "*t*" refer to figures and tables, respectively.

C

F

G

N

T

U

V

W

X

Y

Z

Printed in the United States
by Baker & Taylor Publisher Services